Statistics in Environmental Monitoring and Assessment

Statistics in Environmental Monitoring and Assessment details how statistics plays an essential role in every step of an environmental monitoring and assessment program, from the sampling design to the data management and data analysis methods that are used to extract meaningful information for environmental quality management.

Large temporal and spatial variations of the environmental variables present a great challenge for producing a representative characterisation of the status of any given environmental media, water, air, biota, or solid. This book explains these issues in an easy-to-understand manner and illustrates using actual applications in real-life case studies. Each case study comes with a clear explanation for every related fundamental concept and methodology that details the statistical methods used in the analysis and considerations taken when using a particular method.

Primarily intended for graduate students in environmental sciences who wish to use statistics as a research tool, it will also benefit researchers who are interested in applications of statistical methods for designing environmental monitoring programs and analysis of environmental data to assess the status of various environmental matrices. This book is organised in three parts with Part I providing a review of the basics of statistics, Part II focusing on sampling designs, and Part III presenting applications of statistical methods for data analysis.

Key Features:

- Contains real-life case studies
- Includes practical statistical designs for environmental sampling
- Analyses monitoring data for environmental assessment
- Comprehensively explains theoretical parts related to the statistical analysis methods

Shin'ichi Okamoto is a professor emeritus of Tokyo University of Information Sciences, holding a doctorate degree from Waseda University. He worked for the Industrial Pollution Control Association of Japan (currently the JEMAI) for nine years, then joined the Tokyo University of Information Sciences as a professor, and retired in 2014, having served as the dean of the undergraduate and graduate schools. He has published 36 books or book chapters, and about 160 academic papers, offered intensive lectures on statistics in a sampling design course once a year since 2000 as a visiting professor at the Asian Institute of Technology (AIT). In 2004, he received the Teaching Award for this course from the School of Environment, Resources and Development of AIT.

Nguyen Thi Kim Oanh is a distinguished professor at the Asian Institute of Technology (AIT) with over 35 years of working experience in research and education. She has been affiliated with the Environmental Engineering and Management Program of AIT since 1997 and is the founding director of the Air Quality Nexus Centre at AIT. Recognised for her efforts in characterising environmental pollution issues in Asian developing countries, Prof. Kim Oanh has been selected to serve in several professional organisations and has been credited with over 200 peer-reviewed scientific papers and book chapters, two books (ed.), and many important development reports.

Statistics in Environmental Monitoring and Assessment

Shin'ichi Okamoto and Nguyen Thi Kim Oanh

CRC Press
Taylor & Francis Group
Boca Raton London New York

CRC Press is an imprint of the
Taylor & Francis Group, an **informa** business
A CHAPMAN & HALL BOOK

Designed cover image: Shin'ichi Okamoto and Nguyen Thi Kim Oanh

First edition published 2026
by CRC Press
2385 NW Executive Center Drive, Suite 320, Boca Raton FL 33431

and by CRC Press
4 Park Square, Milton Park, Abingdon, Oxon, OX14 4RN

CRC Press is an imprint of Taylor & Francis Group, LLC

ISBN: 978-1-032-98136-9 (hbk)
ISBN: 978-1-032-98137-6 (pbk)
ISBN: 978-1-003-59721-6 (ebk)

DOI: 10.1201/9781003597216

Typeset in Palatino
by Apex CoVantage, LLC

Contents

Preface xiii
About the Authors xv

1. **Introduction** 1
   1.1 Overview of Monitoring for Environmental Quality Management 1
      1.1.1 Relevant Terminologies and Definitions 1
      1.1.2 Environmental Monitoring Framework 4
      1.1.3 Development Trend in Environmental Monitoring 6
   1.2 Application of Statistics in Environmental Sciences 7
   1.3 Data Quality Management 10
      1.3.1 Data Quality 10
      1.3.2 Data Management and Its Quality Issues 11
      1.3.3 Concerns Related to the Adherence to International Quality Standards 13
   1.4 Summary 14

Part I Fundamentals of Statistics

2. **Descriptive and Inferential Statistics** 19
   2.1 What Is Statistics? 19
      2.1.1 Descriptive Problems 20
      2.1.2 Inferential Problems 20
   2.2 Descriptive Statistics 20
      2.2.1 Picturing the Distribution 20
      2.2.2 Sample Statistics for the Location (Centre of Data) 21
      2.2.3 Sample Statistics for the Variability (Scatter of Data) 22
      2.2.4 Moment 23
      2.2.5 Temporal and Spatial Resolution of Data 24
   2.3 Other Important Characteristics 26
      2.3.1 Law of Large Numbers 27
      2.3.2 Central Limit Theorem 27
      2.3.3 Bayesian Statistics 28
   2.4 Summary 28

3. **Sample and Distribution** 30
   3.1 Population and Sample 30
   3.2 Distribution of Random Variables 31
      3.2.1 Discrete Probability Distribution 31
      3.2.2 Continuous Probability Distribution 31
      3.2.3 Probability Density Distributions 32
      3.2.4 Expected Value and Variance of a Random Variable 33
      3.2.5 Relation between the Sum of Squares and Degrees of Freedom 34

3.3 Sampling Distributions ... 36
3.3.1 Relation between the Population and Sample ... 36
3.3.2 Normal Distribution ... 37
3.3.3 *t* Distribution ... 38
3.3.4 Chi-Square Distribution ... 38
3.3.5 *F* Distribution ... 39
3.3.6 Relations between Some Distributions ... 39
3.4 Summary ... 40

4. Inferential Statistics ... 42
4.1 Sampling and Statistical Inference ... 42
4.2 Statistical Test ... 42
4.2.1 Concept of Statistical Test ... 42
4.2.2 Terminology Used in Hypothesis Testing ... 43
4.2.3 Procedure for Statistical Tests ... 44
4.2.4 Hypothesis Testing ... 45
4.2.5 Significance Testing ... 46
4.2.6 Power of Hypothesis Testing and Sample Size ... 47
4.2.7 One-Tailed Test and Two-Tailed Test ... 48
4.3 Estimation ... 49
4.3.1 Point Estimate ... 50
4.3.2 Interval Estimate ... 50
4.4 Test of Population Means ... 51
4.4.1 Comparing Means (Variances Equal) ... 52
4.4.2 Comparing Means (Variances Unequal) ... 54
4.4.3 Comparing Two Means – Paired Data ... 54
4.5 Testing on Proportion ... 56
4.6 Comparing Variances: The *F* test ... 58
4.7 Summary ... 59

5. Analysis of Variance ... 60
5.1 Relationship between the Test for Differences in Population Means and Analysis of Variance ... 60
5.2 One-Way ANOVA and Randomised Block Design ... 62
5.3 Factorial Design and Some Related Issues ... 68
5.3.1 Comparison with One-Way ANOVA ... 70
5.3.2 Comparison with Two-Way ANOVA (RBD) ... 71
5.3.3 Handling of Random Factors and Their Relationship with Mixed Models ... 72
5.4 Other Useful Experimental Designs ... 74
5.4.1 Split Design ... 74
5.4.2 Orthogonal Design ... 74
5.5 Summary ... 76

6. Correlation and Regression Analysis ... 78
6.1 Correlation Analysis – Variable Relationship ... 78
6.2 Regression Analysis – Cause–Effect Relationship ... 79
6.2.1 Estimation of Regression Coefficients ... 80
6.2.2 Evaluation of Prediction Accuracy ... 83
6.2.3 Decomposition of the Variance and Degrees of Freedom ... 84
6.2.4 Relation between Correlation Analysis and Regression Analysis ... 84
6.2.5 Test for Regression Coefficient ... 85
6.2.6 Regression Analysis for Repetitive Data ... 88
6.2.7 Stratification of the Dataset and Its Relationship to the Mixed Model ... 92

6.3 Multiple Regression Analysis ... 94
6.3.1 Meaning of Partial Regression Coefficient ... 95
6.3.2 Evaluation of Multiple Regression Equation ... 97
6.3.3 Variable Selection Method ... 100
6.4 Summary ... 100

7. Multivariate Analyses ... 102
7.1 Some Important Multivariate Analysis Methods ... 102
7.1.1 Principal Component Analysis ... 103
7.1.2 Factor Analysis ... 104
7.1.3 Discriminant Analysis ... 105
7.2 Additional Topics Related to Advancements in Statistical Analysis ... 106
7.2.1 Relation between Experimental Design and Multivariate Analysis ... 106
7.2.2 Statistical Analysis Software ... 106
7.3 Summary ... 107

Part II Applications for Monitoring and Measurement

8. Monitoring Tool in Environmental Quality Management ... 111
8.1 Environmental Monitoring Parameters and Measurement Units ... 111
8.1.1 Monitoring Parameters for Different Environmental Matrices ... 111
8.1.2 Measurement Units ... 113
8.2 Quality Assurance and Quality Control in Environmental Monitoring ... 114
8.2.1 Quality Assurance and Quality Control ... 114
8.2.2 Sources of Uncertainty of Monitoring Results ... 115
8.2.3 Types of Monitoring Errors ... 117
8.2.4 Principal Indicators of Monitoring Data Quality ... 118
8.3 Monitoring Data Management ... 120
8.3.1 Framework of Environmental Monitoring Data System ... 120
8.3.2 Development in Data Management Methods ... 122
8.4 Development of Environmental Monitoring Techniques ... 123
8.4.1 Manual Monitors ... 123
8.4.2 Continuous Automatic Monitoring Instruments ... 124
8.4.3 Low-Cost Sensors ... 124
8.4.4 Remote Sensing ... 125
8.5 Summary ... 127

9. Sampling Design from Theory to Practical Applications in Environmental Monitoring ... 130
9.1 Concept of Sampling Design ... 130
9.1.1 Sampling Design ... 130
9.1.2 Population ... 130
9.1.3 Population Size ... 131
9.1.4 Target Population ... 131
9.1.5 Sampling Frame ... 131
9.1.6 Sampling Unit ... 131
9.1.7 Sample Size ... 132
9.1.8 Sample vs. Population ... 132
9.2 Statistical Sampling Methods ... 132
9.2.1 Non-Probability Sampling ... 134
9.2.2 Probability Sampling ... 135

9.3 Multistage Sampling Design 138
9.3.1 Multistage Probability Sampling 139
9.3.2 Multistage Mixed Sampling Designs 139
9.4 Calculation of the Sample Size in Simple Random Sampling 142
9.4.1 Sample Size for an Infinite Population 142
9.4.2 Sample Size for a Finite Population 143
9.5 Application of Sampling Designs for Environmental Measurements 144
9.5.1 Specific Considerations for Representative Sampling 144
9.5.2 Types of Environmental Samples and Collection Techniques 146
9.6 Summary 149

10. Industrial Pollution Monitoring 151
10.1 Industrial Pollution Management 151
10.1.1 Industrial Pollution 151
10.1.2 Approaches to Reduction of Industrial Pollution 151
10.2 Broad Categories of Industrial Pollution Monitoring 152
10.2.1 Monitoring for Industrial Pollution Auditing 153
10.2.2 Pollution Monitoring for Providing Information on the Operation Conditions of Production Process 154
10.2.3 Regulatory Monitoring 155
10.3 Case Study with a Chlorine-Bleached Kraft Pulp and Paper Mill 156
10.3.1 Monitoring for Environmental Performance Auditing 157
10.3.2 Pollution Monitoring to Provide Information for Optimising Production Control 160
10.4 Summary 166

11. Monitoring to Assess the Impacts of Industrial Discharges on the Environment 168
11.1 Overview of Monitoring for Potential Harms of Industrial Discharges to the Environment 168
11.2 Monitoring for Assessment of Environmental Impacts from Chlorine-Bleached Kraft Pulp and Paper Mill 169
11.2.1 Environmental Discharges from the Mill 169
11.2.2 Assessment of Potential Harms of the Mill Discharges to the Environment 170
11.3 Summary 180

12. Design and Implementation of Ambient Air Pollution Monitoring Programme in Asian Regional Project of AIRPET 182
12.1 Introduction: Overview of AIRPET 182
12.2 Design of PM Monitoring Programs with QA/QC 184
12.2.1 Sampling Design 184
12.2.2 QA/QC for Chemical Analyses 187
12.2.3 QA/QC for Data Quality 189
12.3 Key Findings of the AIRPET Monitoring Programme 190
12.4 Summary 193

13. Sampling and Analysis for Particulate Matter: Kashima SPM Study 195
13.1 Overview of the Kashima SPM Study 195
13.1.1 Sampling Design 196
13.1.2 Analytical Plan and Related Composite Sampling 197
13.2 Field Survey Program Designs and Quality Assurance in the Kashima Study 198
13.2.1 Sizing Efficiency and Instrumental Differences Effects 198

13.2.2 Flow Rate and Flow Volume Calculation ... 202
13.2.3 Continuous Monitoring Instrument for Particulate Matter ... 204
13.2.4 Effect of Humidity on Measuring Instruments ... 205
13.2.5 Issues Related to Analysis for PM Chemical Composition ... 207
13.3 Key Findings of the Kashima Study ... 212
13.4 Summary ... 215

14. Temporal Resolution and Averaging Time of Concentrations ... 218
14.1 Spatiotemporal Scope of Air Pollution Problems ... 218
14.2 Averaging Time of Concentrations in Air Quality Management ... 219
14.3 Atmospheric Concentration with an Averaging Time Shorter Than One Hour ... 222
14.4 Other Statistical Analyses on Concentration Data ... 223
14.4.1 Goodness-of-Fit Test ... 224
14.4.2 Time-Series Analysis of Concentration Data and Its Relationship to the Law of Large Numbers ... 227
14.5 Summary ... 228

15. Spatial Resolution of Concentration Data and Optimal Placement of Monitoring Stations ... 230
15.1 Concentration Measurements at Air Quality Monitoring Stations ... 230
15.1.1 Analysis of Differences due to Measuring Instruments ... 231
15.1.2 Analysis of Differences due to Measuring Sites ... 232
15.1.3 Statistical Explanation of the Additive Nature of Variance ... 233
15.1.4 Concentration Variability around the Same Building ... 234
15.1.5 Concentration Variability in a Residential Area ... 238
15.2 Optimal Placement of the Monitoring Stations for Producing Representative Data ... 240
15.2.1 Basic Concept of Representativeness of the Monitoring Station ... 240
15.2.2 Estimating the Concentration Spatial Distribution in the Target Area ... 241
15.2.3 Representativeness of a Monitoring Station ... 242
15.2.4 Evaluation of a Monitoring Network ... 245
15.3 Summary ... 246

16. Quality Assurance of the IPCAJ's Acid Rain Study ... 248
16.1 Overview of the IPCAJ's Acid Rain Study ... 248
16.2 Measures for Quality Assurance ... 249
16.2.1 Storage Condition of Rain Samples ... 249
16.2.2 Inter-Laboratory Comparison of the Analytical Results ... 251
16.2.3 Precipitation Sampler and Its Instrumental Error ... 253
16.2.4 Evaluation of the Obtained Data and Data Screening ... 254
16.3 Key Findings of the Acid Rain Study ... 255
16.4 Summary ... 256

17. Quality Assurance of Air Tracer Dispersion Experiments for Roadways ... 258
17.1 Methodology of Air Tracer Experiments for Roadways ... 258
17.1.1 Overview of Air Tracer Experiments ... 258
17.1.2 Air Tracer Releasing (or Disseminating) Plan for Roadway Experiments ... 259
17.1.3 Sampling Plan for Roadway Experiments ... 261
17.1.4 Design of Air Tracer Sampling Grid and Measurement of Turbulence Induced by Running Vehicles ... 264
17.2 Database Development for Highway Line Source Modelling ... 266

17.2.1 Experimental Method and Design ... 267
17.2.2 Quality Assurance by Tracer Flux Calculation ... 270
17.2.3 Key Findings of the Dispersion Experiments ... 273
17.3 Summary ... 277

18. Characteristics of Concentrations Calculated by Air Quality Models ... 281
18.1 Air Quality Model ... 281
18.2 Statistical Characteristics of the Concentration Values as the Output of the Air Quality Model ... 283
18.3 Case Study for the Kashima Area ... 285
18.3.1 Air Quality Model Used in the Case Study ... 285
18.3.2 Issues Related to Categorical Data as Input Conditions ... 286
18.3.3 Issues Regarding the Prediction Target, Average vs. Maximum Values ... 290
18.4 Air Quality Model for Tall Stacks ... 293
18.4.1 Early Studies in the United States ... 293
18.4.2 Analysis of Air Tracer Dispersion Experimental Data in Japan ... 294
18.4.3 Comparison of Japanese Data and US Analysis Results ... 295
18.4.4 Needs for Further Considerations ... 296
18.5 Summary ... 297

19. Evaluation of Air Quality Model Performance ... 300
19.1 Challenges in Model Performance Evaluation ... 300
19.2 Fundamentals of Air Quality Model Performance Evaluation ... 301
19.2.1 Robust Highest Concentrations ... 302
19.2.2 Fractional Bias ... 302
19.3 Performance Evaluation for 3D Photochemical Smog Models ... 302
19.4 Features of the Statistics Used for Evaluating the Model Performance ... 304
19.4.1 Criteria for Performance Evaluation ... 305
19.4.2 Inter-Comparison of Evaluation Criteria ... 305
19.5 Reproducibility of Hourly Concentrations ... 307
19.6 Summary ... 310

Part III Applications for Data Analysis

20. Types of Statistical Data and Analysis Methods Suitable for Their Analysis ... 315
20.1 Types of Data ... 315
20.2 Variables and Data ... 316
20.3 Analytical Methods for Various Types of Data ... 317
20.3.1 Logistic Regression ... 320
20.3.2 Generalised Linear Model ... 322
20.3.3 Perceptron ... 322
20.4 Analysing an Interrelation among Categorical Variables ... 323
20.5 Measuring Scale for Subjective Sensuous Factors ... 326
20.5.1 Level of Measurement ... 326
20.5.2 How to Measure the Scale ... 327
20.6 Summary ... 329

21. Applications of Principal Component Analysis and Exploratory Factor Analysis ... 331
21.1 Important Issues and Analytical Procedures for PCA and EFA ... 331
21.1.1 Terminology ... 331
21.1.2 Factor Extraction Process ... 332

21.1.3 Determination of the Number of Factors (Components) to Retain 332
21.1.4 Rotation 332
21.2 A New Way of Thinking about PCA 333
21.3 Relationship between PCA and EFA 333
21.4 Case Studies on the Application of PCA and EFA 334
21.4.1 Case 1: Analysis of Air Quality Data in Tokyo in the 1960s 335
21.4.2 Case 2: Classification of Meteorology Patterns Affecting Air Quality in a Mountainous Area 336
21.4.3 Case 3: Application of PCA for Analysis of Roadside Air Quality 341
21.4.4 Case 4: Analysis of Spatial Homogeneity in Wind Field by Cosine Similarity 345
21.4.5 Case 5: Analysis of the Spatial Distribution of Photochemical Air Pollution 350
21.4.6 Case 6: Application of a Binary Dataset to Analyse the Biogeographic Distribution of Butterflies 353
21.4.7 Case 7: Application of PCA and EFA to the Analysis of PM Elemental Composition for Identifying Contributing Sources 357
21.5 Summary 362

22. Statistical Analyses for Air Quality Predictions 366
22.1 Challenges with Statistical Air Quality Predictions 366
22.2 Canonical Correlation Analysis and Its Application 368
22.2.1 Algorithms 368
22.2.2 Application 369
22.3 Cluster Analysis 370
22.3.1 Hierarchical Method 371
22.3.2 Non-Hierarchical Method 372
22.4 Artificial Neural Network as Compared to Linear Regression Model: Case Study of Prediction of Maximum Hourly Ozone in an Urban Area 373
22.4.1 Characteristics of Input Data 374
22.4.2 Performance of the Multilayer Perceptron (MLP) Model in Comparison with the Linear Regression (LR) Model 375
22.5 Application of Discriminant Analysis for Forecasting Photochemical Air Pollution in Tokyo 377
22.5.1 Input Data and Its Preliminary Analysis 377
22.5.2 Prediction Results: Comparison between Discriminant Analysis Model and Multiple Regression Model 378
22.6 Case Study of Predicting SO2 Concentration in Tokyo in the 1960s 381
22.6.1 Characteristics of Data Used and Setting of Explanatory Variables 381
22.6.2 Modelling Results and Evaluation 382
22.7 Haze Prediction Using Meteorology Clarification and Stepwise Regression Modelling Approach 384
22.7.1 Characteristics of Haze Associated with Different Synoptic Patterns 384
22.7.2 Construction and Evaluation of Statistical Prediction Model 384
22.7.3 Early Warning of Haze 386
22.8 Summary 386

23. Receptor Models 389
23.1 Basic Concept of Receptor Models 389
23.1.1 What Is a Receptor Model 389
23.1.2 Basic Concept of a Chemical Mass Balance (CMB) Model 390
23.2 Feature of Receptor Models 393
23.2.1 Least Squares Model 395

23.2.2 Effective Variance Model ... 397
23.2.3 Ridge Regression Model ... 397
23.2.4 Linear Programming Model ... 398
23.2.5 Factor Analysis Model ... 398
23.2.6 Positive Matrix Factorisation (PMF) ... 398
23.2.7 Structural Equation Modelling (SEM) ... 398
23.3 Application Procedure of the CMB Model Calculations ... 399
23.3.1 Calculation Procedure ... 399
23.3.2 Case Studies of Result Evaluation and Handling of Abnormal Values ... 402
23.4 Evaluation of the CMB Calculations ... 403
23.4.1 Consistency of Each Piece of Data ... 405
23.4.2 Consistency for Each Element ... 406
23.5 Factor Analysis–Multiple Regression (FA-MR) Model and Its Application ... 408
23.5.1 Formulation of the FA-MR Model ... 408
23.5.2 Data Used in the FA-MR Modelling ... 410
23.5.3 Source Apportionment Result ... 411
23.5.4 Evaluation and Discussion of FA-MR Model Results ... 413
23.6 Positive Matrix Factorisation (PMF) and Its Application ... 415
23.6.1 Basic Concept and Algorithm of the PMF ... 415
23.6.2 Application to a Multi-Site Dataset ... 416
23.6.3 Discussion of Source Profile Estimation Results ... 418
23.6.4 Comparison with the Source Apportionment Results by Different Receptor Models ... 419
23.7 Summary ... 423

24. Statistical Analysis Methods for Questionnaire Survey Data: Case Studies on Environmental Valuation ... 426
24.1 Conjoint Analysis ... 426
24.2 Structural Equation Modelling (SEM) ... 428
24.2.1 SEM Concept ... 428
24.2.2 Path Analysis ... 429
24.3 Measuring Consumer Perceptions of Environmentally Friendly Products ... 431
24.3.1 Consumer's Recognition of Green Products – Perception Mapping by PCA ... 431
24.3.2 Estimating the Consumer's Attitude by Conjoint Analysis ... 432
24.3.3 Causal Model Analysis on Purchase of Green Products ... 435
24.4 Estimating the Economic Value of the Environment ... 437
24.4.1 Preference for Environmental Goods and Attitude toward Environment-Related Products ... 437
24.4.2 Measuring Willingness-to-Pay ... 439
24.4.3 Hedonic Pricing ... 441
24.4.4 Economic Evaluation of Environmental Projects, Policies and Related Initiatives ... 441
24.4.5 Case Study ... 443
24.5 Summary ... 449

Index ... 453

Preface

Environmental monitoring is defined as a systematic process of observation and measurement to obtain data for characterising the state and trend of environmental quality and provide relevant information to stakeholders. The application of statistics is essential not only to ensure the generation of data of the required quality cost-effectively but also to use proper data analysis methods to extract meaningful information for environmental quality management. This book is primarily intended for graduate students in environmental sciences who wish to use statistics as a research tool. Nevertheless, the case studies have been compiled to meet the expectations of many other interested researchers as well. We have tried to keep the theoretical parts related to the statistical analysis methods as simple as possible to benefit readers from a wide range of disciplines. This book is organised in three parts with Part I providing a review of the basics of statistics, followed by Part II and Part III, which provide examples and case studies of using statistical tools for environmental monitoring and data analysis.

As a supplement to other books related to statistical applications in environmental monitoring, this book particularly enriches its readers with real-life case studies illustrating various applications of statistical methods in environmental monitoring and assessment. In each case study, a clear explanation for every related fundamental concept and methodology is provided that details the statistical methods used in the analysis and considerations taken when using a particular method.

Part I (Fundamentals of Statistics) is an edited version of the lecture notes used for an intensive course of sampling design taught to graduate students from the School of Environment, Resources, and Development (SERD) at the Asian Institute of Technology (AIT), Thailand. The lectures focused on providing the statistical background to improve the students' ability to apply statistical methods to design laboratory experiments and field surveys and to analyse obtained data in their research for masters' theses and PhD dissertations. Many examples provided in this part use hypothetical data with easy-to-calculate numerals so that readers can focus on better understanding the methods.

Part II introduces the background on the sampling design illustrated with the development and implementation of statistical sampling designs in real-life case studies. Part III focuses on how to handle the obtained data using statistical analysis methods to extract useful information for environmental quality management purposes. In environmental monitoring, often the measurement method and the data analysis method are highly related; hence they should be planned together. Both Part II and Part III provide case studies to show the various statistical methods used for data analyses. These case studies cover a wide range of environmental monitoring programs for different environmental matrices, i.e., ambient air, industrial pollution discharges and effects on receiving water bodies, along with biological species distribution and market surveys. Additional theoretical background is also included in the explanation of case studies where relevant. Some content of Part II and Part III was also included in the sampling design course taught at AIT.

We would like to thank AIT for the opportunity to offer the sampling design course at the Institute. This was an intensive time-limited course that started as Akira Harada Lectures on Environmental Statistics in 2000. Mr. Akira Harada, the donor for this course implementation, was a counsellor of Oken Associates and had requested AIT to create this endowed course as a gesture of appreciation for the goodwill of those involved while he was assigned in Bangkok. The course coordinator then was Assoc. Prof. Dr. Preeda Parkpian. After the endowment ended, it continued as a regular course until 2016 under the coordination of Prof. Nguyen Thi Kim Oanh, who further enriched the lecture content.

In the preparation of this book, we cited many research results conducted by the Industrial Pollution Control Association of Japan (IPCAJ), where Shin'ichi Okamoto was once affiliated. We are grateful for the approval for the citation of these results from the Japan Environmental Management Association for Industry (JEMAI), the successor organisation of IPCAJ.

Mr. Nguyen Minh Thang is thanked for his assistance in improving the quality of several figures and tables in the book. The research staff Air Quality Nexus Centre at AIT are thanked for their assistance in the formatting of references in several key chapters. Last but not least, we would like to dedicate this book to the beloved families of both authors. Without their understanding and constant support through the years, this book would not have been completed.

About the Authors

Shin'ichi Okamoto is Professor Emeritus of Tokyo University of Information Sciences. He received a doctorate degree from Waseda University in 1977 and worked for the Industrial Pollution Control Association of Japan (currently the JEMAI) for nine years. After that, he joined the Tokyo University of Information Sciences as a professor, and retired in 2014, having served as the dean of the undergraduate and graduate schools. He has published 36 books (including writing only book chapters) and about 160 academic papers. Since 2000, as a visiting professor at Asian Institute of Technology (AIT), he has offered intensive lectures on statistics in a sampling design course, once a year. In 2004, he received the Teaching Award for this course from the School of Environment, Resources and Development of AIT. He was a member of several academic societies, including AWM&A, until his retirement and is currently a fellow of the Japanese Society for Quality Control (JSQC).

Nguyen Thi Kim Oanh is Distinguished Professor at the Asian Institute of Technology (AIT) with over 35 years of working experience in research and education. She has been affiliated with the Environmental Engineering and Management Program of AIT since 1997 and is the founding director of the Air Quality Nexus Centre at AIT. Prof. Kim Oanh enriched her experience with sabbaticals at the University of Washington in Seattle, the University of Illinois at Urbana–Champaign in the USA (2004), and the Institute for Advanced Sustainability Studies in Germany (2012). Recognised for her efforts in characterising environmental pollution issues in Asian developing countries, Prof. Kim Oanh has been selected to serve in several professional organisations. She has been credited with over 200 peer-reviewed scientific papers and book chapters, two books (ed.), and many important development reports.

1

Introduction

Environmental monitoring is defined as the process of generating data on environmental quality by continuous observations and measurements to timely provide relevant information to stakeholders. Environmental monitoring is an essential technical tool to identify environmental problems for decision-making so that appropriate management measures can be taken to regulate the amount, duration, and location of pollutants discharged into the environment. Such measures aim to effectively reduce the pollution to acceptable levels, thereby minimising the environmental and health risks. Producing valid data that can be used to assess the state and trend of environmental quality is the goal of a monitoring program. Statistics plays essential roles in every step of an environmental monitoring and assessment program, from the sampling design to the data management and data analysis. This chapter introduces a general framework of environmental monitoring along with an overview of various statistical applications and data quality management, which will be illustrated with case studies throughout this book.

1.1 Overview of Monitoring for Environmental Quality Management

Monitoring is a technical tool for observing and characterising the environmental quality to identify environmental problems with the aim of preventing environmental degradation. To continuously assess the environmental impacts and control pollution situations of the development activities, "the need for environmental monitoring is as great as ever" (Artiola et al., 2004). In environmental monitoring, standardised methods of measurements and observations are conducted over the long term for one or more environmental matrices. The obtained datasets are analysed to extract information on the status and trend of the environment quality. The monitoring parameters differ between the environmental matrices, e.g., between air, soil, water, and biota. Environmental monitoring has been continuously advanced with new monitoring methods and techniques to measure conveniently and accurately not only the traditional pollutants but also a growing list of emerging pollutants.

1.1.1 Relevant Terminologies and Definitions

This section explains several important technical terminologies needed when discussing environmental quality management concepts.

The environment: Several definitions exist for "environment", but generally it can be defined as the surroundings or conditions in which a person, animal, or plant lives or operates. Thus, for the human being, the environment is the complex totality of physical, chemical, biotic, and anthropic factors that represent the total of all living and non-living elements and their interactions that influence our life.

There are three key components of the Earth system: (1) physical/abiotic (atmosphere, hydrosphere, lithosphere), (2) biological/biotic (biosphere: plants, animals), and (3) human-made (economic, social, cultural), which are interacted to drive the dynamic change of the system. Although the natural components of the earth system have effective functions to handle the stress of pollution, excessive amounts of contaminants released into the environment may

DOI: 10.1201/9781003597216-1

collapse this self-purifying mechanism and result in adverse effects. This may eventually lead to environmental disasters that affect agriculture, biodiversity, human health, and the economic development of a region or a country. For example, too many nutrients (nitrogen and phosphorus compounds) discharged in large quantities into a lake would cause eutrophication, the conditions that enhance the growth of aquatic plants, which in turn depletes the dissolved oxygen level in water hence threatening the survival of aquatic animals. Likewise, the atmospheric processes may, to a certain extent, dilute pollutants released from emissions sources, e.g., road traffic or industry, and remove pollutants from the air by deposition to the earth's surface. However, excessive air pollution loads, such as that from a large forest fire event or a highly polluting factory, discharged into stable atmospheres with limited vertical and horizontal mixing (stagnant atmospheres, in other words) would cause build-ups to extremely high or episodic levels of air pollution.

The term 'air pollution episode' refers to the occurrence of high pollutant concentrations in densely populated areas during persistently stagnant atmospheric conditions that may result in serious health effects with a sharp increase in illnesses and/or deaths from respiratory and related diseases. Air pollution episodes may be caused by natural emissions (volcano eruption, dust storm), or by human-made emissions, or both, and normally last for several days. Historical air pollution episodes have been well documented, with the famous examples of the Great Smog of London and the series of photochemical smog incidents in Los Angeles during the 1940s and early 1950s. The Great Smog of London was caused by excessive emissions from dirty coal burning in residential and industrial combustions in London coupled with extremely stagnant atmospheric conditions in the winter of 1952 (UK-Metoffice, 2021). This smog is also called the London sulphurous winter smog due to the extremely high levels of sulphur dioxide (SO_2) and particulate matter (soot) observed in the air. On the other hand, photochemical smog is prevalent during hot and sunny summer months. It is formed in the photochemical reactions involving the precursors of VOCs (volatile organic compounds) and nitrogen oxides ($NO_x = NO + NO_2$) in the presence of sunlight. Photochemical smog is visible as a yellow-brownish haze in the sky and is a mixture of high levels of surface ozone, fine particles, and several other toxic pollutants. The phenomenon, first identified in the city of Los Angeles in 1944, is nowadays widespread over many cities around the world where intensive emissions of the precursors, i.e., from fuel combustion (vehicle engines, industry, residential), biomass open burning, and biogenic activities, are present. Ozone is a major component of photochemical smog, which can be found in the air of many places around the world. The tropospheric ozone is a strong greenhouse gas and a harmful pollutant that affects both human health and ecosystems. For example, exposure to ozone causes health effects leading to sickness and 365,000 premature deaths in 2019 worldwide (HEI, 2020). Modelling studies showed a few percentage points of yield reductions for major Asian crops (rice, maize, soybean) are associated with exposure to ambient ozone levels (UNEP, 2019). Experimental studies confirmed that exposure to ozone concentrations of 10 ppb above the ambient levels caused significant reductions (10–20%) of crop yield for mung bean, soybean, and sunflower (Van Dinh and Kim Oanh, 2023). In Southeast Asia, haze episodes with extremely high levels of fine particles recur due to emissions from large forest fires annually, causing widespread regional effects. For example, an extensive haze episode was reported in the El Niño year of 1997 when Indonesian forest fires spread smoke across the region (Heil, 2007).

Environmental quality: This is defined as "a measure of the condition of an environment relative to the requirements of one or more species and/or to any human need or purpose" (Johnson et al., 1997). The terms "air quality", "soil quality", "water quality", etc. are used for the corresponding environmental matrix.

It should be noted that when attempting to clarify the definition of the word "quality", various difficulties arise. One can think of the quality of a thing as its appropriateness for its intended use. Accordingly, it is reasonable to consider air quality as the degree to which the air conforms to its purpose, implying that the air is in a desirable state for us to breathe. The UK Department for Environment, Food and Rural Affairs (accessed 2022) has defined "Air quality is the term we use to describe how polluted the air we breathe is. When air quality is poor, pollutants in the air may be hazardous to people, particularly those with lung or heart conditions".

As mentioned, the environment is the surrounding situation as seen from the object. There are also aesthetic aspects of the environment quality such as recreational noise, odour, and landscape scenes. Several characteristics of the environmental quality related to sustainability, such as the conservation of ecosystems and mineral resources, may not be directly related to the pollutants released into the environment. It is worth emphasising that the term "environmental quality" is not necessarily to be interpreted in the same way as the quality of a product or a service. Nevertheless, data quality management in environmental monitoring is as important as in any other field, which is further discussed in Section 1.3.

The environmental quality in a location at a specific time is described using a set of properties and characteristics. A wide array of physical, chemical, and biological parameters is monitored to assess the status and trend of the environmental quality. Environmental quality is highly dynamic and exhibits large temporal and spatial variations. The environmental quality may be described as "good", "satisfactory" or "bad" relative to the requirements of one or more species and or to any human need or purpose. The environmental quality in a location may be "improving" or "deteriorating" as a result of human-made activities and natural events. For example, the environmental quality, notably the quality of ambient air and surface water, in urban areas of many developing countries is deteriorating due to rapid industrialisation, motorisation, and population growth in tandem with insufficient environmental management capability.

Environmental pollution: This may be defined as the addition of any substance to the environment at a rate that is faster than or that exceeds the rate at which the system can handle within its purifying capacity, thereby negatively impacting the quality of the environment, hence resulting in adverse effects on the environmental quality. These substances are referred to as environmental pollutants and include any chemical compounds (solid, liquid and gas), biological organisms (debris, pollen, fungi, bacteria, virus, etc.) or physical substances (heat, sound/noise, radioactivity) that is released intentionally or accidentally by human-made activities and by natural events. Pollution deteriorates environmental quality and causes negative effects on human health and well-being. For example, air pollution in the form of fine particulate matter or $PM_{2.5}$ (particles with an aerodynamic diameter ≤ 2.5 μm) present in ambient air and indoors caused over 6 million premature deaths globally in 2019 (HEI, 2020).

Environmental quality management: This usually refers to the systematic and methodical efforts to bring environmental quality to the desired state. In addition, it may include setting its desired state (e.g., environmental standards) and evaluating the efforts. More specifically, it can also be interpreted as an activity to regulate the duration and location of pollution discharge into the environment to eliminate or reduce the concentration of pollutants to acceptable levels with the aim to minimise environmental and health risks. The "acceptable levels" of pollutants can be regarded as the goals related to environmental quality and are commonly expressed in terms of environmental quality standards. The standards are the legally specified maximum permissible concentrations of contaminants in the respective environmental matrix, e.g., air, soil, sediment, water, or biota, in a specified geographical location during a stipulated period.

Environmental parameter: In the environmental monitoring context, a parameter is a factor which is characterised/measured to assess the environmental quality. An environmental parameter can be defined as a variable, encompassing a spectrum from physical, chemical to biological properties, that pertains to the environment and is necessary to comprehend environmental quality. Selection of monitoring parameters is made based on the environmental matrix (air, water, soil and biota), monitoring objectives and available resources, among others, which is further detailed in Chapter 8. The word "parameter", however, is used with slightly different meanings in mathematics, statistics, and certain engineering disciplines. For example, in statistics, "statistic" refers to the properties of a sample, whereas "parameter" refers to those used to characterise a population from which the sample is taken. In particular, Chapters 2–7 of Part I use the term "parameter" in this sense.

Environmental standards: Standards are the expression of public policy; hence they are adopted by a regulatory authority as enforceable. However, the strength of enforcement varies from country to country. The standards are established based not only on the available scientific information on the dose–response relationship but also by considering other factors related to socio-economic and technological conditions. These include, for example, prevailing exposure levels, other competing environmental problems (air, water, solid wastes), background pollution levels, and availability of resources for monitoring and control. The national standards therefore depend on the country's level of development and environmental management capability; hence the specific values for a particular environmental contaminant differ among countries. Within a country, the values may be different between states or regions or even between different locations: e.g., ambient air quality standards for industrial vs. residential areas, or between sub-categories of a matrix, such as surface water vs. drinking water.

Environmental standard values are not the absolute safe pollution levels for the environment and human health. Therefore, these standards are revised when new scientific knowledge is available or in response to changes in other socio-economic conditions. The revision is done to provide increased protection against adverse effects associated with long- and short-term exposures to the pollution; hence it is done with new pollutants added to the regulated list while more stringent standards are set for the already regulated pollutants. For example, the US National Annual Ambient Air Quality Standard (NAAQS), primary (health-based), for $PM_{2.5}$ is 12 μg/m^3, which was effective as of March 2013, and this is more stringent than the previous level of 15 μg/m^3. Note that before 1997 there was no NAAQS standard for $PM_{2.5}$ promulgated in the country (U.S. EPA, 2013).

The national environmental standards of different countries are not necessarily the same as the WHO guideline values; e.g., the WHO guideline released in 2021 for the annual $PM_{2.5}$ is 5 μg/m^3. This is because the WHO guidelines are derived from the existing evidence of epidemiological and toxicological studies in such a way as to minimise health risks and environmental impacts. Countries set environmental standards for air pollutant concentrations using such statistics as mean, maximum value, second highest value, or percentile values for a given averaging time. Understanding the meaning of these standards is essential for promoting appropriate environmental quality management actions. It is important to note that standards should be enforced. Although the strength of enforcement varies depending on the country and other requirements once a standard is set for a pollutant, its levels should be properly monitored to check compliance.

1.1.2 Environmental Monitoring Framework

The overall monitoring framework presented in Figure 1.1 shows the link between different activities of an environmental monitoring program. It is important to note that monitoring is a resource-consuming process; hence all the steps are designed to ensure that the data generated are of required quality and are used meaningfully for effective environmental quality management. Furthermore, legal backing may be required to justify the expenditure of vast resources. Conversely, there are cases where monitoring is mandatory under laws and regulations for environmental administrations. The monitoring results should be valid and timely communicated to decision-makers, scientific communities, and the public. The monitoring process is a closed loop, and a long-term monitoring program requires periodic review and revision so that necessary changes can be made when the monitoring results or new information from other sources justify the needs.

Monitoring objectives: The overall goal of monitoring is to provide information that serves as the basis for decision-making to those who have the authority to make decisions regarding necessary measures to be taken to safeguard the environmental quality. Concrete monitoring objectives are set for each monitoring program. These objectives define the spatial extent

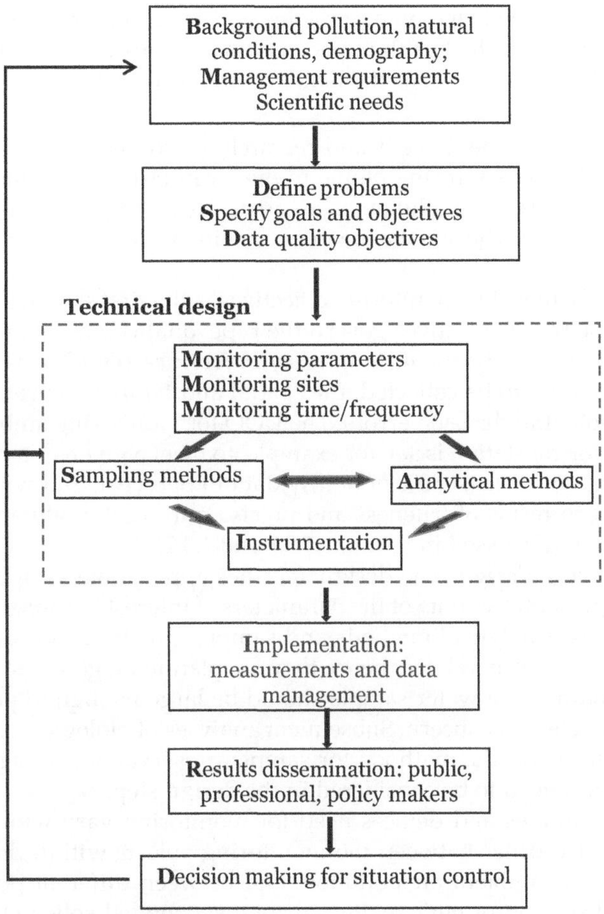

FIGURE 1.1 Framework of environmental monitoring and the link to the decision-making process.

and timescale of a monitoring program, as well as the parameters to be measured, which in turn dictate the monitoring equipment to be selected to produce measurement data. The objectives are set considering the background of important regulatory and management goals or research needs and should be achievable (scientifically, technologically, logistically, and financially). Some key objectives of environmental monitoring are detailed next.

- To assess compliance with regulations and standards: the monitoring results are compared with the regulatory values, such as national ambient air quality standards, surface water quality standards, etc. in order to assess compliance.
- To establish the baseline/existing environmental conditions: for example, the baseline monitoring used in the environmental impact assessment (EIA) is done to understand the existing environmental conditions before starting a development project.
- To evaluate the effectiveness of pollution abatement programs and other interventions: for instance, to demonstrate how the water quality in a receiving river changes after a wastewater treatment plan starts operation or how air quality in a location improves after enforcing a ban on rice straw field burning.
- To identify and quantify existing environmental problems: an example is to assess the nature and extent of pollution to provide early warnings of potential future problems and to prioritise pollution problems to be solved in a given area.

- To enhance general knowledge on ecosystems and environmental health: a case in point is to conduct research on the health impact assessment or investigate the spatial pollution distribution to identify hotspots or the temporal trends of pollution effects on a species or an ecosystem.
- To answer other management needs and research questions: for example, to examine the aging effects of a biomass-burning plume in the air in order to conduct source apportionment studies for $PM_{2.5}$ (an air pollutant) in an urban area, or to assess the effects of industrial wastewater discharge on the aquatic life in a receiving water body.

Setting up clear and achievable monitoring objectives is the first step in designing a monitoring program because these objectives govern the type, quality, and quantity of the data that should be produced to satisfy the ***data quality objectives*** **(DQOs)**. The DQOs define the appropriate types of data to be collected, the spatial and temporal coverage of data, and the tolerable levels of potential decision errors. The DQO for monitoring ambient SO_2 and NO_2 at an air quality monitoring station is set, for example, to meet a given criterion on the temporal coverage of the data, say at least 80% of hourly data in one year, and with a given measurement of uncertainty in terms of trueness and precision for each pollutant. The topic of data quality will be further discussed in Sections 1.3.1 and 1.3.2.

Technical monitoring design: The design specifies appropriate sampling times and locations to generate representative data of the parameters of interest in conformity with the stated monitoring objectives. Analytical methods and sampling methods selected for the environmental parameters are closely related; hence they are planned together to ensure consistency. For instance, the volume of a water sample should be large enough to perform the planned analyses of all parameters of concern. Subsequent analyses of biological, physical, and chemical parameters require specific methods for sample preservation, storage, and preparation; hence all these aspects need to be considered in the design step.

The sampling techniques and devices used for monitoring vary with the environmental matrices, and they also differ between the monitoring objects within the same matrix, e.g., stack gas emissions vs. ambient air pollution, and between different pollutants within the same matrix, e.g., dissolved organic matter vs. total suspended solids (TSS) in water. Many considerations are necessary in designing a monitoring program to capture representative characteristics of the study environment matrices, and these are further detailed in Chapter 9. A quality assurance and quality control (QA/QC) protocol must be developed and implemented to ensure the quality of the data. An overview of the data quality management is given in Section 1.3, whereas the specific QA/QC measures applicable for the environmental monitoring are detailed in Chapter 8.

1.1.3 Development Trend in Environmental Monitoring

Environmental monitoring has been continuously advanced from the traditional to new methods, including those using sensing technologies such as highly accurate sensors and satellite products. Both sampling and analytical techniques of pollutants are continuously evolving to measure conveniently and accurately not only the traditional pollutants but also a growing list of emerging pollutants. The latter include the synthesised chemicals, which are called contaminants of emerging concern of perfluorinated compounds, chlorinated paraffins, and polybrominated diphenyl ethers, among others. These chemicals are potentially toxic and are used in routine products and hence would eventually end up in the environment to cause adverse effects on human health (U.S. EPA, 2023; Environment-Canada, 2017). The so-called microplastics (MP) and nanoplastics (NP) pollution has recently been identified as an increasing threat to the environment; hence monitoring techniques are being developed to characterise the amounts, types, and sizes of the micro- and nanoplastics (MNP) particles in different environmental matrices. Analytical techniques are also continuously developed to accurately measure trace pollutants, e.g., chlorinated pesticides and dioxins in sediments and biota, at parts per trillion (ppt, 10^{-12}) levels or lower.

Continuous development of environmental monitoring techniques allows the generation of massive datasets of multiple pollutants which is a challenge to handle using conventional data management methods. Big Data Technology is used to analyse, process, and interpret the massive amount of structured and unstructured data that could not be processed manually or traditionally. Further details of the development in environmental monitoring are given in Chapter 8.

1.2 Application of Statistics in Environmental Sciences

Various analytical methods based on statistical knowledge have been applied in a wide range of fields to provide a basis for designing experiments and observations and for summarising and interpreting data. In this section, statistical issues in the environmental field are considered. As outlined in Section 1.1, "environment" may be defined as the surroundings from the perspective of the subject or the interaction with these surroundings; however, if defined more specifically, various difficulties arise. When trying to find such a vague definition of environment and to measure its situation, it can be challenging to determine what and how to measure, when and where to measure it, and what data should be collected. These technical terms of environmental monitoring are further detailed in Chapter 8. This section describes only the outline of environmental statistics.

United Nations organisations have provided a broad scope of the environmental statistics. First, the framework of environmental statistics published by the United Nations, namely, the Framework for the Development of Environment Statistics (FDES, 2013) clarifies the scope of environmental statistics and provides an organisational structure to guide the collection and compilation of environmental statistics at the national level. The document is inherently broad and comprehensive, covering issues and aspects of the environment relevant to policy analysis and decision-making, with application to cross-cutting issues such as climate change (UN Statistics Division, 2017). The FDES defines the scope of environmental statistics that covers biophysical aspects of the environment and those aspects of the socio-economic system that directly influence and interact with the environment. The United Nations Statistics Division (UNSD) collects data on environmental statistics from countries using questionnaires; the complete UNSD/UNEP questionnaire on environmental statistics includes statistics on air, water, land and waste. Since 2004, data collection has been limited to the areas of water and waste, and statistics on other topics are collected from other international organisations. The UNSD works closely with partners in the UN family and beyond, coordinating efforts to effectively aid countries in strengthening their national statistical capacities to meet the data demands of the Sustainable Development Goals (SDGs) agenda. This framework aims to assist countries in reporting environmental data of consistent quality. For this reason, it does not necessarily go into technical details such as individual monitoring and analytical methods.

One perspective of environmental statistics is to describe the state and trends of the physical matrix of the environment and the biota within it. The aim is to provide scientific evidence and for appropriate policy responses to environmental problems such as air, water and soil pollution. The goal is to understand the causal relationship between pollution sources and environmental quality. Another important aspect of environmental statistics is to address the societal aspect of the environment by investigating and analysing them, such as public understanding of environmental policies like climate change, and consumer response to environmentally products, including the circular economy (recycling).

Statistics, which will be elucidated in Chapter 2, is a science that extracts samples from the population of subjects of interest, analyses the data, and makes inferences concerning the population. Thus environmental statistics can be understood as statistics where the target population is the environment. The statistical methods used are varied and are continuously expanding, especially in the era of the Big Data science and artificial intelligence.

In statistical analysis, the types of data handled vary depending on the subject matter to which the analysis is applied, and even if there is not much difference in the basic concepts, there are differences in the analysis methods that are frequently used. This is why there are different genres of statistics. Statistical analysis is used in a variety of academic disciplines. For example, the fields of the humanities and social sciences, such as economics, sociology and psychology; the fields of the natural sciences, such as physics, chemistry and biology; and the applied sciences, such as engineering and medicine, making them essential in various scientific fields involving empirical analysis. Many efforts in demographics and biostatistics have greatly contributed to the development of statistics itself. Several fields of applied statistics are active in their own academic societies and publish their scholarly journals.

Biostatistics is closely related to the development of genetics. Since its inception, genetic research has used statistical concepts to make sense of observed experimental results. For example, Mendel began genetic studies investigating segregation patterns of genes in pea families and actively used statistics to explain the data collected (Science Learning Hub, 2023). Medical statistics deals with the application of statistics to the medical and health sciences, including epidemiology, public health, and clinical research. Medical statistics is sometimes considered part of so-called biostatistics, although it is sometimes recognised as a branch of statistics separate from other applied statistics. It is the science of summarising, collecting, presenting, and interpreting data in medical practice and using that data to estimate the magnitude of associations and to test hypotheses. Pharmaceutical statistics is closely related to medical statistics, but it also has its own academic societies and journals. This can range from experimental design issues to analysis of pharmaceutical trials, to pharmaceutical product commercialisation issues. In epidemiology and the evaluation of drug efficacy, such topics as inference on proportion are also important. Such statistical analysis methods are important for monitoring health damage and morbidity due to environmental pollution. The topic of the analysis of discrete random variables used in such cases is discussed in Chapter 20.

Economic statistics is a subject of applied statistics and applied economics concerned with the collection, processing, compilation, dissemination and analysis of economic data. It is closely related to business statistics and econometrics. Recently, game theory and stochastic process theory have been included in this category and considered. Psychological statistics deals with information and data in the field of psychology. Statistical methods in this field include both developed and applied statistical theory and methods for modelling psychological data. Important topics concerning these methods include psychometrics, factor analysis, experimental design, and Bayesian statistics. In monitoring our emotional movements and behavioural triggers, many innovations are needed in terms of their measurement methods (e.g., measurement scale, etc.) and methods of analysis. Some of them are presented in Chapters 20 and 24 in Part III.

Several reports are found in the literature dealing with statistical analyses related to environmental monitoring. For example, Gilbert (1987) describes a method of analysis that focuses on the log-normality of pollutant concentrations. Determining whether the concentrations of pollutants in the environment matrices, such as air or water, follow a log-normal distribution is one of the significant issues in environmental science. This topic, along with related subjects, will also be addressed in Chapter 14 of Part II. Berthouex and Brown (1994) have written a statistics textbook for environmental engineers, but there may be room for improvement using new study materials covering multiple environmental media. To partly fill in the gaps, in this book we present real-life case studies of the applications of statistics on environmental monitoring with a focus on air, water and biota.

Monitoring of the environmental quality of air and water, essential to our lives, and the associated statistical analysis issues are the main subjects in this book. Greenhouse gases (GHGs) such as CO_2 concerning climate change can be discussed within the scope of this category. In monitoring related to climate change, atmospheric temperature, precipitation and the atmospheric concentrations of GHGs are important items. Since a GHG is considered to be a solute dispersed in the solvent of air, its monitoring can be carried out similarly to other air pollutants. The monitoring characteristics of the concentrations of atmospheric constituents are covered in Chapters 12–16 of Part II. In addition, regarding temperature monitoring, problems such as the uneven distribution of observation points have been raised, but it can also be regarded as a problem of how to grasp the mean value of the entire population. Along with climate change, biodiversity is probably the environmental issue that has attracted the most attention in recent years. The statistical analysis of data handled therein can be considered to fall within the category of biostatistics previously mentioned, but some points differ from conventional biology. Since it is not feasible to present a sufficient explanation of statistical analysis on biodiversity in this book, only a brief overview is given here with an example on analysing the biogeographic distribution of butterflies provided in Chapter 21.

The term "biodiversity" was coined by ecologist Rosen, who promoted the National Forum on Biological Diversity (held in 1986), based on the term "biological diversity". The term is widely used today, although it is said that the term was not based on a clear academic definition and had significant policy connotations (Ikeda, 2012). However, various discussions have since taken place, and today the term is generally considered to encompass the following three intentions:

- Genetic diversity,
- Species diversity, and
- Ecosystem diversity.

There are two general indicators of species diversity: species richness and evenness. Species richness refers to the number of species present in a community. In general, the greater the number of species, the more diverse the community. However, even if the number of species in a community is comparable, diversity will be lower if the population of a particular species is high, and the population of other species is low. To express species diversity, a diversity index has been devised that considers both species richness and evenness. The relative abundance curve is a graph that represents the diversity (relative abundance and its uniformity) of species within an ecological community. It plots the ratio of the number of individuals of a species to the total number of individuals on the vertical axis and the ranking of individuals ranked by abundance on the horizontal axis. An example of such a graph is shown in Figure 1.2.

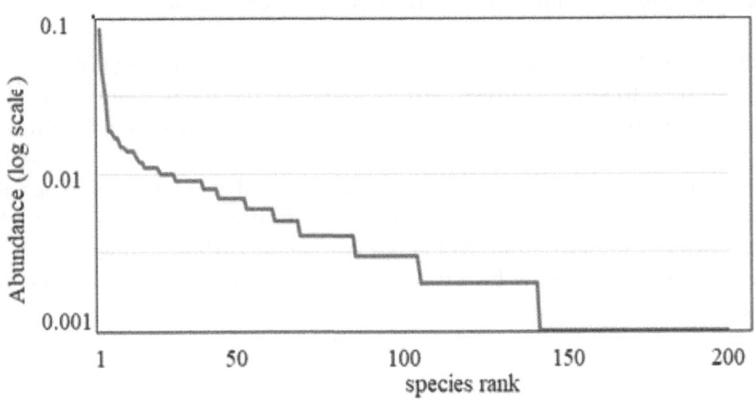

FIGURE 1.2 Species abundance curve for butterflies in the vicinity of the Mae Klong watershed research station, Thailand.

Source: Data were retrieved from Okamoto et al., 2013.

By drawing such a curve for each region and comparing them, one can consider the species diversity of each region. If the species diversity is greater, the graph will extend to the right; if the diversity is less, the hem on the right side will be shorter.

Many ecologists have pursued the factors that influence community diversity. The most representative idea is the model of niche partitioning. It is also a strategy to avoid interspecific competition and coexist under the constraints of finite resource amounts (predators, etc.). However, faced with various challenges, Hubbell (2001) proposed an attempt to use random processes (stochastic processes) to explain patterns of species diversity in a unified manner under minimal constraints (explanatory variables). To test the model, he displayed the results in a graph like this and considered them. Hubbell (2001) also mentioned the importance of sampling in estimating model parameters.

In summary, this section highlights that a variety of statistical analysis methods are used in different research areas. The most crucial aspect of monitoring is to identify the monitoring objectives, which clearly define the required output information with the necessary data quality (DQOs). Following this, it is essential to establish appropriate sampling methods from the population, determine the data to be collected through experiments and observations, ensure the uniformity of the experimental field.

1.3 Data Quality Management

In inferential statistics, to estimate population parameters (e.g., population mean), data are first obtained by extracting a sample from the population and measuring the sample or directly measuring a portion of the population. The next step is to perform a statistical analysis based on the obtained data to estimate the required parameters. This procedure is described in Chapter 4. Sampling theory is essential knowledge when considering how to extract samples from a population, which will be detailed in Chapter 9.

In such cases, the problem of whether the obtained data are appropriate as a basis for critical decision-making, rather than only estimating parameters, is an important issue. For example:

- In the development of new drugs, are there any obvious significant differences between subjects receiving the real drug and placebo?
- When establishing environmental standards for water quality, is there a significant difference in the effects on test organisms when they are cultured in concentrations exceeding the standards compared to when they are cultured in a clean environment?

When considering such issues, various challenges may arise, including not only how to extract samples but also how to store samples, calibrate and adjust measurement equipment and so on. It is a question of the representatives of the obtained data, whether the data are adequate in estimating population parameters and whether they are appropriate as a basis for the necessary decision-making. The degree of appropriateness of the data can be expressed in terms of "data quality". Constructing and operating the system for data acquisition of appropriate quantity, appropriate quality and appropriate cost is called data management. The management of data acquisition systems for environmental monitoring are detailed in Chapter 8, specifically in Section 8.3.

1.3.1 Data Quality

In the international standards on quality, there is a definition of the term. ISO 9000 standard (ISO, viewed in 2022) defines quality as the "degree to which a set of inherent characteristics of an object fulfils requirements".

A more refined definition has been provided by the American Society for Quality (ASQ, viewed in 2025b), that explains quality as "a subjective term for which each person or sector has its own definition". In technical usage, quality can have two meanings:

1. the characteristics of a product or service that bear on its ability to satisfy stated or implied needs;
2. a product or service free of deficiencies.

From this explanation, it can be concluded that data quality refers to its fitness for the purpose for which the data are collected. For example, if the objective is to estimate the population mean, the challenge lies in how to extract the sample and whether the equipment used for measurement is appropriate. If they are appropriate, the quality of the data can be expected to remain in good condition. These efforts are considered quality management in data acquisition. Additionally, these activities may include organising the obtained data and making the data available to those who wish to use them.

1.3.2 Data Management and Its Quality Issues

The concept of data management may appear abstract and difficult to conceptualise. Therefore, first a few examples of data management are presented to explain how data management is done in some institutions and projects where data acquisition is a key issue.

Example 1–1: GE-BICH (Group of Experts on Biological and Chemical Data Management and Exchange Practices) is a working group formed by the International Oceanographic Data and Information Exchange (IODE) programme of the Intergovernmental Oceanographic Commission (IOC). GE-BICH organised a workshop on quality control (QC) and quality assurance (QA) of chemical oceanographic data collections in 2010 (IODE, 2010). They focused on four main areas:

1. Quality control checks and procedures,
2. Common quality control issues,
3. QC flags schemes and strategies, and
4. Units and metadata requirements for data reporting and exchange.

Example 1–2: Castillo et al. (2011) revised the first edition of the World Health Organisation (WHO) manual *Laboratory Methods for the Diagnosis of Meningitis* and published the second edition, including Chapter 2: Epidemiology; Chapter 3: Results management and data reporting; and Chapter 13: Quality control/Quality assurance. Chapter 13 provides a relatively straightforward introduction to quality control and quality assurance.

According to Castillo et al. (2011), quality control/quality assurance (QC/QA) can be defined as the set of planned and systematic activities focused on providing confidence that quality requirements will be fulfilled. It covers a wide range of matters that influence the quality of a product or service. In a medical laboratory, quality can be defined as the accuracy, reliability, and timeliness of the reported test results. QC refers to those measures that must be included in each assay to verify that the test is working properly. QA is defined as the overall program ensuring that the results reported by the laboratory are as correct and accurate as possible.

Example 1–3: The United States Geological Survey (USGS) has detailed information on data management on its website. The USGS (2022, viewed) stated, "Data quality management is the process of adopting protocols and methods to ensure that data are properly collected, handled, processed, used, and maintained at all stages of the scientific data lifecycle". Furthermore, good data management enables the location, sharing, and reuse of data, and reduces the redundancy of data. These attributes of good data management reduce costs in terms of time and money. The following explanations have been extracted from USGS (2022) website:

- ***Why manage your data:*** When data are well documented, you know how and where to look for information and the results you return will be what you expect. Accurate data are legally and scientifically defensible and may aid the agency by reducing litigation and appeals.
- ***Value of data management:*** Poor data quality, redundant data, and lost data can cost companies 15 percent to 25 percent of their operating budget. What would a 15 percent cost reduction be worth to your Project or Program?
- ***Data lifecycle overview:*** Our data are corporate assets with value beyond our immediate need and should be managed throughout the entire data lifecycle. Questions of documentation, storage, quality assurance, and ownership need to be answered for each stage of the lifecycle.

Example 1–4: The California State Water Resources Control Board and the regional water quality control boards (Regional boards, collectively the water board) (2022, viewed) manage and collect vast amounts of data and information on water quality, water quantity, use and agency operations in the state. The state water board adopted an Open data resolution in 2018 to increase attention to data management and accessibility at the water board. The QA/QC details of the management of these data, a central issue, are also provided on the web page.

A common issue in all four of the aforementioned cases is the management and assurance of data quality. When it comes to commodities or products, their quality is relatively clear, making it easy to discuss. However, when considering data quality, there is a challenge due to the wide range of purposes for which data will be used, making it difficult to reach a common understanding. Discussions about the quality of such test data began in medical laboratories as highlighted next.

In the past, the variability of laboratory data among hospitals was so large that Levey and Jennings (1950) proposed the application of Shewhart's control chart to improve the variability of laboratory data. The modified control chart method introduced by Levey and Jennings (1950)demonstrated that variability in laboratory data occurs not only between laboratories but also within laboratories. This chart is also known as the Levey and Jennings chart. This is the origin of what is now called laboratory quality control.

Laboratory quality control (lab QC): Lab QC covers a wide variety of instrumental analyses, but often, due to its historical background, it is assumed to be an analytical laboratory, such as for clinical laboratories. There it is designed to detect, reduce, and correct deficiencies in the laboratory's internal analytical processes to improve the quality of results reported by the laboratory before patient results are published.

Analytical quality control (AQC): AQC refers to all processes and procedures designed to ensure that the results of laboratory analyses are consistent, comparable, accurate and within specified precision limits. This concept has evolved as quality control in analytical chemistry. Analytical techniques, sampling techniques and data processing techniques are key issues here.

Quality control in laboratories evolved uniquely for such clinical testing and came to have several characteristics that differ from those of quality control for the management of manufacturing processes of industrial products. Later, it was applied to various fields, such as environmental data, data from geological surveys and so on, leading to the development of management methods for data acquisition systems discussed here. In Shewhart's control chart, the probability of type I error is set at a very small 0.3%. Note that these control charts, proposed by Shewhart, are tools for process control. These charts utilise the principle that the characteristic values of products produced in a properly controlled process follow a normal distribution, with a 0.3% probability of a value exceeding three times the standard deviation (SD). If a time-series graph displaying quality values of products made in a process shows a value deviating more than three times of the SD from the mean value, it is determined that an abnormality has occurred in the process. In this scenario, there is a 0.3% probability of incorrectly determining that an abnormality has occurred in the process even when no abnormality exists (NIST, 2012).

This is also to avoid confusing the process by erroneously judging that "the process has changed". On the other hand, in laboratory data control, type II errors must also be sufficiently small to detect questionable data reliably. The problem of error, type I and II errors, in such hypothesis testing is discussed in Chapter 4.

The presented examples of data management and data quality initiatives in several organisations in this section provide details on issues such as environmental sample extraction and storage methods, maintenance of measuring equipment, and personnel competence. However, there are still issues to be considered, such as how to define the data quality. As for environmental monitoring, the specific application of QA/QC to ensure data quality will be further discussed in Chapter 8.

1.3.3 Concerns Related to the Adherence to International Quality Standards

It is worth noting here that these organisations list individual QA/QC issues related to data management and data quality management but do not consider them as a total quality management system. A characteristic of the data quality issues in many measurement projects, including the preceding four cases, is that they make no reference to ISO 9001, the well-known international standard for quality management.

Considering the relationship between the quality management of data acquisition systems and the ISO9001 standard may provide more insights. The first edition of ISO 9001–9003 (later integrated into ISO 9001), issued in 1987, was positioned as a system standard for "quality assurance" and ISO 9004 as a system standard for "quality management". Subsequently, it was revised several times, greatly changing its character. The latest version of ISO 9000, which provides the definitions of QMS terms, states that QMS includes not only QC and QA but also QI (quality improvement). Such operations do not seem to be suitable for many data management systems, especially for one-off project management.

The 9001 standard is not a requirement for the quality of a product but rather a requirement for the process of continuously producing a product. Therefore, while it may be beneficial for organisations that continuously provide test and inspection results for their customers, it may not be suitable for one-off projects. This may be one of the reasons why it is difficult for those organisations or projects, such as those in the four examples, to adapt to ISO 9001. For organisations that continuously provide high-quality measurement data to their customers, such as testing centres attached to medical institutions, it is logical to set goals for the quality of the data they provide and to strive to achieve those goals. However, in many projects, efforts are made to obtain data of the quality necessary for important decision-making. In such cases, there cannot be data quality goals that are separate from the project's own goals. This may also be the reason why the example organisation previously mentioned had difficulty implementing a QMS that complies with ISO 9001 for data acquisition.

According to ASQ (2025a):

> Quality assurance and quality control are two aspects of quality management. While some quality assurance and quality control activities are interrelated, the two are defined differently. Typically, QA activities and responsibilities cover virtually all of the quality system in one fashion or another, while QC is a subset of the QA activitises. Also, elements in the quality system might not be specifically covered by QA/QC activities and responsibilities but may involve QA and QC.
>
> (https://asq.org/quality-resources/quality-assurance-vs-control)

In this ASQ definition, the relationship between QA and QC based on ISO 9000 is illustrated, which is quoted and shown in Figure 1.3.

FIGURE 1.3 Quality system, quality assurance and quality control relationships.

Source: Adapted from AQC (viewed in 2025a) and modified.

ASQ considers QC to be a subset of QA, but in some organisations, the two are clearly defined separately, while in others they are combined and lumped together. First, compare the meaning of the words "control" and "assurance". Control has the connotation of governing, controlling, or regulating, and it has the connotation of adhering to established work standards, as well as laws and regulations. On the other hand, assurance has a strong connotation of giving trust to specific or potential stakeholders. QA in ISO 9000 basically means "confidence in the ability to provide products and services of quality as per agreed specifications through demonstration". The key to understanding the essence of this meaning is "as per specification" and "through demonstration". In other words, one can see that "specification" and "demonstration" are the two words that indicate the actual image of trustworthiness (Iizuka, 2018). In addition, historically there have been many twists and turns in the development of quality control and quality management. This can be understood from the fact that even the content in ISO/TC176, which discusses the ISO 9000 standards, has changed over time (Iizuka, 2018).

1.4 Summary

The foremost concern in environmental monitoring is producing the data with the required quality to characterise the state and trend of environmental quality; hence appropriate measures are applied in every step of a monitoring framework. To ensure the data quality, the use of appropriate sampling methods is of importance, i.e., how a sample is extracted from the population under the study, which should be planned together with the measurement methods.

Statistical analysis techniques are utilised in various fields, with applications differing across disciplines. In any case, it is crucial to employ statistical analysis methods that yield valuable insights for intended decision-making and to prepare data with adequate quality and quantity for such analysis. Consequently, implementing a suitable data management system is essential.

This book is organised into three parts with Part I providing a review of the basics of statistics, Part II focusing on sampling designs and Part III focusing on applications of statistical methods for data analysis. The content of all three parts is enriched with examples and real-life case studies.

REFERENCES

American Society for Quality (ASQ), 2025a, *Quality assurance vs control*, retrieved on 10 September 2025 from https://asq.org/quality-resources/quality-assurance-vs-control

American Society for Quality (ASQ), 2025b, *Quality glossary – Q*, retrieved on 10 September 2025 from https://asq.org/quality-resources/quality-glossary/q

Artiola, J.F., Pepper, I.L. and Brusseau, M.L., 2004, *Monitoring and characterization of the environment, chapter 1 in the book environmental monitoring and characterization*, ISBN 978-0-12-064477-3.

Berthouex, P.M. and Brown, L.C., 1994, *Statistics for environmental engineers*, Lewis Publishers, Boca Raton, FL.

California State Water Resources Control Board and Regional Water Quality Control Boards, 2022, *Water boards data and databases*, retrieved on 28 November 2022 from www.waterboards.ca.gov/resources/data_databases/

Castillo, D., Harcourt, B., Hatcher, C., Jackson, M. et al., 2011, *Laboratory methods for the diagnosis of meningitis caused by Neisseria meningitidis, Streptococcus pneumoniae, and Haemophilus influenza, WHO manual, WHO/IVB/11.09*, 2nd edition, World Health Organisation, Centres for Disease Control and Prevention, https://stacks.cdc.gov/view/cdc/11632

Environment Canada, 2017, *Environmental monitoring and surveillance: Chemicals management*, retrieved on 25 September 2023 from www.canada.ca/en/environment-climate-change/services/science-technology/programs/monitoring-surveillance-chemicals-management.html

Gilbert, R.O., 1987, *Statistical methods for environmental pollution monitoring*, Pacific Northwest National Laboratory (PNNL), United States, ISBN 0442230508, www.osti.gov/biblio/7037501

Health Environmental Institute (HEI), 2020, *State of global air 2020, special report*, Health Effects Institute, Boston, MA, ISSN 2578-6873

Heil, A., 2007, *Indonesian forest and peat fires: Emissions, air quality, and human health*, Max Planck Institute for Meteorology, Hamburg, Germany, ISSN 1614-1199, https://pure.mpg.de/rest/items/item_994286/component/file_994285/content

Hubbell, S.P., 2001, *The unified neutral theory of biodiversity and biogeography*, Princeton University Press, Japanese translation by Hirao, T., Shimatani, K. and Murakami, M., 2009, Bun-ichi Sogo Shuppan, Co. Ltd., Tokyo, Japan.

Iizuka, Y., 2018 July, *Quality assurance, tools for achieving sustainable success-learning the basics of quality management-4th lecture, ISOS, no. 248* (In Japanese), www.isosms.info/page/151

Ikeda, K., 2012, *Think about biodiversity, Chukousensho series 009*, Chuokoron-Shinsha, Inc., Tokyo, Japan (In Japanese), ISBN 978-4-12-110009-2 C1345

International Oceanographic Data and Information Exchange (IODE), 2010, *IODE Workshop on QC/QA of Chemical Oceanographic*, Oostende, Belgium, https://goosocean.org/event/496

International Organisation for Standardization (ISO), 2022, *ISO 9000:2015(en), Quality management systems – fundamentals and vocabulary*, www.iso.org/obp/ui/#iso:std:iso:9000:ed-4:v1:en

Johnson, D.L., Ambrose, S.H., Bassett, T.J., Bowen, M.L., Crummey, D.E., Isaacson, J.S., Johnson, J.S., Lamb, P., Saul, M. and Winter-Nelson, A.E., 1997, Meanings of environmental terms, *Journal of Environmental Quality*, 26(3), 581–589.

Levey, S. and Jennings, E.R., 1950, The use of control charts in clinical laboratory, *American Journal Clinical Pathology*, 20(11), 1059–1066, https://doi.org/10.1093/ajcp/20.11_ts.1059

NIST (National Institute of Standards and Technology), 2012, *Chapter 2.2.2.1, Shewhart control chart, in engineering statistics handbook*, https://doi.org/10.18434/M32189, www.itl.nist.gov/div898/handbook/mpc/section2/mpc221.htm

Okamoto, S., Panuthai, S., Aoki, T. and Yamaguchi, S., 2013, Species diversity of butterflies around the Mae Klong watershed research station, Kanchanaburi province, Thailand, *Evolutionary Sciences*, 18, 3–20.

Science Learning Hub, *Article Mendel's principles of inheritance*, viewed on 17 December 2023, www.sciencelearn.org.nz/resources/2000-mendel-s-principles-of-inheritance

UK Department for Environment Food & Rural Affairs, 2022, *Air quality: Explaining air pollution – at a glance*, updated 14 January 2019, retrieved on 2 December 2022 from www.gov.uk/government/publications/air-quality-explaining-air-pollution/air-quality-explaining-air-pollution-at-a-glance

UK-Metoffice, 2021, *Great Smog of 1952*, retrieved on 21 September 2023 from www.metoffice.gov.uk/weather/learn-about/weather/case-studies/great-smog

UN Statistics Division, 2017, *Framework for the Development of Environment Statistics (FDES 2013)*, United Nations Publication, ST/ESA/STAT/SER.M/92, ISBN 978-92-1-161582-1, https://unstats.un.org/unsd/environment/FDES/FDES-2015-supporting-tools/FDES.pdf

UNEP, 2019, *Air pollution in Asia and the Pacific: Science-based solutions*, published by UNEP, Asia Pacific Clean Air Partnership and Climate and Clean Air Coalition, https://wedocs.unep.org/handle/20.500.11822/26861

U.S. EPA, 2013, National ambient air quality standards for particulate matter, *Federal Register*, 78(10), 3086–3287, U.S. Environmental Protection Agency, retrieved on 21 September 2023 from www.govinfo.gov/content/pkg/FR-2013-01-15/pdf/2012-30946.pdf

U.S. EPA, 2023, *PFAS explained*, U.S. Environmental Protection Agency, retrieved on 25 September 2023 from www.epa.gov/pfas/basic-information-pfas#health

U.S. Geological Survey (USGS), 2022, *Manage quality*, retrieved on 2 December 2022 from www.usgs.gov/data-management/manage-quality

Van Dinh Thi Hai and Nguyen Thi Kim Oanh, 2023, Effects of ozone on plant and crop yield: Case studies with rice and peanut. In *Advances in botanical research*, vol. 108 (pp. 83–107), https://doi.org/10.1016/bs.abr.2023.02.002

Part I

Fundamentals of Statistics

Part I describes the basics of descriptive and inferential statistics to enable readers who need such background to follow the applications of statistics in the sampling design and data analysis in environmental monitoring presented in Part II and Part III. The content of Part I includes the basic concepts of statistical hypothesis testing, analysis of variance, and regression analysis, among others, to the extent possible within the context of this book. The purpose is to provide a glimpse of what statistical analysis methods are available and which considerations should be considered when planning various surveys and experiments. Several other concepts are also relevant in the case studies presented in Part II and Part III. References are provided, which enable readers to access more advanced materials.

DOI: 10.1201/9781003597216-2

2

Descriptive and Inferential Statistics

This chapter begins with a brief introduction to the position of statistics. There are two types of statistics: descriptive and inferential statistics. The former is outlined in this chapter. The latter will be discussed in the next and later chapters. This chapter also introduces the law of large numbers and the central limit theorem, which are important theorems that link probability theory and statistics. Also, without going into too much depth, Bayesian statistics, which has recently attracted much attention, is discussed here.

2.1 What Is Statistics?

Statistics is the science of collecting, aggregating, analysing and interpreting quantitative and qualitative data in such a way that the reliability of data-based conclusions is assessed by probability. Statistical techniques can be used to draw reasonable conclusions from limited information in various fields, such as economics and engineering. Statistics requires a lot of mathematical knowledge as a theoretical background. To some extent, knowledge of mathematics is necessary to understand the various ways of statistics. Recently, however, advances in information technology have developed tools to treat statistical software like black boxes, but it is still better to understand some theoretical background. It is important to understand the idea of probability. Probability theory is effective in theoretically dealing with various decision-making problems. The essence of probability theory may also be found in various decisions in gambling. To understand this, it may be beneficial to refer to such examples. In this text, please approach this assignment with the following considerations.

- It is important to understand numerical information in statistics.
- Statistical thinking is important when dealing with uncertainty; we can say, "Statistics is the science of uncertainty".
- In statistics, the ultimate aim is knowledge of the so-called **population**. Extracting a sample from that population is called **sampling**. Then it is possible to generalise about the population based on information obtained from the sample.
- The **sample** is a clue that provides information on the state of the population. In other words, it can be regarded as the only window through which you can see inside the room called the population.
- Using statistical methods, one can analyse the relationship between multiple variables. Analysis of the past behaviour of a predictor also helps to predict future situations and provide information on future trends.

Statistics can be roughly divided into two types. Statistics that collect certain data, make tables and graphs, and understand the characteristics of data by looking at averages, variability and trends is called descriptive statistics. On the other hand, a statistic that takes samples from a population infers the characteristics of the population from the characteristics of the sample or tests whether a hypothesis is correct or not is called inference statistics.

DOI: 10.1201/9781003597216-3

2.1.1 Descriptive Problems

This section explains descriptive statistics. As mentioned earlier, if there is no variation in the sample data, the sample characteristics will represent everything if one value is shown. However, due to the universal variability of the data, different statistical indicators are needed to characterise the state of the population. The most used is the mean value. The mean value is the value that represents the central tendency of the overall data with fluctuations. For example, "The average concentration of a contaminant in Lake A is 60 (of an arbitrary unit but, say, mg/L) the average in Lake B is 55, so it is expressed that Lake A is more contaminated than Lake B". However, is it possible to judge whether the whole lake is polluted only with this single value? For example, if there is an extremely contaminated area in Lake A, and if one calculates the average concentration excluding this highly contaminated sample, the result would be 53. It may also be the case that the samples extracted may not represent the water quality in the whole lake (population). Thus it is helpful to extract enough samples to give information on the state of the population. It would be ideal if a histogram could be constructed that shows the variation (distribution) of the data. By knowing the state of variability, one can learn more about the overall characteristics.

2.1.2 Inferential Problems

As an active study of statistical inference starting from the 20th century, it has a much shorter history than descriptive statistics. Before that, statistics were 100% surveys, and there was no idea about the concept of a population and samples. As the number of survey targets increases, it becomes difficult to conduct an exhaustive survey, so a theory of sampling has become necessary. Statistical inference enables us to obtain the maximum amount of accurate information from a given effort of testing. The sampling survey theory came to be considered for securing representativeness in the questionnaire survey, and the experimental design method was also required for the study of the cultivation method at the experimental farm. These are data collection methodologies for a more accurate understanding of population characteristics, even if the number of samples drawn from the population is small. Typical examples of use are exit poll surveys during elections and TV ratings surveys. In statistics, this inferred statistic occupies a very important place and has evolved in recent years. Furthermore, an overview of sampling theory and its applications will be explained in detail in Chapter 9. However, in the age of Big Data, the position of this inference statistic may be greatly transformed.

The rest content of this chapter primarily discusses issues related to descriptive statistics. Further, issues related to inferential statistics will be discussed in the other chapters, especially in Chapter 4.

2.2 Descriptive Statistics

2.2.1 Picturing the Distribution

If one can collect some data and the number of data increases to some extent, the data can be counted separately for the separated classes, for example, large, medium, and small. If the amount of data further increases, the data can be divided into appropriate intervals and counted. The table aggregated in this way is called a frequency table, and the histogram visualises this table.

When a certain amount of data is obtained, it may be better to create a frequency distribution chart to understand the situation. This figure is called a histogram. Histograms are used to look at the distribution of quantitative data. Divide the data into several classes, create a frequency distribution table, and draw a picture (or a graph). The horizontal axis is the data

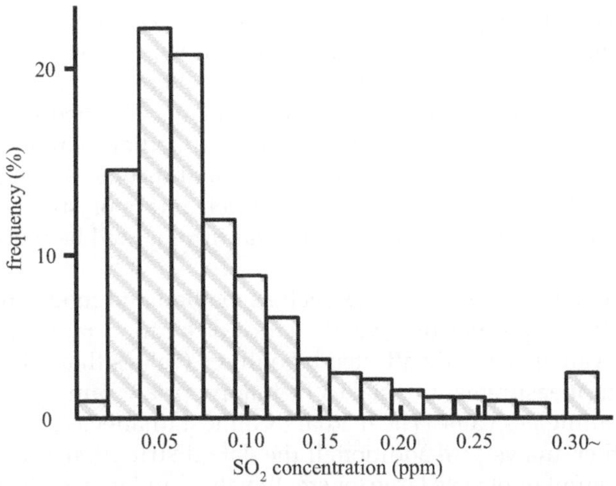

FIGURE 2.1 Example of the histogram. Distribution of hourly sulphur dioxide concentration. The data was measured at the Chuo monitoring station of Kawasaki city from January to March 1969.

Source: Shiozawa et al. (1974).

value, and the vertical axis is the frequency. The histogram looks like a bar graph at first glance, but its area represents the frequency, so be careful with the height if the width of the class is different.

It can therefore be viewed as a data density diagram. Rather than just looking at the data in the frequency table, looking at the visualised histograms gives us much more useful information, such as distribution features. For example, a graph of the appearance frequency of the concentrations of sulphur dioxide in the atmosphere is shown in Figure 2.1. Many air pollution concentrations are distributed with tails on the right (higher concentration side).

2.2.2 Sample Statistics for the Location (Centre of Data)

If a lot of data is available, we would like to know where the centre of the distribution is. In such a case, three parameters that measure the centre of the distribution are important. These parameters, called location parameters or measures of central tendency, include mean, median and mode.

- Mean
- Median
- Mode

The mode value of the discrete probability distribution is a value that maximises the probability mass function. In other words, it is the value that is most likely to appear as a specimen. The mode value of the continuous probability distribution is a value at which the probability density function is maximised and is roughly a peak value. When the probability density function has a plurality of extreme values, each may be the mode of the distribution. Such a continuous probability distribution is called a multimodal distribution.

In a line-symmetric monomodal distribution such as a normal distribution (It will be explained later. Please see Figure 3.1), the mean, median and mode are all the same. For example, if it is found that the distribution follows a line-symmetric distribution, the average of the sample group can be used as an estimate of the mode of the population.

The median is one of the representative values and is a value that is in the centre when a finite number of data are arranged in ascending order. For example, if there are five pieces of data, the third size is the median value. However, if there is an even number of data, the arithmetic average of two values close to the centre is taken. The median is used for purposes like the mean, but, depending on the application, the median may have better properties than the mean. For example, if you consider the annual income of the people, it is easy to understand. In a country with a large gap between rich and poor, some wealthy people raise their average annual income, so the average annual income will be much higher than most people.

In addition, the following characteristics such as median and mode are often used as representative values that represent the overall centre of location rather than average values. The mean (average value) is greatly affected by outliers (values that differ significantly from other values) due to measurement mistakes or inappropriate sampling and may become a questionable or meaningless value. The median is almost unaffected by outliers, so no action is required for further analysis. In addition, if the data distribution is asymmetric, the mean is drawn to a small number of very large (or small) values and may shift more than the centre of the distribution where a lot of data is gathered. However, the median is not significantly affected.

Example 2–1: Five pieces of data were collected by a researcher. The values of the data are 1100, 1200, 1300, 1400 and 5000. The mean is 2000, and the median is 1300. The question is which is more acceptable to most people as a way of displaying the centre of distribution.

Note that the mean is better than the median for statistically estimating the centre of the population. One can see that by finding the variance of the sample means. If the distribution of the data is a symmetrical bell-down shape, that is, it follows a normal distribution, the values of the mean, median and mode are all the same. However, the median is sometimes better when the shape of the distribution is distorted. In data such as national income, it is possible that in countries where some of the richest people make up most of the wealth in the country, people with higher incomes than the mean may be only the royal family and some related parties. In such cases, the magnitude of the statistic will be in the order of mean, median and mode.

2.2.3 Sample Statistics for the Variability (Scatter of Data)

The data obtained by measurement often has variability. There are various measures to express the degree of this variation. The statistical variation is minimal when all the data are the same value and larger, as the difference among the data is larger. Important values that indicate variability include range, variance and standard deviation.

- range
- variance
- standard deviation

The range is calculated from the maximum value in the dataset minus the minimum value. It can easily represent the degree of scattering of data.

The average of the squares of the values obtained by subtracting the average value from the value of each data point is called variance. When the population and the sample are considered separately, they are called population variance and sample variance respectively. In descriptive statistics, the variance is denoted as s^2. Its square root s (positive square root) is called the standard deviation.

Since the standard deviation has the same unit dimension as the original data, it is easier to understand when expressing the extent of dispersion of actual data. On the other hand, the variance is the "second moment around the mean", which is suitable for mathematical logical thinking.

2.2.4 Moment

When X is a random variable and α is a constant, the n-th order moment concerning α is defined as

$$\mathit{Moment}\,(X) = \langle (X-\alpha)^n \rangle n = 1, 2, \ldots \tag{2–1}$$

Here, $\langle \ldots \rangle$ represents an operation that takes an expected value (see more detail in Chapter 3). The moment is one of the characteristics of distribution. The moment is used to quantify the average value, variation, distortion and kurtosis of the distribution. Distribution moments are useful to explain distribution characteristics when the density function is unknown. In general, the first to fourth moments are fundamental.

Note that the explanation of moment is briefly included here because it can be used to describe the characteristics such as the kurtosis and skewness of the distribution of the data. Section 3.2 of Chapter 3 provides the related information, e.g., random variables and expected values.

Especially when $\alpha = 0$, the moment is written as m_n.

$$m_n = \langle X^n \rangle \qquad n = 1, 2, \ldots$$

The expected value μ is equal to the first moment m_1. The variance σ^2 can be expressed by the second moments m_1 and m_2. That is,

$$\mu = m_1,$$

$$\sigma^2 = m_2 - m_1^2.$$

The n-th moment about m_1 is represented by μ_n and is called the n-th order centre moment or the n-th centring moment.

$$\mu_n = <(X - m_1)^n > n = 1, 2,$$

Here, the second-order centring moment μ_2 agrees with the variance. Next, if we use centring moment, the values of n from 1 to 4 have the following meanings.

$n = 1$	1st moment	... location (centre of distribution)
$n = 2$	2nd moment	... variability (scatteredness)
$n = 3$	3rd moment	... skewness
$n = 4$	4th moment	... kurtosis (sharpness)

The skewness can be calculated by the following equation.

$$\alpha 3 = \langle (X - \bar{x})^3 \rangle / \sigma^3 \tag{2–2}$$

$\alpha 3 < 0$		Long hem is on the left.
$\alpha 3 = 0$		Left and right are symmetrical.
$\alpha 3 > 0$		Long hem is on the right.

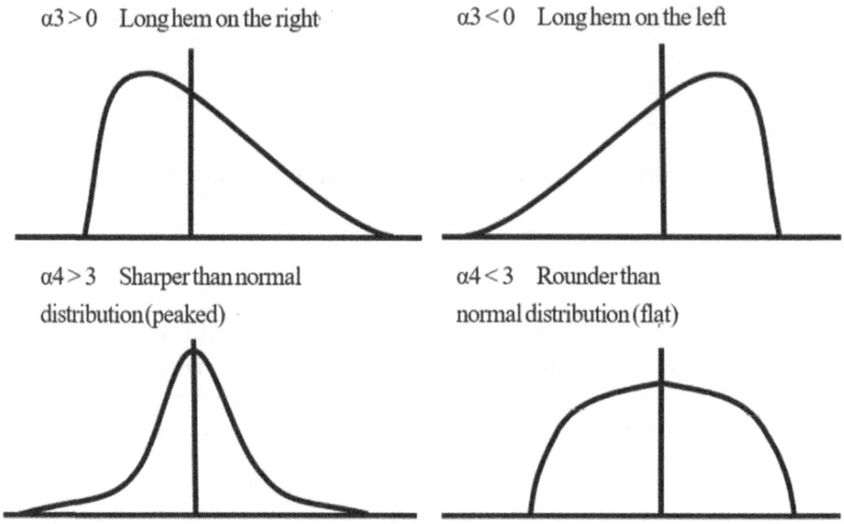

FIGURE 2.2 Shapes of distributions.

The kurtosis can be calculated by the following equation.

$$\alpha 4 = \left\langle (X - \bar{x})^4 \right\rangle / \sigma^4 \tag{2–3}$$

α4	<	3	Rounder than normal distribution	(flat)
α4	=	3	Same kurtosis as normal distribution	
α4	>	3	Sharper than normal distribution	(peaked)

Figure 2.2 shows the typical distribution shapes associated with different levels of skewness and kurtosis.

2.2.5 Temporal and Spatial Resolution of Data

This section describes the characteristics of the data that are obtained by measuring or counting. Important ideas about the basic treatment of these data can be found in many statistics textbooks, Data on products in factories and data obtained through marketing surveys can also be handled in this way. However, when it comes to air and water quality monitoring data, it is necessary to resolve issues such as the spatial and temporal resolutions of the data, which until now has not been considered very important by researchers in the field of statistics. The issue regarding data acquisition will be discussed in Chapters 8 and 9.

The concentration of some constituents in air, water, soil or other environmental matrices varies from place to place and by time. Therefore, when the concern is whether the air or water pollution concentration in a certain place exceeds the environmental standard, the question is what kind of average value of space and what kind of average value of time should be used, which should be specified. Naturally, the concentration of substances differs widely with location and time of measurements. The spatial and temporal scales required for the data presentation also differ between the problems of concern. For example, when discussing issues such as global warming and the depletion of the ozone layer, it is necessary to consider the average concentration of the whole Earth. On the other hand, the average indoor concentration is also important when considering the quality of air in our daily lives. A proper understanding of these concentrations at these different spatial scales and at different timescales is essential for considering environmental science.

For example, air pollution from automobiles is a problem near roads. That is, it will be in the range of about 1 m to 100 m. In addition, air pollution by automobile exhaust gas in the whole city will be a problem on a scale of about 1–10 km. In the discussion of fine particles related to the acid rain phenomenon, the related scale to be considered would be 10–1000 km. For many air pollutants for which many environmental standards have been established, the concentration for 1 hour is the basic unit of measurement.

In addition, the probability distribution of the annual mean and the daily mean over a year may be the subject of consideration. Local CO_2 concentration distribution is a less important issue when it comes to the matter of climate change by reducing CO_2 emissions. Because climate change is a problem of the whole Earth, if its horizontal distance is measured, it is an average value on a scale of 1000 km or more. In addition, the average concentration of 1 day or 1 hour does not matter very much. Rather, the annual average is more important. This section only discusses the basics of data handling. Further detailed discussion will be found in Chapter 14.

By collecting a large amount of data (for example, 8760 hourly CO data points per year), it is possible to determine how often the concentration would exceed a certain level, say 5 ppm measured at a roadside, in a year. Next, by taking the frequency on the vertical axis and the concentration on the horizontal axis to create a histogram, one can understand the characteristics of the distribution. It is possible to also check if the concentration distribution can be

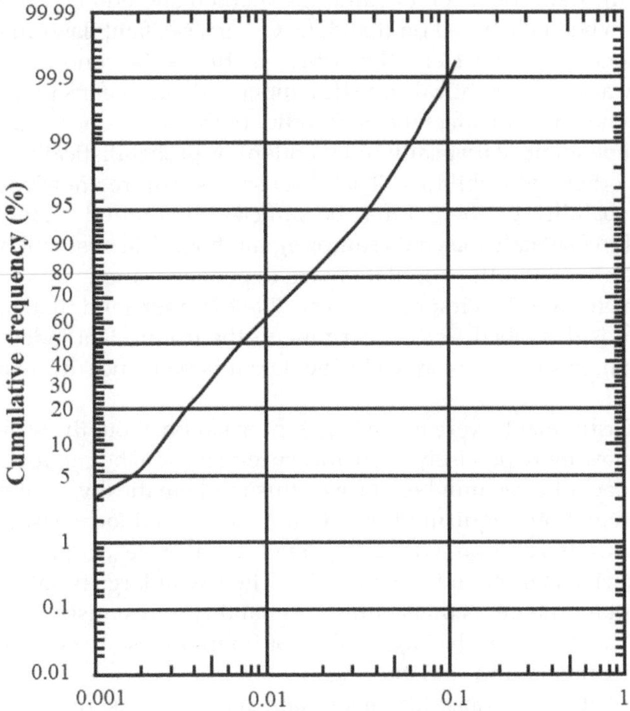

FIGURE 2.3 Example of PC curve (percentile concentration curve). The hourly concentration of nitrogen oxides at the No. 108 monitoring station in the Kashima area from April 1, 1980, to March 31, 1981. The number of data was 8215, and the annual average concentration was 0.012 ppm.

Source: Industrial Pollution Control Association of Japan (IPCAJ, 1985).

approximated by a normal distribution. Furthermore, a graph can be created by taking concentration on the horizontal axis and the cumulative appearance probability corresponding to the cumulative frequency of the normal distribution on the vertical axis. Since the concentration of many air pollutants has a distorted distribution towards higher concentrations, it is convenient to represent the concentration on a logarithmic scale. Such graph paper is called log-normal probability paper, and if the plot on this figure is close to a straight line, this distribution can be approximated by means of log-normal. Such a graph (curve) is called a PC curve (percentile-concentration curve). (See Figure 2.3.) More discussion on this issue can be found in Chapters 14 and 18.

When collecting samples to measure the concentration of pollutants in air and water, it is very difficult and almost impossible to accurately extract the sample at the location and time indicated by the values in the random number table. This is because these trace components are transferred by the flow of air and water and diffuse in it. When samples are taken continuously at the same location, they are always influenced by conditions just before and slightly upstream from that location. This can also be understood from the physical laws (diffusion equations for substances) that describe the behaviour of substances moving in fluids such as water and air. Therefore, it is important to understand its characteristics before analysing the concentration data. The next section provides more details on this issue.

2.3 Other Important Characteristics

In statistics, first, the data collection or sampling method should be appropriate so that various discussions can be made based on that data. Otherwise, fatal flaws in statistical interpretation and judgment can occur. Here, the critical matter is the concept of **independence**. In probability theory and mathematical statistics, independence means that, in multiple events, the occurrence of one does not affect the probability of the occurrence of the other. Two events are said to be independent, statistically independent or probabilistically independent if they do not affect each other's probabilities. Strict discussions require knowledge of mathematical set theory and probability, but here it may be sufficient if you understand its basic idea. It is strictly prohibited to use malicious data sampling methods. The rest will be explained where necessary. Some elements in the population are dependent on one another and may not be independent. In such cases, it is important to be able to understand that. Many of the statistical analysis methods described in this text rely on the use of data independently extracted from the population. In such cases, avoid losing data independence due to improper sampling techniques.

For example, measurements were made to get information about the water quality in a lake. If it is desired to know more precisely, more measurements will be conducted. This is an intuitive decision. The law of large numbers proves this mathematically. A related theorem is the central limit theorem, both requiring the data to be extracted (or measured) independently, and are discussed in this section. For further reading on the exact proof, please refer to the technical books, such as Orloff and Bloom (2014). The law of large numbers and central limit theorem are important theorems that connect probability and statistics. These theorems provide important ideas that form the basic pillars of inferential statistics and serve as a bridge between probability theory and statistics.

The relationship between probability theory and statistics is useful in this context. In probability theory, the probability of an event occurring is fixed, and things are considered based on that probability. Statistics, on the other hand, attempts to draw some conclusion related to the probability of an event from the facts and data obtained. While probability theory is a deductive approach, statistics is an inductive approach that starts from observed events and data and tries to explain the real situation behind them and what is happening.

There has been a long debate about what probability is and how to define it. It has had a profound impact on statistical thinking and methods beyond the scope of probability theory. A typical example of this is the Bayesian school of statistics, shown next. A system of probability spaces composed of axiomatic probabilities is the theorem that guarantees that it is consistent with statistical probability. This is the concept of frequentism in statistics, which is one of the definitions of probability. It defines the probability of an event as the extreme value of the frequency of the event when the number of trials is increased without limit. This statistical probability is obtained as an approximation by increasing the number of trial iterations.

2.3.1 Law of Large Numbers

The law of large numbers states: "As the sample size increases, its average approaches the average for the entire population, that is population mean". For example, let us consider rolling dice to record the number of eyes that come out. If the number of trials is infinitely increased, the probability that the sample mean of the number of eyes will deviate from a neighbourhood of 3.5, the population mean of the eyes, will become small. This is a typical example of a consequence derived from the law of large numbers.

Strictly speaking, there are the weak law of large numbers and the strong law of large numbers, as detailed in technical books such as Taboga (2021).

2.3.2 Central Limit Theorem

The central limit theorem discusses the error between the sample mean and the population mean. In most cases, regardless of the shape of the distribution of the population, the distribution of the sample means of the data independently extracted from the population approaches a normal distribution as the sample size increases.

The distribution of the mean value obtained from many data is explained here using an easy-to-understand example, the average value of the number of dice. First, the distribution of the number of eyes when throwing one die would have the odds of 1 to 6 of the dice of 1/6, and the average value of 3.5.

$$\overline{x} = \frac{1}{6}(1+2+3+4+5+6) = 3.5$$

What happens if one throws two dice at the same time and takes an average of the number of eyes? It is easy to make a two-dimensional table of the number of two dice. What happens if one throws three dice at the same time? If one throws four or more dice, a distribution with a shape that closely resembles the normal distribution will be obtained, as shown in Figure 2.4.

Thus, with this kind of random data, if a lot of data points are collected to take the average, one can see that the dataset has the property of approaching a normal distribution. This nature is very important and allows statistical analysis of many data, in which a normal distribution can be assumed. In this case, the data must be extracted properly.

Note that, in air pollution monitoring, the distribution of the air pollutant concentration data is empirically known to follow a log-normal distribution. Moreover, this property is maintained for hourly data, daily average data, and further for longer averaging times. This seems to be an event that violates the central limit theorem described earlier. The reason is that the air pollutant concentration data are not independent of one another in time-series. In the case of air pollution concentration data, if a concentration with a previous time point is shown, there is a high probability that the next time will be a value close to that concentration. Therefore, the concentration of air pollution is not independent in time-series, so even if one gathers consecutive 24 pieces and averages them, its distribution will not approach the

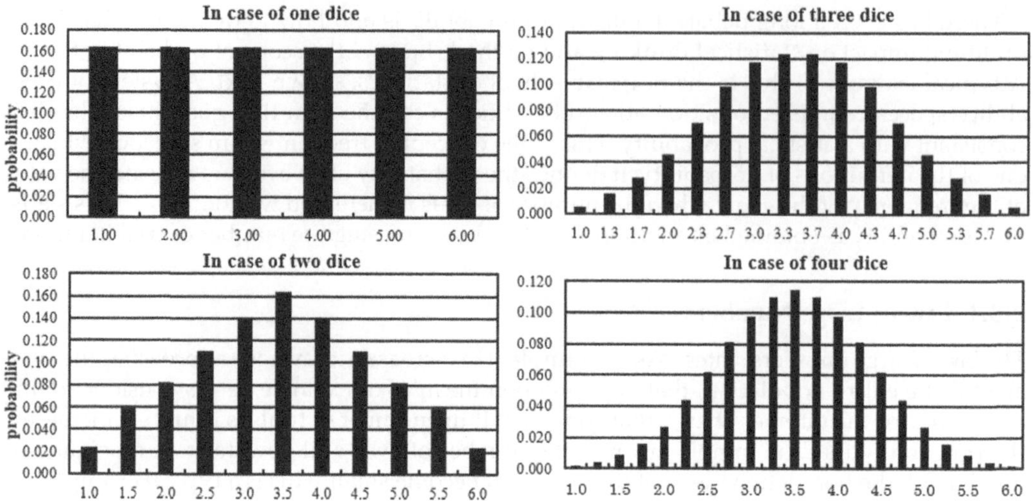

FIGURE 2.4 Distribution of average values, the average value of the number of eyes that appear when throwing 1 to 4 dice.

normal distribution. More details on this issue will be provided along with the case studies in Chapter 14.

2.3.3 Bayesian Statistics

Bayesian statistics refers to statistics based on the idea of Bayes' theorem proposed by Thomas Bayes in the 18th century (Cowles et al., 2019; SAS, 2019). It is mainly used for "estimating the unknown and uncertain". There is a way to think about the position of Bayesian statistics in statistical classification as follows: statistics can be broadly divided into three categories: descriptive statistics, inferential statistics, and Bayesian statistics. The position that emphasises the former two is called frequentist.

Bayesian statistics is a theory in the field of statistics based on the Bayesian interpretation of probability. Probability represents the degree of confidence in an event. It may be based on previous experience or the results of previous experiments and is called prior probability. This differs from many other interpretations of probability, such as the frequentist interpretation, which views probability as a limit on the relative frequency of events after many trials.

In Bayesian statistics, even with insufficient sample data, the prior probability that a certain event will occur is set, and the posterior probability that the prior probability changes with each additional piece of information is updated to derive the subjective probability of the event that would originally occur. This model is said to be compatible with machine learning because it is a model that improves its accuracy by adding and relearning data.

2.4 Summary

In this chapter, we introduced the fundamental concepts of statistics and explained descriptive statistics. However, we only provided a conceptual introduction to inferential statistics, which will be elaborated on in the subsequent chapters. We also discussed that calculating the average value of a large concentration of data in the environment, it rarely follows a normal distribution. This can be related to data independence and the central limit theorem; we also addressed such issues that have not been extensively discussed in the past.

Descriptive statistics is a method for objectively and efficiently describing the characteristics of acquired data. For example, it is possible to calculate the average value, variance, standard deviation, etc. of the obtained data, and then graph it to make the results easier to see.

REFERENCES

Cowles, K., Kass, R. and O'Hagan, T., 2019, *What is Bayesian Analysis*? International Society for Bayesian Analysis, retrieved on 2 December 2022 from https://bayesian.org/what-is-bayesian-analysis/

Industrial Pollution Control Association of Japan (IPCAJ), 1985, *The study on the behaviour of particulate matter, vol. 8: Report on dispersion simulation* (In Japanese), Japan Environmental Management Association for Industry, Tokyo.

Orloff, J. and Bloom, J., 2014, *Central limit theorem and the law of large numbers, MIT open courseware 18.05 introduction to probability and statistic*, Massachusetts Institute of Technology, https://ocw.mit.edu/courses/18-05-introduction-to-probability-and-statistics-spring-2014/73a6c3dad25e892d0d3582122e1c6c01_MIT18_05S14_Reading6b.pdf

SAS Institute Inc., 2019, *Introduction to Bayesian analysis procedures*, SAS/STAT User's Guide, SAS Institute Inc., Cary, North California, United States, http://documentation.sas.com/doc/ja/pgmsascdc/9.4_3.4/statug/statug_introbayes_toc.htm

Shiozawa, K., Ohtaki, T., Okamoto, S. and Yamane, J., 1974, A mathematical model for air pollution control and its application, *Operations Research*, 19(8), 36–41 (In Japanese).

Taboga, M., 2021, *Law of large numbers, Lectures on probability theory and mathematical statistics, online appendix*, Kindle Direct Publishing, Washington, United States, retrieved on 2 December 2022 from www.statlect.com/asymptotic-theory/law-of-large-numbers

3

Sample and Distribution

This chapter explains the concepts of population and sample and some important statistical distributions. The entire research target is called the population. The process of selecting a part of the population as a sample is called sampling. Inferential statistics is the process of estimating the properties of a population using samples drawn from the population. Understanding the relationship between the population and the sample is the basis of inferential statistics.

3.1 Population and Sample

As global warming is a concern, it is important to have an idea of the concentration of CO_2 in the atmosphere. The mass of air on Earth is estimated to be 5.3×10^{18} kg, so it is not realistic to measure all the air. Therefore, measurements can only be made on the air sucked into some measuring devices. and the data obtained is used to estimate the average concentration in the atmosphere. At this point, the whole object of interest is called the population, which in this example is the air on Earth. The volumes of air selected to measure CO_2 are called samples.

Selecting a part of the population as a sample is called extraction or sampling. Inferring the nature of a population using samples extracted from the population is called inferential statistics. On the other hand, it is called descriptive statistics in that the characteristics of data are expressed in an easy-to-understand manner using the acquired data at hand. The relationship between the two has been already explained in Chapter 2.

There are two types of populations: finite population and infinite population. A finite population is a population of finite elements. Although the amount of air on Earth is enormous, it is strictly not unlimited because it is finite. However, in many cases, even if regarded as a true finite quantity, changes in the population after sampling can be ignored, so it can be treated as an infinite population. On the other hand, an infinite population is a population consisting of infinite elements, for example, the data of the eye that throws out dice. The experiment of throwing the dice can be repeated indefinitely, and the data obtained are infinite. Take, for example, a dice and think about recording the eyes that came out. Repeating such a trial enormously, it is possible to reduce the probability that the average of the resulting eyes (sample average) deviates from the vicinity of 3.5 of the average of the eyes. This is a classic example of the consequences derived from the law of large numbers. The law of large numbers means that the mean of a sample following an independent distribution converges to the mean of the population, as discussed in the previous Section 2.3.1.

The mean of the population is called the population mean, and the mean calculated from the sample data is called the sample mean. Thus statistical terms such as "mean" and "standard

TABLE 3.1

Notation of Parameters and Statistics

Statistics	Population	Sample
Mean	μ	$\bar{x}$
Variance	σ^2	V (or s^2)
Standard deviation	σ	s

DOI: 10.1201/9781003597216-4

deviation" are used for both the population and the sample. To distinguish between these two terms, Greek letters are used for the population, and Roman letters are used for the sample. Typical examples are shown in Table 3.1.

3.2 Distribution of Random Variables

A random variable is a value assigned to the outcome of the trial that is to be conducted. It can be a characteristic value that one can determine by measurement or trial (e.g., length, weight, eyes of a dice). Its exact definition requires knowledge of probability theory, set theory etc. which will not be further detailed in this chapter

Generally, the data obtained by counting are integers, and such data become discrete random variables. On the other hand, the data obtained by measurement with various measuring instruments are real numbers, and such data are continuous random variables. The types of data corresponding to these random variables and their characteristics will also be discussed in Chapter 20.

3.2.1 Discrete Probability Distribution

As mentioned earlier, a random variable is a variable that has a specific probability when a variable takes a specific value. When you throw the dice, the eye you see is one of the {1, 2, 3, 4, 5, 6}, and the probability of each eye appearing is 1/6. Therefore, it can be said that the number of eyes that come out when throwing the dice is a random variable.

In this case, if the value of the random variable (= the number of eyes of the die) is X, it can be expressed as the following equation. The parentheses on the right side indicate the range of values that X takes, and in this example, it indicates that "a random variable takes an integer value of 1 to 6".

$$P(X) = \frac{1}{6} \; (X = 1,2,3,4,5,6)$$

For example: "The probability of an event that throws a dice and the result is 'third' is 1/6" can be written as:

$$P(X = 3) = \frac{1}{6}$$

$P\,(X = x)$ = Probability that the value of X is x

The probability distribution when a random variable is a discrete type is called a discrete probability distribution or a discrete distribution.

3.2.2 Continuous Probability Distribution

If a probability exists that corresponds to the possible values of a continuous variable, then this variable is called a continuous random variable. The probability distribution when the random variable is continuous is called the continuous probability distribution or continuous distribution. The data measured by various instruments is usually a continuous random variable. If many such data points are obtained, then a histogram can be prepared. If the amount of data is very large, the graph connecting the tops of the histogram can be approximated by a smooth curve. This figure can be considered as the probability distribution function of a continuous random variable. That is, the horizontal axis represents the random variable, and the vertical axis represents the probability density.

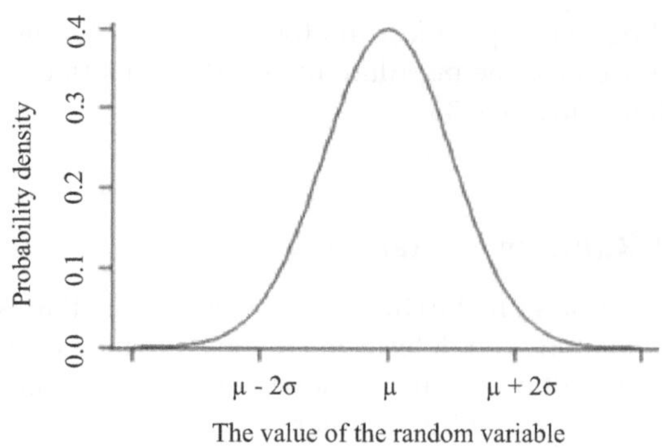

FIGURE 3.1 Probability density function of normal distribution.

The most typical continuous random variable probability distribution is the normal distribution. Its functional shape resembles a bell or mountain shape, as shown in Figure 3.1, which describes the basic nature of the normal distribution.

- Mean, mode and median are the same value.
- It is symmetrical about the mean value.
- The *x*-axis is an asymptote.

Normal distribution with mean μ and variance σ^2 is denoted by N (μ, σ^2) and can be expressed mathematically as follows:

$$f(x)=\frac{1}{\sqrt{2\pi}\sigma}e^{-\frac{(x-\mu)^2}{2\sigma^2}} \quad (-\infty < x < \infty) \tag{3–1}$$

3.2.3 Probability Density Distributions

For continuous data such as length and weight, the characteristics of the distribution can be understood by looking at the histogram when the number of data is large. If the number of data is in the tens to hundreds, we can easily create a histogram using software such as Excel. At that time, the number of classes is set automatically. If the accuracy of the measuring instrument is sufficiently high and the number of data is increased to 1000, 10,000, 100,000 or more, the number of classes can be further increased. The resulting histogram looks like a smooth curve. You can see that this curve approaches the probability density function of a continuous random variable, as shown in Figure 3.2.

When the sample size becomes very large, the frequency distribution becomes smoother and approaches the shape of population frequency distribution. The mathematical function used to represent the population frequency distribution of a continuous random variable is called the probability density function. Probability is always given by the area under the probability density function, and the laws of probability require that the total area under this function equals one (1.0). The probability that a continuous random variable *x* is larger than *a* and smaller than *b* is an (approximate) trapezoidal area with base (*b*-*a*) and height *f* (*x*).

In the histogram, the probability is expressed as a rectangular area. For example, consider a histogram of a person's natural height of 100 students. If 18 students are between 170 and 180 cm in height, the probability of a class between 170 and 180 cm is 18%. However, the probability of a height of 170 cm is zero. For continuous quantities, there is no data for a specific value. For continuous random variables, this means that the data are measured by a measuring

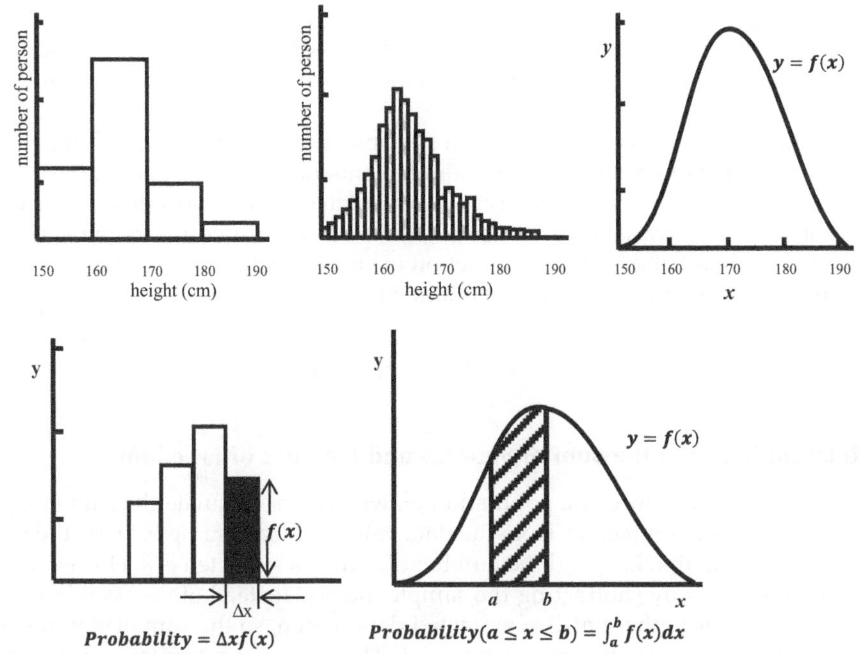

FIGURE 3.2 Histograms and probability density function of continuous distribution.

instrument with an infinite number of significant digits of measurement accuracy. A person who is 170 cm tall does not mean a person who falls between 169.5 and 170.5 cm but should be a person of 170.00000 . . . (infinite number of zeros) cm.

3.2.4 Expected Value and Variance of a Random Variable

The expected value of a random variable is the sum of the products of the value taken by the random variable and its probability, and it represents the average value of the random variable. The expected value is also called expectation. The expected value is one piece of information used to grasp the characteristics of the distribution and is expressed using the initial E for expectation. For example, the expected value of the random variable X is expressed as $E(X)$.

Note that, to clearly distinguish between the random variable and its possible values, the random variable is denoted by upper case letters and its possible values by lower case letters. The probability that a discrete random variable X has the value x is denoted by $P(X = x)$.

The expected value for a discrete random variable can be calculated by multiplying each value x that can be taken by the random variable by the probability p corresponding to the value and adding all the results.

$$E(X) = \sum_i xi\, P(X = xi) \tag{3–2}$$

The expected value of a continuous random variable can be calculated by integrating the product of the value of the variable and the probability density function. In this case, $f(x)\,dx$ is considered to be the probability of x.

$$E(X) = \int_{-\infty}^{\infty} x f(x)dx \tag{3–3}$$

The expected value corresponds to the average of the entire population, that is, the population mean. Therefore, once we know the probability density function of the random variable, we can express the expected value using Equation (3–3), and the population mean is also understood.

Next, consider the variance of the random variable. It is the same as population variance, which is an indicator of variability when all elements of the entire population are known. When it is a discrete random variable, it can be calculated if the probabilities associated with the values of all random variables are known. When it is a continuous random variable, it can be calculated if the probability density function is known. In both cases, the variance of random variables $V(X)$ can be calculated by Equation (3–4).

$$V(X) = E\left(X^2\right) - \left(E(X)\right)^2 \tag{3–4}$$

3.2.5 Relation between the Sum of Squares and Degrees of Freedom

When estimating the variance of a population σ^2, we need to calculate the sum of squares by subtracting the population mean μ from the data values for each sample. Here, the number of elements of the population is N, and the number of samples extracted is n. However, we calculate the sum of squares by subtracting the sample mean $\bar{x}$ instead of the population mean μ. In this way, $\bar{x}$ is closer to the centre of extracted data than μ, so the sum of squares is always calculated to be smaller, as shown in Figure 3.3. The reason the sum of squares calculated in this way is smaller is that it introduces a formula for calculating the sample average as a constraint condition. With the introduction of this conditional expression, we believe that the number of data required to calculate variation has decreased by one. In calculating the sum of squares, even if n data are used, it is considered that the effective data have decreased by one by incorporating the formula for calculating the sample average, resulting in $(n - 1)$. This is the concept of the degrees of freedom. A detailed theoretical explanation is not given here, but this concept is outlined next.

The degrees of freedom have a meaning that it is the amount of data minus several constraints. In this case $(x_1 + x_2 + \ldots + x_n)/n = \bar{x}$ is a constraint for the calculation of the variance. The expected value of the sum of squares (SS) divided by the degrees of freedom (df) provides an estimate of the population variance.

$$E\,(\text{Sum of squares}/\text{Degrees of freedom}) = \sigma^2 \tag{3–5}$$

The average value of the sampled observation data varies. Here, the observed average value is called the sample mean. However, the population mean μ does not vary.

Consider a square whose sides are the size of the deviation (equal to the effect of the variation in individual data). The square of the effect is called the amount of information. Since the area of a square is a squared value, it becomes the value of the information amount of dispersion. For this reason, the observation data x is allocated to the information amount of the population mean value, the information amount of the sample mean and the information of other variations, as illustrated in Figure 3.4. Consider a square with the magnitude of the deviation (equal to the effect part of the variation of the individual data) as one side. All the data have squares of various sizes. The area of the square is the variance, and the length of one side is the standard deviation (Ishida, 2020).

An example of information decomposition is shown for four measured values in Figure 3.4. The population mean μ, the variation of the sample mean and the amount of information of the remaining variation are color-coded in squares. If the sum of squares is calculated with μ, there is no problem, but since it is calculated with the sample mean, it shows that the variation part (darker grey) of the average value is missing from the calculated sum of squares.

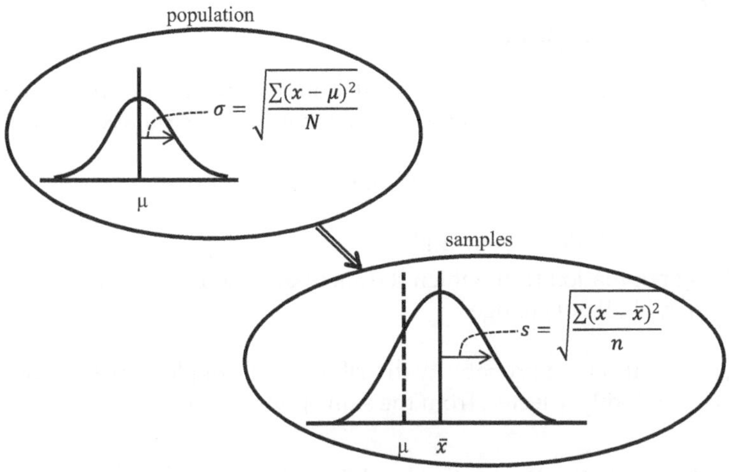

FIGURE 3.3 Variability in population and sample.

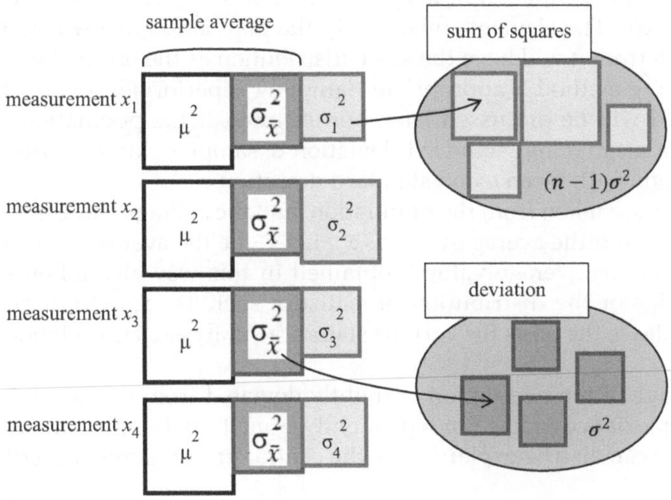

FIGURE 3.4 Schematic explanation of degrees of freedom and sum of squares.

Source: Adapted and modified from Ishida (2020), http://home.a02.itscom.net/coffee/tako08Annex2.html

Since the variation part of the average value (sample mean) is σ^2/n, the total amount of missing information is σ^2 in total. For this reason, the variance σ^2 should be calculated by dividing the sum of squares by $n - 1$ degrees of freedom. Σx_i^2, which is the sum of the squared values of the observation data, is the total amount of information. The total amount of information is distributed as follows.

Total information amount = Total information amount of population mean value + Total information amount of variation part of sample mean value + Remaining total variation information amount

$$\Sigma x_i^2 = n\mu^2 + \sigma^2 + (n-1)\sigma^2 \tag{3–6}$$

3.3 Sampling Distributions

The sample must be representative of the population from which it is drawn. It must therefore be drawn randomly, which means that every member of the population has an equal chance of being drawn in every trial.

A random sample has to satisfy two conditions:

- All sample values are independent, and
- The underlying population from which the successive samples are drawn remains shall be the same, namely shall not change.

The sampling distribution is a probability distribution of sample statistics calculated from all possible samples randomly selected from the same population.

3.3.1 Relation between the Population and Sample

As mentioned earlier, when two events are independent, it means that even if it is known that one event has occurred, the probability of the other event does not change. That is, if two random variables are independent, knowing the value of one variable does not affect the value of the other variable. The elements included in the population follow a certain distribution. The sample taken from it will have the same distribution as the original population distribution if the sampling method is appropriate. Sampling is performed so that it will be, but this sampling method will be discussed later. For example, if the population follows a normal distribution with mean μ and standard deviation σ, samples extracted from it will follow a normal distribution with mean μ and standard deviation σ.

Here, n samples are taken from the population, and the average value is $\bar{x}_1$. Next, n samples are also taken out, and the average value is $\bar{x}_2$. Let $\bar{x}_m$ be the average value of samples taken out the m-th time. The average value $\bar{x}$ obtained in this way also follows a certain distribution. Knowledge of the distribution of statistics such as mean and variance calculated from these samples is the basis for various statistical analyses. This relationship is shown in Figure 3.5.

Even if the shape of the population is slightly deviated from the normal distribution, the average of multiple data extracted independently from it can be approximated by the normal distribution. This can also be explained by the central limit theorem described earlier.

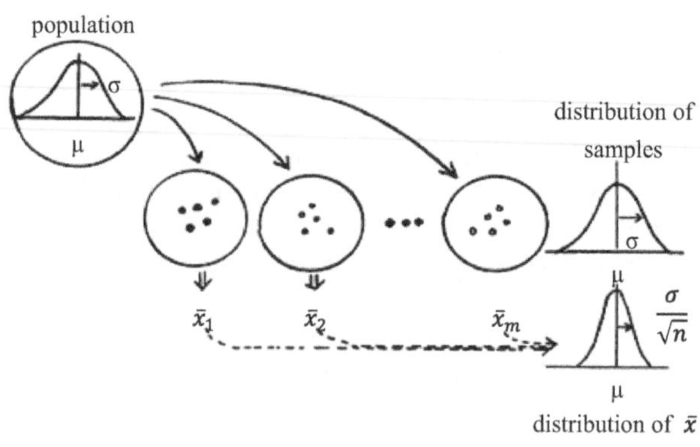

FIGURE 3.5 Relation between the population and samples.

3.3.2 Normal Distribution

The basic idea of the normal distribution has already been mentioned in the section on random variables. Here, let us explain the points to note when performing various analyses using the statistical table of the normal distribution. As the variance (or standard deviation) increases, the peaks of the curve become lower and spread out to the left and right and become flat. As the variance (or standard deviation) decreases, the peaks increase and become sharper. Next, the expected value $E(X)$ and variance $V(X)$ of continuous random variable X can be obtained by integrating the probability density function $f(x)$ of the normal distribution shown in Equation (3–1).

$$E(X) = \int xf(x)dx = \mu \tag{3–7}$$

$$V(X) = \int \{x - E(X)\}^2 f(x)dx = \sigma^2 \tag{3–8}$$

A graph of the probability density function of a random variable following a normal distribution is shown in Figure 3.6. For data following a normal distribution, there is approximately a 5% probability of data appearing that is more than twice the standard deviation from the mean. In addition, there is about a 0.3% chance that data will appear that is more than three times the standard deviation from the mean. This is a handy number to remember.

The probability that the data following the normal distribution will be greater than a certain value can be determined by integrating the probability density function of the normal distribution. However, since the number of the combination of mean μ and standard deviation σ is infinite, it is inconvenient to perform the integral calculation to obtain the probability every time. We consider a variable transformation from X to Z such that the following two probabilities can be considered the same as shown in Figure 3.7. One is the probability that a random variable X that follows the normal distribution with mean μ and standard deviation σ is greater than x, and the other is the probability that the random variable Z that follows a normal distribution with mean 0 and standard deviation 1 is greater than z. This operation is called standardisation. The following equation determines this:

$$Z = (X - \mu)/\sigma \tag{3–9}$$

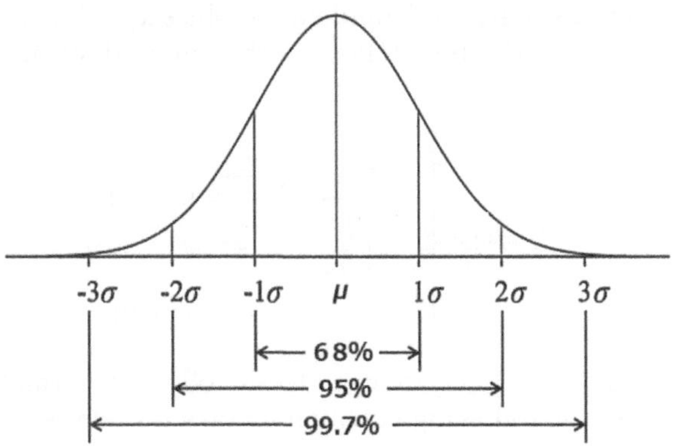

FIGURE 3.6 Probability that data will fall within that range in case the data follow a normal distribution.

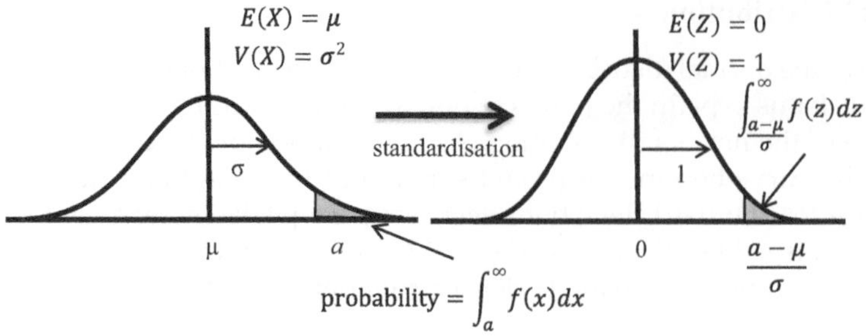

FIGURE 3.7 Standardisation of normal distribution.

3.3.3 *t* Distribution

The mean value of samples extracted independently from an arbitrary population follows a normal distribution by the central limit theorem. The expectation of the sample mean is the same as that of the original population. The variance of the sample mean is the population variance divided by the square root of the number of samples.

The standardised sample mean $\bar{x}$ follows the standard normal distribution. Here, the expected value of variance is the deviation sum of squares divided by $(n - 1)$, but the expectation of standard deviation cannot be described with a simple function. Therefore, instead of dividing by the standard deviation, we consider another way to standardise the sample mean $\bar{x}$; it is possible to consider dividing it by the expectation. However, the value obtained does not follow the normal distribution, but the following *t* distribution:

$$f(t) = \frac{\Gamma((\nu+1)/2)}{\sqrt{\nu\pi}\Gamma(\nu/2)}\left(1+t^2/\nu\right)^{-(\nu+1)/2} \tag{3–10}$$

where degrees of freedom $\gamma = n - 1$, Γ is the gamma function. The shape of the probability density function of the *t* distribution is very similar to that of the normal distribution, as shown in Figure 3.8.

3.3.4 Chi-Square Distribution

The chi-square distribution is one of the most widely used statistical distributions in inferential statistics. The sum of squares of *k* random variables $x_1, \ldots, x_k$, which independently follows a standard normal distribution, follows a chi-square distribution with *k* degrees of freedom.

$$f(x;k) = \begin{cases} \dfrac{1}{2^{\frac{k}{2}}\Gamma\left(\frac{k}{2}\right)} e^{-\frac{x}{2}} x^{\frac{k}{2}-1} & (0 \le x) \\ 0 & (x < 0) \end{cases} \tag{3–11}$$

When each random variable $x_1, \ldots, x_k$ independently follows a normal distribution $N(\mu, \sigma^2)$, the value calculated from the following equation follows a χ^2 (chi-square) distribution with degrees of freedom *k*.

$$\sum_{i=1}^{k}\left(\frac{x_i - \mu}{\sigma}\right)^2 \sim \chi^2(k)$$

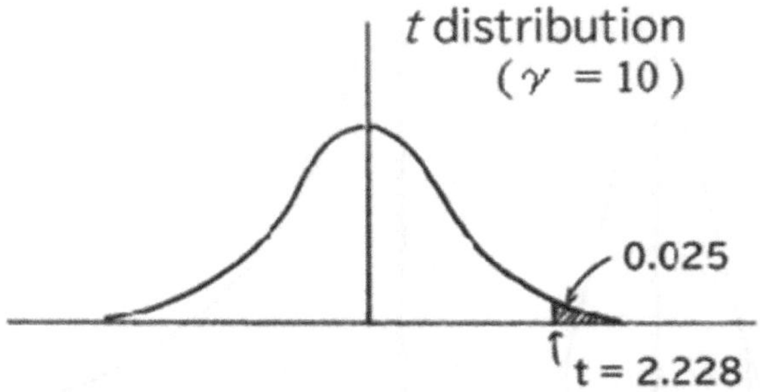

FIGURE 3.8 Approximate shape of t distribution with $\gamma = 10$. In this t distribution, the probability that the value of t will be greater than 2.228 is 0025.

The value of the following equation obtained by expanding this equation follows the χ^2 distribution of degrees of freedom $k - 1$. Where x is the value of the sample extracted from a normal population, $\bar{x}$ is the sample mean, and s^2 is the unbiased variance.

$$\sum_{i=1}^{k}\left(\frac{x_i - \bar{x}}{\sigma}\right)^2 = \frac{(k-1)s^2}{\sigma^2} \sim \chi^2(k-1)$$

3.3.5 *F* Distribution

Consider two populations that follow a normal distribution. Let s^2_1 and s^2_2 be the unbiased variance when sample sizes n_1 and n_2 are extracted from these populations, respectively. Currently, the following relationship holds in the formula for obtaining F. This F follows the F distribution with degrees of freedom (k_1, k_2), where $k_1 = n_1-1$, and $k_2 = n_2-1$.

$$F = \frac{\chi_1^2/k_1}{\chi_2^2/k_2} = \frac{s_1^2/\sigma_1^2}{s_2^2/\sigma_2^2} \tag{3–12}$$

When we can assume that the population variances of the two populations are equal, the preceding equation can be transformed as follows:

$$F = \frac{\chi_1^2/k_1}{\chi_2^2/k_2} = \frac{s_1^2/\sigma_1^2}{s_2^2/\sigma_2^2} = \frac{s_1^2}{s_2^2}$$

Therefore, it is used for an equal variance test in which the null hypothesis is "the variances of the two populations are equal".

3.3.6 Relations between Some Distributions

This section describes the relationship between the normal distribution, t distribution, χ^2 distribution and F distribution. The t distribution, χ_2 distribution and F distribution are important distributions that are often used in inferential statistics, but their probability density functions are complicated, and you do not need to remember their functional forms. These distributions are derived from random variables that follow a normal distribution. Understanding these relationships is also useful for deepening the understanding of the content described in the following chapters.

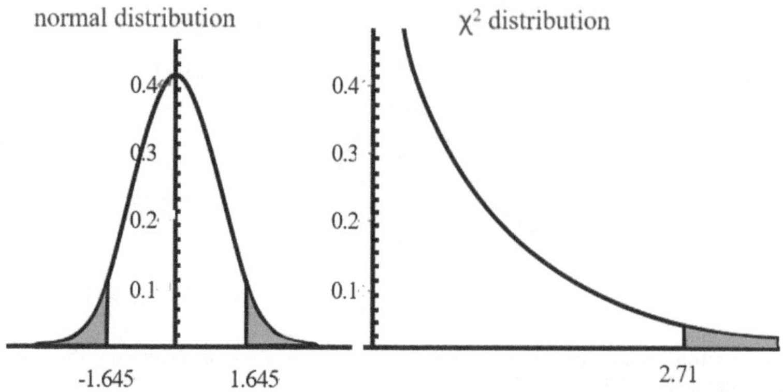

FIGURE 3.9 Two-tailed rejection area of standard normal distribution (each side 5% rejection area), z (0.05) = 1.645, and corresponding rejection area of χ^2 distribution of $df = 1$, $\chi^2(0.10, df = 1) = 2.71$. (*Note:* In this figure, small probabilities are exaggerated slightly to make it easier to see.)

- When $x_1, \cdots, x_n$ independently follow the standard normal distribution $N(0,1)$, the distribution $\chi^2 = x^2_1 + \cdots x^2_n$ follows a χ^2 distribution with n degrees of freedom. Using this nature, we can see the following relation:

$$\{z(\alpha)\}^2 = \chi^2(2\alpha, df = 1) \tag{3–13}$$

Namely, the square of the statistic of standard normal distribution $N(0,1)$ is the same as the statistic of χ^2 distribution with 1 degree of freedom. This relationship is shown in Figure 3.9.

- Let χ^2_1 and χ^2_2 be independent random variables and follow a χ^2 distribution with γ_1 and γ_2 degrees of freedom, respectively. At this time,

$$F = (\chi_1^2/\gamma_1)/(\chi_2^2/\gamma_2) \tag{3–14}$$

follows the F distribution with (γ_1, γ_2) degrees of freedom. Write this as $F(\gamma_1, \gamma_2)$.

- When the "degrees of freedom for the numerator" is 1 in Equation (3–12), the F statistic is consistent with the square of the t statistic. Like the relationship between the normal distribution and the χ^2 distribution, the following relationship holds.

$$\{t(\alpha/2, \gamma_2)\}^2 = F_\alpha(1, \gamma_2) \tag{3–15}$$

Here, the t value is represented by a single-sided probability, but if expressed by both-sided probability, it should be $\{t(\alpha, \gamma_2)\}^2$.

3.4 Summary

This chapter explained the relationship between population and sample statistics and important statistical distributions often used in inferential statistics. Particularly important distributions are the normal distribution, the chi-square distribution, the t distribution and the

F distribution. The concept of degrees of freedom, which is essential to understanding these distributions, was also explained.

REFERENCE

Ishida, H., 2020, *The degrees of freedom and unbiased variance that can be seen in the illustration* (In Japanese), retrieved on 20 January 2020 from http://home.a02.itscom.net/coffee/tako08 Annex2.html

4

Inferential Statistics

This chapter addresses the statistical inferences employed when information about a population is sought. Statistical testing and estimation are crucial aspects of inferential statistics. In statistical test, a hypothesis is first formulated for the population, data are then collected, and ultimately a decision is made to accept or reject the hypothesis. Estimation, on the other hand, is estimating the parameters of a population from the acquired data.

4.1 Sampling and Statistical Inference

Statistics developed before the 19th century are so-called descriptive statistics. The purpose of statistics was to find the regularity of a group. Still, in descriptive statistics, it was thought that the regularity of a group could only be discovered by observing many samples. For this reason, descriptive statistics cannot determine the regularity of the population to which only a small number of samples can be obtained due to limited resources. Inferential statistics was developed to overcome these problems.

In inferential statistics, the main propositions are how to extract a sample from a population and how to estimate a required population parameter from the obtained data.

4.2 Statistical Test

The test (statistical test) is a statistical procedure for deciding whether to reject the hypothesis or not to reject the hypothesis based on the observed value. The procedure is determined so that the probability of rejecting is less than or equal to α even though the hypothesis holds. This α is called a significance level. To understand the concept of statistical testing, it may be a good idea to consider what decisions can be made when participating in gambling (an activity largely dependent on luck).

4.2.1 Concept of Statistical Test

Before we explain the specific test method, let us consider the concept. See the following example.

Example 4–1: A game of dice: Consider a gambling game using dice. In this game, when your opponent throws the dice three times and all the eyes of 1 appear, you will have to pay a certain amount of money. If 1 is displayed three times, you can perform one of the following two actions:

1. pay money, or
2. say the claim.

If you took action 2, it would be because you judged that such a rare phenomenon does not occur so often. Perhaps such an idea may rely on the following hypothesis:

DOI: 10.1201/9781003597216-5

$$(1/6)^3 = 1/216$$

This probability is calculated based on the assumption that the dice is correct. In other words, the chance (probability) is 1/6 for pip 1.

Procedure for solving Example 4–1: Consider the following statistical methods.

Step 1: Enter the underlying hypothesis H_0: the dice are correct. According to this hypothesis, the probability of pip 1 appearing is 1/6, and we can calculate the probability of the event.

Step 2: The level of significance probability α is set to 0.01. This significance level is the probability that a hypothesis will be rejected even though it is correct.

Step 3: Probability calculation, $p = (1/6)^3 = 1/216 = 0.00463$.

Step 4: Comparing the calculated probability p with the significance level α. Since $p < \alpha$, that is, the calculated probability is smaller than the significance probability, the following conclusion is drawn. H_0 should be rejected. That is, the dice are not correct. Judge that something is wrongly crafted.

In this example, it was hypothesised that the dice were made correctly. Based on this hypothesis, the probability can be calculated assuming that the probability of seeing a 1 is 1/6. The probability of 0.00463 obtained as a trial result was smaller than the probability of the criterion of 0.01 that it is not a "rarely occurring phenomenon", so the provisional was discarded. Whether this judgment is correct or not can be considered in a binary table, as shown in Table 4.1. If the hypothesis is false, it is correct to reject it. If the hypothesis is correct, the correct answer is not to reject it. On the other hand, it is an error to reject the hypothesis that is correct; do not reject the hypothesis even though it is false is also an error. Here, the former is called the first type of error, and the latter is called the second type of error. The probability of the first type of error is called α, and the probability of the second type of error is called β. In general, when careful judgment is required, the probability α of the first type of error is reduced.

- A small value of α means prudently, carefully, cautiously, or conservatively.
- A large value of α means boldly or fearlessly.

4.2.2 Terminology Used in Hypothesis Testing

This section explains the technical terms related to the important matters discussed in this chapter.

- Null hypothesis: The hypothesis you want to verify with the hypothesis test. Usually expressed as H_0.

TABLE 4.1

Two Types of Errors in Statistical Tests

		Real situation	
		H_0 is true	**H_0 is false**
Decision	Reject H_0	Type I error (probability = α)	Correct decision
	Do not reject H_0	Correct decision	Type II error (probability = β)

Note: The hypothesis is displayed as H_0. For more information, see the following glossary.

Note that, in many cases, we do some work, expect to see some results and then perform a statistical test. In doing so, we make the hypothesis, "There is no result, and it is the same as the previous state" and expect it to be rejected. In other words, it is called the null hypothesis because it is a hypothesis that is expected to return to nothing.

- ***Alternative hypothesis:*** Describes the state where the null hypothesis is not valid. If it is judged that the null hypothesis is wrong, this alternative hypothesis is adopted. Usually, it is expressed as H_1.
- ***Rejection:*** To determine that the hypothesis is false and eliminate it.
- ***Type I error:*** The error that rejects the null hypothesis, even if the null hypothesis is correct.
- ***Type II error:*** The error that cannot reject the null hypothesis even if the alternative hypothesis is correct.
- ***Significance level:*** Criteria for rejecting the null hypothesis. The null hypothesis is rejected when the first type error probability is smaller than the value of this significance level. In letters, it is expressed as α (alpha), often taking 5% or 1%.
- ***P value:*** Probability that the actual value of observation data or more extreme observation data can be obtained when the null hypothesis is correct. When this is below the significance level, the null hypothesis is rejected.

4.2.3 Procedure for Statistical Tests

The specific procedure of the test is explained using the following example.

Example 4–2: The average weight of the products was 10.0 g. Recently, 11 samples of our products were taken and weighed; the average of the data is 10.5 g. The sample standard deviation s is 1.0 g. Determine whether the weight of the product has changed or not.

Procedure for solving Example 4–2:

Step 1: Set the hypothesis.

$$H_0: \mu = 10.0$$
$$H_1: \mu \neq 10.0$$

Step 2: Set the level of significance, $\alpha = 0.05$.

Step 3: Determine the appropriate distribution. This is a test for a population mean where σ is unknown, and therefore the t distribution is selected.

Step 4: Determine the area in which H_0 can be rejected. (See Figure 4.1.)

$$t \leq -2.228 \text{ and } t \geq 2.228$$

Step 5: Calculate the t value.

$$t = (\bar{x} - \mu) / \left(s/\sqrt{n}\right) = (10.5 - 10.0)/(1.0/\sqrt{11}) = 1.66$$

Step 6: Conclusion.

Since the calculated t value is 1.66, that is less than 2.228. Therefore, the H_0 cannot be rejected; namely, one cannot say that the weight of the product has been changed.

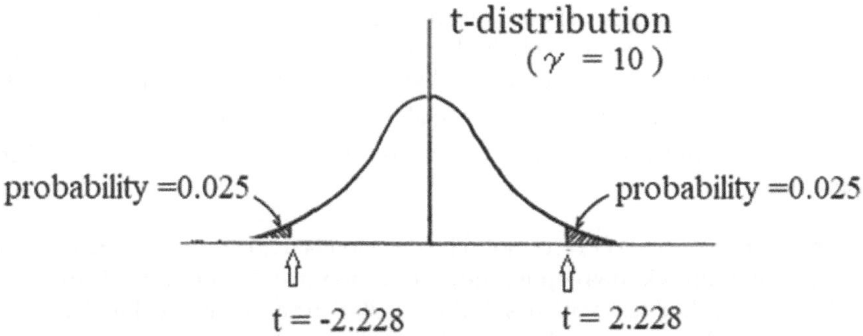

FIGURE 4.1 *t* distribution with degrees of freedom $\gamma = 10$.

Note that, in a typical hypothesis test, only the probability of type I error is set up for testing. If it is desired to consider both type I and type II errors, it is necessary to determine the amount of data in the procedure, as explained in Section 4.2.6 of this chapter. If the test is conducted using the available data, setting only the probability of a type I error (significance level), the probability of a type II error will have been ignored. In this case, if the null hypothesis can be rejected, there is no problem. However, if it cannot be rejected, then it is dangerous to say that the null hypothesis is correct actively. That is because, even though the null hypothesis is wrong, the probability that it cannot be rejected has not been specified. The conclusion in such a case would be a roundabout way of saying, "It cannot be said that . . .".

4.2.3.1 Two Types of Testing

There are two types of statistical tests:

1. Hypothesis testing: Set up a population parameter θ before conducting a test.
2. Significance testing: Decision-making based on *P* value.

For a statistical test, two hypotheses are involved, **alternative hypothesis H_1** and **null hypothesis H_0**. An alternative hypothesis is the hypothesis being proposed by the experimenter according to the objective of testing. The null hypothesis is the negation of H_1. It is hoped that the evidence will lead us to reject H_0 and thereby accept H_1. The significance level is set to prove the hypothesis.

4.2.4 Hypothesis Testing

Consider that a typical estimation problem has a population parameter θ whose values are approximated based on a sample. Usually, there is no preconception about the actual value of this parameter. We are trying to verify its value as much as possible. In contrast, when testing the hypothesis concerning θ, there is a preconception about its value. This means that two theories or hypotheses are involved in this kind of statistical research. One is the hypothesis proposed by the experimenter, and the other is the denial of the hypothesis. The former is denoted by H_1 and is called the alternative hypothesis or research hypothesis. The latter is indicated by H_0 and is called the null hypothesis. The purpose of the experiment is to determine whether the evidence tends to refute the null hypothesis. The following three guidelines help in deciding how to state H_0 and H_1.

1. When testing a hypothesis about the value of a certain parameter θ, the statement is always included in hypothesis H_0. In this way, H_0 identifies a specific number that can be the actual value of θ. This value is called the null value and is indicated by θ_0.
2. Whatever is detected or supported is the alternative hypothesis H_1.
3. Since our research hypothesis is H_1, we expect the evidence to reject H_0 and thereby accept H_1.

The decision will either to reject H_0 or not. The decision is made by observing statistics whose probability distribution is known under the assumption that the null value is the true value of θ. Such statistics are called test statistics. If the test statistic assumes a value that is rarely seen when $\theta = \theta_0$ and tends toward credence in the alternative hypothesis, then H_0 is rejected in favour of H_1. If the observed values are those that typically occur under the assumption that $\theta = \theta_0$, the null hypothesis cannot be rejected.

Here, the null value can be a measured (quantitative) value or a counted (qualitative) value. The number of people is a counted value because it is not measured but counted, and its ratio also becomes a counted value. For example, suppose the average person's cancer rate was 2.0% and the survey result showed that of smokers was 4.0%. In this case, it is desirable to know whether smoking was the cause of the increased incidence. In the statistical test currently, the hypothesis is "the carcinogenesis rate of smokers is the same as the carcinogenesis rate of all people", and the test is whether this hypothesis can be rejected. Furthermore, since it is not conceivable that the rate of cancer by smoking will decrease, the alternative hypothesis to the null hypothesis is "the carcinogenesis rate of smokers is higher than that of ordinary people". In an expression, this is:

- H_0: $p_0 = 2.0$
- H_1: $p_0 > 2.0$

This means that, at the end of the study, the situations summarised in Table 4.1 will occur, as detailed here.

- Rejected H_0 when it was true and shall have committed what is known as a type I error.
- Made the correct decision of rejecting H_0 when the alternative H_1 was true.
- Failed to reject H_0 when the alternative H_1 was true. In this case, this is what is known as a type II error.
- Made the correct decision of failing to reject H_0 when H_0 was true.

4.2.5 Significance Testing

Significance testing is conceptual and a little difficult to understand. Therefore, let us first give an easy-to-understand case and then give a theoretical explanation later.

Example 4–3: The average weight of the product was 10.0 g. Recently, 36 samples were collected and measured, and the average of the data is 10.5 g, and the sample standard deviation s is 1.0 g. Determine whether the weight of the product has been increased or not.

Procedure for solving Example 4–3: Since the sample size is 36 and there are more than 30, the statistics can be calculated assuming the population variance is known. Since the number of data is large enough, the sample variance can be approximately used as the value of the population variance. Assume that σ is not affected by this change. Because the research hypothesis is taken as the alternative hypothesis, the test will be:

- H_0: $\mu = 10.0$
- H_1: $\mu > 10.0$

The sample mean for these data is $\bar{x}$ = 10.5 g (sample size = 36). This value is larger than the null value for μ of 10.0 g. To see if there is enough difference to cause us to reject H_0, we find the *P* value for the test. That is, we compute the probability of observing a sample mean of 10.5 or larger if μ = 10.0 and σ = 1.0. This is done by noting if μ = 10.0 and σ = 1.0; then the test statistic $\bar{x}$ is, by the central limit theorem, at least approximately normally distributed with mean μ = 10.0 and standard deviation $\sigma/\sqrt{(n)}$ = 1/6. Therefore:

$$P(\bar{x} > 10.5 \mid H_0) = P\big(z > (10.5 - 10.0)/(1/6)\big) = P\big(z > 3.0\big) = 0.00135$$

There are two explanations for this very small probability. The null hypothesis is true, and we have observed a very rare case that by chance has a large sample mean; the null hypothesis is not true, and it is reasonable to think that something has changed in the production process and the weight of the product has increased. We may prefer the latter explanation. That is, we shall reject H_0 and report that the *P* value of our test is sufficiently small, namely the probability of wrongly rejecting it is small.

It is important to detail here the concept of "significance testing". In hypothesis testing, a method for deciding whether to reject a null hypothesis is used. In the method of deciding whether to reject H_0, the H_0 and H_1 must be set as just presented. However, in the significance testing, α is not set in advance to specify a strict critical area. Rather, after evaluating the test statistic, the probability is determined for observing a value of the test statistic that is at least as extreme as the value recorded under the assumption that $\theta = \theta_0$. This probability is referenced as the test probability or *P* value. The *P* value is the minimum level at which α can be preset and H_0 can be rejected. If this *P* value is small, H_0 is rejected. In other words, even if the hypothesis is correct, the probability of rejecting it is sufficiently small that it can be interpreted as urging that the null hypothesis be positively denied.

4.2.6 Power of Hypothesis Testing and Sample Size

As a way of example, the following case is considered. A researcher is studying a reagent that is to be used to improve the quality of water used in an industrial facility. The random variable considered is the change in contaminant concentration before and after using the reagent. Data obtained in this experiment is obtained by subtracting the concentration after the reagent addition from that measured before the addition. If the reagent is effective, the average difference in readings should be a positive value. The hypothesis to be tested is:

- H_0: μ = 0 (The reagent is ineffective.)
- H_1: μ > 0 (The reagent is effective in improving the water quality level.)

Next, the researcher needs to determine how large the sample size should be collected. In this case, it can be interpreted as how many measurements need to be made to have enough data for the test.

Consider a test to detect whether there is a difference between the mean values of two populations with approximately the same standard deviation. Suppose that *n* samples are taken from each population, and the average of each sample $\bar{x}_1$ and $\bar{x}_2$ is calculated. The variance of the sample mean is σ^2/n if the population variance is known. Therefore, as shown in Figure 4.2, the larger the number of samples, the easier it is to detect the difference between two means. Therefore, as the number of data increases, the power of the test becomes higher, as seen in Figure 4.3.

Recall that, in testing a hypothesis, two errors are possible. One might reject H_0 when in fact H_0 is true, thus committing a type I error. In this example, it leads to the use of an ineffective agent. The probability of making this mistake can be controlled by agreeing to preset α at some acceptably low value. A type II error occurs when it fails to reject H_0 even though the

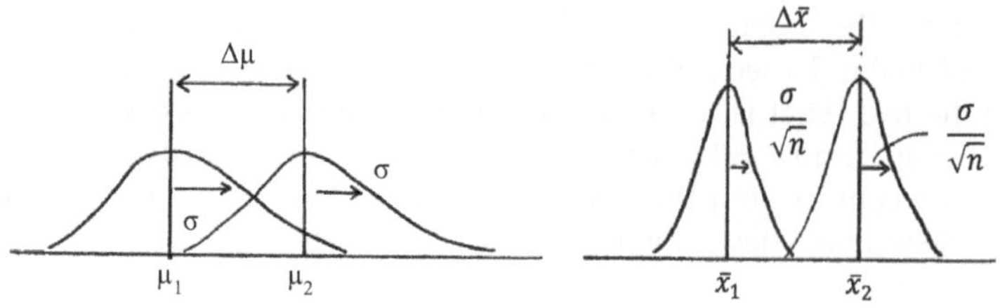

FIGURE 4.2 As the number of data increases, the overlapping area of the two distributions decreases.

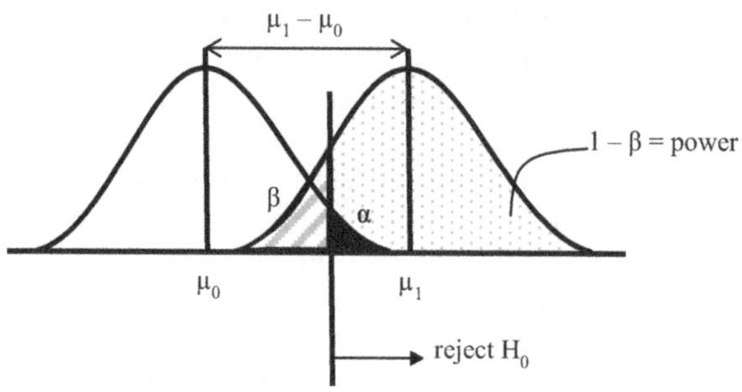

FIGURE 4.3 Power of testing in which the hypothesis is H_0: $\mu = \mu_0$ and H_1: $\mu > \mu_0$.

research hypothesis is true. In this example, a type II error leads to the inability to recognise the effectiveness of the agent being studied. The probability of committing a type II error is beta (β). It is evident that, ideally in any hypothesis testing setting, β should be sufficiently small. The probability that one will be able to detect a true research theory is called power of testing, and computationally power = $1 - \beta$, which is explained in Figure 4.3.

Since the power is the probability of supporting the research hypothesis when it is right, experiments should be designed so that power is high. If the type I error and the type II error sizes are specified in the test for the population mean, the required sample size is obtained from Table 4.2. As shown in Figure 4.3, the power of the test can be considered as the probability to be able to be detected that the population mean has become μ_1 instead of μ_0 set by the hypothesis.

In hypothesis testing, type I errors are often of more concern. In other words, it can be interpreted that the risk of making mistakes by the person in a hurry is more emphasised. But what if the type II errors are also equally important? As seen from Figure 4.2, the standard deviation of the sample means decreases as the number of samples increases. Therefore, even if there is the same difference between the two means, there is a higher probability that the difference can be detected. In other words, the probability of committing errors in both type I and type II can be reduced at the same time as the sample size increases. Table 4.2 shows the required number of samples when the predetermined probability of type I and type II errors is specified using this relationship.

How to use Table 4.2 is explained by the following example. The test is single-sided (a one-tailed test) with $\alpha = 0.05$, $\beta = 0.10$ (power = 0.90). In this sample case, it is desired to identify the difference $\Delta = (\mu - \mu_0)/\sigma = 0.50$, and Table 4.2 shows that the sample size $n = 36$.

4.2.7 One-Tailed Test and Two-Tailed Test

The results of experimental studies are evaluated based on the data obtained. In many cases, the hypothesis corresponding to the expected outcome is the alternative hypothesis because

TABLE 4.2

Minimum Number of Samples for Identifying the Difference (Δ) with Designated α and β Values (Single-Side Testing)

	$\alpha = 0.05$	$\alpha = 0.05$	$\alpha = 0.05$	$\alpha = 0.05$	$\alpha = 0.05$
$\Delta = (\mu - \mu 0)/\sigma$	$\beta = 0.010$	$\beta = 0.05$	$\beta = 0.10$	$\beta = 0.20$	$\beta = 0.50$
0.20					70
0.30		122	97	71	32
0.40	101	70	55	40	19
0.50	65	45	**36**	27	13
0.60	46	32	26	19	9
0.70	34	24	19	15	8
0.80	27	19	15	12	6
0.90	21	15	13	10	5
1.00	18	13	11	8	5

Source: Modified from Table 5.4 of Milton et al. (1997).

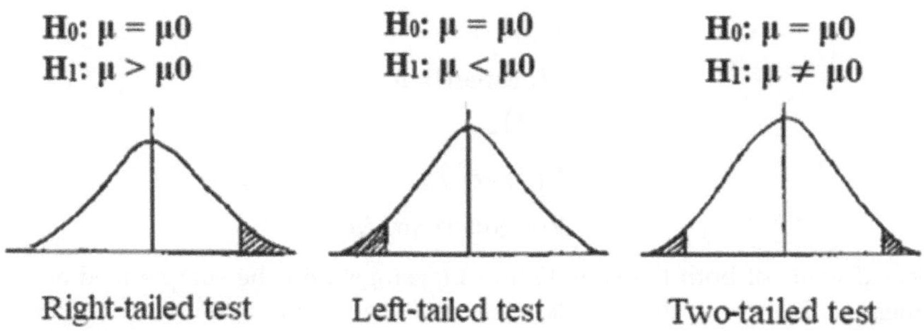

FIGURE 4.4 One-tailed test and two-tailed test. (*Note:* the shaded area in the figure is the probability of the rejection region.)

the null hypothesis is that there is no outcome. For example, suppose a student studies before an examination and statistically tests whether the examination score becomes higher as an outcome. In this case, the null hypothesis is "the examination score does not change", whereas the alternative hypothesis is "the examination score becomes higher". Logically, the outcome of interest is achieving a higher score. It is not presumed that studying before the exam will result in a lower score. That is, only the higher tail of the distribution is considered. This type of test is known as a one-tailed test.

On the other hand, if the results of an experiment of interest may be larger or smaller than the compared value, then it is necessary to consider both sides in the alternative hypothesis. The relation between the one-tail test and two-tail test is shown in Figure 4.4.

4.3 Estimation

Estimating a population parameter such as the mean or variance of a population based on some samples is called estimation. There are two types of estimation:

1. Point estimation is to obtain a single numerical value as an estimate of a parameter (such as population mean or population variance), and
2. Interval estimation is to estimate that a parameter is included in a certain interval.

The point estimation of the population mean uses the characteristics that the expected value of the sample mean matches the population mean. This estimation is possible because the population mean estimator satisfies two characteristics: consistency and unbiasedness.

When estimating parameters such as population mean or population variance based on estimators, the estimation must be accurate. The law of large numbers is that the sample mean approaches the population mean as the sample size n increases. In this way, when n increases, the property that the estimator gradually approaches the true parameter is called coincidence. When estimating a parameter from an estimator, it is meaningless if the estimator deviates significantly from the true parameter. In other words, the expected value of the estimator must match the parameter. This property is called unbiased. "The expected value of the estimator is equal to the parameter regardless of the number of samples n". In other words, it shows that there is no difference in how much the estimated amount deviates when n is small or large.

4.3.1 Point Estimate

If $E(\hat{\theta}) = \theta$, $\hat{\theta}$ is an unbiased estimator of θ, where θ is a population parameter.

$$
\begin{aligned}
&E(\bar{x}) = \mu \\
&E(\text{median}) = \mu \\
&E(s^2) = \sigma^2 \\
&V(\bar{x}) = \sigma^2/n \\
&V(\text{Median}) > \sigma^2/n
\end{aligned}
$$

The expected value of both the sample mean (average) and the sample median is the population mean μ. Thus both the sample mean and the sample median are unbiased estimators of the population mean. However, the variance of the sample mean is smaller than that of the sample median. Therefore, the sample mean is a better estimator from a consistency perspective.

4.3.2 Interval Estimate

Interval estimation is the estimation of population parameters such as population mean with widths (intervals) based on sample statistics. This estimated width is called the confidence interval.

Here, to estimate the population mean, n samples are taken, the sample mean $\bar{x}$ is calculated and an estimation interval containing the population mean with a probability of 95% or higher is obtained.

In the standard normal distribution, the probability point $z(\alpha/2)$ where the probability on both sides α is obtained from the normal distribution table. At this point, the probability that the random variable according to the standard normal distribution is larger than $-z(\alpha/2)$ and smaller than $z(\alpha/2)$ is $1 - \alpha$, as shown in Figure 4.5.

If the population standard deviation is known, the confidence limits for the estimate of the population mean can be calculated using the relationships shown in Figure 4.5, as follows:

$$
\begin{aligned}
&P\{-z(\alpha/2) \le z \le z(\alpha/2)\} = 1-\alpha \\
&P\{\bar{x} + z(\alpha/2)(\sigma/\sqrt{n}) \ge \mu \ge \bar{x} - z(\alpha/2)(\sigma/\sqrt{n})\} = 1-\alpha
\end{aligned}
$$

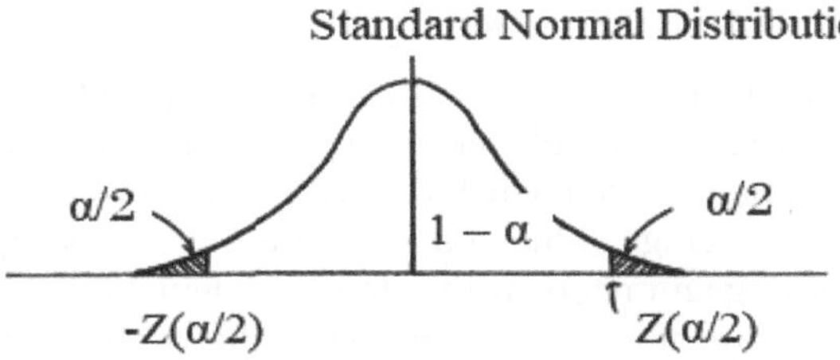

FIGURE 4.5 In the standard normal distribution, the probability inside the area between $\pm z(\alpha/2)$ is $1 - \alpha$.

For example, $\alpha = 0.05$, $z(0.05/2) = z(0.025) = 1.96$

$$P\{\bar{x} - 1.96(\sigma/\sqrt{n}) \le \mu \le \bar{x} + 1.96(\sigma/\sqrt{n})\} = 0.95$$

These values, $\bar{x} - 1.96\ (\sigma/\sqrt{(n)})$ and $\bar{x} + 1.96\ (\sigma/\sqrt{(n)})$ are the lower and upper end-points, respectively, of the confidence interval. These end-points are also called confidence intervals. In case that σ is unknown, we can take a similar procedure, in which σ change into s, and $z(\alpha/2)$ change into $t(n - 1, \alpha/2)$.

4.4 Test of Population Means

A sample from a population is taken to know about the properties of the population of concern. Then the parameters of the population are estimated from the measurements of the sample. In most cases, the population mean is of interest. In addition, if one has some information about the population in advance, it is desirable to confirm its validity. Often, that information is set as a hypothesis, and a decision is made based on the data obtained from the sample. There are two types of hypotheses in this population mean test. One is when the population mean is a specified value. The other is when there are two populations and population 1 has the same mean as population 2. Both types of the population mean test have two cases, namely, the case when the variance of the population is known, and the case when it is unknown. The types of tests for population mean are given next.

Classification of the testing means.

- $H_0{:}\mu = \mu_0$ — σ known — standard normal
- $H_0{:}\mu = \mu_0$ — σ unknown — *t* distribution
- $H_0{:}\mu_1 = \mu_2$ — σ known $\sigma_1^2 = \sigma_2^2$ — standard normal
- $H_0{:}\mu_1 = \mu_2$ — σ unknown $\sigma_1^2 = \sigma_2^2$ — *t* distribution
- $H_0{:}\ \mu_1 = \mu_2$ — σ known $\sigma_1^2 \neq \sigma_2^2$ — standard normal
- $H_0{:}\ \mu_1 = \mu_2$ — σ unknown $\sigma_1^2 \neq \sigma_2^2$ — *t* distribution (Satterthwaite)
- $H_0{:}\ \mu_1 = \mu_2$ ($H_0{:}\ \mu_d = 0$) paired data — *t* distribution

If the variances of the two populations are unknown, the sample size (number of elements to be extracted from each population) is preferably equal, i.e., $n_1 = n_2$. Even if $\sigma_1^2 \neq \sigma_2^2$, the conclusion may not be so different from the case of $\sigma_1^2 = \sigma_2^2$. Strictly speaking, it is not possible to test the mean of two populations that have different population variances. However, in the approximate calculation, it is known that the power of testing decreases when there is a large difference between the sample sizes extracted from the two populations. Thus it is safer that both of the sample sizes are nearly the same. The theoretical proof of this statement can be found in statistical textbooks, for example, Kondo and Ando (1967).

4.4.1 Comparing Means (Variances Equal)

The z or t statistic can be used for the test of the difference between the two population means. The reason is that the difference of sample means ($\bar{x}_1 - \bar{x}_2$) follows a normal distribution. The background is that the population from which the original data were extracted follows a normal distribution or that the average value of a certain number of samples follows a normal distribution by the law of large numbers. If the standard deviation of ($\bar{x}_1 - \bar{x}_2$) is known, the next statistic follows a standard normal distribution. If the variance (standard deviation) of two populations is known, the variance of the difference between the means of the samples extracted from that population is the sum of the variances of sample means due to the additive nature of variance.

$$z = \left((\bar{x}_1 - \bar{x}_2) - (\mu_1 - \mu_2)_0\right) / \text{standard deviation of } (\bar{x}_1 - \bar{x}_2)$$
$$= \left((\bar{x}_1 - \bar{x}_2) - (\mu_1 - \mu_2)\right) / \sqrt{\left(\sigma_1^2/n_1 + \sigma_2^2/n_2\right)},$$

Assuming that the variances of the two sets of populations are not different, one can estimate the variance by adding the data extracted from the two sets of populations. The method of testing using the sample variance calculated in this way is called a pooled test. In case the primary objective of a study is to compare means, and one has no reason to believe that population variances are not equal, a procedure called the pooled t test to compare μ_1 to μ_2 can be used. Here, let s_p^2 be the estimated variance obtained by summing the two sets of samples. That is the sum of the two sets of sums of squared deviations divided by the sum of the two sets of degrees of freedom. The following t statistics serve as a test statistic for testing any of the usual hypotheses.

$$\left((\bar{x}_1 - \bar{x}_2) - (\mu_1 - \mu_2)_0\right) / \sqrt{s_p^2 (1/n_1 + 1/n_2)} = t \tag{4–1}$$

where $(\mu_1 - \mu_2)_0$ denotes the hypothesized difference in population means.

The hypothesised difference can be of any value, but the most commonly appearing value is zero. If the purpose is to determine whether the population means differ and, if so, which is the larger or smaller, such hypotheses take these forms:

- H_0: $\mu_1 = \mu_2$, H1: $\mu_1 > \mu_2$ . . . right-tailed test
- H_0: $\mu_1 = \mu_2$, H1: $\mu_1 < \mu_2$ . . . left-tailed test
- H_0: $\mu_1 = \mu_2$, H1: $\mu_1 \neq \mu_2$ . . . two-tailed test

Next, consider the case where the standard deviation of ($\bar{x}_1 - \bar{x}_2$) is unknown. If the variances of the two populations are different, one will rely on the method explained in the next section (comparing means when variances are unequal). However, if there is no significant difference between the population variances based on experience or preliminary survey, then the pooled t test, just explained, can be used. That is, assuming that the two sets of population variances are equal, the two sets of data are added together to estimate the population variance. This makes it possible to test with the t distribution. The variances estimated from the aggregated data are called the pooled value.

$$t = (\bar{x}_1 - \bar{x}_2) - (\mu_1 - \mu_2)_0) / \sqrt{\text{estimated variance of } (\bar{x}_1 - \bar{x}_2)}$$

$$\text{estimated variance of } (\bar{x}_1 - \bar{x}_2) = \frac{\widehat{\sigma^2_{\text{pooled}}}}{n_1} + \frac{\widehat{\sigma^2_{\text{pooled}}}}{n_2} = \widehat{\sigma^2_{\text{pooled}}}\left(\frac{1}{n_1} + \frac{1}{n_2}\right)$$

$$\text{where } \widehat{\sigma^2_{\text{pooled}}} = s_p^2 = \frac{SS_1 + SS_2}{(n_1 - 1) + (n_2 - 1)}$$

SS_1 and SS_2 are the sum of squares of two sets of samples. n_1 and n_2 are the number of data extracted from each population.

The validity of this method when the two sets of population variances are not equal has been discussed by many researchers. They also reported that matching both sample sizes contributes to increasing the robustness of this method (Kirk, 1999).

Example 4–4: One bought mechanical parts from two companies and surveyed them to see whether there were any differences in the length of the parts. A sample of ten pieces was taken for each purchased item, and the length was measured. The average values $\bar{x}$ and the sum of squares SS are determined. The significance level of the test is 5%. From past purchases, one can consider that there is no significant difference between the two companies' parts in terms of variation.

$$n_1 = n_2 = 10$$
$$\bar{x}_1 = 100.4\text{ mm}, \bar{x}_2 = 101.1\text{ mm}$$
$$SS_1 = 10.0, SS_2 = 17.0$$

Answer for Example 4–4:

Step 1: Set the null and alternative hypotheses.

$$H_0 : \mu_1 = \mu_2$$
$$H_1 : \mu_1 \neq \mu_2$$

Step 2: Specify the values given in the problem.

$$n_1 = n_2 = 10, \bar{x}_1 = 100.4, \bar{x}_2 = 101.1, SS_1 = 10.0, SS_2 = 17.0, \text{and } \alpha = 0.05$$

Step 3: Determine the appropriate distribution and rejection area of the test.
In this problem, the t distribution is appropriate. Draw the t value from a statistical table.

$$df = (n_1 - 1) + (n_2 - 1) = (10 - 1) + (10 - 1) = 18$$
$$t(\alpha/2, df) = t(0.025, 18) = 2.101$$

The rejection area of this test is $t \leq -2.101$ and $t \geq 2.101$.
Step 4: Calculate the statistic for the testing.

$$s_p^2 = (SS_1 + SS_2)/\{(n_1 - 1) + (n_2 - 1)\} = (10.0 + 17.0)/18 = 1.5$$
$$t = \{(\bar{x}_1 - \bar{x}_2) - (\mu_1 - \mu_2)_0\}/\sqrt{s_p^2 (1/n_1 + 1/n_2)}$$
$$= (100.4 - 101.1)/\sqrt{1.5}\ (1/10 + 1/10)$$
$$= -0.7/0.55$$
$$= -1.27$$

Step 5: Conclude.
Since the calculated t value is not in the rejection area, we cannot reject H_0. Therefore, we cannot say that there is a significant difference between the parts of the two companies.

In Example 4–4, the difference in population means was tested using the fact that ($\bar{x}_1 - \bar{x}_2$) follows the normal distribution. Because the population standard deviation is unknown, the t-test was performed using the estimated population variance. Here, if the number of each sample is sufficiently large, it can be considered that the sample standard deviation can be approximately used as an estimator of the population standard deviation.

Note that, comparing the statistic obtained from the standard normal distribution with that from the *t* distribution, if the degrees of freedom are sufficiently large, both will have almost the same value. Furthermore, the two coincide perfectly in the case of infinite degrees of freedom. Therefore, in the test of the difference between the two population means, if the sample size of each is sufficiently large, testing is possible by means of the normal distribution instead of the *t* distribution. At that time, each sample size should be 30 or more (Clark, 1991). For example, in Example 4–3, there are 36 samples, so the variance calculated from the samples can be used to obtain the test statistic for the normal distribution.

4.4.2 Comparing Means (Variances Unequal)

If a difference is found when comparing the population variances, it is inappropriate to use a pooling procedure. It may still be possible to compare means using an approximate *t* statistic. The desired statistic is found by logically modifying the following *z* random variable:

$$\left((\bar{x}_1 - \bar{x}_2) - (\mu_1 - \mu_2)\right) / \sqrt{\left(\sigma_1^2 / n_1 + \sigma_2^2 / n_2\right)}$$

Because there is evidence that $\sigma_1^2 \neq \sigma_2^2$, each population variance is estimated separately; i.e., these estimators are not combined. Instead, the population variances in the *z* random variable are replaced by their respective estimators, s_1^2, and s_2^2, to obtain the random variable:

$$\left((\bar{x}_1 - \bar{x}_2) - (\mu_1 - \mu_2)\right) / \sqrt{\left(s_1^2 / n_1 + s_2^2 / n_2\right)}.$$

Some difficulties arise due to an approximation from a random variable of normal distribution to the *t* distribution. This time, however, the number of degrees of freedom must be estimated from the data. Several methods have been suggested for doing this. Here we recommended the Smith–Satterthwaite procedure. However, more detailed explanations will be left to other higher-level textbooks.

4.4.3 Comparing Two Means – Paired Data

In many instances, problems arise in which two random samples are available, but they may not be independent; rather, each observation in one sample is naturally or by design paired with an observation in the other. Note that when data are paired, we can define a new random variable *d* by $d = x - y$. The *n* differences $di = xi - yi$, $i = 1, 2, 3 \ldots, n$ constitute a set of observations on *d*; that is, they constitute a random sample of size *n* drawn from the population of differences. Therefore, according to the rules for expectation:

$$\mu_x - \mu_y = E(x) - E(y) = E(x - y) = E(d) = \mu_d$$

The original question, "What is $\mu_x - \mu_y$?" is equivalent to the other question, "What is μ_d?" One can reduce from the original two-sample problem to the one-sample problem by making an inference on the mean of the population differences. In particular, the formula for the 100 $(1 - \alpha)$ % confidence intervals on $\mu_x - \mu_y = \mu_d$ is shown by the formula:

$$\left\{\text{Confidence intervals on } \mu_x - \mu_y \text{ for paired data: } \bar{d} \pm t\,(\alpha/2)\, s_d / \sqrt{n}\right\}$$

In this formula, $\bar{d}$ and s_d are the sample mean and sample standard deviation of the sample of difference scores, respectively. The $t\,(\alpha/2)$ is the appropriate point relative to the *t* distribution with $n - 1$ degrees of freedom. The null hypothesis $\mu_x = \mu_y$ is equivalent to the hypothesis $\mu_d = 0$. The test statistics for testing this hypothesis based on the sample of difference scores is

$$(\bar{d} - 0)/(s_d/\sqrt{n})$$

which follows the t distribution with $n - 1$ degrees of freedom if H_0 is true. The use of this statistic is illustrated using Example 4–5.

Example 4–5: The concentration of contaminant X in the lake was measured. Water samples were collected from sites A and B every day during a five-day monitoring program. It is desired to compare the concentrations at sites A and B.

Answer for Example 4–5: For this example, it is necessary to test whether there is a significant difference in the concentration between sites A and B, and, for example, the significance level is set, α = 5%. Because daily samples were taken, it is better to think of these data as a pair rather than as if they were obtained independently.

The null and alternative hypotheses are:

- H_0: $\mu_A = \mu_B$
- H_1: $\mu_A \neq \mu_B$

This is equivalent to the following testing:

- H_0: $\mu_d = 0$
- H_1: $\mu_d \neq 0$

The observed value of the test statistic is calculated as follows:

$$\bar{d} = (-8)/5 = -1.6$$

$$s_d = \sqrt{22 - ((-8)^2/5)/(5-1)} = \sqrt{(9.2/4)} = 1.5$$

$$(\bar{d} - 0)/(s_d/\sqrt{n}) = -1.6/(1.5/\sqrt{5}) = -2.36$$

With the number of data of 5, the degree of freedom is 4. Even if the hypothesis that there is no difference between the population means of the two, that is, the difference between the population means is zero is true, the probability of rejecting it is 5%. The values in the rejection area are shown in Figure 4.6. The critical value of the area of the rejection area is obtained from the t distribution table with four degrees of freedom. The calculated absolute t value (2.36) for this test is less than 2.776 obtained from t table. Since this t value is not sufficiently large, one cannot reject H_0 and conclude and hence cannot say that the concentrations in the lake at two sites show a significant difference.

TABLE 4.3

Data for the Paired Data Test (Concentration in Arbitrary Units)

Day/site	Site A	Site B	Difference	Square
6 Mar. 06	9	9	0	0
7 Mar. 06	2	6	−4	16
8 Mar. 06	4	5	−1	1
9 Mar. 06	6	8	−2	4
10 Mar. 06	7	8	−1	1
Total			−8	22

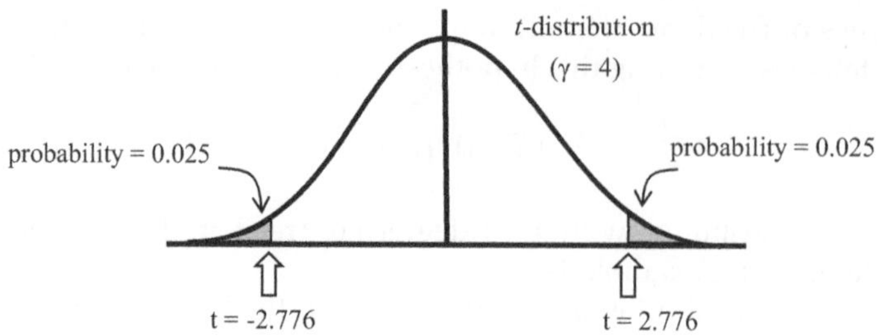

FIGURE 4.6 Rejection area of the t distribution with the degrees of freedom $\gamma = 4$.

4.5 Testing on Proportion

Consider proportional data, such as the percentage of people who can pass the entrance examination or the percentage of people who have lung cancer due to smoking. The analysis of such categorical variables is discussed in detail in Chapter 20; only the test of the proportion is included here.

If n samples are extracted from the population and the number of samples with a certain trait is X, the proportion in the sample is X/n. From that data, consider estimating or testing the proportion of elements with that trait across the population. In general, data for the number of objects in a sample with such specified traits follows a binomial distribution. The number of dice introduced in Section 3.2.1 is a typical example of the binomial distribution. This is a typical discrete probability distribution. More generally, it can be expressed as follows; the binomial distribution is based on the following assumptions:

- There is a fixed sample size of n separate trials.
- The trials are independent. Each trial has two possible outcomes. One has some specified trait, and the other does not.
- Each trail has the same probability p as being in the class of the outcome being counted.

The values of n (some positive integer) and p (a real number between 0 and 1) determine the full distribution (list of possible values and associated probabilities). For a random variable with a binomial distribution, its expected value and variance are:

$$E(X/n) = np$$
$$V(X/n) = np(1-p)$$

A random variable with a binomial distribution approaches a normal distribution approximately as the number of samples increases. According to Kondo and Ando (1967), the normal approximation is possible when $np \geq 5$ and $n(1-p) \geq 5$. Example 4–6 next is provided to explain this concept.

Example 4–6: It is said that the carcinogenesis rate of X (X denotes the organ) in the public area is 2.0%. A recent survey of 300 smokers found that 12 people had X cancer. Use this data to determine whether smoking is the cause of the increased morbidity.

Answer for Example 4–6: There were 300 people surveyed and 12 people suffering from X cancer. The X cancer rate p can be estimated as:

$$\hat{p} = x/n$$

where x is the number in the sample who have the trait of interest, namely have the X cancer, and n is the sample size.

$$\hat{p} = 12/300 = 0.04$$

The following hypothesis is set up, and the test is performed. This hypothesis considers whether the population proportion is equivalent to the pre-specified value, p_0. Statistical tests will be used to examine whether this morbidity differs from the numerical values conventionally told to the public. The hypothesis here is that the carcinogenic rate is 2.0%, namely, the same as that of the general population in this case. If this hypothesis is rejected, the carcinogenic rate of smokers will differ from that of the general population. Furthermore, if one thinks that smoking never reduces the incidence of cancer, the alternative hypothesis is a single-side; and this is the one-tailed test.

- Null hypothesis H_0: $p = p_0 = 0.02$
- Alternative hypothesis H_1: $p > p_0 = 0.02$

The proportion follows a binomial distribution. The expected value and variance of a random variable following a binomial distribution are as follows, and, under the assumption of the null hypothesis, they can be calculated.

$$E(p) = np_0 = 300 \times 0.02 = 6$$
$$V(p) = np_0(1 - p_0) = 300 \times 0.02 \times (1 - 0.02) = 5.88$$

To perform this test, one first checks whether the data satisfy the rule of thumb used to approximate the binomial distribution with a normal distribution:

$$np_0 = 300 \times 0.02 = 6, \text{ and}$$
$$n(1 - p_0) = 300 \times (1 - 0.02) = 294$$

Both these values are greater than 5. Therefore, one can use the test statistics of the normal distribution. In other words, this test statistic approximately follows a standard normal distribution under the null hypothesis.

$$z = \frac{\hat{p} - p_0}{\sqrt{p_0(1 - p_0)/n}} = \frac{(0.04 - 0.02)}{\sqrt{0.02(1 - 0.02)/300}} = 2.47$$

According to the decision rule, if the z value is larger than 1.645 (one-tailed test), one can reject the null hypothesis.

In the survey, the percentage of X cancers among smokers was 0.04. The calculated value of the z statistic is 2.47, which is larger than the critical value of the rejection area of 1.645 associated with a significance level of $\alpha = 0.05$. Therefore, one rejects the null hypothesis and concludes that the prevalence of cancer in smokers differs from 0.02 in the public.

Example 4–7: Town A has severe air pollution, and town B has clean air. Respondents were randomly extracted to find out how many people complained of some form of poor physical condition. The number of subjects in the sample was 50 for each town. As a result, the number of people who complained of feeling unwell was 15 and 5, respectively. The question: is it possible to say that the proportion of people in town A who are in poor physical condition is higher than in town B?

Answer for Example 4–7: Hypotheses of this sort can be tested using the test statistics with $(p_1 - p_2)_0$ set equal to zero. In this problem, the null hypothesis is that the prevalence is the same. On the other hand, the alternative hypothesis is $p_1 - p_2 > 0$ because it is stated that town A is badly polluted and town B is clean. This procedure makes use of the fact that if H_0 is true, $\hat{p}_1$ and $\hat{p}_2$ are both estimators for the same proportion, which are denoted by p. To see how to use this information, note that the variance of $\hat{p}_1$ - $\hat{p}_2$ can be written as $p(1 - p)(1/n_1 + 1/n_2)$.

Consider that the following random variable in Equation (4–2) has a distribution that is approximately standard normal.

$$\frac{\hat{p}_1 - \hat{p}_2}{\sqrt{p(1-p)\left(\frac{1}{n_1} + \frac{1}{n_2}\right)}} \tag{4–2}$$

If one can accept the null hypothesis, then the pooled estimator for p can be used and can be replaced with $\hat{p}$ instead of p in Equation (4–2).

In this example, the sample size for each town is the same, $n_1 = n_2 = 50$. The estimator $\hat{p}_1$ and $\hat{p}_2$ is 0.30 and 0.10, respectively. The pooled $\hat{p}$ is:

$$\hat{p} = (0.30 \times 50 + 0.10 \times 50)/(50 + 50) = 0.20$$

Therefore, the test statistics are:

$$z = \frac{\hat{p}_1 - \hat{p}_2}{\sqrt{p(1-p)\left(\frac{1}{n_1} + \frac{1}{n_2}\right)}} = \frac{0.30 - 0.10}{\sqrt{0.02(1-0.20)\left(\frac{1}{50} + \frac{1}{50}\right)}} = 2.5$$

Since the calculated z value is larger than 1.645 (one-tailed test with a significance level of $\alpha = 0.05$), one can reject the null hypothesis.

In this case, the calculated z statistic under the hypothesis that the incidences of sick people in the two towns are the same. It is greater than the critical value of the rejection area 1.645 associated with the significance level $\alpha = 0.05$. Therefore, one rejects the null hypothesis and concludes that the prevalence of the sickness in the two towns is different, and the prevalence in town A is higher than in town B.

Note that this test utilises the fact that the difference between the two proportions approximately follows a normal distribution under the null hypothesis H_0. However, this difference in proportion is not a continuous quantity. A histogram can represent the distribution of discrete random variables. When applying this to the probability density function of the normal distribution, it corresponds to the centre value of each class of the histogram. However, there will always be a difference of half of the class width from the boundary value of the class. For a stricter test, one can consider the correction method.

The chi-square test utilising the contingency table, as will be discussed in Chapter 20, also employs a corrective method known as Yates' correction. This technique is elaborated upon in technical texts, including those by Kondo and Ando (1967) and Haviland (1990).

4.6 Comparing Variances: The *F* test

First of all, it is noted that the test for variance using the F distribution is called the F test. Similarly, the test for population means using the t distribution is sometimes called the t test.

In this case, it is desired to find out whether there is a difference in the variances of the two populations. Samples are extracted from each population, and the sample variance is calculated. To test whether the variances of the two populations are equal, the samples of size n_1 and n_2 were extracted from each population, respectively, and the statistic was calculated. The null and alternative hypotheses are as follows:

- H_0: $\sigma_1^2 = \sigma_2^2$ H_1: $\sigma_1^2 > \sigma_2^2$ . . . right-tailed test
- H_0: $\sigma_1^2 = \sigma_2^2$ H_1: $\sigma_1^2 < \sigma_2^2$ . . . left-tailed test
- H_0: $\sigma_1^2 = \sigma_2^2$ H_1: $\sigma_1^2 \neq \sigma_2^2$ . . . two-tailed test

It is easy to find the logical statistic for comparing variances. Here, V_1 and V_2 (or expressed s_1^2, s_2^2 in some textbooks) are unbiased estimators for the population variances σ_1^2, and σ_2^2, respectively. If $\sigma_1^2 = \sigma_2^2$, then the statistics s_1^2/s_2^2 (or V_1/V_2) follows the *F* distribution with $n_1 - 1$ and $n_2 - 1$ degrees of freedom.

4.7 Summary

This chapter explained inferential statistics with a focus on statistical tests and estimation. The concept of hypothesis testing introduced here is essential knowledge for various statistical analyses. A simple explanation has been provided by preparing an example using relatively simple numbers. When testing the population mean and its difference, the analysis method differs depending on whether the population variance is known or unknown. It is also important to understand that this will result in a difference in the distribution of the test statistics used.

As explained here, it is crucial to use appropriate statistics for our analysis based on an informative sample from the population we are interested in. Additionally, it is important to understand that the power of the test depends on the amount of data available for our analysis.

REFERENCES

Clark, L.J., 1991, *The essentials of business statistics II* (p. 109), Research & Education Association, Piscataway, NJ, United States.

Haviland, M.G., 1990, Yates's correction for continuity and the analysis of 2 × 2 contingency tables, *Statistics in Medicine*, 9(4) 363–367, https://doi.org/10.1002/sim.4780090403

Kirk, R.E., 1999, *Statistics – an introduction*, 4th edition (p. 393), Harcourt Brace College Publishers, Fort Worth, TX, United States.

Kondo, Y. and Ando, T., 1967, *Statistical methods – hundred questions and hundred answers* (pp. 25–27, 60, 75), JUSE Publishing, Co., Tokyo (In Japanese).

Milton, J.S., McTeer, P.M. and Corbet, J.J., 1997, *Introduction to statistics*, WCB/McGraw-Hill Co., Inc., Boston, United States.

5

Analysis of Variance

The analysis of variance and its use are explained in this chapter. There is a variety of data from field observations and environmental monitoring stations and data from the laboratory under planned conditions. In the latter case, it is necessary to plan the survey or experiment in a way that most effectively captures the data that researchers want to know. Analysis of variance is a form of statistical hypothesis testing often used in the analysis of experimental data. A typical use of analysis of variance is the hypothesis that "all groups are random samples from the same population". The basic idea was established by Fisher in the 1920s and is sometimes called Fisher's analysis of variance. The term in analysis of variance is often derived from experimental design. While many methods are used to analyse experimental data and develop models, knowledge of analysis of variance is essential when planning such experiments.

5.1 Relationship between the Test for Differences in Population Means and Analysis of Variance

As seen from Chapter 4, the two-population means test uses the t distribution. It is explained in Chapter 3 that the square of the statistic that follows the t distribution becomes the statistic followed by F distribution (see Equation [3–15]); the numerator then corresponds to the variance corresponding to the mean difference between the two populations, and the denominator corresponds to the variance within each population, as shown in Figure 5.1. In other words, the test can be performed by comparing the variance corresponding to the difference between the means of the two populations with the variance within each population. This is the same concept as testing the difference between two population variances. Of course, the variances for the two populations must not be significantly different.

Within treatment: $(SS_1 + SS_2)/(n_1 + n_2 - 2)$, where SS = sum of squares
Between treatment: $(\bar{x}_1 - \bar{x}_2)^2$

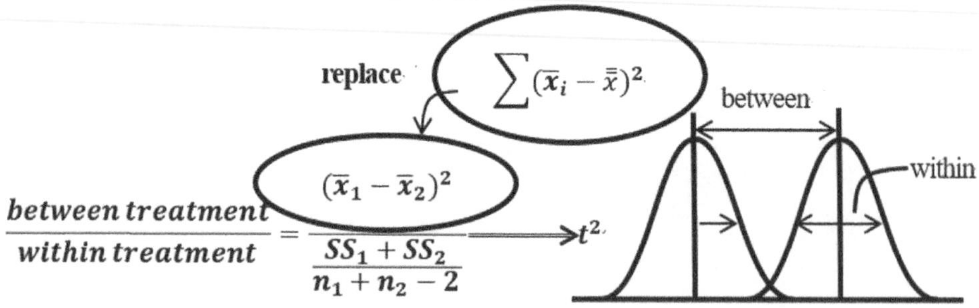

FIGURE 5.1 Comparison of variability between treatments and variability within treatments.

DOI: 10.1201/9781003597216-6

Here, the numerator corresponds to variance between treatments, and the denominator corresponds to variance within treatment. This is the basic idea of analysis of variance, which is a technique that considers the difference in averages between groups as deviations and tests whether it is significant. It is commonly referred to as ANOVA. This analysis can be extended also to three or more treatments.

First, let us consider the method of using the *F* statistics described as the square of the *t* statistics for testing two population means. The sum of squares of the deviations, namely variance components of the data, is decomposed to separate the fluctuation due to the treatment effect and the fluctuation due to the error. Next, a mean square is calculated in which we divide the sum of squares by the degrees of freedom. Then, the *F* value is calculated by using the mean square explained by the treatment effect as the numerator and the mean square explained by the error as the denominator. The significance of the effect of the treatments is determined by an *F* test with a specific significance level. In the case of a comparison between the means of two sets of populations, the square root of this *F* value can be the *t* distribution.

$$z \text{ (or } t) = \text{difference of } \bar{x}/\text{standard deviation of } \bar{x}$$
$$\text{Square of } z \text{ (or } t) = \text{between-group difference/within-group difference}$$

In this formula, the larger the numerator or the smaller the denominator, the larger the value of *z* or *t*. Therefore, the following can be said:

The within-group variability is smaller, and the power of the test is higher.
The between-group difference is larger, and the power of the test is higher.

This idea can be easily extended. For example, suppose it is desired to experiment at four levels and see whether there are any differences between the levels. There are multiple data for each of the four levels, and the average value is $\bar{x}_1$, $\bar{x}_2$, $\bar{x}_3$ and $\bar{x}_4$. If one collects these averages, it should have a distribution. The variation in the distribution of the average values is between-group variability. In addition, since there are multiple data within each level; variations in the distribution are considered as within-group variability. This relationship is illustrated in Figure 5.2. In addition, strictly speaking, the variance of each within-group variability must not be significantly different, as discussed when testing the population mean with the *t* distribution.

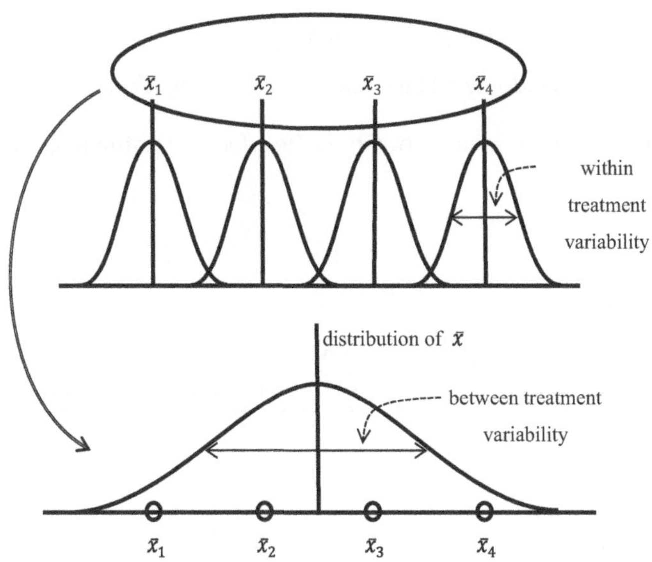

FIGURE 5.2 Variability within and between the treatments.

5.2 One-Way ANOVA and Randomised Block Design

An experiment is considered to see whether there are differences in the efficacy of the three types of reagents. A total of 12 experiments were conducted four times for each of the three types of reagents. If you organize this data, it will look like Table 5.1. To analyse this data, statistical application software can be used, and one may select a method of one-way ANOVA. In this case, the number of treatments is 3, and the number of repetitions is 4, and it is necessary to consider whether all 12 experiments had been conducted in randomised order or not. Ideally, one would establish the sequence for the experiment by selecting numbers from a random number table. If not, what should we do? Read the following explanation, and then think about it.

For example, the effects of the reagents are sensitive to changes in temperature and humidity, and it turns out that it is difficult to complete all experiments in a single day. Accordingly, each day, one experiment using one reagent is performed; hence, in total, the experiments were conducted in 4 days or they are divided into 4 groups. In this case, the data will also be organised in the same format as Table 5.1. However, the data number is not just a sequence but has the meaning of a date. The analysis of such data requires the ANOVA of the randomised block design.

In the one-way analysis of variance and the randomised block design, the raw data table format is the same. However, there are differences in the meaning of data numbers. In the former, it simply represents the sequence of the data order. However, the latter shows the correspondence of a specific place, time, processing method, or something else. In case the number of samples only shows the number of the data, for example, repetition of analysis or sequential order of the data, this is the one-way ANOVA. In case the number of samples shows the site number, sampling date or something related to the condition of treatment, this is the ANOVA of randomised block design.

The role of the analysis of variance is to find out whether there is a difference between the means of multiple populations or multiple groups within the population. It is also important to consider why the difference occurred. A useful concept at this time is the data structure. In the data shown in Table 5.1, if the data number has no meaning, the data structure will be Equation (5–1). If the data number is a date and there are daily effects (such as temperature and humidity directly), it is called a block effect, and the data structure is as shown in Equation (5–2):

$$\text{One-way ANOVA: } x_{ij} = \mu + tr_i + e_{ij} \tag{5–1}$$

$$\text{Randomised block design: } x_{ij} = \mu + tr_i + b_j + e_{ij} \tag{5–2}$$

where x_{ij} is the j-th data of the treatment i, tr_i is the effect of treatment i, b_j is the block effect j, and e_{ij} is the error term.

TABLE 5.1

Data Format for a Typical Case of ANOVA

No. of sample	Treatment 1	Treatment 2	Treatment 3
1	Xxxx	Xxxx	Xxxx
2	Xxxx	Xxxx	Xxxx
3	Xxxx	Xxxx	Xxxx
4	Xxxx	Xxxx	Xxxx

TABLE 5.2

Data of the Contaminant Concentrations (unit is arbitrary scale)

Date	Site A	Site B	Site C
6 Feb. 2016	3	0	0
7 Feb. 2016	3	0	2
8 Feb. 2016	8	4	3
9 Feb. 2016	4	1	1
10 Feb. 2016	2	0	1
11 Feb. 2016	5	0	3

To understand the concept of analysis of variance, let us go through the steps of a calculating procedure with a simple numerical example by using hypothetical Example 5–1.

Example 5–1: To investigate the lake water quality, concentrations of Contaminant X were measured at three sites A, B and C. This monitoring was conducted for 6 consecutive days, and the results are shown in Table 5.2. Is the water quality different among the three sites based on the data?

ANOVA is a technique that allows us to test two or more population means simultaneously. For example, if the test is for 3 population means, then the null hypothesis would be stated as:

- H_0: $\mu_1 = \mu_2 = \mu_3$
- H_1: There are some inequalities.

Determine the values of k, b, and N (number of treatment k, block b, and total data N).

$$k = 3, b = 6, \text{and } N = 18$$

Using the following auxiliary tables, we can calculate the sum of squares (SS). Next, mean squares (MS) are calculated from SS divided by degrees of freedom (df). The calculated SS, df, MS, and F values are summarised in the ANOVA table. Here, the mean squares are another name for variance, and this term is generally used in the analysis of variance.

One can calculate the sum of squares SST, the sum of squares of concentration differences among the sites SSTR, the sum of squares of block (day) SSB, and the sum of squares of residuals (or errors) SSE using the intermediate values in the following auxiliary table. A more detailed explanation is given later, including the relationship with the structure of the data.

Calculation auxiliary table for finding the sum of squares

3	0	0	3		9		3		9	
3	0	2	5		25		5		25	
8	4	3	15		225		15		225	
4	1	1	6		36		6		36	
2	0	1	3		9		3		9	
5	0	3	8		64		8		64	
25	5	10	40				40		368	
625	25	100			1600		625	25	100	750

9	0	0	9		9					
9	0	4	13		13					
64	16	9	89		89					
16	1	1	18		18					
4	0	1	5		5					
25	0	9	34		34					
127	17	24			168					
127	17	24	168							

$$SST = 168 - (40 \times 40)/18 = 79.1$$
$$SSTR = 750/6 - (40 \times 40)/18 = 36.1$$
$$SSB = 368/3 - (40 \times 40)/18 = 33.8$$
$$SSE = 79.1 - 36.1 - 33.8 = 9.2$$

By summarising these calculation results, the ANOVA table in Table 5.3 can be obtained. In the case of this problem, the mean values at the three sites are $\overline{x}_1 = 4.17$, $\overline{x}_2 = 0.83$, and $\overline{x}_3 = 1.67$, respectively. The variance of these means, that is. the variability between these groups, is 18.05 in unbiased variance (mean square). The within-group variability also includes daily fluctuation, so if we subtract that amount, the mean square is 0.92. Therefore, the *F* value, which is the ratio of unbiased variance (mean square), is 19.6.

Based on the null hypothesis, the *F* value was calculated as shown in Table 5–3, and the *F* table value is as follows:

$$F \text{ table} = F_{0.05}(2,10) = 4.103$$

Since $F_0 = 19.6 > 4.103$, H_0 can be rejected. Therefore, it can be concluded that there are significant differences between the sites.

In general, even if the block effect is significant, usually the effect is not estimated. This is a measure for removing the separable dispersion component included in the error term and increasing the power of testing. The test of the difference between two sets of population means can be analysed by analysis of variance with two levels. Here, a simple population difference test corresponds to a one-way analysis of variance, and a paired data difference test corresponds to a randomised block design (RBD). Now, to help understand this relationship more clearly, two examples, Example 5–2 and Example 5–3, using simple numbers that are easy to calculate, are presented in this section.

TABLE 5.3

ANOVA Table with the Result of Example 5–1

	Sum of squares, SS	Degrees of freedom, df	Mean square MS	*F* value
Site	36.1	2	18.05	19.6
Block(day)	33.8	5	6.76	7.3
Error	9.2	10	0.92	
Total	79.1	17		

Example 5–2: A study was conducted to compare the water quality of two rivers, A and B. From the two rivers, five samples were collected by a researcher, and the concentration of a toxic substance was analysed for each sample. Determine whether the water qualities of rivers A and B are different.

A: 106, 103, 104, 108, 104 (ng/L)
B: 105, 102, 104, 106, 103 (ng/L)

Answer for Example 5–2: In this problem, five samples from two rivers were taken for analysis. Each of the five samples was collected at a different time, so it is considered that the five samples were collected separately. In this case, it should be usually analysed by testing the difference between the two population means, and it can also be analysed by one-way analysis of variance which is equivalent to testing the following hypotheses:

- H_0: $\mu_A = \mu_B$
- H_1: $\mu_A \neq \mu_B$

The result is shown in Table 5.4.

Based on the one-way ANOVA (Table 5.4), one cannot say that there was a significant difference between the two rivers because the obtained *F* value is too small and H_0 cannot be rejected.

The data structure in a one-way ANOVA is as follows:

$$x_{ij} = \mu + tr_i + e_{ij}$$

The breakdown of the data according to this structure is seen in Figure 5.3.

TABLE 5.4
ANOVA Table with the Result of Example 5–2

	SS	df	MS	*F* value
Site	2.5	1	2.5	0.77
Error	26.0	8	3.25	
Total	28.5	9		

Xij			μ			Tr1	Tr2		eij	
106	105		104.5	104.5		0.5	-0.5		1	1
103	102		104.5	104.5		0.5	-0.5		-2	-2
104	104	=	104.5	104.5	+	0.5	-0.5	+	-1	0
108	106		104.5	104.5		0.5	-0.5		3	2
104	103		104.5	104.5		0.5	-0.5		-1	-1

FIGURE 5.3 Data structure for the data of Example 5–2. (*Note:* Tr1 is the effect of site 1, and Tr2 is the effect of site 2.)

The basic concept of analysis of variance is the decomposition of the sum of squared deviations, which for the data from Example 5–2 is:

$$SS = \Sigma\left(x_{ij} - \bar{\bar{x}}\right)^2$$
$$= \Sigma\left(x_{ij} - \bar{x}_i + \bar{x}_i - \bar{\bar{x}}\right)^2$$
$$= \Sigma\left(\left(x_{ij} - \bar{x}_i\right) + \left(\bar{x}_i - \bar{\bar{x}}\right)\right)^2$$

Now let us conduct a simple operation before calculating the variance. For example, there are two sets of data, (104,105,106) and (4,5,6). The variance for these two sets of data is the same. Here, considering the decomposition of data using the values obtained by subtracting 104.5 (overall average value) from the original data of Example 5–2, the result is as follows:

$$\begin{bmatrix} 1.5 & 0.5 \\ -1.5 & -2.5 \\ 0.5 & 0.5 \\ 3.5 & 1.5 \\ 0.5 & 1.5 \end{bmatrix} = \begin{bmatrix} 0.5 & -0.5 \\ 0.5 & -0.5 \\ 0.5 & -0.5 \\ 0.5 & -0.5 \\ 0.5 & -0.5 \end{bmatrix} + \begin{bmatrix} 1 & 1 \\ -2 & -2 \\ -1 & 0 \\ 3 & 2 \\ -1 & -1 \end{bmatrix}$$

Now, using the decomposed values of these individual data, calculate the sum of squared deviation SST corresponding to the total variance, the sum of the squared deviation of the level effect SSTR, and the sum of the squared deviation of residual SSE.

- $SST = 1.5^2 + (-1.5)^2 + 0.5^2 + 3.5^2 + 0.5^2 + 0.5^2 + (-2.5)^2 + 0.5^2 + 1.5^2 + 1.5^2 = 28.5$
- $SSTR = (0.5)^2 \times 5 + (-0.5)^2 \times 5 = 2.5$
- $SSE = 1^2 + (-2)^2 + (-1)^2 + 3^2 + (-1)^2 + 1^2 + (-2)^2 + 0^2 + 2^2 + (-1)^2 = 26$

Thus the basic function of ANOVA is to decompose the sum of the squares of the deviations and compare and test the size of the variation due to the effect of the treatment and the residual variation. For this analysis, it is more convenient to use the calculation method presented for Example 5–1, if no suitable statistical analysis software is available.

Example 5–3: The concentration of heavy metal in particulate matter is analysed by two measuring methods, and it is suspected that the results obtained by the two methods might be different. Consequently, the researchers collected five particulate matter filter samples, divided each one into two, and subjected them to analysis using both methods. Table 5.5 presents the values acquired through Method A and Method B.

Answer for Example 5–3: In this case, the number of samples is significant (That is, the data numbers are the individual sample numbers, not just sequential order.) because five samples were utilised to demonstrate the differences in methods. It is not only a sequential order.

TABLE 5.5

Data of the Experiment (unit: ng/m^3)

PM filter No.	Result by method A	Result by method B
1	106	105
2	103	102
3	104	104
4	108	106
5	104	103

Therefore, this is analysed by a test of the paired data. In the analysis of variance, it becomes a randomised block design, which is presented here.

- H_0: $\mu_A = \mu_B$
- H_1: $\mu_A \neq \mu_B$

Based on the null hypothesis, the *F* value can be calculated as shown in Table 5.6, and the *F* table value is as follows:

$$F \text{ table} = F_{0.05}(1,4) = 7.71$$

In conclusion, it can be said that there is a significant difference between the methods because $F_0 > F$ table.

The data structure in a randomised block design ANOVA is as follows:

$$x_{ij} = \mu + tr_i + b_j + e_{ij}$$

The breakdown of the data in this example according to this structure is shown in Figure 5.4.

Here, as in the case of Example 5–2, subtract 104.5 from the original data of Example 5–3.

$$\begin{bmatrix} 1.5 & 0.5 \\ -1.5 & -2.5 \\ 0.5 & 0.5 \\ 3.5 & 1.5 \\ 0.5 & 1.5 \end{bmatrix} = \begin{bmatrix} 0.5 & -0.5 \\ 0.5 & -0.5 \\ 0.5 & -0.5 \\ 0.5 & -0.5 \\ 0.5 & -0.5 \end{bmatrix} + \begin{bmatrix} 1 & 1 \\ -2 & -2 \\ -0.5 & -0.5 \\ 2.5 & 2.5 \\ -1 & -1 \end{bmatrix} + \begin{bmatrix} 0 & 0 \\ 0 & 0 \\ -0.5 & 0.5 \\ 0.5 & -0.5 \\ 0 & 0 \end{bmatrix}$$

TABLE 5.6
ANOVA Table with the Result of Example 5–3

	SS	df	MS	*F* value
Method	2.5	1	2.5	10.0
Block(sample)	25.0	4	6.25	25.0
Error	1.0	4	0.25	
Total	28.5	9		

Xij			µ			Tr1	Tr2						eij	
106	105		104.5	104.5		0.5	-0.5		1.0	1.0	b1		0	0
103	102		104.5	104.5		0.5	-0.5		-2.0	-2.0	b2		0	0
104	104	=	104.5	104.5	+	0.5	-0.5	+	-0.5	-0.5	b3	+	-0.5	0.5
108	106		104.5	104.5		0.5	-0.5		2.5	2.5	b4		0.5	-0.5
104	103		104.5	104.5		0.5	-0.5		-1.0	-1.0	b5		0	0

FIGURE 5.4 Data structure for the data of Example 5–3. (*Note:* Tr1 is the effect of site 1 and Tr2 is the effect of site 2.)

$$SST = 1.5^2 + (-1.5)^2 + 0.5^2 + 3.5^2 + 0.5^2 + 0.5^2 + (-2.5)^2 + 0.5^2 + 1.5^2 + 1.5^2 = 28.5$$
$$SSTR = (0.5)^2 \times 5 + (-0.5)^2 \times 5 = 2.5$$
$$SSB = \{1^2 + (-2)^2 + (-0.5)^2 + 2.5^2 + (-1)^2\} \times 2 = 25$$
$$SSE = 0^2 + 0^2 + (-0.5)^2 + 0.5^2 + 0^2 + 0^2 + 0^2 + 0.5^2 + (-0.5)^2 + 0^2 = 1$$

Comparing this analysis result with that of Example 5–2, one can see that the numerical values of raw data are the same, but the conclusions are different. Why did this difference occur? Even if the mean square of the treatment effect is the same, the mean square of the residual is different. In this problem, the block effect was subtracted from the residual variation. In other words, it can be said that the power of testing has increased as the variation of residuals has decreased.

5.3 Factorial Design and Some Related Issues

Suppose that multiple factors (also called independent or explanatory variables) affect the objective variable (also called dependent variable). If the influence of each factor can be explained separately, that is, it is additive, interaction is not significant. On the other hand, when the impact of one factor changes depending on the level of another factor, it is said that there is an interaction. One can visualise the effect of an interaction by plotting the data on a graph with the objective variable on the vertical axis and the factor level on the horizontal axis. For an experimental design that includes factors that may have interactions, one should plan to obtain multiple data for all levels of the factors so that the variance of the interaction can be separated from the variance of the residuals. In other words, it is necessary to repeat the experiment for all levels. Such a plan is called a factorial design.

A factorial design is an experimental design that consists of two or more elements, and the experimental unit adopts all possible combinations of these levels for all these factors. These experiments are useful in studies to understand the influence of each factor on the response variable and the interaction between factors on the response variable. In Figure 5.5, the interaction is negligible if the factor effect graph is nearly parallel but becomes significant if the lines intersect or have opposite slopes.

A full factorial experiment is a plan that consists of two or more factors, each with its value or "level", and the experiment is performed on all possible combinations of these levels. Such

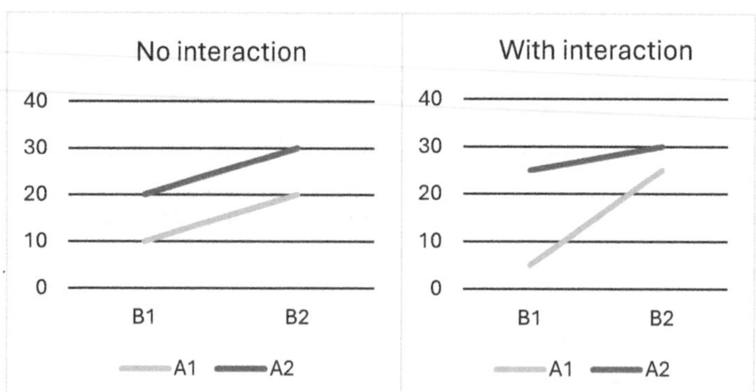

FIGURE 5.5 Graph plots for the effects with or without interaction. In this example, there are two factors, factor A and factor B, each with two levels. The vertical axis is the effect value.

TABLE 5.7

Raw Data of the Water Quality Survey (unit: mg/L)

Date	Concentration in Lake A	Concentration in Lake B
1 March (sunny)	116	126
2 March (sunny)	114	123
3 March (rain)	96	105
4 March (sunny)	115	126
7 March (rain)	96	107
9 March (rain)	93	103

an experiment allows the experimenter to examine the influence of each factor on the response variable and the interaction between the factors on the response variable.

In a complete factorial design, there are too many combinations to conduct, so some of the possible combinations can be omitted, for example by using an orthogonal table to avoid confounding important factors. But for beginners, this procedure is not easy to understand, and one can start with a simpler procedure that is illustrated in this section. More advanced studies can be found in relevant textbooks. Information given here aims to endow readers with the meaning of important technical terms that may be useful.

In the ANOVA, it is necessary to correctly understand how the data was collected and choose an appropriate method. It is also necessary to understand what we want to know and devise a way to collect data. Choosing the wrong method can also lead to unintended and erroneous conclusions. Consider such a problem by referring to Example 5–4.

Example 5–4: Complaints have been lodged regarding the water quality of two small lakes on the outskirts of the city, both of which receive industrial wastewater discharges. It has been observed that the water quality appears to slightly improve with the increased inflow of river water during rainy days. For the survey, three days of clear weather and three days of rainy weather were chosen. The concentrations of pollutants in lakes A and B are detailed in Table 5.7. Prior to commencing this analysis, the researchers were informed of the following:

- Lake B is more contaminated than Lake A.
- The pollutant concentration is higher on sunny days compared to rainy days.

From this survey, researchers want to know the difference in the contaminant levels between places (lakes) and the weather conditions. Here, for the place (lake), there are two lakes, a finite number, which is all one wants to know. On the other hand, for the dates, three days are selected from countless sunny days and rainy days. In such cases, the former is a factor of the parameter model, and the latter is a factor of the random model. There is no need to pay attention at the stage of creating the ANOVA table, but caution is required at the stage of subsequent estimation. In general analysis of variance, it is sufficient if one knows whether there is a significant difference in the factors of the random model. As for the block factors, it is meaningless in most cases to estimate the population means for each level. This is because the level of the factor is not the value of the set target level but a random variable. The topic will be elaborated upon in Section 5.3.3 on mixed models. However, for a more in-depth understanding, consulting specialised textbooks is recommended

The data obtained from this survey are two levels for the place (lake) and two levels for the weather. Therefore, the data can be summarised in a 2×2 spreadsheet with 3 data points in each space as shown in Table 5.8. This data can therefore be analysed using a factorial design two-way ANOVA.

The degrees of freedom for each variable are the number of levels minus one, so both are one. In addition, because of the repeated data, one can calculate the mean square of the

TABLE 5.8

Data Rearranged for Two-Way ANOVA of Factorial Designs

	Lake A	Lake B
Sunny day	116, 114, 115	126, 123, 126
Rainy day	96, 96, 93	105, 107, 103

TABLE 5.9

Result of Two-Way ANOVA (factorial design) for Example 5–4

Factor	SS	df	MS	F
Site (lake)	300	1	300	109.1
Weather	1200	1	1200	436.4
Interaction	0	1	0	0
Error	22	8	2.75	
Total	1522	11		

$F_{0.05}(1, 8) = 5.318$

interaction. Table 5.9 summarises the results of the analysis of variance, which indicates that there is a significant difference in the contaminant concentration between Lake A and Lake B. The next step is to estimate each lake's mean concentration, which is required in environmental monitoring studies but is not shown here.

A significant difference is also found in the contaminant concentration depending on the weather. Moreover, for example, later, it was also found that, although there were no measurements made to confirm quantitatively, the changes in solar radiation intensity and rainfall intensity also have effects on the contaminant concentrations. In this case, the data should be analysed by a randomised block design (RBD). Clear weather and rainy weather are categorical variables. If there is an effect of precipitation and it is not the main purpose to see the effect quantitatively, it can be revealed by the analysis of variance of RBD.

5.3.1 Comparison with One-Way ANOVA

Suppose researchers ignore the influence of the weather and take six samples at random and separately from Lake A and Lake B. In such a case, the data with six repetitions as shown in Table 5.10 will be analysed by one-way analysis of variance.

The results of this analysis of variance are shown in Table 5.11, which can be compared with Table 5.9 to reveal what has changed by ignoring the effects of weather. The sum of the squares of the whole and the site is the same, and one can see that the sum of squares of the weather is added to the error (residual) term.

In conclusion, it cannot be said that there is a significant difference between Lakes A and B. Since this data have two groups to compare, they can also be analysed by testing the difference between the two sets of population means. The analysis result in this case is as follows.

Test as an independent sample: H_0: $\mu_A = \mu_B$

$$t = -1.567, df = 10. \left(t^2 = 2.455\right)$$

One cannot say that there is a significant difference between Lakes A and B. Note that the F value in the analysis of variance table matches the square of the t value in the mean difference test. Both tests are the same for this case.

TABLE 5.10

Rearranged Data for One-Way ANOVA

No.	Lake A	Lake B
1	116	126
2	114	123
3	96	105
4	115	126
5	96	107
6	93	103

TABLE 5.11

Result of One-Way ANOVA for Example 5–4

Factor	SS	df	MS	*F*
Site (lake)	300	1	300	2.455
Error	1222	10	122.2	
Total	1522	11		

$F_{0.05}$ (1, 10) = 4.965

TABLE 5.12

Rearranged Data for Paired-Data Comparison (Example 5–4)

Sample number	Date	Lake A	Lake B
1	1 March (sunny)	116	126
2	2 March (sunny)	114	123
3	3 March (rain)	96	105
4	4 March (sunny)	115	126
5	7 March (rain)	96	107
6	9 March (rain)	93	103

When analysing data in such a way, the dispersion due to the influence of the weather is mixed in with the variation of the residual term, and the variance within the group increases. Recall that, as explained in Chapter 4 for the mean difference test, even if the same mean difference, the greater the variation, the lower the power of testing, making it impossible to identify significant differences.

5.3.2 Comparison with Two-Way ANOVA (RBD)

Next, let us consider the following example in the case that it is not clear whether the weather has effects on water quality or whether there may be something other than the weather or maybe a common factor that changes day by day. In such cases, it is safer to take a sample at the same time and analyse the data as paired data on the same day. If there are two groups to compare, the sample can be analysed with a paired data population means test, but if there are three or more groups, it is analysed with a randomised block design ANOVA. Here, the number of blocks is two, but let's analyse it using a randomised block design. In this case, the data format is the same as for a two-way ANOVA without repetition (Table 5.12) with the two-way ANOVA shown in Table 5.13.

Comparing Table 5–11 and Table 5–13, the sums of squares for the site (lake) are the same, but the sums of squares of the residual terms are different. Most of the residual sums of squares

TABLE 5.13
Result of Two-Way ANOVA (RBD) for Example 5–4

Factor	SS	df	MS	*F*
Site (lake)	300	1	300	750
Block (day)	1220	5	244	610
Error	2	5	0.4	
Total	1522	11		

$F_{0.05}\,(1, 5) = 6.61$

in Table 5.11 are transferred to the effect of the day, which is a blocking factor in Table 5.13. As a result, the residual sum of squares in Table 5.13 is small. And because the residual mean squares were small, the power of the test increased, and the difference in the lake (location) could be determined as significant. According to the result shown in Table 5.13, one can say that there is a significant difference between Lakes A and B.

This dataset can also be analysed by the test of the difference in the population mean because two groups need to be compared. However, since the samples were extracted from each population on the same day, it becomes a statistical test as paired data. In this case, the hypotheses and statistic calculation procedure yielded as follows:

$$H_0 : \mu_A - \mu_B = 0; H_1 : \mu_A - \mu_B \neq 0$$

$$t = \bar{d} / \sqrt{\left(s_d^{\,2}/n\right)} = (-10) / \sqrt{(0.8/-6)} = -10/0.365 = -27.39$$

$$t = -27.386, \text{df} = 5, \left(t^2 = 750.0\right)$$

where d is the difference between paired data.

One thus can say that there is a significant difference between Lakes A and B. It is seen that the randomised block design ANOVA with two factors is the same as the paired t test for population means. This can also be confirmed from the corresponding statistic values of Equation (3–15), as shown next. For paired data, one can calculate each element of the ANOVA table from statistics such as the variance of the difference.

$$\text{MSE} = 0.8/2 = 0.4$$

$$\text{MSTR} = \left(\Sigma d\right)^2 / (2n) = (-60)^2 / (2 \times 6) = 300$$

$$F_0 = \text{MSTR}/\text{MSE} = 300/0.4 = 750$$

5.3.3 Handling of Random Factors and Their Relationship with Mixed Models

A t test is utilised when comparing the means of two populations. While, to examine the differences in means across three or more groups, an analysis of variance (ANOVA) is conducted. If the objects taken up for comparison are all there are, and there is nothing else, it makes sense to compare their mean values. However, if the objects for comparison are representatives selected from a large number of candidates, it is meaningless to estimate each mean value, and it may be sufficient to understand the degree of variation. In this context, the former are referred to as fixed factors, while the latter are known as random factors. The method used to analyse both simultaneously is known as a mixed model.

A mixed model includes both fixed and random factors and is frequently used to assess mean value differences in data analyses, such as with hierarchical or longitudinal data. For example, in Example 5–4 (Table 5.7), to determine whether differences in monitored values are attributable to variations in "location (lake)" and "day", "location" is regarded as a fixed factor, and "day" as a random factor. The primary effect of a fixed factor is termed a fixed effect and that of a random factor is a random effect. Fixed effects are interpreted based on the average value differences across levels, whereas random effects concentrate on factor variability, assessing the impact of this variability on the dependent variable. Typically, factors used for level comparisons or effect verification are labelled fixed factors, and those that are uncontrolled or significantly influential and challenging to account for in the error term are labelled random factors. However, the choice between fixed and random effects should be based on the specific context and hypothesis of the model's application.

Analysis of variance can be employed to examine data involving both fixed and random factors. In many instances, as discussed in this text, ANOVA is applied to fixed factors (parameter model) and random factors (variate model) alike. The methods for estimating optimal levels are different, but the differences are not significant until one constructs an ANOVA table. However, this process is based on certain assumptions:

- Errors are assumed to be independent.
- Uncontrollable factors beyond fixed effects are accounted for in the error term.
- The dependent variable is presumed to follow a normal distribution.

ANOVA necessitates a study design that mitigates the impact of certain assumptions when they are not satisfied. In contrast, mixed models integrate these uncontrollable factors as random effects across various models. Differing from ANOVA, which presumes error independence, mixed models accommodate varying error assumptions. Thus mixed models enable flexible hypothesis-based modelling. Assuming a normal distribution for the dependent variable, a linear mixed model is applicable; for non-normal distributions, a generalised linear mixed model (GLMM) is suitable. Generalised linear mixed models accommodate various probability distributions, such as normal and binomial distributions. A thorough understanding of the mixed model requires familiarity with certain complex theories that are beyond the scope of this text and are thus omitted. For further understanding, textbooks such as Seltman (2014) may be consulted.

5.3.3.1 Analysis for Repeated Measurement Data

In a study design involving repeated measures data, each experimental unit is observed multiple times, with dependent variable measurements taken at various time points. Repeated measures data are frequently synonymous with longitudinal data. In such analyses, the experimental group variable is considered the primary independent variable. Statistical analyses for these designs should include all repeated observations to test the research hypotheses. Nonetheless, it is common for conventional statistical analyses to summarise repeated measures data by calculating an average for each experimental unit.

When aggregating repeated measurements, ANOVA is typically used. However, applying ANOVA to aggregated repeated measurements may breach the fundamental assumption of independence that ANOVA requires. Repeated measurements are correlated observations, given that they are observed from the same experimental unit. Repeated measurements collected closer in time are more correlated than measurements collected further in time. Correlation within an experimental unit is ignored when repeated measurements are aggregated. This violation of independence may lead to biased analysis results and incorrect interpretations. For the analysis of such repeated measurement data, see Muhammad (2023) and others

In econometrics, the analysis involves data on various economic indicators across multiple time points. Linear mixed models are commonly employed for this purpose, and the process is typically referred to as "panel data analysis" in this field.

5.4 Other Useful Experimental Designs

ANOVA is an effective method for analysing the influence of multiple factors. However, conducting all experiments in a random sequence can be challenging. Valuable techniques are available, even when numerous factors need investigation or specific conditions are present, but they demand considerable caution. In such instances, recalling Fisher's three principles is beneficial: These principles, crucial for experimental design, encompass replication, randomisation and local control (Saxena, 2024).

- ***Replication*** involves repeating each treatment at least twice under identical conditions to discern whether errors are systematic (differences due to treatment variations) or random (errors occurring by chance).
- ***Randomisation*** of the experiment order mitigates the impact of uncontrolled factors on the results.
- ***Local control*** ensures that the experiment is conducted within a specific plot with uniform management of experimental conditions.

In this context, the randomisation of experimental order pertains to the prerequisites for utilising ANOVA, which include the independence of errors. Split design serves as a method to circumvent the requirement of local control. This section provides two examples of such designs, split design and orthogonal design. Further details can be found in relevant advanced statistical textbooks.

5.4.1 Split Design

Various studies that are conducted in a laboratory often do not work under the ideal conditions assumed in a statistical textbook. Commonly, it is not possible to combine all levels of all factors and study them all in the order in which they are determined by a random number table. For example, if one must change the temperature set for a constant temperature bath, it takes several days for the bath temperature to become steady. In this case, one would probably want to run some experiments together under the same temperature conditions and vary other conditions such as the amount of reagent added. However, if the experiments are conducted in this procedure, it is no longer possible to estimate the error term in the way that has been described so far, since not all experiments are performed in the randomised order. That is, the format of the ANOVA table will change. This method is called split design, which is detailed in the advanced statistical textbooks such as that of Pennsylvania State University (2020). This section provides a brief description of this issue.

A split or split-plot design is an experimental method used when two or more levels of experimental units cannot randomise all experimental orders. In a split-plot design, two (or more) levels of randomisation are applied, and experimental units are assigned to the process. The first level of randomisation is applied to the entire plot and is used to assign experimental units to the level of the first treatment. The whole plot is divided into several subplots, and a second level of randomisation is used to assign the experimental units of the subplot to the next level of treatment. Because there are two levels of experimental units in the design of a split-plot, there is a separate experimental error throughout the plot and subplot portions. In split design, the ANOVA table is split, so one should adopt the corresponding error as the denominator in the F test.

5.4.2 Orthogonal Design

Next, let us consider an experimental method when there are many factors to be investigated. For example, in an experiment in which the composition of many raw material reagents will

TABLE 5.14

Example of the Orthogonal Array Table with a Level of Two

	a	b	c	d	f	g	h
1	1	1	1	1	1	1	1
2	1	1	1	2	2	2	2
3	1	2	2	1	1	1	2
4	1	2	2	2	2	2	1
5	2	1	2	1	2	1	2
6	2	1	2	2	1	2	1
7	2	2	1	1	2	2	1
8	2	2	1	2	1	1	2

be changed, one would like to find the best mixing condition to attain the highest efficiency of the mixed chemical product. Even if two levels are set for all ten kinds of raw material reagents, a total of 1024 (2^{10} = 1024) experiments are required in the factorial design experiment. Furthermore, to confirm all interactions, repetition will be necessary, so 2048 experiments will be performed. It would be impossible to carry out so many experiments with stable conditions with the laboratory and equipment kept constant. Here, if one knows in advance that almost all interactions can be negligible, then using the orthogonal array design experiment is a good alternative. In this case, 16 experiments may be enough, as explained.

The orthogonal array design experiment is to decide the level of the factor in each experiment according to the orthogonal array table and conduct the experiment according to this table. An example of a simple orthogonal table is shown in Table 5.14.

Note that, for an orthogonal table, the title of the row is the experimental number, and the title of the column is the factor. The number in each cell is the level number of the factor in that column. Therefore, 8 experiments (rows) are conducted according to this table. Each factor is set to the level specified in each cell. In this table, to avoid the mistake of an error term, the letter "e" was not used with only a, b, c, d, f, g and h (7 factors) specified in the columns. In this orthogonal array table, the column titles are a, b and c in lowercase letters. In actual experiments, enter the experimental conditions, i.e., amount of reagents added, set temperature, working conditions etc. here. In this textbook, these experimental conditions are indicated by capital letters A, B, C.

An orthogonal table assigns factors to each column of the spreadsheet and gives each row an experiment number. Then assign a level number for each factor to each cell in the spreadsheet. It is important to note that, no matter which two columns are taken from the orthogonal table, the inner product should be zero. The inner product of the vectors is equivalent to the cosine similarity, which means that the correlation coefficient is zero, that is, uncorrelated. In the orthogonal array experiment, as a rule, it is advisable to arrange the levels of all factors to be 2 or 3. Of course, more factor levels can be used, but it is not recommended because it requires a sophisticated technique.

To illustrate the application of this design, a simple case is used in this section. Consider the experimentation on five factors: A, B, C, D, and F, with all factors having two levels. In the factorial design experiment, 32 experiments are required, because 2^5=32. If it seems that interactions can be ignored, one can use an orthogonal array experiment. Assign factors A, B, C, D and F to columns a, b, c, d and f in Table 5.14, respectively. The first experiment, according to the table, will be conducted at the first level for all factors. In the second experiment, A, B and C are the first levels, and D and F are the second levels. In this way, the level of each factor is determined according to this table, and a total of eight experiments are conducted.

In this way, the effects of A, B, C, D and F can be evaluated from 8 experimental data. If one obtains the average of the experimental results from the first to the fourth and the average of the experimental data from the fifth to the eighth, the effect of factor A can be determined by comparing two average values. In this case, the number of the frequency for each level of B, C, D and F that appears in the two datasets (from the first to fourth, and from fifth to eighth) is quite the same. This means "orthogonal" in the orthogonal table. The same is true for factors B, C, D and F. For more information, read advanced technical books or consult with relevant experts.

Table 5.14 can be used to analyse the effects of up to seven factors. However, if seven factors are adopted, there is no error term. In an orthogonal array table, the error terms are presented by the column that may appear as neglected interactions. In an orthogonal array experiment, one can adopt factors up to the maximum number of columns, but it is recommended to leave some columns to ensure the degree of freedom for the error term. Also, if a particular interaction may be expected, it is necessary to confirm the column in which the interaction appears and to prevent other factors from being assigned to that column. In addition, various calculations are performed using it as an error term. Therefore, it is important to recognise that unexpected interactions can make all results unreliable.

One of the most useful tools for effectively estimating the effects of many factors may be the experimental design with orthogonal arrays. Of course, there are some restrictions, such as the need for prior information about the interactions, but it is hoped that it will be considered to improve the cost-effectiveness of the experimental results. Specific examples can be found in Chapter 24 (Tables 24.4 and 24.5 for examples) in Part III. Further information can be found in specialised books in this field, for example, SAS Institute Inc. (1999).

Data from designed experiments, as will be described in Chapter 6, can also be analysed using regression analysis. Regression analysis of well-organised data has many advantages, but for such analyses see technical works such as Draper and Smith (1981) and Box et al. (1978) for further details on the application of regression analysis in constructing empirical equations.

5.5 Summary

Analysis of variance is employed to analyse various datasets. When examining data from field observations, it is crucial to accurately comprehend the nature of the data and choose a suitable analysis method. Conversely, in numerous laboratory experiments, it is preferable to acquire data under an appropriate experimental design and subsequently analyse that data to obtain more pertinent information. In considering such an experimental design, knowledge of ANOVA is also highly beneficial. For split-plot design experiments and experiments utilising orthogonal arrays, the explanation provided in this chapter is insufficient, and more advanced specialist literature should be consulted.

When planning a significant experiment, understanding available analysis methods and the necessary precautions for each method will enhance the likelihood of obtaining appropriate results.

REFERENCES

Box, G.E.P., Hunter, W.G. and Hunter, J.S., 1978, *Statistics for experimenter, an introduction to design, data analysis, and model building*, Wiley & Sons, Inc., New York, United States.

Draper, N.R. and Smith, H., 1981, *Multiple regression analysis applied to analysis of variance problems, chapter 9 of applied regression analysis*, 2nd edition, John Wiley & Sons, Inc., New York, United States.

Muhammad, L.N., 2023 June 1, Guidelines for repeated measures statistical analysis approaches with basic science research considerations, *J Clin Invest*, 133(11), e171058, https://doi.org/10.1172/JCI171058

Pennsylvania State University, 2020, *The split-plot designs, chapter 14.3 of STAT 503*, Pennsylvania State University, Eberly College of Science, retrieved on 20 September 2023 from https://online.stat.psu.edu/stat503/lesson/14/14.3

SAS Institute Inc., 1999, *Chapter 16 theory of orthogonal designs, version 8* (pp. 501–508), SAS/QC User's Guide, SAS Institute Inc., Cary, NC, United States, http://www.math.wpi.edu/saspdf/qc/chap16.pdf

Saxena, S., 2024, *Basic principles of experimental designs, lecture 15 in research methodology*, viewed on 10 October 2024, https://www.ramauniversity.ac.in/online-study-material/fcm/bba/ivsemester/researchmethodology/lecture–15.pdf

Seltman, H.J., 2014, *Mixed models, a flexible approach to correlated data, chapter 15 in "experimental design and analysis"*, Carnegie Mellon University, https://rauterberg.employee.id.tue.nl/lecturenotes/2014-Seltman-Statistics_Book.pdf

6

Correlation and Regression Analysis

In this chapter, we will explain statistical analysis methods for studying the relationship between two variables: correlation analysis and regression analysis. Regression analysis is applied when we want to estimate the value of one variable from the value of another variable. In many cases, this is a study of the causal and effect relationship between two variables. Correlation analysis is a common method for analysing the relationship between two variables, regardless of whether there is a causal relationship between them.

In correlation analysis, there are two positions. One is descriptive statistics that simply looks at the strength of the relationship between the two variables obtained, and another is inferential statistics that estimate the relationship between the elements of the population based on the extracted samples. Regression analysis is a very practical statistical analysis technique and is widely used in analysing data from various experiments and surveys.

6.1 Correlation Analysis – Variable Relationship

A study on the strength of the relationship between two variables is correlation analysis, and the strength is measured by the correlation coefficient. This measure is limited in the linear relationship. Therefore, to make the scatter diagram, it is useful to know whether a non-linear relationship exists. The correlation coefficient r is calculated by the following formula:

$$r = \frac{\sum (x-\bar{x})(y-\bar{y})}{\left[\sum (x-\bar{x})^2 \sum (y-\bar{y})^2\right]^{1/2}} = \frac{SSXY}{(SSX \times SSY)^{1/2}} \tag{6–1}$$

where SSX and SSY are the sums of the squared deviations of x and y, respectively. SSXY is the sum of the products of x and y. The correlation coefficient ranges between –1.0 and 1.0.

The relationship between two variables can be clarified by visualisation using a scatter plot. If a scatter plot is obtained that shows that the increase in one variable is not related to the increase or decrease in the other, the correlation coefficient between those variables is close to zero. If the scatter plots are distributed from the top right to the bottom left, there is a positive correlation. In such a case, if the correlation coefficient is calculated, it becomes a positive value. On the other hand, if the plots are distributed from the upper left to the lower right, it is said that there is a negative correlation.

On a scatter diagram, if all plots are distributed on one straight line, the correlation coefficient is either plus one or minus one. The more scattered plots deviate from the straight line, the closer the correlation is to zero. Such a trend is expressed as a weakened correlation. A typical scatter diagram showing the correlation from strong to weak is introduced in Figure 6.1.

Regarding the correlation coefficient, when it is necessary to consider the population and the sample separately, they are called the population correlation coefficient and the sample correlation coefficient, and the symbols ρ and r are used respectively.

DOI: 10.1201/9781003597216-7

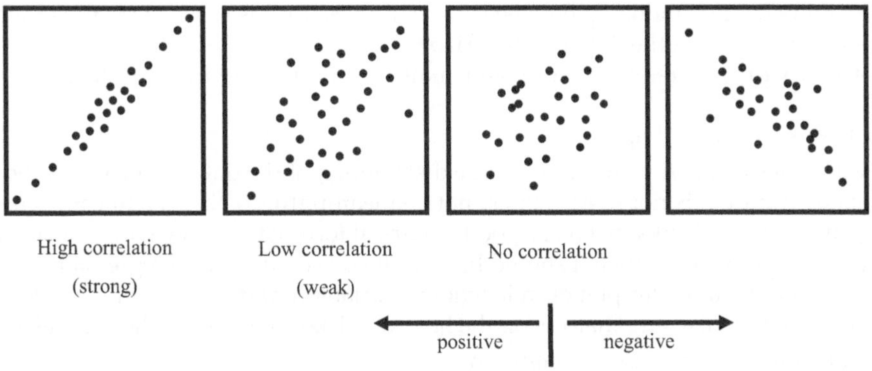

FIGURE 6.1 Example of typical and characteristic scatter plots.

Suppose one extracts some datasets from a population and calculates the correlation coefficient between pairs in the sample. Suppose it is desired to consider whether there is a correlation between the elements of the population. the following hypotheses are set for a statistical test as follows.

$$H_0 : \rho = 0$$
$$H_1 : \rho \neq 0$$

The test for the population correlation coefficient ρ is usually conducted with null hypothesis H_0: $\rho = 0$ and alternative hypothesis H_1: $\rho \neq 0$.

Null hypothesis: there is no correlation between the two variables, $\rho = 0$.
Alternative hypothesis: there is a correlation between both variables, $\rho \neq 0$.

When the null hypothesis is rejected, it can be concluded that there is a correlation between the two variables. When the null hypothesis is not rejected, it implies that we cannot confirm a significant correlation between the two variables. Even when the population correlation coefficient ρ is 0, the calculated correlation coefficient r from a sample taken at random may often be a quite large value of nearly 0.5, if the sample size (number of elements in the sample) is small. Let r be the correlation coefficient of the samples extracted from a population that follows a bivariate normal distribution. Let ρ be the correlation coefficient of the population. If the hypothesis is that there is no correlation in the population, that is, $\rho = 0$, the t statistic in Equation (6–2) follows the t distribution with $n - 2$ degrees of freedom.

$$t = r\sqrt{\frac{n-2}{1-r^2}} \tag{6–2}$$

If the absolute value of the t value calculated by Equation (6–2) is greater than t $(n - 2, \alpha/2)$ for the significance level α, the null hypothesis can be rejected. A further explanation of the procedure of this test can be found in relevant textbooks, e.g., Milton and Arnold (1995).

6.2 Regression Analysis – Cause–Effect Relationship

Regression analysis is a very useful statistical method in which the value of one variable is estimated based on the value of another variable. This method can be applied to the study of the cause-and-effect relationship between two variables. In this case, one is the independent variable, and the other is the dependent variable.

The ordinary procedure of the regression analysis is performed in three steps:

Step 1: Make a scatter diagram.

Draw a scatter plot and visually check the distribution of the data on the plot. When the data distribution is not linear, it may not be meaningful to obtain a linear regression equation. If one thinks that it is a good empirical formula to apply with a quadratic or exponential function, then it should be examined accordingly. In regression analysis, when creating a scatter plot of independent variable x and dependent variable y, it is important to make sure that x is on the horizontal axis and y is on the vertical axis.

Step 2: Estimate the regression coefficient.

Estimate the coefficient of the regression equation from the obtained dataset. That is the slope and y-intercept of the regression line. The least squares fitting is commonly used to estimate the coefficient.

Step 3: Evaluate the prediction accuracy.

Examine whether the dependent variable y can be estimated from the independent variable x using the obtained regression equation. In the case, when x and y are completely irrelevant and it is meaningless to know the value of x when estimating the value of y, the slope of the regression line is zero. Therefore, first consider whether the hypothesis that the regression coefficient (the slope of the regression line) is zero can be rejected. If it can be rejected, further consider the size of the estimation error and so on.

6.2.1 Estimation of Regression Coefficients

In Figure 6.2, the i-th data is displayed as (x_i, y_i), and $\hat{y}_i$ is the value of the dependent variable estimated by the regression equation when the value of the independent variable is x_i. Here $y_i - \hat{y}_i$ is the residual e_i.

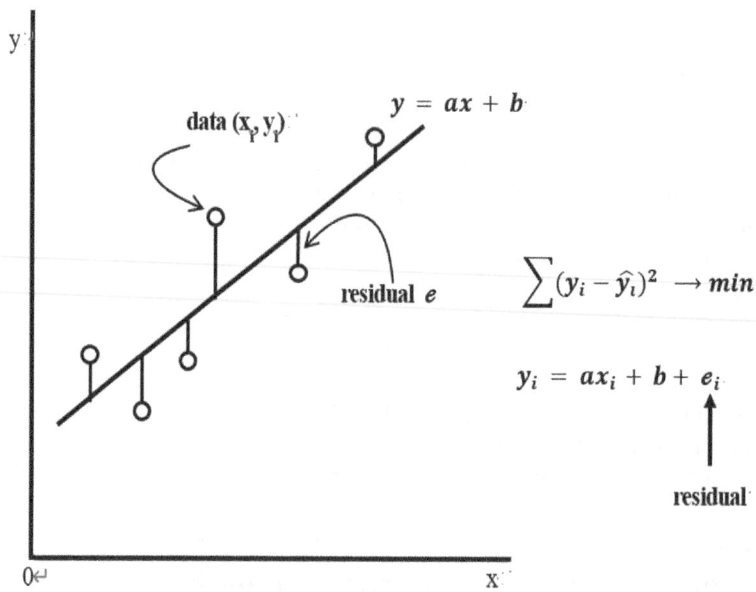

FIGURE 6.2 Regression line on the scatter diagram.

y is the value of the dependent variable or predictor variable.
x is the value of the independent variable or explanatory variable.
a is the intercept, which is the point at which the regression line crosses the y-axis.
b is the regression coefficient, namely the slope of the regression line.
e is a prediction error or residual, defined as the difference between the observed value y and the predicted value.

To determine an optimal empirical formula for estimating y from x, it is essential to minimise the residual sum of squares (also referred to as the sum of squared errors, SSE). The residuals can be positive or negative, so it is meaningless to add them as they are. It is necessary to determine the coefficient of the line (regression coefficient) so that the arrangement of all data is as close as possible to the line. This can be done by minimising the sum of the squares of the residuals.

Here, the equation is expanded by substituting a linear equation including coefficients a and b for the estimated value of y. Then a and b can be differentiated as variables. For the residual sum of squares SSE, the derivative of each of a and b is zero at the local minimum. The value can be determined from the simultaneous equations obtained in this way.

$$\text{residual sum of squares, SSE} = \sum \{y_i - (a + bx_i)\}^2$$

$$\underbrace{\frac{\partial(SSE)}{\partial a} = 0 \quad \frac{\partial(SSE)}{\partial b} = 0}_{\Downarrow} \tag{6–3}$$

$$b = \frac{SSXY}{SSX}, \ a = \overline{y} - b\overline{x}$$

where SSX is the sum of squares, and SSXY is the sum of products, as in Equation (6–1). The regression line always passes the point of the average value of x and y $(\overline{x}, \overline{y})$ due to its nature. Therefore, there are two ways to display it.

$$y = a + b\,x \tag{6–4}$$

$$y - \overline{y} = b(x - \overline{x}) \tag{6–5}$$

The regression line should only be used to estimate the value of the dependent variable y within the range of the observed data. Extrapolating beyond this range for predictions is not advisable. When a scatter plot shows that most data points are distant from the origin, the intercept value, or the regression constant a, becomes less important, as illustrated in Figure 6.3. Consequently, the second equation, Equation (6–5), might serve as a more suitable regression line equation from this perspective.

Let us explain the regression analysis procedure using the hypothetical numerical example provided in Example 6–1. Although the sample size is small, the focus here is to facilitate understanding of the calculation process. In a real-world scenario, a sufficient number of data points for each variable is essential to produce a meaningful analysis result.

Example 6–1: Air quality was monitored at five locations within an urban area. The concentrations of sulphur dioxide and nitrogen dioxide are presented in Table 6.1. The question arises: if the sulphur dioxide concentration is known, can it be used to estimate the nitrogen dioxide concentration, and if so, how?

Answer for Example 6–1: The regression coefficient b can be calculated from the sum of squares SSX and the deviation sum of products SSXY as seen in the calculation auxiliary of Table 6.2.

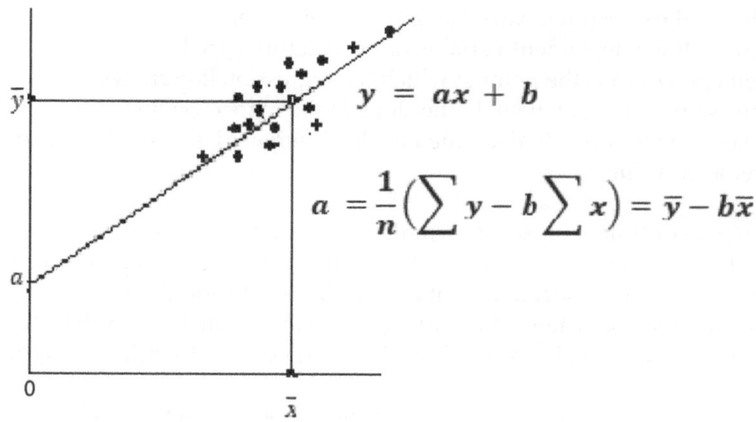

FIGURE 6.3 Regression equation, the relation between the intercept *a* and data.

TABLE 6.1

Air Pollutant Concentrations at 5 Sites (arbitrary unit)

Site	SO_2	NO_2
A	4	5
B	1	3
C	6	10
D	2	1
E	2	6

TABLE 6.2

Auxiliary Table for Calculating Regression Coefficients and Residuals in Regression Analysis

x	y	xy	x^2	y^2	$\hat{y}= a+bx$	$e = y - \hat{y}$	e^2	$\hat{y}^2$
4	5	20	16	25	6.375	−1.375	1.891	40.6408
1	3	3	1	9	2.25	0.75	0.563	5.0625
6	10	60	36	100	9.125	0.875	0.766	83.2656
2	1	2	4	1	3.625	−2.625	6.891	13.1406
2	6	12	4	36	3.625	2.375	5.641	13.1406
15	25	97	61	171	25	0	15.75	155.2501

$SSX = 61 - 15^2/5 = 16$, $SSXY = 97 - 15 \times 25/5 = 22$, $SSY = 171 - 25^2/5 = 46$

Since the average values of x and y are 3 and 5, respectively, the regression coefficient b and intercept a can be calculated by Equation (6–3):

$b = 22/16 = 1.375$

$a = 5 - 1.375 \times 3 = 0.875$

Therefore, the obtained regression equation is

$$y = 0.875 + 1.375x$$

6.2.2 Evaluation of Prediction Accuracy

The predictive accuracy of regression equations is usually evaluated by the magnitude of the residual. This term is called the error. The value of the dependent variable estimated by the regression equation is $\hat{y}$. The variance of the difference between the estimated value and the measured value y is the error variance e^2. The number of elements of the population is N, and the number of samples extracted is n. The square root of the unbiased estimate from the n samples is the **standard error of the estimate** $s_{y.x}$.

$$
\begin{aligned}
&e^2 = \frac{1}{N}\sum (y_i - \hat{y}_i)^2 \\
&\hat{e}^2 = \frac{1}{n-2}\sum (y_i - \hat{y}_i)^2 = s^2_{y.x} \\
&s_{y.x} = \sqrt{\Sigma\,(y_i - \hat{y}_i)^2 / (n-2)} \\
&\text{where, } \hat{y} = a + bx.
\end{aligned}
\tag{6–6}
$$

In the manual calculation, it is convenient to use the following procedure:

$$s^2_{y.x} = \mathrm{SSE}/(n-2),\ \mathrm{SSE} = \Sigma y^2 - a\Sigma y - b\Sigma xy$$

When the i-th value of the independent variable is x_i, the i-th value of the dependent variable is y_i, and the estimated value y_i from x_i is $\hat{y}_i$. The error (or residual) is e_i.

$$e_i = y_i - \hat{y}_i$$

The standard deviation of e_i is σ_e. The square of σ_e, namely variance of error σ_e^2 can be estimated by the sum of squares of error (or residual) divided by its degrees of freedom. The square root of this value is called the standard error of the estimate and has $s_{y.x}$ as the notation.

$$
\begin{aligned}
&Ve = \Sigma\,(y_i - \hat{y}_i)^2 / (n-2) = \mathrm{SSE}/\mathrm{df} \\
&s_{y.x} = \sqrt{(Ve)}
\end{aligned}
$$

The $s^2_{y.x}$ for the case study Example 6–1 can be calculated as follows:

$$
\begin{aligned}
\mathrm{SSE} &= \Sigma\, y^2 - a\,\Sigma y - b\Sigma\, xy \\
&= 171 - 0.875 \times 25 - 1.375 \times 97 = 15.75
\end{aligned}
$$

The degrees of freedom (df) are 3 because the number of data is 5 and df = 5 – 2 =3.

$$s^2_{y.x} = 15.75/3 = 5.25$$

In this example, the sum of squares error SSE is 15.75, and it can be found in Table 6.2 (last row of e^2 column). The calculation of the degrees of freedom for the SSE may be interpreted as having two constraints, that is, the average value of the objective variable and the deviation from the regression equation, which are further explained in the following section.

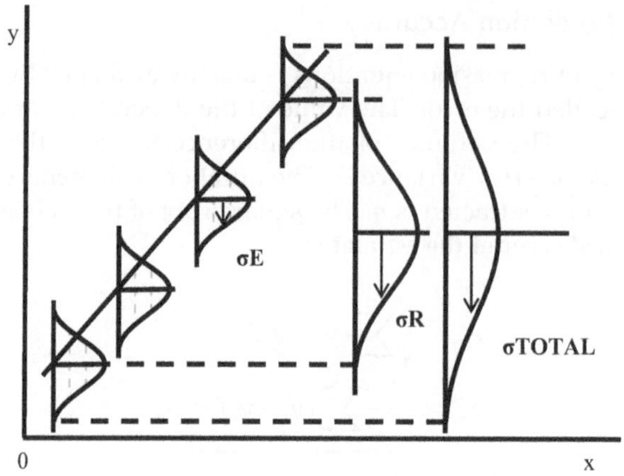

FIGURE 6.4 Relationship between the variability of data from the regression line and the variability that can be explained by the regression equation.

6.2.3 Decomposition of the Variance and Degrees of Freedom

The effectiveness of the regression equation can also be measured as follows. Let us consider how much of the variation of the dependent variable (objective variable) can be estimated by the regression equation.

If all the plots in the scatter diagram are on the regression line and the variation from the regression equation is zero, 100% of the variability can be explained by the regression equation. If the plots of the data are scattered around the regression equation, one can compare the two, as shown in Figure 6.4. The relation between the variations of the objective variable due to regression (regression σ_R) and variation within regression (error σ_E) is shown in Figure 6.4.

Figure 6.4 illustrates that the variance of the dependent variable (objective variable) can be decomposed into two parts. One is that can be estimated by the regression equation, and the other is the variability around the regression equation. This can be explained using a mathematical formula. The sum of squares of y can be decomposed as follows:

$$
\begin{aligned}
SSY &= \Sigma(y_i - \hat{y})^2 = \Sigma(y_i - \hat{y}_i + \hat{y}_i - \overline{y})^2 \\
&= \Sigma\{(y_i - \hat{y}_i) + (\hat{y}_i - \overline{y})\}^2 = \Sigma(\hat{y}_i - \overline{y})^2 + \Sigma(y_i - \hat{y}_i)^2 \\
&= SSR + SSE
\end{aligned}
$$

$$
\begin{array}{ccccc}
(n-1) & = & 1 & + & (n-2) \\
\text{df for SSY} & & \text{df for SSR} & & \text{df for SSE}
\end{array}
$$

As for the case of Example 6–1, one can confirm that 155.25 plus 15.75 equals 171, which is also seen in Table 6.2.

6.2.4 Relation between Correlation Analysis and Regression Analysis

Correlation analysis examines the relation between two variables, e.g., x and y. In this case, one usually tests for no correlation, i.e.:

$$
\begin{aligned}
H_0 &: \rho = 0 \\
H_1 &: \rho \neq 0
\end{aligned}
$$

(ρ: population, r: sample)

Regression analysis examines the relationship between the dependent and independent variables. In this case, one usually tests for the regression coefficient.

$$H_0 : \beta = 0$$
$$H_1 : \beta \neq 0$$
(β: population, b: sample)

Regression analysis and correlation analysis are different concepts of statistical analysis. However, the calculation results are closely related. If a regression analysis is performed on a dataset with a correlation coefficient of 0, the regression coefficient will be 0. It is also a good idea to calculate the correlation coefficient between variables in advance when performing a regression analysis on two variables that are considered to have a causal relationship.

The sum of squares of the objective variable is SSY. The sum of squares of the variation of the objective variable explained by regression is SSR. The ratio SSR/SSY is considered to be an evaluation scale for seeing to what extent an objective variable can be estimated from an explanatory variable by a regression equation. It is called the coefficient of determination or contribution rate. The estimated value $\hat{y}_i$ of the objective variable can be rewritten as follows:

$$\hat{y}_i - \overline{y} = b\,(x_i - \overline{x})$$

Therefore, the sum of squares of fluctuations due to regression SSR is written as follows:

$$\text{SSR} = \Sigma(\hat{y}_i - \overline{y})^2 = b^2\,\Sigma\,(x_i - \overline{x})^2 = b^2\,\text{SSX}$$

Using Equation (6–3) and this relationship, the following equation is obtained.

$$\begin{aligned}\text{SSR/SSY} &= b^2\,\text{SSX/SSY} = (\text{SSXY/SSX})^2\,\text{SSX/SSY}\\ &= \text{SSXY}^2/(\text{SSX SSY}) = r^2\end{aligned}$$

It can be seen that the coefficient of determination is the square of the correlation coefficient. That is, the higher the correlation coefficient between the objective variable and the explanatory variable, the better the fit of the obtained regression equation. Some statistical analysis software displays the coefficient of determination instead of the correlation coefficient in the scatter plot accompanying the regression analysis.

6.2.5 Test for Regression Coefficient

If the variation explained due to the regression is sufficiently larger than the variation deviated from the regression (scattering around the regression line), it can be considered meaningful to estimate y from x by using the regression equation. Now let us consider the expected value of the sum of squared deviations. The expectation of SSR can be expressed as follows:

$$\begin{aligned}E(\text{SSR}) &= \sigma_e^2 + \beta^2\,\text{SSX}\\ E(\text{SSE}) &= E(\text{SSY} - \text{SSR}) = E(\text{SSY}) - E(\text{SSR})\\ &= \left\{(n-1)\,\sigma_e^2 + \beta^2\,\text{SSX}\right\} - \left(\sigma_e^2 + \beta^2\,\text{SSX}\right) = (n-2)\sigma_e^2\\ E\left(\text{SSE}/(n-2)\right) &= \sigma_e^2\end{aligned}$$

If $\beta = 0$, the expected value of SSR is identical to σ_e^2, namely, the same as the expected value of (SSE/(n – 2)). Therefore, the hypothesis of H_0: $\beta = 0$ can be evaluated by comparing two population variances, namely the F test. Another explanation is as follows:

If $\beta = 0$,

$$SSR/\sigma_e^2 \sim \chi^2(1)$$
$$SSE/\sigma_e^2 \sim \chi^2(n-2)$$

and both are independent, then $F_0 = (SSR/\sigma_e^2)/(SSE\ /\sigma_e^2)$ follows F distribution. Based on the concept, ANOVA for regression analysis can be shown in Table 6.3.

The ANOVA table for Example 6–1 is shown in Table 6.4.

As seen in Table 6.4, the F_0 value is less than $F_{0.05}(1,3)$. Therefore, one cannot say that this regression equation has significant meaning. The sum of squares of regression SSR can be directly calculated as shown in Table 6.5.

This ANOVA result is identical to the t test for regression coefficient (slope).

$$H_0: \beta = 0$$
$$H_1: \beta \neq 0 \text{ (two-tailed test)}$$

In this example, $n = 5$, $b = 1.375$, and s_b is calculated as follows:

$$s_b = \sqrt{}(Ve/SSX) \text{ or } s_{y.x}/\sqrt{}(\text{SSX})$$

TABLE 6.3

ANOVA Format for Regression Analysis

Source	Sum of squares	Degrees of freedom	Mean square	F-value
Regression	$SSR = SSXY^2/SSX$	1	$V_R = SSR$	$F_0 = V_R/Ve$
Residual	SSE = SSY-SSR	n – 2	Ve = SSE/(n – 2)	
Total	SST = SSY	n – 1		

TABLE 6.4

ANOVA of Regression Analysis for Example 6–1

Source	SS	df	MS	F_0
Regression	SSR = 30.25	1	$V_R = 30.25$	$F_0 = 5.762$
Residual	SSE = 15.75	3	Ve = 5.25	
Total	SST = 46.00	4		

Note: $22^2/16 = 30.25$; The F value of df1=1, df2=3 is $F_{0.05}(1,3) = 10.1$;

TABLE 6.5

Auxiliary Table for Calculating SSR

No.	$\hat{y}$	$\bar{y}$	$\hat{y} - \bar{y}$	$(\hat{y} - \bar{y})^2$
1	6.375	5	1.375	1.890625
2	2.25	5	−2.75	7.5625
3	9.125	5	4.125	17.015625
4	3.625	5	−1.375	1.890625
5	3.625	5	−1.375	1.890625
Total	25	25	0	30.25

The value of b is the least squares estimate of β, and this b follows the normal distribution N (β, σ_e^2/SSX). Therefore, the standardised value of b follows the standard normal distribution. Note that the notation "~" means "follow".

$$(b-\beta)/\sqrt{(\sigma_e^2/\mathrm{SSX})} \sim \mathrm{N}\left(0,1^2\right), \text{standard normal}$$

Here, σ_e^2/SSX is replaced by the estimated value, Ve/SSX.

$$(b-\beta)/\sqrt{(Ve/\mathrm{SSX})} \sim t\,(n-2).$$

This property can be used to test whether the regression coefficient is significant.

6.2.5.1 Hypothesis Test for Regression Coefficient (Slope)

The regression coefficient of the population is β, and the regression coefficient calculated from the data of the sample extracted from there is b. We consider the population regression coefficient from this data. The hypothesis testing is performed by following these five steps:

Step 1: Set the hypotheses.

$$\mathrm{H}_0\colon \beta = 0$$
$$\mathrm{H}_1\colon \beta \neq 0$$

Step 2: Clarify the value, n, b, and s_b.

$$s_b = s_{y.x}/\sqrt{(\text{sum of squares of } x)}$$

Step 3: Determine the rejection region using t table.
Step 4: Calculate the t value.

$$t = b/s_b$$

Step 5: compare the t value with the marginal value of the rejection region, and draw the conclusion

The steps mentioned are applicable to Example 6–1 for testing the regression coefficient, yielding the following result.

Step 1: Set the hypotheses.

$$\mathrm{H}_0\colon \beta = 0$$
$$\mathrm{H}_1\colon \beta \neq 0$$

Step 2: Clarify the value, $n = 5$, $b = 1.375$, and $s^2_{y.x} = 15.75/3 = 5.25$.

$$s_b = s_{y.x}/\sqrt{\mathrm{SSX}} = \sqrt{(5.25/16)} = 0.5728$$

Step 3: Determine the rejection region using t-table.

The t value of the both-tailed test for df = 5 − 2 = 3 is 3.182.

Step 4: Calculate the t value.

$$t = b/s_b = 1.375/0.5728 = 2.400$$

Step 5: Compare the t value with the marginal value of the rejection region, and draw the conclusion.

The calculated t value is less than the t table value of df = 3 and $\alpha = 0.05$. Therefore, one cannot say that this regression equation is significant.

This t value is compared with the F value in the corresponding ANOVA table shown in Table 6.4.

$$2.4^2 = 5.76$$

Recall that $F_\alpha(1, \gamma_2) = \{t\ (\alpha/2, \gamma_2)\}^2$ in Equation (3–15), which also confirms this relation. The results show the following key features: rejecting the hypothesis that the regression coefficient is zero is equivalent to rejecting the hypothesis that the variation due to regression is the same as the variation deviated from regression.

6.2.6 Regression Analysis for Repetitive Data

The case discussed in this section is related to performing a regression analysis on repeated data at each level of an explanatory variable. For example, one may have some experimental datasets for each level of the explanatory variable and would like to find an empirical equation after the analysis of variance. In laboratory experiments, one can use experiment designs to systematically set levels of explanatory variables. In a laboratory experiment, collecting data in this way is not so difficult and enables various useful analyses. Note that regression analysis based on data obtained from field surveys does not allow such settings.

Given such data, it is also possible to analyse whether there is a significant difference between the levels of the factors mentioned in Section 5.1 of Chapter 5 before regression analysis. If necessary, it may be possible to examine whether the effects of unexpected factors are encountered in the field of experiments. Consider this issue through a hypothetical but concrete numerical case of Example 6–2.

Example 6–2: Experiments were conducted to determine the optimal dosage of additives to purify water. The amount of chemical reagent was changed, and three measurements were carried out for the same conditions. A predetermined amount of purifying agent was added to the contaminated factory wastewater, stirred well and the concentration of the contaminated substance was measured after a lapse of a predetermined time. Data is shown in Table 6.6. Note that in this experiment, all 12 runs were performed in random order. It is desired to obtain the regression equation from these data.

Note: The data presented were derived from the experiment conducted in a completely random sequence. However, if the experiment had been performed according to the sequence shown in this table, refer to Section 5.3.3 in the preceding chapter.

Answer for Example 6–2: In this case, with a total of 12 data prepared, one can simply conduct regression analysis with these 12 data points. However, there are 3 repeated data at the same level. In other words, there are three cases for each explanatory variable hence various analyses can be made. Let us start with a simple regression analysis. First, create an auxiliary calculation table to find the sum of deviation products and the sum of squares of deviations (Table 6.7).

The average value and the sum of products are calculated as follows.

$$SSX = 375, SSY = 155, SSXY = -195, \bar{x} = 17.5 \text{ and } \bar{y} = 13.5.$$

TABLE 6.6

Data for Example 6–2: Concentrations Measured in the Experiments

Case number (reagent level)	10 mg/L	15 mg/L	20 mg/L	25 mg/L
1	16	16	13	7
2	20	11	14	12
3	18	15	9	11

Note: The unit of the contaminant concentration is μg/L.

TABLE 6.7

Calculation Auxiliary Table

No.	agent (x)	Conc. (y)	x^2	y^2	xy
1	10	16	100	256	160
2	10	20	100	400	200
3	10	18	100	324	180
4	15	16	225	256	240
5	15	11	225	121	165
6	15	15	225	225	225
7	20	13	400	169	260
8	20	14	400	196	280
9	20	9	400	81	180
10	25	7	625	49	175
11	25	12	625	144	300
12	25	11	625	121	275
total	210	162	4050	2342	2640

The regression coefficient and y-intercept are calculated:

$$b = \mathrm{SSXY}/\mathrm{SSX} = -0.52$$

$$a = \overline{y} - b\overline{x} = 22.6$$

The obtained regression equation is:

$$y = -0.52\,x + 22.6$$

Next, an evaluation of this regression equation is performed:

$$\mathrm{SSR} = (\mathrm{SSXY})^2/\mathrm{SSX} = 101.4,\ \mathrm{SSE} = \mathrm{SSY} - \mathrm{SSR} = 53.6$$

$$s_b = \sqrt{}(V_e/\mathrm{SSX}) = 0.120$$

Using these values, one conducts the test of whether the regression coefficient is zero with the t value calculated as follows.

$$t = b/s_b = -4.33$$

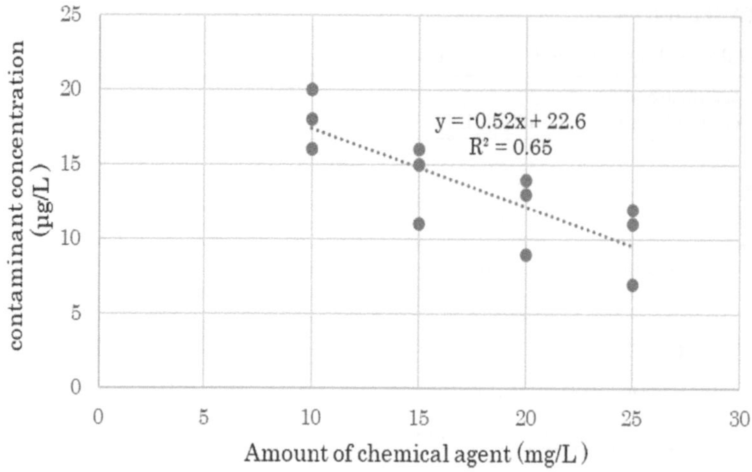

FIGURE 6.5 Scatter plots for the data of Example 6–2 using Excel.

The absolute value of the calculated t value is larger than the value of the t-table with a significance level α of 0.05, degrees of freedom (n–2). So, the hypothesis (H_0: $\beta = 0$) can be rejected. Next, examine the scatter plot of this data in Figure 6.5.

When finding empirical formulas from such experimental data, a lot of useful information can be used if one performs multiple experiments at each level. First, a one-way analysis of variance can be used to determine whether there is a significant difference between levels. Next, as an empirical formula, one can see if linear regression lines are appropriate or not. In addition, the confidence interval of the target value obtained by the empirical formula can be estimated.

In this example, it is also stated that all experiments were conducted in randomised order which is commonly assumed in exercise questions that often appear in statistical textbooks. However, it is a great challenge in reality when experiments are conducted. For example, changing some experimental conditions is too often too difficult compared with changing other factors. In such a case, one can adopt the split design of the experiment shown in Section 5.4.1 of Chapter 5. Before carrying out an experiment, it is beneficial to quantitatively assess the impact of other unforeseen factors and to conduct a preliminary experiment to reduce their influence. If unavoidable factors still exist, they can be partially addressed, if not entirely, by including them as block factors in the experiment, as elaborated in Chapter 5. Furthermore, while not discussed here, analysis using a mixed model may be feasible. For more information, consult a specialised advanced text.

Using the results performed under these controlled experimental conditions, it is also possible to check whether there is a significant difference between each level by ANOVA before performing regression analysis as seen in Figure 6.6.

In this step, the statistical analysis software of SPSS was used to conduct the analysis and make an ANOVA table for regression analysis. First, let's change the amount of reagent input from 10 to 25 mg/L and see if there is a significant difference in experimental results at each level. The var0001 in Table 6.8 is the objective variable. This data was analysed with one-way ANOVA, in which the number of levels is 4, and the number of repetitions is 3. As shown in Table 6.9, there is a difference at the 0.05 significance level. Therefore, the hypothesis that there is no difference between the levels at a significance level of 0.05 is rejected, and it can be said that the concentration of treated water varies depending on the amount of reagent added.

Further, regression analysis was performed with VAR001 as the dependent variable y and VAR003 as the independent variable x in Table 6.8. The results are shown in Table 6.10. Next, calculate the average value for each level and replace all data with this average value. The

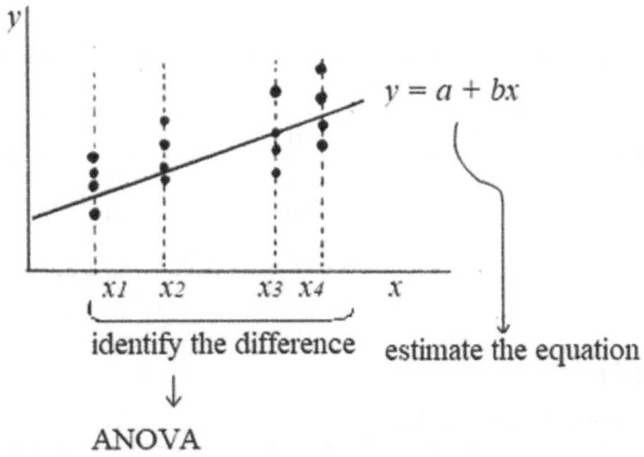

FIGURE 6.6 Experimental result by multiple trials for each condition.

TABLE 6.8
Data File of Example 6–2 in SPSS Format

No.	Var001	Var002	Var003	Var004
1	16.00	1.00	10.00	18.00
2	20.00	1.00	10.00	18.00
3	18.00	1.00	10.00	18.00
4	16.00	2.00	15.00	14.00
5	11.00	2.00	15.00	14.00
6	15.00	2.00	15.00	14.00
7	13.00	3.00	20.00	12.00
8	14.00	3.00	20.00	12.00
9	9.00	3.00	20.00	12.00
10	7.00	4.00	25.00	10.00
11	12.00	4.00	25.00	10.00
12	11.00	4.00	25.00	10.00

TABLE 6.9
Result of the One-Way ANOVA

	SS	df	MS	*F*
Agent (between-group)	105.0	3	35.0	5.6
Error (within-group)	50.0	8	6.25	
Total	155.0	11		

$F_{0.05}(3, 8) = 4.07$

TABLE 6.10
ANOVA for Regression Analysis (Independent Variable: VAR003, Dependent Variable: VAR001)

Source	SS	df	MS	*F*
Regression	101.4	1	101.4	18.9
Residual	53.6	10	5.36	
Total	155	11		

TABLE 6.11

ANOVA for Regression Analysis by Using Average Data within the Group (Independent Variable: VAR003, Dependent Variable: VAR004)

Source	SS	df	MS	*F*
Regression	101.4	1	101.4	281. 7
Residual	3.6	10	0.36	
Total	105	11		

TABLE 6.12

Composite ANOVA Table

Source	SS	df	MS	*F*
Linear Regression	101.4	1	101.4	
Lack of fit	3.6	2	1.8	<1.0
Between-group	105	3	35	5.6
Within-group	50	8	6.25	
Total	155	11		

data obtained in this way are VAR004 in Table 6.8. Regression analysis was performed with VAR004 as the dependent variable y and VAR003 as the independent variable x, and the results are seen in Table 6.11.

In Tables 6.10 and 6.11, the sum of squares of variation due to regression is the same. To suppress the fluctuation within each level of the factor, individual data were replaced with the average value for each level. This shows that the residual sum of squares has decreased from 53.6 to 3.6. Also, comparing Tables 6.9 and 6.10, the sum of squares between groups in one-way ANOVA is 105, whereas the sum of squares in regression is 101.4, and this difference can be considered a deviation from a linear relationship. In other words, the sum of squares of deviation from linearity is considered to be 3.6 (105.0–101.4). Finally, combine the results in Tables 6.10 and 6.11 to make the composite result in Table 6.12. In this composite ANOVA table, if the variance of the lack of fit is not significant, one can pool this variation into the total error, namely "within the group" and Table 6.10 can be used as the final result. If the variance of the lack of fit is significant, one should consider a more sophisticated regression, maybe a higher-order regression equation.

6.2.7 Stratification of the Dataset and Its Relationship to the Mixed Model

In correlation and regression analysis, it is advisable to begin by plotting a scatter plot to understand the data characteristics before analysing the data. Should there be a potential issue, e.g. unexpected causes, consider constructing a stratified scatter diagram utilising various symbols to differentiate the datasets. This step is crucial to ascertain if there are unexpected factors, aside from the independent variables (explanatory variables), influencing the dependent variable. Should the segregated scatter plot reveal any trends, it is essential to stratify the data within the dataset and verify these trends prior to conducting the detailed analysis.

For example, it was said that there was a correlation between pollutant components X and Y in industrial wastewater. Consequently, certain tests were carried out. For these tests, two devices, A and B, were chosen from the various measuring devices employed on the production line. Observing the overall scatter plot in Figure 6.7, a distinct correlation is not evident.

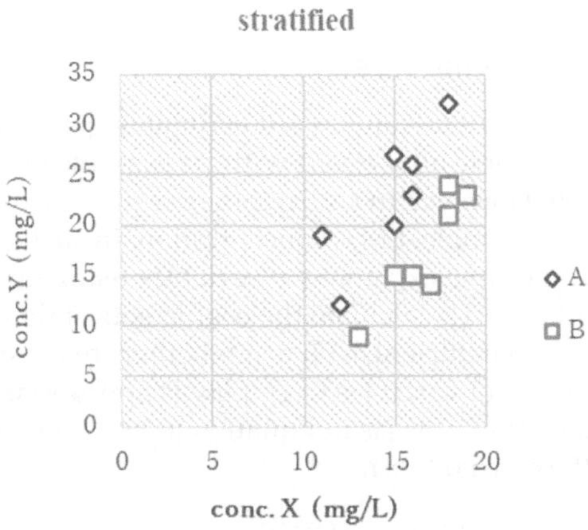

FIGURE 6.7 Relationship between pollutant concentrations *X* and *Y* in wastewater: stratification by measuring devices A and B. The correlation coefficient for all the data is 0.53, but when stratified by measurement devices A and B, the correlation coefficients are 0.83 and 0.91 respectively.

Nevertheless, when the scatter plot data is categorised by the measuring device, a relatively strong correlation emerges for each divided dataset.

The scatter plot indicates that the discrepancy between measuring devices A and B significantly impacts the correlation between pollutants X and Y. This discrepancy suggests a potential issue with one or both devices. Consequently, in formulating a regression equation to predict the dependent variable Y from the independent variable X, it may be essential to stratify the data by the measuring device to ensure an accurate prediction equation. The question is: are these the only two measuring devices used in this test, and there are no other options, or are they two that just happened to be taken out of a machine warehouse where a large number of devices are stored? Should the former be the case, it would be reasonable to consider them as fixed factors among the independent variables. Conversely, if they are two examples extracted from several candidates, they should be considered variable factors. The latter case is considered to correspond to the randomised block design discussed in the Chapter 5. This blocking effect corresponds here to the variable factor.

Recall the data structure of RBD as explained in the previous example in Chapter 5. In Figure 5.4, the block effects (b) and the data values (*y*) are independent of each other, which is a prerequisite for analysing this data with ANOVA. Generally, this experimental design aids in distinguishing uncontrollable factors that significantly affect the dependent variable from the error term, thereby enhancing the power of the test to detect the effect of the factor of interest. Moreover, comparing the equation of the RBD data structure to the linear regression equation is insightful. If the blocking effect is linear and independent of all factors and dependent variables, it can be included in the model as an explanatory variable. However, if it is not, we must then rely on the concepts of mixed models. Now we will briefly explain it, referencing Kawakubo (2021); the linear mixed model can typically be expressed as follows:

$$y = X\beta + Zb + \varepsilon, \quad b \sim (0,G), \quad \varepsilon \sim (0,R) \tag{6–7}$$

The term $\boldsymbol{X\beta}$ represents the fixed effect, while the term $\boldsymbol{Zb}$ represents the random effect. It is termed a linear mixed model because it combines these two effects. Furthermore, **G** and ***R*** are the variance–covariance matrices for $\boldsymbol{b}$ and ϵ, respectively. Both $\boldsymbol{b}$ and ϵ are commonly assumed to follow normal distributions. The design of **Z** allows for the modelling of individual effects, spatial effects and more. For further details, refer to technical publications like the UCLA, Statistical Consulting Group (2024).

6.3 Multiple Regression Analysis

Section 6.2 described regression analysis, which estimates the value of an objective variable from a single explanatory variable, illustrated with an example of estimating NO_2 concentration from SO_2 concentration. However, in reality, other available explanatory variables need to be added to the analysis, for example, $PM_{2.5}$ concentration. In such cases, one can consider the regression equation using multiple explanatory variables using multiple regression analysis. In this analysis, it is important to (1) calculate the coefficients of this multiple regression equation (called partial regression coefficients), (2) test whether this is statistically significant and (3) consider whether the choice of variables to employ is appropriate. The multiple regression equation using m explanatory variables is in Equation (6–8), where bi is the i-th partial regression coefficient and b_0 is the constant term.

$$y = b_0 + b_1 x_1 + b_2 x_2 + \cdots + b_m x_m \tag{6–8}$$

As in the case of single regression analysis, the partial regression coefficient of the multiple regression equation is estimated by minimising the deviation sum of squares of the dependent variable y and its estimated value $\hat{y}$, that is, by the least squares method.

$$\text{SSE} = \Sigma(y - \hat{y})^2 \rightarrow \text{minimise} \tag{6–9}$$

A detailed explanation of the theory of multiple regression analysis is omitted here. How to find the partial regression coefficient of a multiple regression equation will be explained using Example 6–3. Table 6.13 shows composite data in which $PM_{2.5}$ concentration data are added into Table 6.1.

Example 6–3: The $PM_{2.5}$ and sulphur dioxide concentrations are monitored at five sites in the city. Make an empirical equation to estimate the nitrogen dioxide concentration based on the available monitoring data (Table 6.13).

In such a multiple regression analysis, statistical analysis software is commonly used. The results are summarised in Table 6.14.

So the regression equation with the constant term is written in Equation (6–10).

$$y = 0.65x_1 + 1.05x_2 - 1.4 \tag{6–10}$$

TABLE 6.13

Atmospheric Concentrations at Five Sites (arbitrary unit)

Site	x_1 ($PM_{2.5}$)	x_2 (SO_2)	y (NO_2)
A	6	4	5
B	3	1	3
C	6	6	10
D	2	2	1
E	8	2	6

TABLE 6.14

Regression Coefficients for the Example 6–3

Regression coefficient	b_1	b_2	b_0
Estimated value	0.65	1.05	–1.4

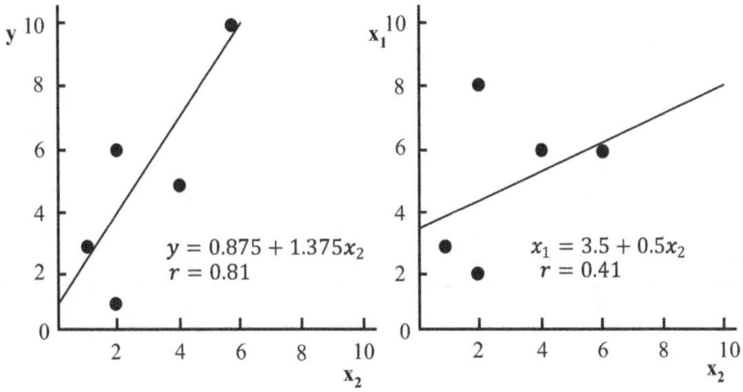

FIGURE 6.8 Scatter diagrams, between the objective variable y and x_2 and between x_1 and x_2.

6.3.1 Meaning of Partial Regression Coefficient

When there are two or more explanatory variables, it is more difficult to interpret the meaning of the regression coefficient as compared to the case of simple regression analysis with only one explanatory variable. In particular, even when the correlation coefficient with the objective variable is positive, the partial regression coefficient may take a negative value. Let us explain the meaning of the partial regression coefficient of the multiple regression equation of Example 6–3 shown in Table 6.13. First, let's consider the meaning of the partial regression coefficient 0.65 for the explanatory variable x_1 in Equation (6–10). There is another explanatory variable x_2 in Equation (6–10), so it is conceivable to eliminate the fluctuation explained by that beforehand. The relationship between the objective variables y and x_2 and the relationship between x_1 and x_2 is visualised in the scatter diagrams in Figure 6.8.

Here, Equation (6–11) is the regression equation for estimating y from x_2 and the estimated value is $\hat{y}$. Likewise, the regression equation for estimating x_1 from x_2 is Equation (6–12), and the estimated value is $\hat{x}_1$. If $p = y - \hat{y}$ and $q = x_1 - \hat{x}_1$, then p is the residual of y from which the variation explained by x_2 is removed, and q is the residual of x_1 from which the variation explained by x_2 is removed.

Therefore, if the regression equation for estimating p from q is obtained, this is the regression equation for estimating y from x_1 after eliminating the influence of x_2.

$$y = 1.375x_2 + 0.875 \tag{6–11}$$

$$x_1 = 0.5x_2 + 3.5 \tag{6–12}$$

Table 6.15 is an auxiliary table created to estimate p from q, which shows the mean and sum of squares of residuals q and p.

$$\overline{p} = 0, \overline{q} = 0, \text{SSp} = 15.75, \text{SSq} = 20.00, \text{SSpq} = 13.00$$

The regression coefficient for estimating p from q is as follows.

$$b = \frac{\text{SSpq}}{\text{SSq}} = \frac{13.00}{20.00} = 0.65, \quad a = \overline{p} - b\overline{q} = 0 - 0.65 \times 0 = 0$$

Therefore, the regression equation obtained is Equation (6–13).

$$p = 0.65q \tag{6–13}$$

TABLE 6.15
An auxiliary table to estimate p from q.

y	$\hat{y} = 0.875 + 1.375x_2$	$p = y - \hat{y}$	x_1	$\hat{x}_1 = 3.5 + 0.5x_2$	$q = x_1 - \hat{x}_1$	p^2	q^2	pq
5	6.375	−1.375	6	5.5	0.5	1.891	0.25	−0.6875
3	2.250	0.750	3	4.0	−1.0	0.563	1.00	−0.750
10	9.125	0.875	6	6.5	−0.5	0.766	0.25	−0.4375
1	3.625	−2.625	2	4.5	−2.5	6.891	6.25	6.5625
6	3.625	2.375	8	4.5	3.5	5.641	12.25	8.3125
25	25	0	25	25	0	15.75	20.00	13.00

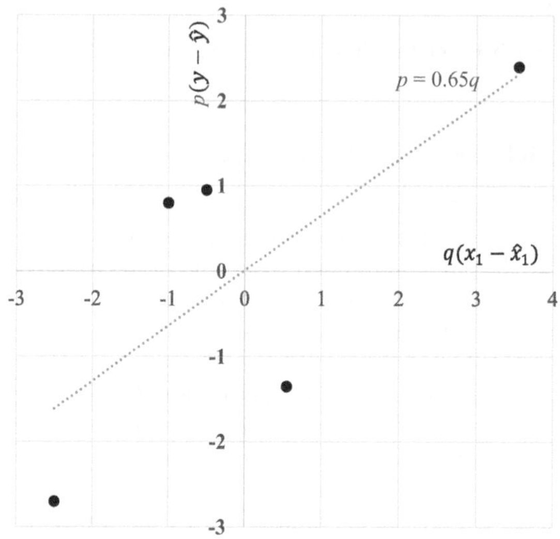

FIGURE 6.9 Scatter diagram of residuals p and q.

The scatter plot of the residuals p and q is shown in Figure 6.9. The value 0.65 of the coefficient b is the same as the value of the partial regression coefficient of the multiple regression equation shown in Equation (6–10).

Next, equations, $p = y - \hat{y}$ and $q = x_1 = \hat{x}_1$ are substituted into Equation (6–13).

$$y - \hat{y} = 0.65\,(x_1 - \hat{x}_1)$$
$$y - (1.375\,x_2 + 0.875) = 0.65\{x_1 - (0.5\,x_2 + 3.5)\}$$
$$y = 0.65\,x_1 - 0.65 \times 3.5 - 0.65 \times 0.5x_2 + 0.875 + 1.375x_2$$
$$y = 0.65x_1 + 1.05x_2 - 1.4$$

It can be seen that this equation is the same as Equation (6–10) obtained in the first place for the two explanatory variables. The correlation coefficient between x_1 and x_2 is 0.41, and the difference $q(q = x_1 - \hat{x}_1$, where $\hat{x}_1$ is estimated from $x_2)$ is used as the explanatory variable in the next stage. Now consider the case if the difference q is zero. A difference q of zero means that the correlation coefficient between x_1 and x_2 is 1.0. In such a case, it is not necessary to use both x_1 and x_2 as explanatory variables in the regression equation for estimating y; either one is sufficient.

Furthermore, when the number of explanatory variables increases, it is necessary to consider whether it is better to use all explanatory variables or to select and use several explanatory variables. Some statistical software includes a variable selection algorithm. In such a case, it is necessary to consider the meaning of residual variance, coefficient of determination, partial regression coefficient and the like, which will be illustrated using Example 6–4 in the next section.

6.3.2 Evaluation of Multiple Regression Equation

The correlation coefficient between the observed value y of the objective variable and its estimated value $\hat{y}$ is called the multiple correlation coefficient. The multiple correlation coefficient R is calculated by Equation (6–14).

$$R = CY\hat{y} / \sqrt{(\text{SSY} \times \text{SS}\hat{y})} \tag{6–14}$$

where $CY\hat{y}$ is the sum of (cross) products Y and $\hat{y}$. SSY and $\text{SS}\hat{y}$ are the sum of squares.

As explained in the case of the single regression analysis, the variation SSY of the objective variable represented by the sum of squares can be decomposed into two parts: the variation SSR that can be explained by the regression equation and the variation SSE of the deviations around the regression equation.

$$\text{SSY} = \text{SSR} + \text{SSE} \tag{6–15}$$

When the equation of the multiple correlation coefficient is squared and transformed, Equation (6–16) can be obtained.

$$\begin{aligned} R^2 &= \text{SSR}/\text{SSY} = (\text{SSY} - \text{SSE})/\text{SSY} \\ &= 1 - \text{SSE}/\text{SSY} \end{aligned} \tag{6–16}$$

The value of R^2 represents the ratio of variation explained by regression and the overall variation of the objective variable y and is called the coefficient of determination. This property is the same as that of a single regression analysis. The effectiveness of multiple regression equations can be assessed using the F test, which compares the variation of s^2_R (or SSR) and s^2_e (or SSE). The idea is the same as the analysis of variance for single regression analysis with one explanatory variable shown earlier in Table 6.3.

Here, by dividing the numerator and denominator of Equation (6–16) by the number of data n, the variance of the residual and the variance of the objective variable can be obtained, so let us consider the meaning of this ratio.

$$R^2 = 1 - s_e^2 / s_y^2 \tag{6–17}$$

If the coefficient of determination is defined by Equation (6–16), the estimated value of the variance will be biased when the number of data is small. That is, when the variance of the population (the entire set behind the data of the extracted sample) is σ^2, dividing the sum of squared deviations by the number of data results in an underestimation of the population variance. The unbiased estimator of the variance is the sum of squares divided by the degrees of freedom. The sum of squares SSY of the objective variable has a degree of freedom of $n - 1$, and the residual sum of squares has a degree of freedom of $n - m - 1$. Here, m is the number of explanatory variables of the multiple regression equation.

$$\begin{aligned} Vy &= \text{SSY}/(n-1) \\ Ve &= \text{SSE}/(n-m-1) \end{aligned}$$

The coefficient of determination can be calculated from the ratio of the unbiased variance. This coefficient of determination is called a degrees-of-freedom-adjusted coefficient of determination and is represented by R^{*2}. When selecting a variable of the regression equation, which will be described later, the regression equation can be evaluated by R^{*2}.

$$R^{*2} = 1 - Ve/Vy \tag{6–18}$$

Example 6–4 is used to explain how to select the explanatory variables to be included in the multiple regression equation.

Example 6–4: The data on consumer prices and salary were collected from 15 cities around the world (Table 6.16); one wants to find a regression equation that estimates salary from prices. What prediction formula is appropriate?

Although the data are not new, they contribute to a suggestive analysis example. Here, a regression equation is considered to estimate salary from the prices of three major commodities. The correlation coefficient between beef and salary is positive, but the partial regression coefficient of beef in the regression equation is negative. Let us examine why this happened.

Regression analysis was conducted using all explanatory variables by using statistical software. First, the correlation matrix is constructed as shown in Table 6.17, which shows the

TABLE 6.16

Consumer Prices and the Annual Income of Office Starting Salaries in Each City in the World in Fiscal Year 1988

	x_1: Plain bread (Kyen/kg)	x_2: Beef (Kyen/kg)	x_3: Electricity charges (Kyen/200kWh)	y: Tax-included annual income (M yen)
Tokyo	0.363	3.57	4.372	2.655
Taipei	0.404	1.68	2.175	1.598
Hong Kong	0.117	1.50	2.250	1.517
Seoul	0.298	2.155	3.105	1.275
New Delhi	0.139	0.603	1.428	0.213
Sidney	0.189	1.363	2.306	2.621
Paris	0.593	1.67	2.510	3.341
Milan	0.609	2.344	3.000	2.344
London	0.198	1.571	4.407	2.203
Dusseldorf	0.300	1.175	4.149	4.106
Prague	0.051	0.657	3.000	0.314
Toronto	0.352	2.112	1.835	2.717
New York	0.330	1.430	3.405	2.860
Mexico City	0.099	1.108	0.921	0.904
Sao Paulo	0.097	0.349	1.337	0.488

Source: This example is quoted from Okamoto and Shinmen (2012).

TABLE 6.17

Correlation Matrix

	Plain bread	Beef	Electricity	Annual income
Plain bread	1	0.612	0.293	0.634
Beef	0.612	1	0.503	0.476
Electricity	0.293	0.503	1	0.562
Annual income	0.634	0.476	0.562	1

TABLE 6.18

Regression Coefficients for the Regression Equation in Which All Variables Were Used

Independent variable	Partial regression coefficient	Standard partial regression coefficient	t value	
x_1	3.75	0.558	2.206	$R = 0.749$
x_2	−0.127	−0.088	−0.314	SSE = 8.284
x_3	0.471	0.443	1.916	$R^{*2} = 0.441$
(constant)	−0.156			

TABLE 6.19

Regression Coefficient, Where Only Selected Independent Variables Were Used

Independent variable	Partial regression coefficient	Standard partial regression coefficient	t value	
x_1	3.451	0.513	2.552	$R = 0.746$
x_3	0.438	0.412	2.051	SSE = 8.358
(Constant)	−0.182			$R^{*2} = 0.483$

positive values for all combinations of variables. In other words, the higher the price of the three commodities, the higher the salary.

Next, the regression coefficients were calculated, and the results are shown in Table 6.18.

R^{*2} is the adjusted coefficient of determination. Since the non-adjusted R^2 automatically and spuriously increases when extra explanatory variables are encountered in the model, the use of an adjusted R^{*2} is an attempt to account for this phenomenon to avoid these obstacles. Using the results in Table 6.18, the obtained regression equation is shown in Equation (6–19).

$$y = 3.750x_1 - 0.127x_2 + 0.471x_3 - 0.156 \quad (6\text{–}19)$$

As for the partial regression coefficient of the explanatory variable x_2, the absolute value of the t value is very small. So let us remove x_2 (the beef) and find another regression equation that estimates salary from only bread and electricity. The result is shown in Table 6.19 and Equation (6–20).

$$y = 3.451x_1 + 0.438x_3 - 0.182 \quad (6\text{–}20)$$

Comparing the two regression equations, there is not so much difference in the partial regression coefficients of x_1 and x_3, and the value of the multiple correlation coefficient is nearly the same. It was found that the beef data are useless to improve the estimation accuracy of the regression equation. Here, even if the correlation coefficient between y and x_2 is positive (0.476), the partial regression coefficient for x_2 is a negative value (−0.127), as shown in Equation (6–19). Furthermore, assuming that the values of y and x_2 are estimated from x_1 and x_3, it is quite possible that the correlation coefficient between $(y - \hat{y})$ and $(x_2 - \hat{x}_2)$ is negative. In particular, if $(x_2 - \hat{x}_2)$ is close to zero, the calculation results will become unstable. And when $(x_2 - \hat{x}_2)$ becomes completely zero, it becomes impossible to calculate the regression coefficient.

Compared with Equation (6–20), Equation (6–19) seems to be better from the viewpoint of a multiple correlation coefficient R (and the coefficient of determination that is the square of R). However, Equation (6–20) is better for the adjusted coefficient of determination. Thus it cannot be said that it is better to have as many explanatory variables as possible in the multiple regression equation. In particular, it is important to check for highly correlated combinations among explanatory variables.

Multicollinearity is such a situation where there is a correlation between explanatory variables. This is a characteristic of the obtained data rather than a deficiency in the setting of the multiple regression equation. To prevent this situation, there is no other way than adding new data collected from a wide range or deleting some explanatory variables from the multiple regression equations already obtained. Such a phenomenon often occurs when analysis with a receptor model, which is often used in atmospheric science is conducted using a large dataset having similar elemental compositions, which is detailed in Chapter 23. In particular, the variance inflation factor (VIF) quantifies the severity of multicollinearity in the least squares regression analysis and can be calculated by Equation (23–14).

6.3.3 Variable Selection Method

If there are too many explanatory variables in a multiple regression analysis, not only does it become complicated to interpret the meaning of the partial regression coefficients in the multiple regression equation and to predict the value of the target variable, but it can also lead to problems such as a decrease in the reliability of the results due to multicollinearity between the explanatory variables. In such cases, it may be necessary to select key explanatory variables and eliminate those that are not useful. Several methods for automatically selecting explanatory variables will be introduced here.

- ***Forward selection method:*** First test the addition of each variable using the criterion of the variable selection, and then add the variables that provide the most statistical improvement to estimate the objective variable. Repeat this process until no improvement is seen.
- ***Backward elimination method:*** Starting with all the candidate variables selected, test the deletion of each variable using the selected model conformance criterion, and delete the variable with the lowest loss in the model conformance statistically. This means that when the multiple regression equation is calculated by expelling the variables, the decrease in the multiple contributions is minimal. Further, repeat the process until the variable cannot be deleted.
- ***Stepwise method:*** There are two types of this method: a stepwise forward selection method and a stepwise backward selection method. The former is more common and hence will be explained here. It is a method that combines an increasing method and a decreasing method. First, as in the variable forward selection method, variables are selected according to the selection criterion. Then, each time one variable is imported, it is examined to see if any of the already imported variables satisfy the same eviction criteria as the backward selection method, and, if so, it is expelled. And then the import of more variables continues.

An illustration of the application of the stepwise method for variable selection in multiple regression analysis is provided in Chapter 22, Section 22.7.

6.4 Summary

In this chapter, we have explained correlation analysis and regression analysis in detail using easy-to-understand examples. Regression analysis and correlation analysis are analytical methods with different concepts, but they are closely related, and understanding this relationship is useful when examining many analytical results. In regression analysis, multiple explanatory variables can also be introduced. This type of method is called multiple regression analysis and is also part of multivariate analysis methods.

REFERENCES

Kawakubo, Y., 2021, *Linear mixed model and its applications, Qiita presentations* (In Japanese), https://qiita.com/ykawakubo/items/f57c38b72d2ddc7798fd

Milton, J.S. and Arnold, J.C., 1995, *Introduction to probability and statistics*, 3rd edition (p. 430), McGraw-Hill, Inc., New York, NY.

Okamoto, S. and Shinmen, K., 2012, *Fundamentals of multivariate statistics*, 2nd edition, Soseisha Publishing Co., Tokyo (In Japanese).

UCLA, Statistical Consulting Group, 2024, *Introduction to linear mixed models*, https://stats.oarc.ucla.edu/other/mult-pkg/introduction-to-linear-mixed-models/, accessed 27 September 2024.

7

Multivariate Analyses

Multivariate analysis is a general term for various statistical methods used to analyse the relationships between multiple variables. Representative examples include multiple regression analysis, principal component analysis, factor analysis, and discriminant analysis. For instance, the values of multiple test items in a health check can be used to understand trends in the signs and symptoms of a certain disease. In such cases, it is useful to know which variables are related to one another, and multivariate analysis is applied. Here, we will briefly introduce the characteristics of a few representative methods. A more detailed explanation will be provided in the appropriate chapters in Part III.

Typically, multivariate analysis is not feasible to perform manually. While the theoretical concepts have existed for quite some time, their practical implementation has advanced alongside the evolution of computers and analytical software. The enhancement of computing capabilities has further spurred the creation of innovative analytical techniques. This chapter will provide a concise overview of these modern developments

7.1 Some Important Multivariate Analysis Methods

The analysis of a large number of variables can be divided into cases where there are objective variables and cases where there are no objective variables. If one wants to estimate the value of the variable of interest (objective variable) from many other variables, then this is further classified by the type of objective variable. In this instance, if the variables to be estimated are metric values (data obtained by measuring length, weight, electrical properties etc.), multiple regression analysis, as detailed in Chapter 6, is beneficial. Conversely, if the variables to be estimated are categorical measures such as (pass/fail) or (healthy/sick), then discriminant analysis proves to be valuable.

If there is no target variable (which one wants to be estimated), but instead it is desirable to know which, as well as which are deeply related among many variables. One may also want to know which variables are convenient to use when classifying a large number of samples. A typical method used in such a case is principal component analysis and cluster analysis. Multivariate analysis methods can be classified according to whether or not it has a dependent variable. Representative examples of each type are shown in Table 7.1.

Details of these methods are available in more advanced textbooks. This chapter will provide the basic knowledge relevant to the case studies presented in Part II and Part III of this book. Note that multivariate analysis cannot be processed without a computer and analysis software such as SAS, SPSS and R. It is important to read the description in the software manuals and understand the algorithm when software is selected to use. Some knowledge of linear algebra is also required to understand the multivariate analysis algorithm. More details of the method can be found in the relevant references such as Harris (1985). Beginners struggling with mathematical details might benefit from reviewing books that present statistical analysis within marketing research for non-engineers, like those by Aaker et al. (1995) and Parasuraman (1991). Additionally, a wide array of web-based educational resources have become available recently, providing more advanced examples.

DOI: 10.1201/9781003597216-8

TABLE 7.1

Classification of Multivariate Analysis

Dependent variable (external criteria)		Method
Yes	Continuous value, measured value, indiscrete value	Multiple regression analysis
	Category (classification)	Discriminant analysis
None		Principal component analysis Factor analysis Cluster analysis

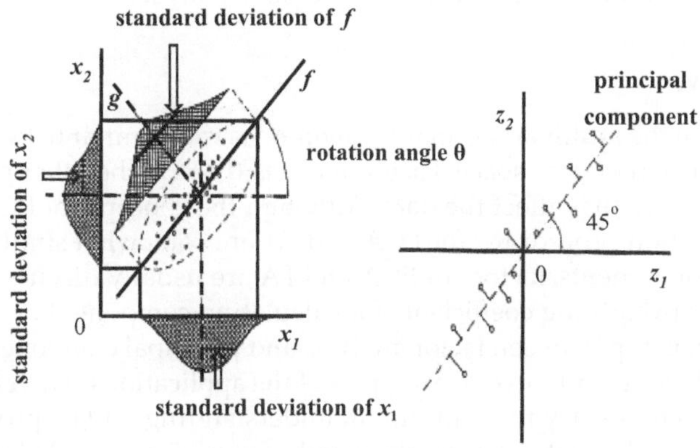

FIGURE 7.1 Geometric description of principal component analysis.

7.1.1 Principal Component Analysis

Principal component analysis (PCA) is a method of multivariate analysis that finds a few composite variables that can best represent the most variation of all original variables, that is, transformation to the smaller number of uncorrelated variables from a large number of correlated original variables. Typically, this is used to reduce the dimension of the dataset.

Since the principal component is a linear combination of the original variables, obtaining the weighting coefficient is the first step of principal component analysis. At that time, the variance of the first principal component is maximized, and the subsequent principal component is selected to maximize the variance under the constraint that it is orthogonal to the principal component formerly determined so far. Principal component analysis starts by obtaining eigenvalues and eigenvectors of a covariance matrix or correlation matrix.

In general, principal component analysis is usually applied to datasets composed of a large number of variables, but to make it easier to understand the basic concept, this section uses the case of two variables for explanation. The arrangement of data in variables x_1 and x_2 is shown in Figure 7.1. Rotate the coordinate system of x_1 and x_2 by angle θ, and make the new coordinate system of f and g. Here, this angle may be determined to maximize the variance on the axis f. If the data of x_1 and x_2 were standardized, the rotation angle is always 45° (when the correlation between x_1 and x_2 is positive). Here, z_1 and z_2 are the standardised data of x_1 and x_2.

$$f_i = w_{11}\, z_{1i} + w_{12}\, z_{2i}$$
$$g_i = w_{21}\, z_{1i} + w_{22}\, z_{2i}$$

Here, the value of the synthetic variables, f and g is called the principal component score (PC score), and the weighting coefficient w for the calculation is called the principal component loading (PC loading).

Due to the ability of principal component analysis to transform large amounts of data into a set of PC scores for a smaller number of uncorrelated new variables, this technique can be used in combination with other analytical procedures such as multiple regression and discriminant analysis.

Suppose a large number of closely related explanatory variables are included in a multiple regression equation. The key issues here are the multicollinearity problem and the lack of a sufficient number of observations. An effective way to avoid both problems is to reduce the number of independent variables entered. Performing principal component analysis properly does not significantly reduce the overall amount of information and allows input data to be limited to a few numbers of uncorrelated new variables. One can use this small set of independent variables as input data for multiple regression analysis.

7.1.2 Factor Analysis

The factor analysis in the multivariate analysis method is very similar to PCA. PCA is a method to reduce the dimension of data space. Factor analysis (FA), on the other hand, is a technique for finding latent factors that affect the data. Although the concepts of PCA and FA are quite different, the calculation procedures for PCA and FA are somewhat similar. The characteristics of individual components/factors in PCA and FA are usually discussed in terms of their scores and loadings (weighting coefficients for calculating component or factor scores). More details on the relationship between factor analysis and principal component analysis are provided in Chapter 21 along with several examples of the applications. Therefore, in this section, only the information necessary for a minimum understanding will be provided.

Factor analysis is used to extract a smaller number of variables, called factors or latent factors, from a dataset consisting of many variables. There is also the idea of testing an a priori model (hypothesis) for the latent factors based on the obtained data by using factor analysis. However, at the stage of starting analysis, it is not possible to set a clear hypothesis to be tested, and one may want to consider such a hypothesis that which a latent factor should be adopted. In this way, it also allows consideration of what latent factors should be employed to increase the amount of information (variance) that can be explained by the factors. In other words, factors are extracted from the original dataset so that as much information as possible can be retained. This concept is called exploratory factor analysis (EFA).

This operation can also be understood as follows: suppose that all data is described as a linear combination of latent factors, in which the number of latent factors is sufficiently few compared with the number of original variables. For example, a spreadsheet-like dataset with 20 data variables is all described by a linear combination of a small number (maybe 3 or 4) of latent factors. The number of these factors can be dealt with by arranging the variance of latent factors in descending order and truncating when the amount of information is less than the amount of information (variance) of the original single variable. For example, if there are 3 latent factors, the i-th data of the j-th variable can be expressed as:

$$x_{ij} = w_{1j} f_{i1} + w_{2j} f_{i2} + w_{3j} f_{i3}$$

where w is the factor loading (weighting coefficient), and f is the factor score. See the textbook for multivariate analysis such as that by Harris (1985) for further explanation.

Exploratory factor analysis (EFA) is used to identify complex interrelationships between multiple variables, and researchers do not make rigorous assumptions about the relationship between factors in advance. While confirmatory factor analysis (CFA) is an approach aimed at testing the hypothesis that an item is associated with a particular factor. CFA is often conducted in structural equation modelling (SEM) to test the preset models. This allows us to evaluate the relationship between the observed variables and the expected factors. For a more detailed explanation, see Section 24.2 in Chapter 24.

7.1.3 Discriminant Analysis

Discriminant analysis is a method for estimating which group an unclassified sample should belong to when there are some samples about which the group they belong is already known. The sample is composed of many variables. Several methods have been proposed for discrimination, but two typical methods are described next.

Mahalanobis distance: Assume that the number of variables is m. First, standardise all variables (subtract the mean and divide by the standard deviation). Next, find the position of the centre of gravity of each group. Find the distance between each specimen/sample and the centre of gravity of each group. Further, assume that the new specimen belongs to the group that has the shortest distance from the centre of the group. This virtual distance in m-dimensional space is called the Mahalanobis distance.

Multiple correlation ratio: Let us consider a method of separating data by the total score obtained by multiplying each variable by the weight. First, divide the total score distribution into intragroup (within-group) and intergroup (between-group) distributions. From the perspective of partitioning data, it is most desirable if the variance between groups is large and the variance within the group is small. The ratio of the variance between groups and the variance within groups is called the multiple correlation ratio. Therefore, a weighting coefficient for each variable that maximises the multiple correlation ratio can be obtained.

This method can be applied even if the number of variables is large, but it can be visualised on a plane with two variables x_1 and x_2 (bivariate) as shown in Figure 7.2. Two groups of data are drawn on the diagram in Figure 7.2.

In the diagram, the data for both variables show partial overlap between the two groups. Consequently, a single variable is insufficient to differentiate the two groups. Nonetheless, by appropriately rotating the coordinate axes, it is possible to minimise or completely remove the overlap between the groups. Here, determining the weighting coefficient of the discriminant function is equivalent to determining the weighting coefficient for coordinate transformation according to this rotation angle. An application example of this discriminant analysis can be found in Chapter 22, Section 22.5.

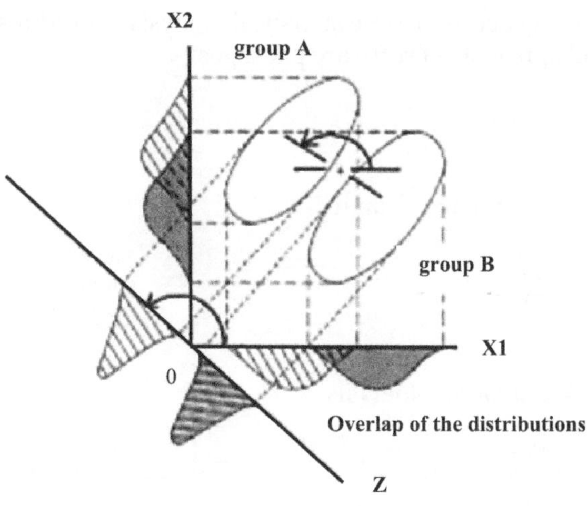

FIGURE 7.2 Geometric description of discriminant analysis.

7.2 Additional Topics Related to Advancements in Statistical Analysis

7.2.1 Relation between Experimental Design and Multivariate Analysis

When a huge amount of effort is spent for collecting small amounts of data, such as industrial or agricultural experiments using large-scale equipment, research on how to acquire data efficiently is the most important concern in statistics, and therefore analytical technologies such as experimental design have advanced. Although advances in automated measurement technology and data transfer technology have enabled the collection of large amounts of data with much less effort, optimum experimental designs are still of foremost importance. Nevertheless, with the large monitoring datasets becoming more available, the application area of statistics has expanded to the proposition of how to extract useful information buried in a huge amount of data. Experimental design is a typical example of the former statistical analysis method that efficiently collects a small number of valuable data, and multivariate analysis is a representative example of the latter statistical analysis method that extracts important information buried in a huge pile of data. The relationship between the two is schematically illustrated in Figure 7.3.

In the environmental monitoring field, with the development of sensor technologies, large environmental datasets are generated which are not possible to deal with using manual calculations. Big Data Technology is a software tool to analyse, process and interpret the massive amount of structured and unstructured data that could not be processed manually in the traditional way. As these technologies evolve, statistical analysis may also advance rapidly. More details of the development in environmental monitoring can be found in Chapter 8.

7.2.2 Statistical Analysis Software

Computers likely started being utilised for data analysis in earnest in the 1960s. Specifically, the advent of analysis software made it feasible to perform complex multivariate analysis methods, which involve matrix calculations too intricate for manual computation. Subsequently, the evolution of hardware saw a transition from bulky mainframes to personal computers (PCs), with statistical software adapting accordingly. Notably, established programs like SAS and SPSS were released as Windows-compatible PC versions, reflecting this shift. Initially, researchers probably developed analysis programs using compiled languages by themself. However, by the 1970s, standard analysis software started to become available for frequently used analysis methods. SAS and SPSS were the forerunners in this development. Subsequently, a range of general-purpose statistical analysis algorithms have been integrated into mathematical and spreadsheet software packages.

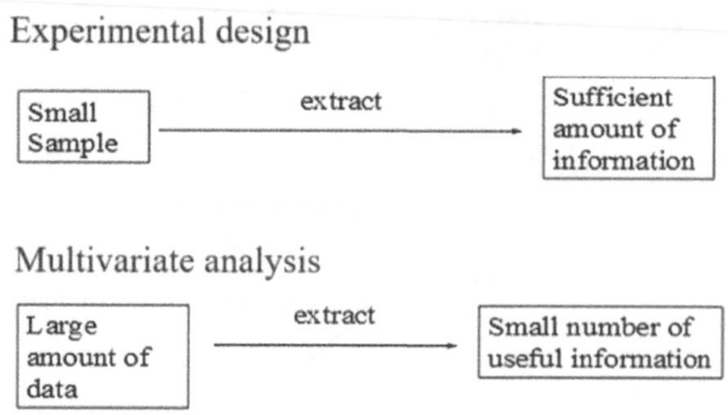

FIGURE 7.3 Relationship between experimental design and multivariate analysis.

Historically, the processing of statistical data for critical outputs like academic papers depended on specialised software, necessitating continuous support and updates, which proved cumbersome for users. Consequently, there was a significant demand for the development of open-source, accessible, and affordable statistical analysis tools. The R programming language emerged in response to this demand. Similar to languages like Python and Matlab, R excels in these areas. However, R has the advantage of allowing the writing of statistical analysis programs with relatively concise syntax. The R programming language is overseen by an entity known as the Comprehensive R Archive Network (CRAN). This website allows for the installation of R across different operating systems. Moreover, it offers a range of information for its users (The CRAN Team, 2024).

Python, along with the R language, is a programming language that is widely used for statistical analysis. It is object-oriented and known for its simplicity and readability, making it accessible for beginners. Python's popularity soars in the realm of machine learning, where it is a favoured tool among data scientists and AI researchers. Recently, several statistics textbooks that include statistical analysis tools have been published. These books are frequently engaging, providing examples of data preparation and analysis result interpretation. For instance, the textbook by Agresti and Kateri (2021) is one such example.

7.3 Summary

This chapter provided an overview of several important multivariate analysis methods. It introduced the basic concepts of discriminant functions as a representative example of cases where there is an external criterion and of principal component analysis and factor analysis as analysis methods when there is no external criterion. The main focus here is to help readers understand what types of data these methods can be applied to analysing. Therefore, when actually applying these methods, consult a more detailed specialist book. Additionally, explanations of other analysis methods not introduced here and points to note when analysing will be discussed in several chapters in Part III.

REFERENCES

Aaker, D.A., Kumar, V. and Day, G.S., 1995, *Marketing research*, 5th edition, John Wiley & Sons, Inc.

Agresti, A. and Kateri, M., 2021, *Foundations of statistics for data scientists with R and Python*, Chapman & Hall, ISBN 9780367748456

The CRAN Team, 2024, *The comprehensive R archive network*, viewed on 10 October 2024, https://cran.r-project.org/

Harris, R.J., 1985, *A primer of multivariate statistics*, Academic Press Inc., Orlando, FL.

Parasuraman, A., 1991, *Marketing research*, Addison-Wesley Publishing Co., Reading, MA.

Part II

Applications for Monitoring and Measurement

The monitoring process generates data on environmental quality by systematic observations and measurements for different environmental components of air, water, soil/land, and biota, as well as for the pollution discharged into the environment. Well-designed protocols of sampling and sample analysis, in conjunction with appropriate data management processes, are crucially required to ensure the collection of valid data in conformity with the monitoring objectives. Such data are essentially required to correctly identify and effectively solve environmental pollution problems. Part II introduces the statistical methods primarily useful in the designing of environmental monitoring programs with a focus on quality assurance and quality control to obtain data with the required quality in terms of accuracy (trueness and precision), as well as spatial and temporal coverages. The challenges to obtaining representative monitoring data for highly dynamic environmental matrices with large temporal and spatial variations (Chapters 14 and 15) call for thorough considerations of the temporal and spatial scales in designing monitoring programs, e.g., for ambient air (Chapters 12 and 13), wastewater (Chapter 10), receiving environmental bodies (Chapter 11) or acid deposition (Chapter 16). The more theoretically oriented content of Chapters 8 and 9 is illustrated with rich case studies of the monitoring programs for industrial pollution and its environmental effects in Chapters 10 and 11 and particulate air pollution at regional (Chapter 12) and local scales (Chapter 13). Chapter 17 underlines the need for well-designed monitoring programs in air tracer release experiments to produce systematic data for the validation of air quality models. Understanding the statistical characteristics of the model-simulated air concentrations is essential for model evaluation and applications (Chapters 18 and 19).

DOI: 10.1201/9781003597216-9

8

Monitoring Tool in Environmental Quality Management

Environmental monitoring is a closed-loop process, and a long-term monitoring program requires periodic review and revision so that necessary changes can be made to correctly characterise the environmental quality in a timely manner. This process mainly involves systematic observations and measurements of the presence and levels of pollution in different forms in the environment and associated impacts. The foremost concern in such a monitoring program is to cost-effectively obtain the data that meet the data quality objectives for the intended uses. This chapter primarily deals with the technical aspects of environmental monitoring with the aim of minimising errors and increasing the representativeness of the data. The technical design core of a monitoring program includes the selection of monitoring parameters, sites, time and frequency, which in turn define the monitoring methods to be used to meet the monitoring objectives (see Figure 1.1, Chapter 1). Key contents of this chapter include the sources and types of monitoring errors, QA/QC procedures and the data management process. Highlights are provided on the development of monitoring techniques to measure conveniently and correctly not only the traditional pollutants but also the newly emerged, along with the development in data management methods to manage big datasets.

8.1 Environmental Monitoring Parameters and Measurement Units

8.1.1 Monitoring Parameters for Different Environmental Matrices

Physical, chemical, and biological parameters are used to characterise the environmental matrices of air, water, soil and biota (vegetation, animal etc.), which should be specified in the technical design to achieve the objectives of a monitoring program. Different parameters are used for monitoring different environmental matrices, also called the environmental media, e.g., air (ambient and indoor), water (surface water, groundwater and wastewater), soil, solid waste, vegetation, animal, sediment etc. The selected parameters also differ between subcategories of an environmental matrix, e.g., ambient air quality vs. indoor air quality, surface water vs. groundwater, etc.

8.1.1.1 Monitoring Parameters for Air Pollution

To assess compliance with the National Ambient Air Quality Standards (NAAQS), the monitoring parameters include the regulated pollutants that are specified in the standards. These normally include the particulate matter (PM) of different size ranges and the content of lead (Pb) in PM, and gases of sulphur dioxide (SO_2), nitrogen dioxide (NO_2), carbon monoxide (CO) and ground-level ozone (O_3). In several countries, the regulated parameters also include specific volatile organic compounds (VOCs) such as benzene, ammonia (NH_3), as well as heavy metals and polycyclic aromatic hydrocarbons (PAHs) contents in PM, among others. The regulated PM parameters normally include PM_{10} (respirable particles, ≤10 μm in aerodynamic diameters) and $PM_{2.5}$ (inhalable particles, ≤2.5 μm in aerodynamic diameters), but in a few countries, TSP (total suspended particulates) is also regulated.

DOI: 10.1201/9781003597216-10

For research purposes, smaller PM sizes are also of interest, e.g., the sub-micron particles of $PM_{1.0}$ (≤1.0 μm in aerodynamic diameters) or ultrafine particles (UFPs) (≤ 0.1 μm in aerodynamic diameters). UFP are ubiquitous in the air with high count concentrations but with very tiny mass; hence the number concentration (counts per unit of volume) is of importance. Monitoring may include heavy metals and trace organic compounds in the air, for example, PAHs, dioxins and chlorinated pesticides. The latter are semi-volatile organic compounds (SVOCs); hence the measurements are made for both particulate and gas phases. Different compositions of PM, i.e., black carbon (BC), organic carbon (OC), water-soluble ions, organic ions, elements and a range of specific organic compounds are monitored to provide input data for receptor modelling analysis, as further detailed in Chapter 23. Case studies of PM monitoring programs at urban, regional and multi-country scales are further detailed in Chapters 12 and 13.

For indoor air quality, besides those just listed, other pollutants are also of interest, such as radon, asbestos particles, specific VOCs, e.g., benzene, aldehydes, ketones, along with NH_3, hydrogen sulphite (H_2S) and selected biological parameters. The latter include the airborne viable or no-longer viable particles, such as bacteria, fungus, animal dander, house dust mites etc. For personal exposure monitoring, a combination of several of the listed parameters can be monitored and commonly include $PM_{2.5}$, CO, VOCs, NO_2 and O_3. These pollutants are measured using small portable sensors, which are placed at the breathing zone of the device wearer.

In monitoring atmospheric emissions from a source, such as from an industrial stack, parameters are selected based on specific characteristics of the emission source, which in turn depend on the types of products, manufacturing processes, and combustion technologies used in the industry. For combustion sources, such as a coal-fired boiler, emission monitoring parameters normally include PM, CO, CO_2, SO_2 and NO_X etc., along with temperature, pressure and oxygen content in the flue gas. Emissions from waste incineration facilities should also be monitored for heavy metals, dioxins and PAHs. Flow-rate of the emissions is monitored and is used to estimate the pollution load, which is the product of concentration (kg/m^3) and flow-rate (m^3/h), from the emission source to the air, e.g., in kg/h or kg/day.

8.1.1.2 Monitoring Parameters for Water Pollution

Compliance monitoring for surface water quality (in a river or lake) is done to compare with water quality standards. The most common physicochemical parameters in this monitoring include pH, dissolved oxygen (DO), electrical conductivity (EC), turbidity, biochemical oxygen demand (BOD), chemical oxygen demand (COD) and total suspended solids (TSS). Phosphorus and nitrogen compounds are also monitored, especially when eutrophication is of concern, and may include the total phosphorus (TP), total Kjeldahl nitrogen or TKN (is the sum of nitrogen bound in organic substances, nitrogen in ammonia and nitrogen in ammonium), and inorganic nitrogen (nitrates and nitrites).

For groundwater monitoring, besides these parameters, depending on the monitoring objectives, other parameters are also covered, such as salinity (to assess salt intrusion), heavy metals and organic compounds, e.g., pesticides. In most cases, the groundwater level is of interest and is recorded accordingly.

The monitoring parameters for industrial wastewater are selected based on the industry type and manufacturing processes. Besides the common water pollution parameters just listed, specific parameters of concern may be included, for example, oil and grease, various nitrogen compounds (from food processing), and sulphates (Kraft pulping industry). The flow-rate of wastewater discharged from each production unit (m^3/h) is measured to assess the pollution load discharged from the unit. A case study of monitoring implemented for a chlorine-bleached Kraft pulp and paper mill is detailed in Chapter 10.

8.1.1.3 Monitoring Parameters for Soil and Sediment Pollution

Soil and sediment sorb the contaminants from native materials and from the pollution discharges that are built up and present for an extended time in the environmental matrices. The pollutants of interest normally include heavy metals and persistent organic compounds but may also cover any others depending on the monitoring objectives. Conventional soil quality parameters include soil density, pH, fertility (NPK), EC, organic matter (OM) etc. Examples of monitoring for contamination in sediments in the receiving environmental bodies of industrial discharges are included in Chapter 11.

8.1.1.4 Biological Monitoring Parameters

Besides physicochemical parameters, biological parameters are monitored in all the environmental matrices to assess biological effects at various levels, from molecule, cell, organ, organism and population to the ecosystem. For example, monitoring may be done to assess the change in the population of specific groups of fish or plankton in a river receiving wastewater from an industry or following the construction of a dam. In bioaccumulation studies, monitoring can be done to assess, e.g., the levels of specific pesticides or heavy metals in animals or plants. Pollen and other viable or no longer viable biological parameters, listed earlier, are monitored for ambient and indoor air quality assessment. Microbial biomass and earthworms are monitored for soil quality assessment. Toxicity testing is commonly conducted to assess the potential biological effects of a discharge. Concrete examples of biological monitoring including toxicity testing, heavy metals and organochlorines in the biota in the receiving bodies of industrial discharges are provided in Chapter 11.

8.1.2 Measurement Units

Monitoring results are expressed in terms of the amount of the monitored contaminant detected in a unit of the environmental matrix, which is referred to as the concentration. The concentration is expressed in several measurement units, such as mass to mass (mg/kg), mass to volume (mg/m^3), molar to volume (mol/m^3), number to volume (number/m^3) or volume to volume (m^3/m^3).

For water pollution, common concentration units are the mass of contaminant (g, mg, µg) in a unit of water volume (1 L). One liter (L) of water has an approximate weight of 1 kg at room temperature (note that only at 4°C, 1 L of water weighs exactly 1 kg) hence the units in mass-to-volume are readily converted to the units in mass-to-mass, e.g., ppm (parts per million), ppb (parts per billion) etc.

Contaminants in solid matrices (soil, sediment, biomass) have the concentration commonly expressed in the mass of contaminants in a unit mass of matrix, i.e., ppm, ppb etc. For the trace compounds, such as dioxins or chlorinated pesticides, smaller concentration units are used, e.g., the dioxins contamination in soil is expressed in parts per trillion (ppt: 10^{-12}) which is equivalent to one nanogram (ng) per kilogram (kg) of soil.

In air pollution monitoring, the units for particulate pollutants, e.g., $PM_{2.5}$, and particulate components, such as BC and OC, are in a mass of the pollutant in a unit of air volume, e.g., µg/m^3, or in number to volume unit. The latter, as mentioned previously, is conveniently used for UFPs, which are better quantified by counting than by weighing.

The concentration units of gaseous pollutants may be expressed either in the mass or the volume of the pollutants in a unit of air volume. The volume-to-volume concentrations of gases are abbreviated as ppm (parts per million), ppb (parts per billion) etc., and normally with "v" added to indicate the "by volume" units, i.e., ppmv. The mass of 1 m^3 of air varies widely with temperature and pressure and is not equal to 1 kg at normal ambient conditions. Therefore, when the pollutant concentrations are expressed in mass-to-volume units, the air temperature and pressure must be reported along with, e.g., at 25°C and 1 atmosphere. With the ideal gas assumption and as long as all the gases in the air remain in the gaseous state, the concentration values of gaseous pollutants in the volume-to-volume units will not change with temperature and pressure. Therefore, if the concentrations in volume-to-volume units are reported, there is no need to indicate the temperature and pressure at which the measurements have been taken.

The conversion from the mass-to-volume unit (mg/m^3) to the volume-to-volume unit (ppmv) for a gaseous pollutant is done considering the gas molecular weight, temperature and pressure of the monitored air, following Equation (8–1).

$$C[mg/m^3] = 10^{-3} * C[ppmv] * MW_{pol} * P/(R_u T) \tag{8–1}$$

where

C[mg/m^3] is the concentration in the mass unit (mg/m^3), and C[ppmv] is the concentration in the volume unit of parts per million (by volume);

MW_{pol} is the molecular weight of the gas (g-mole);

P is the atmospheric pressure, atm;

T is Kelvin temperature, K (K = 273 + °C);

R_u is the universal gas constant, R_u= 0.08208 *10^{-3} m^3. atm/g-mole.°K.

Equation (8–1) is simplified for the standard temperature and pressure (0°C, 1 atm):

$$C[mg/m^3] = C\,[ppmv] * MW_{pol}/22.41 \tag{8–2}$$

and for 25°C and 1 atm:

$$C[mg/m^3] = C\,[ppm] * MW_{pol}/22.45 \tag{8–3}$$

These equations can be used to conveniently convert the units for gaseous pollutants. For example, the measured hourly concentration of CO at a roadside is 12 ppm. It is required to check whether the level has exceeded the NAAQS for hourly CO of 30 mg/m^3 at 25°C and 1 atmosphere. The MW of CO is 28 g-mole. Equation (8–3) is used to convert 12 ppm to the equivalent concentration in mg/m^3, which yields 13.7 mg/m^3; hence it did not exceed the NAAQS.

Due to the inherent large variations in the concentrations of air pollutants in time, it is important to know the averaging time over which the measurements are integrated, e.g., 1-hour, 8-hour, 24-hour, monthly or annual average. The averaging time (or sampling time) is the foremost essential information of the air quality measurement data, and it should be indicated along with the measurement results. Further details on the spatial and temporal scale of air pollutants are included in Chapters 14 and 15, respectively.

8.2 Quality Assurance and Quality Control in Environmental Monitoring

8.2.1 Quality Assurance and Quality Control

Quality assurance and quality control (QA/QC) in environmental monitoring are independent functions of the monitoring process that are essential for providing monitoring data with sufficient quality (detailed in Section 8.2.4) for the intended uses. A QA/QC plan must be established and incorporated in every step of a monitoring program, i.e., from the program design to its implementation and data management. Further details can be found, for example, in Chow et al. (2012). An overview of data quality management has been given in Chapter 1 (Section 1.3), which outlines general concepts of QA/QC in a broad sense. This section provides concrete examples of QA/QC measures that are applicable to environmental monitoring.

The QC measures in environmental monitoring include those technical activities taken to control the quality for meeting the defined standards and/or specifications. The goal is to ensure that the uncertainty of the measurement data remains within the tolerance range required for the intended data uses. For example, the QC measures require that both sampling and analytical devices are operated following the specific standard protocols, respectively. Further, QC measures are implemented to ensure the selection of representative sampling sites and sampling times, the use of the correct sample preservation method, periodic calibration of equipment, the use of different types of blanks (QC samples), analysis of split and spiked samples etc. When monitoring trace levels of chemical parameters, e.g., pesticides in the lake sediment or PAHs in the air, the blank samples are used for both sampling (to assess bias, and imprecision resulting from the collection, transportation, and storage of samples) and analysis (to assess associated bias and imprecision of the analytical methods). An important QC component is capacity-building, which should be incorporated into the monitoring program to ensure the necessary skills of the involved personnel.

The QA measures in environmental monitoring cover, for example, the independent performance audit, external assessment, or inter-laboratory calibration to ensure that the monitoring activities are being performed according to the standards or specifications at a stated confidence level. QA is an activity that expects to gain the trust of stakeholders. Thus, in many cases, it may be considered an activity to ensure the quality of the data that the stakeholders who may use the acquired data will consider appropriate. Implementation of QA/QC measures thus enhances ascertaining the accuracy and validity of measurements in conformity with the data quality objectives (DQOs) set by a particular monitoring program.

8.2.2 Sources of Uncertainty of Monitoring Results

Monitoring data containing measurement errors and knowing the sources of the errors helps minimize them. The sources of measurement uncertainty exist in both sampling and analytical stages, which are illustrated in Figure 8.1, where the area of the oval shape is schematically made proportional to the magnitude of the potential errors. Sampling activities have more inherent uncertainties due to the diverse spatial and temporal distributions of the contaminants, along with the uncontrolled environmental conditions that exist in sampling. The

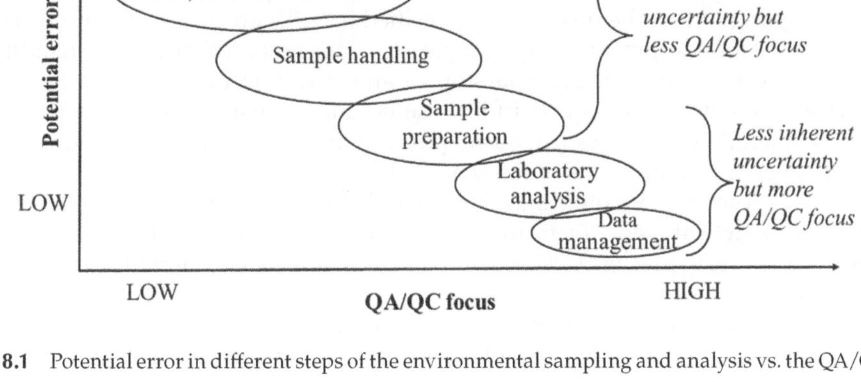

FIGURE 8.1 Potential error in different steps of the environmental sampling and analysis vs. the QA/QC focus.

sampling activities, however, normally get less QA/QC focus as compared to the laboratory analysis activities. The latter are commonly performed in more controlled conditions and have a good QA/QC plan with more detailed guidelines for standard operation procedures/ protocols (SOP). It is important to emphasize that the monitoring results are only valid when a representative sample has been collected in the first place. Otherwise, regardless of how sophisticated and accurate analytical procedures are applied, the results still fail to representatively characterise the environmental conditions being monitored.

The largest source of measurement uncertainty is associated with the heterogeneous spatial and temporal variations of the pollutants in the complexly interacted environmental matrices as further detailed in Chapters 14 and 15. The pollutant's spatial and temporal variations in an environment matrix are mainly influenced by (1) the variation in pollution discharge rates and locations; (2) the change in the dilution and removal capacity of pollution by the environmental matrix, e.g., transport by wind and mixing by turbulence in the air; (3) exchanges with other matrices, e.g., settling of suspended solids in water, atmospheric deposition to the earth surface, or uptake by biota; and (4) physicochemical transformation of pollutants, e.g., loss of volatile compounds from liquid and solid media or formation of secondary air pollutants in the atmosphere. The environmental matrices also contain stochastic factors such as the turbulent motions in air and water that induce random variability of the pollutant concentrations. These large spatial and temporal variations of pollutants in the environment are the key challenges for collecting representative samples.

Understanding the environmental matrix being studied is important to prepare a suitable sampling design, e.g., to select sampling sites and times, with the aim of obtaining representative pollution levels. As a way of illustration, large spatiotemporal distribution of air pollution levels in an urban area is a great challenge for producing representative monitoring results. $PM_{2.5}$ levels, for example, vary spatially due to the influence of the emission sources, i.e., roadside vs. residential site, and temporally due to the change in the emission rates of primary $PM_{2.5}$ (rush hours vs. non-rush hours). Phuc and Kim Oanh (2022) reported significant variations of pollutants in an urban area with heavy traffic, the difference between the roadside and an ambient site located about 150 m apart was 4–5 times for BTEX compounds (benzene, toluene, ethylbenzene and xylenes) and 1.2–1.3 for $PM_{2.5}$. This further indicates the need for monitoring data with sufficiently refined spatial resolution, especially required for health effect studies. The atmospheric processes cause both spatial and temporal variations of air pollutants and include, for example, transport by mean wind, mixing by turbulence, removal (wet and dry deposition) and chemical transformation that leads to the formation of secondary particles (sulphates, nitrates, secondary organic particles, tropospheric ozone) from respective precursors.

Field monitoring activity encounters uncontrolled conditions that potentially introduce errors to the monitoring results. These include, for example, power failure, limited accessibility to the intended sites, unstable operation of equipment etc., which are some of the challenges commonly faced. Further, field sampling activities are affected by weather conditions (stormy, rainy, windy or hot), which in turn influence the quality of samples collected. To minimize sampling errors, the instructions specified in SOPs are strictly followed. The SOPs provide guidance for equipment set-up, sample container preparation, sample handling (taking subsamples, compositing etc.), sample storage, and transport, among others. Suitable sample preservation methods are required for those pollutants that cannot be measured onsite to retard the alteration processes (chemical, physical and biological), hence minimising the potential changes of contaminants in the samples before analysis.

Contaminants from the surroundings may be added to the samples (gaining) during both sampling and analytical steps (from the ambient air, contaminated glassware, equipment, injection syringes, reagents etc.). These additive amounts are commonly assessed using QC blank samples of several types (field blanks, transport blanks, laboratory blanks etc.). Contaminants in a sample may also be lost, e.g., by evaporation or by adsorption on the sample container, during the sampling, sample storage and analysis. Various types of spiked QC samples, e.g., matrix spike, surrogate spike etc., are used to assess the loss or the gain of the contaminants (USEPA, 2022a). A known amount of the contaminant or the suitable surrogate compound is added to a sample, and the measured (recovered) amount is compared with the added amount to quantify the recovery rate. A recovery rate close to 100% is ideal, but, in reality, a certain range of loss and gain is normally accepted.

With the development of the monitoring equipment, many parameters can be measured online continuously using the instrumentation method, as further detailed in Section 8.5. For example, automatic analysers are widely used for air pollution monitoring to provide continuous real-time measurements of PM ($PM_{2.5}$, PM_{10}) and gases (CO, SO_2, NO_x, O_3 etc.). Using the automatic analysers avoids the need for sample preservation, transport etc., but the need remains for taking representative samples for online measurements. These include, for example, selecting the representative monitoring site to install a monitor and correctly positioning the intake probe of the monitor.

8.2.3 Types of Monitoring Errors

Monitoring data potentially contains errors of two types, systematic and random. The systematic errors are constant, either positive or negative, and are commonly associated with the selection of sampling procedures, i.e., sampling times, points, equipment, sample preservation and transportation, and personnel skill levels. For example, in the Asian monsoon climate, air pollution levels are significantly higher in the dry season as compared to the wet season. Therefore, to represent annual pollution levels, the monitoring period should cover both dry and wet seasons. If the monitoring is done only in the dry season, the average level would be higher than the annual average. Spatially, the levels of CO measured at busy roadsides are higher than in ambient areas located a few hundred meters away. Systematic errors cause "bias", the difference between a measurement result and the true value, which is the statistical measure of the trueness.

In environmental monitoring practice, the true value is largely unknown; hence necessary QC measures are taken to minimise the bias. For example, the errors related to the equipment can be removed by calibration, e.g., to correct the pumping flow-rate of an air sampler or to calibrate a pH or conductivity meter for water quality monitoring. Personnel training is a necessary component of QC to ensure conformity with SOPs, thereby minimising human-related systematic errors. For analytical instrument calibration, the standard reference materials (containing known concentrations of heavy metals or trace organic compounds etc.) can be used to quantify and minimise the bias (NIST, 2016).

The random errors, on the other hand, are variable both in magnitude and sign (negative or positive). Statically, these errors lead to statistical variability or dispersion of the measurement points and are related to the "precision". Some degree of random errors is caused by unavoidable random factors inherent in every measurement procedure, along with various uncontrolled variables of both instrumental conditions and human factors previously discussed. Due to their random nature, these errors principally can be described by the law of probability with approximate normal distributions.

When systematic errors are negligible, then a large number of "imprecise" measurements can still produce an average approaching the true value, i.e., good trueness. This is because the accuracy of estimating the population mean can be improved by adopting the average of the measurements generated by multiple trials, i.e., the sample mean approaches the population mean due to the law of large numbers. If random errors are sufficiently small, the measurement results produced by multiple attempts are close to each other, i.e., they are well repeated, or they are precise. For example, for evaluating a device, several measurements are made on a single sample and the results are precise if they produce an average value with a small standard deviation. Measurements may have high degrees of both precision and trueness, or neither nor both, as schematically presented in Figure 8.2. The trueness of a measurement can be evaluated by using the standard reference sample (its bias towards the known

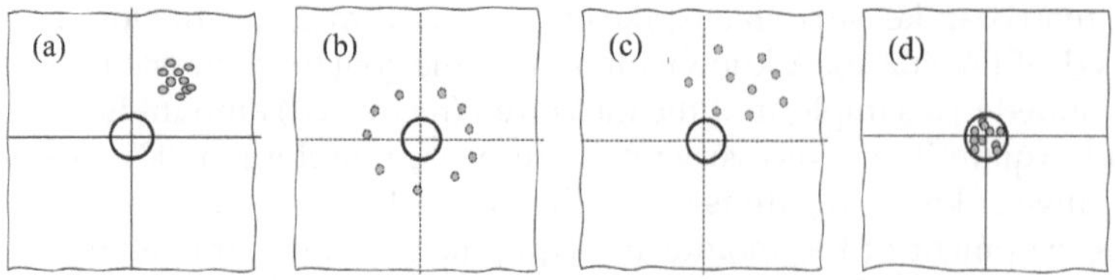

FIGURE 8.2 Schematic presentation of accuracy related to two types of measurement errors (two-dimensional plane view: the centre point is the true value, and the cycle is the specified tolerance range of the measurement): (a) High precision but poor trueness; (b) Low precision but good trueness (considering the mean value); (c) Low precision and poor trueness; and (d) High precision and good trueness.

standard reference value) whereas multiple measurements are needed to evaluate precision. In statistics, bias is the difference between the average of measurements (sample mean) and the population mean. In other words, low bias means high trueness. If the systematic errors are small, then the sample mean is close to the population mean (μ). If the random errors are small, then the dispersion of the measurement points is small, or the standard deviation is small.

8.2.4 Principal Indicators of Monitoring Data Quality

Monitoring data are the final product of any monitoring program; hence the ultimate goal is to produce valid data, i.e., the data with the required quality and quantity in conformity with the DQOs set to meet the monitoring objectives. Practically, due to the constraints related to resources and technology, a monitoring program would not aim at producing "the topmost quality" data but to produce the "best possible data given the constraints", i.e., to cost-effectively produce the data satisfying the DQOs. Principal indicators of monitoring data quality include (1) trueness, (2) precision, (3) representativeness, (4) spatial coverage, (5) temporal completeness, (6) detectability and (7) comparability or harmonisation.

8.2.4.1 Accuracy, Trueness and Precision

The International Organization for Standardization (ISO 5725, 2023) considers "accuracy" to include both "trueness" and "precision". The term "accuracy" at one time was used to cover only the one component that is linked to the statistical parameter of bias is now named "trueness". For a measurement method, "trueness" refers to the closeness of agreement between the arithmetic mean of a large number of test results and the accepted reference or true value. Precision refers to the closeness of agreement between test results or, statistically, the dispersion between the data points, obtained under specified conditions.

Trueness and precision, the two terms used to evaluate the accuracy of measurements, are related to the two types of measurement error previously mentioned: systematic and random errors, respectively. The smaller the bias of a measurement, the better its trueness. As for precision, the smaller the dispersion of the repeated measurement data points, the higher the degree of precision will be. An accurate measurement means the measurement has negligible both systematic and random errors; it should be in the centre of the circle in Figure 8.2. Multiple measurements made at the specified conditions are considered to have a high accuracy if they are both precise and true, as seen in Figure 8.2d.

This section considers the terms "precision" (dispersion) and "trueness" (bias) separately reflecting their relation to the two types of errors. This is because, in environmental monitoring, there is a tendency to emphasize precision, i.e., to obtain an average of the repeated measurements with a small standard deviation. Bias is often overlooked because it is often more difficult to quantify. After all, the "true value" is mostly unknown. For example, the true concentration of $PM_{2.5}$ at an urban site or TSS in a river at a given time is largely unknown. Even quantifying the precision may not be straightforward because the repeated measurements are rarely done for identical environmental samples and under identical conditions to truly produce the "repeatability" or "precision" of the results.

In reality, minimising bias is more important because a large bias makes the monitoring data invalid. The invalid data in turn causes the misclassification of the environmental problems. Consequently, expensive control programs may be initiated based on monitoring results that indicate erroneously high pollution levels, a situation of false positive or over-representation of pollution concentrations. In the opposite case, no control actions would be taken if the monitoring results are erroneously low, a situation of false negative or under-representation of pollution concentrations. Both cases induce an excessive cost due to the consequent effects on society.

8.2.4.2 Representativeness

Representativeness is a measure of the degree to which monitoring data accurately represents the characteristics of environmental quality or conditions, in both time and space. For air quality monitoring, the time of measurement is important; e.g., the hourly CO level at a busy roadside during rush hours would be much higher than that during non-rush hours. It is therefore important to have the monitoring period cover the entire 24 hours to assess the diurnal variation of hourly CO at the site. Likewise, the water quality passing a crowded city varies significantly between upstream and downstream locations of the city; hence representative sites are needed to assess the effects of urban activities on the river water quality.

8.2.4.3 Spatial coverage

Spatial coverage refers to the extent that the monitoring results capture the spatial variation of the considered pollutants. The number and locations of monitoring sites are of importance. For example, when designing a monitoring network to characterize urban air quality, the sites should be selected in such a way as to catch the upwind, downwind, high, medium and low concentration ranges of air pollutants. To assess the exposure and health effects due to air pollution, the sites should be located in the areas where people live and conduct their daily activities.

8.2.4.4 Temporal Completeness

Temporal completeness means that the data collected from a monitoring site should be sufficient to present the temporal variation of pollutants during the monitoring period. For example, the DQOs for ambient air monitoring networks may specify that the monitoring stations should provide at least a certain percentage, for example 80%, of hourly data in a year, with the monitoring period covering all seasons to provide a representative annual average of the pollution levels.

8.2.4.5 Detectability

Detectability refers to the sensitivity of analytical equipment, in laboratories or built-in for automatic analysers, to detect the pollutants of interest. Low levels of a pollutant, e.g., a few ppt of dioxins, can be detected and quantified only by highly sensitive equipment, whereas for less sensitive equipment, such a level would be "not detected". The not-detected values may be assigned as "zero" in the data analysis step; hence leading to erroneous monitoring results. The poor detectability is a cause of the "false negative" situation.

8.2.4.6 *Comparability or Harmonisation*

The data collected by different monitoring methods for a given parameter should be comparatively analysed before being combined in a single dataset for further analysis. In some developing countries, the monitoring data of a parameter in different sites during different periods, and in some cases even at a single site over an extended period, are generated by different techniques, e.g., manual sampling vs. automatic analyser, and with varying QA/QC levels. The data are first harmonised in terms of quality and reliability before being used for extracting useful information for environmental quality management. Collocating different environmental pollution monitors would help reveal the inconsistency in their measurement results. The ambient air quality data, in particular, should be harmonised in terms of the averaging time and the unit, considering the temperature and pressure of measurements.

8.3 Monitoring Data Management

The monitoring results should be properly interpreted and presented in such a way that they can be used in making decisions about potential environmental risks and in prompting necessary corrective actions. The quality of the conclusions derived from monitoring data interpretation can be affected by uncertainty not only related to the sampling and analytical procedures but also in the data management.

8.3.1 Framework of Environmental Monitoring Data System

The overall data system of a monitoring program includes a data gathering sensor/monitor system, a data recording and transmission system, and a data management system. The data system has the roles of collecting, compiling, analysing and finally reporting the data for use. Data management is an essential part of the overall data system and has the goal to ensure that monitoring data are properly validated and managed after being obtained and that the data to be stored must be of good quality, can be easily retrieved and analysed to extract useful information for public dissemination and decision-making. The data flow and data management activities from a monitoring program are schematically shown in Figure 8.3.

8.3.1.1 *Raw Data*

These are the measurements received from monitors that must undergo the "validation" process to produce a valid database that is further analysed/interpreted to extract the information about the trend, the compliance to standards etc. Poor data quality is a key challenge to data management, and data validation is a must step that helps identify suspicious data points and eliminate wrong data points. A robust understanding of the contaminant chemistry, environmental fate (transport, mixing and transformation), local discharge sources and instrument calibration processes, among others, is required in processing the data validation.

8.3.1.2 *Data Validation*

Data validation is done by environmental professionals and is usually conducted at different levels (Chow et al., 2012). The first level (Level I) aims to verify field sampling and laboratory analysis parameters, units and averaging time of pollutant concentrations and to check data transmission and data processing. The invalid data points due to the instrument malfunction and other inoperable periods are eliminated from the dataset. The data are then corrected against the calibration or interference biases based on the QA/QC procedure. Level II is intended to evaluate chemical data for internal consistency. For example, in urban air quality monitoring, this step would examine the diurnal variations of hourly levels of $PM_{2.5}$, CO,

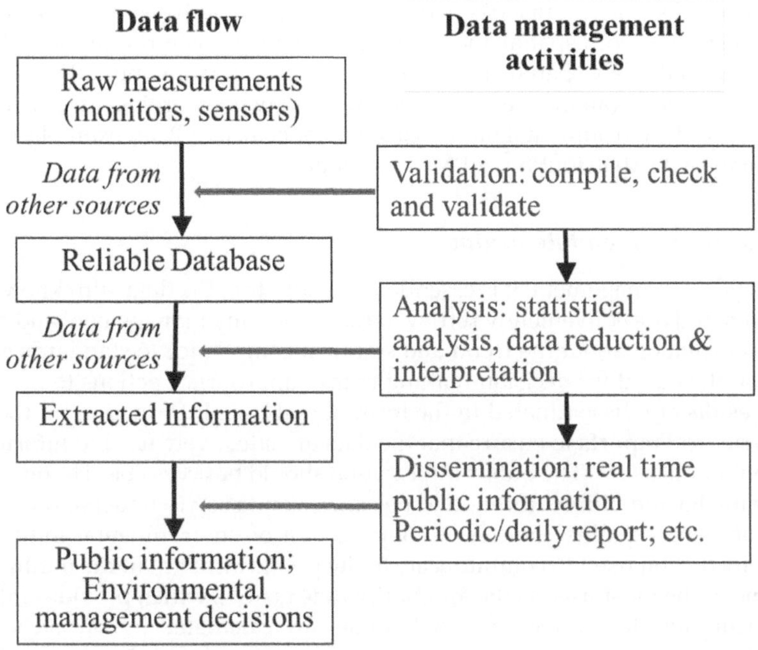

FIGURE 8.3 Data through output and data management activities in the environmental monitoring data system.

NO_x and O_3 to reveal if they reflect the anticipated changes according to the traffic rush hours (peaks in $PM_{2.5}$, CO, NO_x) and the atmospheric photochemistry (peak in O_3 in the afternoon). In Level III, statistical analysis, such as time-series analysis and quantitative statistical analyses, is conducted to illustrate temporal and spatial variations of the monitoring parameters. Available historical data of the pollutants (levels, temporal and spatial distribution) for the monitoring area are useful to verify and explain those obtained in a given monitoring program, e.g., against a change in the environmental emissions, such as an increase in emissions due to new sources or a reduction in emissions linked to a pollution control program. Related data from other sources are useful input for validation of the monitoring results. For ambient air monitoring programs, these include, for example, the results of emission inventory, observed and simulated meteorology, outputs of air quality dispersion modelling, satellite air quality monitoring, data from other monitoring programs and projects etc.

8.3.1.3 Data Analysis

The validated database is mathematically processed using several data reduction methods (average, percentiles, range, variations etc.) to get useful information on the trend and status of environmental quality. The monitoring dataset consisting of several variables can be analysed using multivariate analysis methods to reveal the relationship between the variables. These include, e.g., PCA and FA as the data reduction methods; multivariate analysis of variance (MANOVA); multivariate regression analysis; and others, as detailed in Part I and illustrated with case studies in Part III of this book.

Environmental parameters vary widely in both space and time; hence it is important to understand the change in environmental quality or pollution patterns over time and across different locations in a region. Spatiotemporal analysis can be applied for monitoring data having sufficient temporal (detailed in Chapter 14) and spatial (detailed in Chapter 15) coverages to identify and predict patterns and trend of environmental phenomena over time, for example occurrences of haze episodes in different areas of Southeast Asia in a year. This type of analysis integrating both spatial and temporal dimensions can reveal the features that are hidden in traditional solely spatial or time-series analyses. Spatiotemporal analysis has applications across various domains such as biology, ecology, meteorology, medicine, as well as climate studies and environmental monitoring. Gräler et al. (2013), for example, have used this method to analyse the distribution of PM_{10} in Europe.

8.3.1.4 Dissemination and Reporting

In monitoring data management, it is essential to ensure that the data with known quality are securely stored and are conveniently retrieved for use for any management and research purposes. The information about the trend and status of compliance to standards etc., extracted from the dataset, is used for decision-making to take appropriate actions to manage the situations. The results are disseminated in the form of real-time information for the public or in daily and periodical reports to ensure that the data are effectively used to inform about environmental pollution issues. The public information should be accessible, i.e., on websites and/or broadcast media and billboards etc. and be concise and easy to understand.

The environmental quality indexes, an aggregation of environmental indicators, are the simplified information used to communicate to the public rather than the multiple concentration data. One of the most used is the air quality index (AQI), which provides information on air quality using unit-less values. The AQI values are calculated in consistency with regulated air pollutants with varying measurement units. AQI values are segregated into several ranges, presented with colour codes, following the severity of the pollution effects on different people groups to enable them to potentially minimise the exposure. The presentation of AQI for public information by USEPA has colour codes ranging from green (good), yellow (moderate), orange (unhealthy for sensitive groups), red (unhealthy), purple (very unhealthy) and maroon (hazardous) categories of air quality, which are provided along with the cautionary statements to guide the public for avoiding excessive exposure (USEPA, 2022b). The information for policy making is meant for the government to make decisions on pollution control actions, and hence is presented incorporating relevant environmental quality issues or pollution control scenarios. Professionals and researchers, on the other hand, would prefer that detailed technical data that can be used as references for their studies.

8.3.2 Development in Data Management Methods

Continuous development of environmental monitoring techniques has brought in new monitoring techniques such as low-cost sensors (LCS) and satellite products (see further details in Section 8.4), which allow the generation of massive data. Handling these massive datasets may be a great challenge for the conventional methods of data management. Big Data Technology has become widely used to analyse, process and interpret the massive amount of structured and unstructured data that could not be processed manually or traditionally. The big datasets generated from environmental monitoring can be processed and managed using machine learning, artificial intelligence (AI) and cloud computing for publicising (Laffly, 2020). Statistical analysis methods remain an essential element in the process of handling these massive environmental monitoring datasets.

Particularly, statistical methods are used to accurately build models and develop the estimation algorithm and hence should play key roles in Big Data science (Olhede and Wolfe, 2018). Since 2017, for example, post-deployment calibration techniques in data science have been developed to improve the data quality of an LCS network of more than 2500 sensors used for monitoring $PM_{2.5}$ in Taiwan. A generalised additive model was used to estimate the correction for selected sensors to significantly reduce the root mean square errors (RMSE); machine learning was then applied to obtain the corrections for other sensors and reduce overall RMSE by more than three times, i.e., to about 5.0 μg/m^3 (Lung et al., 2022). In addition, advanced techniques in data science were used to detect malfunctioned sensors in the network using an anomaly detection framework (ADF) developed with statistical and machine-learning methods to identify abnormal sensors and further detect the rate of faulty sensors for improving the data quality (Lung et al., 2022).

8.4 Development of Environmental Monitoring Techniques

Environmental monitoring has been continuously advanced with new monitoring techniques and equipment evolving to measure conveniently and accurately not only the traditional pollutants but also a growing list of emerging pollutants. The analytical techniques in particular are continuously developed to measure trace pollutants at parts per billion (ppb) levels or lower with high accuracy. These include, for example, chlorinated pesticides and dioxins in the air, water, soil and biota.

New monitoring techniques are also developed to respond to the need for monitoring a growing number of synthesized chemicals. These are called "contaminants of emerging concern" due to their potential toxicity and include, for example, perfluorinated compounds, chlorinated paraffins and polybrominated diphenyl ethers, among others. These compounds are used in routine products and hence would eventually end up in the environment to cause adverse effects on human health (USEPA, 2023; Environment Canada, 2017). The new pollutants named microplastics (MP) and nanoplastics (NP), collectively known as MNP, have been identified as an increasing threat to the environment. The term "MP" generally considers all plastic particles with a size range of 1 μm–5 mm. The NP size is commonly considered to be in the range of 1–1000 nm, although some researchers may define NP as those having the size in the nanoparticle range, i.e., 1–100 nm in all three dimensions (Bibi et al., 2023). MNP are detected in air, water, soil, sediment, and biota. MPs enter the food chain through ingestion by aquatic and marine animals and eventually reaches up to humans (Zhang et al., 2020; Bibi et al., 2023).

Environmental monitoring normally uses one or a combination of several methods, which include manual traditional sampling (intermittent sampling for subsequent laboratory analysis), automatic continuous monitoring or automatic analysers (continuous extractive sampling and online automatic analysis), and in situ monitoring (involving no sample extraction) with the aid of remote sensing and online low-cost sensing techniques.

8.4.1 Manual Monitors

Manual monitors have been among the first equipment used in monitoring conventional parameters of air, water, soil, sediment and biota. The manual method has two separate tasks: (1) sampling or sample collection and (2) subsequent laboratory analysis of the sample. The method is well-established, highly reliable and accurate and can measure a wide range of pollutants. However, both sampling and analysis tasks are time-consuming. Besides, this method relies on bulky equipment and skilful human resources.

The manual sampling methods for fluid matrices of air and water may be classified as active sampling, i.e., when a pump is used, and passive sampling. Active sampling takes a sample over a relatively short period, typically between less than 1 hour to a few days, and provides the average concentration over the corresponding period. Depending on the target pollutants and sampling purpose, the sampling devices are designed to collect bulk matrix samples or selective samples that retain only specific compounds. For example, in air pollution sampling, the bulk air samples are collected in a sampling bag, syringe or canister, which takes the actual air mixture at the site. On the other hand, selective air sampling retains only the pollutants of concern in the selected sampling media, such as using a filter to collect particulate matter or an activated carbon sorbent trap to retain VOCs. The pumping rate in the active sampling directly provides information on the sampled volume.

The passive sampling method does not use any pumping device. Instead, the pollutants are collected based on physical and chemical principles, such as diffusion and chemical trapping of target gases in diffusive samplers or settling of particles on a surrogate collecting surface. Passive sampling normally collects a sample over an extended period of a few weeks to months. The volume of sampled matrix (air or water) passed through the collector is not readily known. Instead, it is commonly calculated using the matrix conditions, hence normally inducing considerable uncertainty.

Chapter 9 provides details on the types of environmental samples and the challenges to obtaining representative samples. Active water sampling for industrial discharges and for receiving water bodies is detailed in Chapters 10 and 11, while that of PM is described in Chapters 12 and 13.

8.4.2 Continuous Automatic Monitoring Instruments

Continuous automatic monitors are commonly used to monitor air and water quality. The monitors are online instrumentation providing real-time measurement data, and, with internet access, the data can be conveniently shared. The continuous records of the air and water parameters provide detailed pictures of their temporal variations that help identify in real time the processes leading to high pollution levels. This is an advantage of the method over the manual monitors, which only give a snapshot concentration at the sampling time or only one average value over the integrated sampling period.

The continuous water quality monitors most commonly provide readings of temperature, electrical conductivity, dissolved oxygen, pH and turbidity. However, several other parameters, such as chlorine, fluoride, nitrate, particle count or total organic carbon etc., can be also monitored online, either in situ or by a flow-through system when a pump is used to send extracted samples to sensors (Raich, 2013). The SOP describing procedures to be performed routinely at a continuous water quality station needs to be closely followed to ensure the data quality (Wagner et al., 2006).

For air quality, automatic analysers are available for the most commonly regulated air pollutants, i.e., SO_2, NO_2, CO, O_3, $PM_{2.5}$ and PM_{10}. These instruments are programmed to generate consecutive hourly average pollution levels and are used in automatic air monitoring stations worldwide. Air samples are continuously extracted and sent (pumped) to the analysers. For PM, optical methods, e.g., light scattering and light obscuration, are commonly used for concentration measurement. Beta attenuation monitoring (BAM) is a widely used monitoring technique for PM_{10} and $PM_{2.5}$. BAM relies on the attenuation of the beta ray passing through the particulate layer collected on a filter, and the responses are subsequently converted to PM mass concentrations. For gaseous pollutants, a wide range of automatic analysers are available that measure target gaseous pollutants based on the physical and chemical properties of the monitored gases.

8.4.3 Low-Cost Sensors

The fast-growing low-cost sensor technology provides promising perspectives for environmental monitoring. These small sensors translate the physical environment conditions, in this case, the levels of pollutants, to electrical signals. Output signals from the sensors are calibrated against the reference standard instruments to obtain the measurement data in the relevant concentration unit (Williams et al., 2014). Development in wireless communications and electronics technology has made the wireless sensor network (WSN) into reality. WSN is applied for air quality, water quality, and natural disasters such as landslides, forest fires or volcanic eruptions (Murphy et al., 2015).

Low-cost sensors gain growing applications in air quality monitoring, both for PM and gaseous pollutants. The key advantages of these sensors are their low cost, light weight and small size with low energy consumption. For these reasons, they can be distributed densely in a monitored area. The use of wireless low-cost sensors substantially increases the spatiotemporal density of monitoring data, which would supplement the traditional pollution monitoring in improving exposure estimates and raising community awareness about air pollution. Key drawbacks of the wireless sensors used at present in air quality monitoring include their lower accuracy, being less reliable and having poorer detection range for some pollutants as compared to conventional equipment (Loi et al., 2020). Output signals of LCS are affected by environmental conditions such as humidity, wind, temperature, rain and insects, as well as the signal instability or drift over the operation period. The measurement results produced by these sensors should be carefully evaluated and validated (Lewis et al., 2018). Presently, data quality remains a major concern that hinders the widespread adoption of LCS technology for air quality monitoring, especially for regulatory purposes.

8.4.4 Remote Sensing

Remote sensing techniques are available for monitoring environmental quality. These techniques involve no sample extraction; instead, the online data recording is made in situ. The measurement results are normally spatially integrated: vertically and horizontally. Earth-observing satellite sensors monitor environmental parameters with large spatial coverage, such as land cover, sea surface temperature and colour, solar radiation, meteorological characteristics, water and air quality. Spatial resolution of satellite data varies by the type of instrument, from very fine of less than 10 m to coarser resolution of 1 km or more. Temporal resolution is the time period required for a sensor to observe the same location on Earth, and it depends on the orbit of the satellite. Geostationary satellites make continuous measurements of the same area on Earth, so they have high temporal resolutions (typically minutes to hours). The polar-orbiting satellites have the global coverage and high spatial resolutions but lower temporal resolution, i.e., 1–16 days (Engel-Cox et al., 2012).

Satellite data make it possible to establish the long-term baseline conditions for any region of the Earth. The large spatial coverage of the satellite data is useful for environmental surveys that are conducted prior to technical designing of the ground-level monitoring programs for land use, forest area, air and water pollution. In water quality monitoring, satellite sensing provide data on the sea surface temperature, TSS and chlorophyll-*a* etc. (Wu et al., 2009). The data are useful in characterizing the status and trends of water quality at both local and regional scales and in identifying the water quality changes associated with harmful algae blooms and red tides.

Several satellite sensors have been used for air quality monitoring of PM (aerosol) and gaseous species (Martin, 2008). Aerosol is monitored using the parameter of AOD or AOT product (aerosol optical depth or thickness, unit-less), which is a measure of the scattering and absorption of light by aerosol in a vertical column from the surface of the Earth to the top of the atmosphere. Gaseous pollutants are measured using defined spectral channels of the satellite sensors and include, for example, SO_2, NO_2, O_3, CO, formaldehyde (HCHO) and carbon dioxide (CO_2). Satellite remote sensing provides information about regional air quality and long-distance transport of pollutants, which complements available ground-based data, especially for remote areas. Satellite products have been used to monitor urban areas, large industrial plumes, volcanic events and biomass burning (Engel-Cox et al., 2013).

Satellite monitoring, however, has disadvantages related to the inherent uncertainties of satellite sensors and the dependence on atmospheric conditions, e.g., cloudiness and humidity. In addition, satellite data are columnar measurement results and hence have limited vertical resolutions. When integrating the satellite data with ground-based monitoring results, there is the need to consider the mismatch in temporal and spatial resolutions. The ground-based

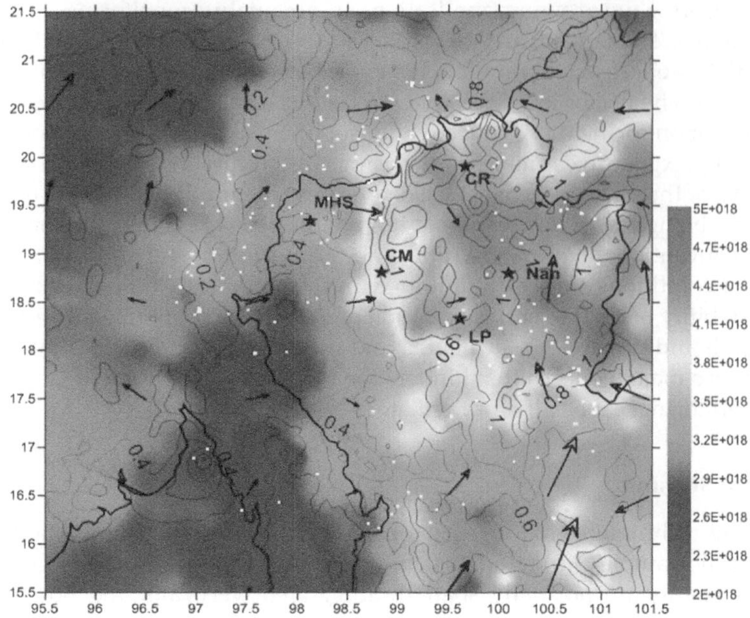

FIGURE 8.4 Satellite monitoring CO, AOD and forest fires in northern Thailand. (*Note:* MODIS AOD contours at interval of 0.2; MOPITT CO total column in mole/cm^2 (grey-colour scales); surface wind vectors; and forest fires (white spots) at 10:00 LST, 7 March 2009.)

Source: Sukitpaneenit and Kim Oanh (2013).

monitors provide the most accurate measurements of air and water quality at a single point (monitoring site), but the data are time-averaged, typically for air quality, on an hourly or daily basis. The satellite sensors provide columnar data with horizontal resolutions of a few square kilometers and for polar-orbiting satellites, the data represent only a snapshot at a single point in time. Regression studies are conducted to convert the columnar satellite data to the actual pollution levels measured near the ground for mapping the regional distribution of pollutants. Figure 8.4 shows an example of the mapping of forest fire hotspots, CO and AOD over a region (Sukitpaneenit and Kim Oanh, 2013). In this case study, the linear correlation coefficients between satellite CO and ground-level CO concentrations measured at different stations were in the range 0.42–0.71, while that between AOD and PM_{10} were 0.50–0.73. Lower correlation coefficients were obtained for those stations that had strong influences from local emission sources in the vicinity.

Other commonly used remote sensing techniques, mainly in atmospheric monitoring, include LiDAR, AERONET and DOAS. The open-path technique using DOAS (differential optical absorption spectroscopy) can measure, simultaneously and in real time, different types of atmospheric molecular compounds. The monitoring principle relies on the relationship between the amount of absorbed light and the number of molecules the light passes through, stated by the Beer-Lambert absorption law. The species monitored by this technique commonly include NO_x, SO_2, O_3, VOCs, NH_3, CO, CO_2, etc. (USEPA, 1998), which are spatially averaged over a long optical path (hundreds of meters). LiDAR (light detection and ranging) is active remote sensing technology that emits pulses of light (i.e., the laser beams) and detects reflected light (NOAA, 2012). LiDAR can be used for observation of the Earth surface and is applied for monitoring ambient air pollution (PM and gases) and emission sources (Gaudio et al., 2015). The AERONET (AErosol RObotic NETwork) is a network of ground-based remote sensing using Sun photometers for measuring atmospheric aerosol properties and provides globally distributed observations of AOD along with other products (AERONET, 2021).

8.5 Summary

Environmental monitoring, an essential technical tool for environmental quality management, may be defined as the process of generating data on the environmental condition that is commonly described by the presence and levels of pollution in different forms along with the associated environmental impacts. Monitoring is conducted for one or more environmental matrices using relevant chemical, physical and biological parameters to characterise and assess the impacts of pollution on the environment. These monitoring parameters as well as their monitoring methods differ largely between the environmental matrices, e.g., air, water, soil/land and biota. Collection of valid data in conformity with specified monitoring objectives is a crucial requirement of environmental monitoring programs. Only such valid data can be used to provide science-based information to establish the status and trend of environmental quality. The valid data thus help to correctly identify environmental problems so that appropriate actions can be taken to control the pollution situation and effectively solve the problems.

Technical design of a monitoring program requires good understanding of the environmental matrices being studied. Such knowledge is important to prepare and implement a suitable monitoring program for obtaining representative environmental monitoring data. Sources of monitoring errors exist; hence a well-established QA/QC procedure is needed to minimise the errors and to cost-effectively produce the data that satisfy the data quality objectives.

Environmental monitoring methods are continuously evolving with the new sampling and measurement techniques that have been developed to accurately measure not only traditional pollutants but also emerging pollutants. Data management is also developed to effectively handle big datasets generated by new monitoring methods. Statistical analysis remains an essential element in the process of handling these massive environmental monitoring datasets. Spatiotemporal models can be developed using the monitoring data to identify and predict patterns or occurrences of the environmental phenomena within a specified region over time.

REFERENCES

AERONET, 2021, *Aerosol robotic network*, retrieved on 25 September 2023 from https://aeronet.gsfc.nasa.gov/

Bibi, Aisha, Can, Ali, Pant, Udit, Hardiman, Gary, Hill, Daniel, Elliott, Christopher, Cao, Cuong, 2023, Chapter six – a review on state-of-the-art detection techniques for micro- and nano-plastics with prospective use in point-of-site detection, *Comprehensive Analytical Chemistry*, 101, 143–196, https://doi.org/10.1016/bs.coac.2022.11.003

Chow, J., Watson, J. and Tropp, R., 2012, Technical challenges in air quality monitoring: Quality control and quality assurance, *Presented at better air quality symposium on improving air quality monitoring systems in Asia*, Hong Kong SAR.

Engel-Cox, J.A., Huff, A.K., Kanabkaew, T. and Kim Oanh, N.T., 2012, *Satellite tools for air quality management with focus on particulate matter, chapter 4 in the book Integrated air quality management: Asian case studies*, Taylor and Francis Group, CRC Press, ISBN 9781439862254

Engel-Cox, J.A., Kim Oanh, N.T., van Donkelaar, A., Martin, R.V. and Zell, E., 2013, Toward the next generation of air quality monitoring: Particulate matter, *Atmospheric Environment*, 80, 584–590, https://doi.org/10.1016/j.atmosenv.2013.08.016

Environment Canada, 2017, *Environmental monitoring and surveillance: Chemicals management*, retrieved on 25 September 2023 from https://www.canada.ca/en/environment-climate-change/services/science-technology/programs/monitoring-surveillance-chemicals-management.html

Gaudio, P., Gelfusa, M., Malizia, A., Parracino, S., Richetta, M., De Leo, L., Perrimezzi, C. and Bellecci, C., 2015, Detection and monitoring of pollutant sources with Lidar/dial techniques, *Journal of Physics: Conference Series*, 658(1), 012004, https://doi.org/10.1088/1742-6596/658/1/012004

Gräler, B., Rehr, M., Gerharz, L. and Pebesma, E., 2013, *Spatio-temporal analysis and interpolation of PM10 measurements in Europe for 2009, ETC/ACM technical paper 2012/8*, revised version, Institute for Geoinformatics, University of Muenster, https://www.eionet.europa.eu/etcs/etc-atni/products/etc-atni-reports/etcacm_2012_8_spatio-temp_pm10analyses

International Organization for Standardization (ISO), 2023, *ISO 5725–1:2023 (en), accuracy (trueness and precision) of measurement methods and results – part 1: General principles and definitions*, https://www.iso.org/obp/ui/#iso:std:iso:5725:-1:ed-2:v1:en

Laffly, D., 2020, *TORUS 3 – toward an open resource using services cloud computing for environmental data*, ISTE Ltd., UK and John Wiley & Sons Inc., New York, United States, ISBN 978-1-78630-599-2

Lewis, A., Peltier, W.R., von Schneidemesser, E. and Peltier, R.E., 2018, *Low-cost sensors for the measurement of atmospheric composition: Overview of topic and future applications*, World Meteorological Organization (WMO), Geneva, Switzerland, ISBN 978-92-63-11215-6

Loi, H.T., Fakprapai, S. and Kim Oanh, N.T., 2020, *Air quality monitoring with focus on wireless sensor application and data management, chapter 2 in the book TORUS 3 – toward an open resource using services cloud computing for environmental data*, ISTE Ltd., UK and John Wiley & Sons Inc., New York, United States, ISBN 978-1-78630-599-2

Lung, S.C.C., Hien, T.T, Cambaliza, M.O.L., Hlaing, O.M.T., Kim Oanh, N.T., Latif, M.T., Lestari, P., Salam, A., Lee, S-Y., Wang, W-C.V., Tsou, M-C.M., Cong-Thanh, T., Cruz, M.T., Tantrakarnapa, K., Othman, M., Roy, S., Dang, T.N. and Agustian, D., 2022, Research priorities of applying low-cost $PM_{2.5}$ sensors in Southeast Asian countries, *International Journal of Environmental Research and Public Health*, 19(3), 1522, https://doi.org/10.3390/ijerph19031522

Martin, R.V., 2008, Satellite remote sensing of surface air quality, *Atmospheric Environment*, 42(34), 7823–7843, https://doi.org/10.1016/j.atmosenv.2008.07.018

Murphy, K., Heery, B., Sullivan, T., Zhang, D., Paludetti, L., Lau, K.T., Diamond, D., Costa, E., O'Connor, N. and Regan, F., 2015, A low-cost autonomous optical sensor for water quality monitoring, *Talanta*, 132, 520–527, https://doi.org/10.1016/j.talanta.2014.09.045

National Institute of Standards and Technology (NIST), 2016, *Standard reference materials*, U.S. Department of Commerce, retrieved on 21 September 2023 from https://www.nist.gov/srm/about-nist-srms

National Oceanic and Atmospheric Administration (NOAA), 2012, *Lidar 101: An introduction to Lidar technology, data, and applications*, NOAA Coastal Services Center, retrieved on 25 September 2023 from https://coast.noaa.gov/data/digitalcoast/pdf/lidar–101.pdf

Olhede, S.C. and Wolfe, P.J., 2018, The future of statistics and data science, *Statistics & Probability Letters*, 136, 46–50, https://doi.org/10.1016/j.spl.2018.02.042

Phuc, N.H. and Kim Oanh, N.T., 2022, Large spatio-temporal variations of size-resolved particulate matter and volatile organic compounds in urban area with heavy traffic, *Environmental Science and Pollution Research*, 29, 21491–21507, https://doi.org/10.1007/s11356-021-16921-9

Raich, J., 2013, *Review of sensors to monitor water quality, European reference network for critical infrastructure protection (ERNCIP) project*, Publications Office of the European Union, Luxembourg, ISBN 978-92-79-34618-7

Sukitpaneenit, M. and Kim Oanh, N.T., 2013, Satellite monitoring for carbon monoxide and particulate matter during forest fire episodes in Northern Thailand, *Environmental Monitoring and Assessment*, 186, 2495–2504, https://doi.org/10.1007/s10661-013-3556-x

U.S. EPA, 1998, *40 CFR part 58 Appendix E – probe and monitoring path siting criteria for ambient air quality monitoring*, U.S. Environmental Protection Agency, retrieved on 25 September 2023 from https://www.epa.gov/amtic/ambient-air-monitoring-quality-assurance

U.S. EPA, 2022a, *Quality assurance (QA) – quality control (QC), hazardous waste test methods SW-846*, U.S. Environmental Protection Agency, retrieved on 21 September 2023 from https://www.epa.gov/hw-sw846/quality-assurance-qa-quality-control-qc

U.S. EPA, 2022b, *AQI breakpoints*, U.S. Environmental Protection Agency, retrieved on 21 September 2023 from https://aqs.epa.gov/aqsweb/documents/codetables/aqi_breakpoints.html

U.S. EPA, 2023, *PFAS explained*, U.S. Environmental Protection Agency, retrieved on 25 September 2023 from https://www.epa.gov/pfas/basic-information-pfas#health

Wagner, R.J., Boulger, R.W., Jr., Oblinger, C.J. and Smith, B.A., 2006, *Guidelines and standard procedures for continuous water-quality monitors-station operation, record computation, and data reporting, techniques and methods 1–D3*, U.S. Geological Survey, retrieved on 25 September 2023 from https://pubs.usgs.gov/tm/2006/tm1D3/

Williams, R., Long, R., Beaver, M., Kaufman, A., Zeiger, F., Heimbinder, M., Hang, I., Yap, R., Acharya, B., Ginwald, B., Kupcho, K., Robinson, S., Zaouak, O., Aubert, B., Hannigan, M., Piedrahita, R., Masson, N., Moran, B., Rook, M., Heppner, P., Cogar, C., Nikzad, N. and Griswold, W., 2014, *Sensor evaluation report, EPA/600/R-14/143 (NTIS PB2015–100611)*, U.S. Environmental Protection Agency, retrieved on 25 September 2023 from https://cfpub.epa.gov/si/si_public_record_Report.cfm?Lab=NERL&dirEntryId=277270

Wu, M., Zhang, W., Wang, X. and Luo, D., 2009, Application of MODIS satellite data in monitoring water quality parameters of Chaohu Lake in China, *Environmental Monitoring and Assessment*, 148, 255–264, https://doi.org/10.1007/s10661-008-0156-2

Zhang, Y., Kang, S., Allen, S., Allen, D., Gao, T. and Sillanpää, M., 2020, Atmospheric microplastics: A review on current status and perspectives, *Earth-Science Reviews*, 203, 103118, https://doi.org/10.1016/j.earscirev.2020.103118

9

Sampling Design from Theory to Practical Applications in Environmental Monitoring

This chapter provides basic concepts of statistical sampling designs and how these designs are applied in environmental monitoring to produce representative data for characterising the status and trend of environmental quality. Knowledge of the population under study, even if it is not complete, can be used to find more efficient methods to extract samples, e.g., to rely more on judgmental sampling or more on random sampling designs. Often, multistage mixed sampling designs are used, and in each stage, a combination of random and/or judgmental sampling techniques may be applied. This chapter elaborates on the need to consider the measurement protocols for the intended monitoring parameters along with the sampling design. Specific considerations in applications of statistical sampling designs for environmental measurements are emphasised, detailing the types of samples and collection techniques required to characterise highly dynamic environmental matrices with large variations in time and space, such as air and water. Examples and case studies are used to explicitly illustrate practical applications of the sampling designs.

9.1 Concept of Sampling Design

Sampling, by definition, is a process of removing a portion, namely a sample, from a population with the aim of using the characteristics obtained from analysing the sample to represent those of the whole population. Therefore, the goal of the sampling is to extract the most representative sample from the population to provide the most appropriate material for the analyses, let them be of physical, chemical, biological or social characteristics, in conformity with the study objectives. Sampling is a universal technology for many kinds of research fields including physics, chemistry, engineering and social sciences. Sampling for a questionnaire survey in marketing research is almost the same as in an epidemiological survey.

9.1.1 Sampling Design

Sampling design is a fundamental part of data collection for science-based decision-making. It is a framework or a plan to collect a sample for obtaining some type of information about the population of interest. A sampling design commonly includes information about the target population, the sampling frame, sampling methods, sampling units, and sample sizes for every sampling stage. A well-prepared sampling design is critical to ensure that the obtained data are representative of the target population and in conformity with the study objectives. Such a sampling design is required to ensure that the data generated are sufficient to draw the conclusions needed with minimum spending of resources (USEPA, 2002).

9.1.2 Population

A statistical population is a complete set of elements (items or events) that share at least one common property that is the subject of a statistical analysis. For example, a supermarket

DOI: 10.1201/9781003597216-11

purchased 200 boxes of eggs on a particular day, and each box contains 10 eggs; hence a total of 2000 eggs were purchased. The population in this case is 2000 eggs that have a common property of being purchased by the supermarket for sale. The concept of "environmental population" of water or air matrices is normally not so well-defined; hence a good understanding of the environmental matrices under the study is required to develop a sampling design, which is further detailed in this chapter.

9.1.3 Population Size

The number of elements or events included in a population is called the population size. When the size is infinite, it is called an infinite population, and when it is finite, it is called a finite population. In the preceding example, the eggs purchased by the supermarket have a finite population of 2000. Note that a population can be regarded as infinite when it is not possible to limit the number. For example, even though the number of products from a factory is finite, if they are continuously manufactured in the future, then the population can be considered infinite.

9.1.4 Target Population

This is the complete collection of items that investigators are interested in. For example, the researchers would like to study the contamination levels of polychlorinated biphenyls (PCBs) in the eggs purchased by the supermarket. The target population in this case is 2000 eggs.

9.1.5 Sampling Frame

This includes the items from the target population that are available for sampling. Ideally, the researchers would need to take a sample from 2000 eggs. However, in the example, 500 eggs were sold, hence leaving only 1500 eggs available for the study. When the investigators collect a random sample from the sampling frame of 1500 eggs, the research results may not be perfectly representative of the target population of 2000 eggs. In other words, the results, in this case, would contain a sampling frame error. Therefore, it is important to make the sampling frame as similar to the target population as possible. When this is not possible, careful considerations need to be made to get the most representative sample from the sampling frame. Matching the sampling frame with the target population is a common challenge in environmental monitoring; for example, the target population is the ambient air at the breathing level in a city but the sampling frame, i.e., the air volume accessible for sampling, is only possible at certain sites. Therefore, the selection of representative sampling sites in the environmental monitoring (air, water, soil or biota) is one of the most crucial activities in the sampling design.

9.1.6 Sampling Unit

A sampling unit, by definition from the *Oxford Dictionary of Statistical Terms* (Dodge, 2003), is "one of the units into which an aggregate is divided for the purpose of sampling, each unit being regarded as individual and indivisible when the selection is made". The sampling units are the elements constituting the population, and in multistage sampling designs the sampling units may vary between the stages as further detailed in Section 9.3. For example, when taking eggs for the PCB study in the example, in the first stage the researchers selected 20 boxes of eggs, then in the second stage they took 3 eggs from each box for various measurements. In this case, the boxes are the primary sampling units while the individual eggs are the secondary sampling units.

9.1.7 Sample Size

A statistical sample size is defined as the number of sampling units to be included in a sample. Determination of the sample size is an important activity of the sampling design. Only with an appropriate sample size, the investigators can make inferences about a population from the results of the sample analysis. In practice, the sample size used in a study is usually determined based on the cost, time or convenience of collecting the data and the need to provide sufficient statistical power of the results. In a census, data are sought for an entire population; hence the intended sample size is equal to the population. However, in most of the empirical studies, the sample size is just a portion of the population, i.e., a few hundred people from a city are selected for a poll. In environmental monitoring, a sample may be some Liters of water taken from a lake or a few cubic meters collected from a continuous stack emission. As discussed further in Section 9.4, in simple random sampling, the calculation of sample size is more straightforward using a well-established formula. Whereas, in quota sampling, the sample size is decided based on the quota given, say 10% of the population in each stratum. In a multistage sampling design, the sample size varies with the sampling method used in each stage.

9.1.8 Sample vs. Population

When investigating a problem, the aim of collecting a sample is not to know the sample itself but to get information about the population under study. Therefore, investigators need to collect samples that allow them to correctly infer the information about the population they want to know. The sample is like an open window for examining the population. Therefore, if the window opened is not connected to the population, investigators cannot know the status of the population.

To understand the two most important foundations of statistics, the population and the sample, two examples are used for illustration:

- Example 1: In order to check the audience rating of a TV, a recorder was installed on the TV receiver in a selected home. The channel of the time zone that the TV is on was recorded. However, no person was sitting in front of the television, only a cat was in the house.
- Example 2: A man bought a package of strawberries from a supermarket. When viewed from above, the strawberries looked very good. But at home when unpacked, the man found that the strawberries in the lower layers were almost rotten.

These are examples of the failure of sampling methods. In example 1, the fact that the TV is on does not necessarily mean that a person is watching the TV. In practice, a similar situation could happen if one tries to measure a substitute characteristic that is easy to measure. In example 2, the population was all the strawberries in the pack, but the man decided on the whole population with only the information from the sampling frame, which is the strawberries on the top row.

9.2 Statistical Sampling Methods

There are two ways to extract samples from a population: random sampling and non-random sampling. Random sampling, also known as probability sampling, has the sampling units selected stochastically, and every element of a population has the same probability to be drawn in the sample. This makes it possible to make statistical inferences about population information based on examining the sample. The non-random method, also known as non-probability, purposive or judgmental sampling, has the sampling units selected based on certain professional knowledge about the population. Therefore, conclusions about the target population are limited and depend entirely on the validity and accuracy of professional judgment. Non-probability sampling contains hidden biases and uncertainties that cannot be quantified by statistical analysis. The biggest problem with this method is that the investigators cannot quantify the error of the estimates of the population parameters based on the information obtained from the sample.

Random sampling, even though named "random", does not mean it is done irresponsibly. Careful considerations should be made in order to obtain a sample that can represent the population for the properties of interest, or, in other words, the sample collected is representative of the studied population. While considered the best sampling method to correctly characterise a population, random sampling is often the most labor-intensive and expensive. Thus, if the status of the population is known to some extent, even though it is incomplete, the information can be used, and a more cost-effective sampling method can be employed. This section discussed the types of random and non-random sampling commonly used in environmental studies that are schematically summarised in Figure 9.1. Multistage sampling designs have the extraction of sampling units performed in more than one stage, and the sampling method often varies vary with stages as detailed in Section 9.3.3.

To illustrate the probability and non-probability sampling methods, two case studies are provided:

- Case study 1: A researcher investigates the attitude of the students in a class of 45 toward learning a statistical course.
- Case study 2: A researcher investigates the effects of the wastewater discharged from a food processing industry on the water quality of a receiving river.

Examples of the sampling design are provided in the following sections to explain the concepts of the sampling techniques for these two case studies and will be referred to as "Case study 1" and "Case study 2".

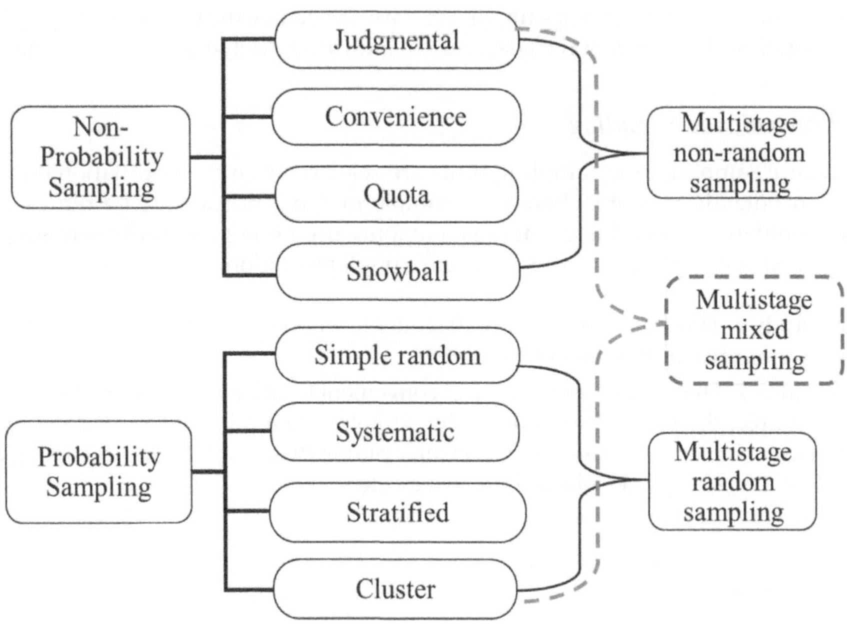

FIGURE 9.1 Overview of statistical sampling methods.

9.2.1 Non-Probability Sampling

The most commonly used non-probability methods include judgmental sampling, convenience sampling, quota sampling, and snowball sampling. The first three are more often used in environmental monitoring when samples are collected for subsequent laboratory analyses. The last type, snowball sampling, finds more applications in social surveys.

9.2.1.1 Judgmental Sampling

In judgmental sampling method, the choice of sampling units is based on the professional knowledge and judgment of investigators about the population under investigation. Based on the researchers' knowledge, they choose only those sampling units that are perceived to fit best for the study purpose. The representativeness of the sample for the population highly depends on the investigators' professional knowledge of the population. The results of this sampling method contain subjective factors and can be influenced by the researchers' prejudiced views. Nevertheless, judgmental sampling is commonly used in environmental monitoring because it is cost-effective, and if the investigators have profound knowledge of the population being studied, the results could be sufficiently accurate.

- Case study 1: Using this sampling method, the researcher selects 10 students that he/she thinks can provide the best representation of the 45 students in the class for interview. The selection is made based on the researcher's judgments using the criteria set for the purpose, e.g., based on the GPA (grade point average) of the students.
- Case study 2: Based on the professional knowledge of the characteristics of the industrial discharge (diurnal variations of the wastewater flow, types of pollutants and variations of pollutants concentrations in the wastewater etc.) and the characteristics of the receiving river (the hydrology, the tidal effects, discharges of other wastewaters into the same river etc.), the researcher selects the sampling locations, sampling time and frequency. For example, the researcher may select one sampling point upstream, one at the industrial wastewater discharge point, and a few points following the wastewater plume downstream in the river. The number of water samples taken per day, the sampling period, and the types of sampling etc. are decided by the researcher to provide the best possible assessment of the effects of the industrial discharge on the river quality.

9.2.1.2 Convenience Sampling

In convenience sampling, the sampling units are selected from the population only because they are conveniently available; hence the sample may not necessarily be representative of the target population. Nevertheless, it is a commonly used non-probability sampling method because of its speed, cost-effectiveness and easy implementation.

- Case study 1: The researcher selects 10 students who volunteered to participate in the survey to represent the class of 45.
- Case study 2: The researcher may select convenient locations, timing etc. for sampling. For example, the researcher takes water samples from a bridge located about 1 km downstream to characterise the wastewater plume dispersion (instead of using a boat to follow the wastewater plume in the river, etc.).

9.2.1.3 Quota Sampling

In quota sampling, researchers decide the sampling units to be included in a sample based on some quota while aiming at selecting a final sample that represents the target population. The researcher first divides the population into strata and takes a number of sampling units from each stratum for examination. The number of sampling units, i.e., the sample size, taken from a stratum is restricted by the given quota.

- Case study 1: The researcher decides to survey only 10 students in the class who should represent different GPA ranges. Accordingly, the class is divided into two strata with stratum 1 consisting of students having a GPA equal to and above the class average and stratum 2 consisting of those with a GPA below the class average. Stratum 1 shared 70% of the class while Stratum 2 shared the remaining 30%. The researcher then selects, based on the criteria set for the study, 7 students from stratum 1 and 3 students from stratum 2 for the survey.
- Case study 2: The researcher divides the river passing the industrial discharge point into three segments: upstream, around the discharge point, and downstream. Subsequently, based on the resources available, the researcher decided that the number of sampling points in the upstream to be 2, surrounding the discharge point to be 3, and in the downstream wastewater plume at different distances to be 3. Common constraints that define the "quota" in this case include the resources for sampling and the cost of laboratory analyses.

9.2.1.4 Snowball Sampling

This sampling technique is normally used in demographic surveys, such as for opinion polls or market research. The researcher first finds suitable subjects and asks them for assistance to seek similar subjects to obtain a sufficient sample size.

- Case study 1: In the first round, the researcher finds a few students who do not like the statistical course. He/she then asks these students if they know any classmates who also do not like the subject. In the second round, the researcher would pose the same question to the new members of the group and continue until no new names appear in the list. The researcher can then interview the students in the final group (who do not like the subjects) to analyse the causes, etc.

9.2.2 Probability Sampling

9.2.2.1 Simple Random Sampling

In simple random sampling, all elements of a population have the same chance of being selected; hence the inference has low bias. To obtain a random sample, all the elements of a target population are numbered; then the numbered elements are selected randomly to form a sample of a given size. The main advantages of this design include:

- Providing unbiased estimates of the mean and standard variation,
- Easy to understand,
- Data analyses are very straightforward, and
- Relatively simple calculation method for determining sample size.

Simple random sampling is most useful in social surveys when the population of interest is relatively homogeneous. For a heterogeneous population, i.e., when the sampling units are widely different from one another, this method normally requires a large sample size to ensure accuracy in characterizing the population. For environmental monitoring, simple random sampling is usually more difficult to implement than the other types of designs due to several reasons, e.g., the strong variations of the monitoring parameters (heterogeneity) in both space and time, and the cost involved when collecting and examining a large number of samples.

In principle, the random locations/sites for environmental sampling can be selected using the random generator, i.e., for x, y coordinates when dealing with two dimensions and for x, y, z coordinates when investigating a three-dimensional matrix. For example, when investigating a waste pile, a three-dimensional coordinate system (x, y, z) can first be superimposed over the waste pile, then x, y and z coordinates are randomly generated to obtain the sampling points. It is conceptually easy to generate random sampling points in both two- and three-dimensional matrices. However, in reality, collecting a sample from every randomly and precisely identified location may be difficult or impossible in some cases. Implementation of a simple random sample, therefore, can be more difficult than some other types of sampling designs, e.g., grid sampling. Generally, the simple random sampling method may be used at the final stage of a multistage design, more for surveys to collect count data than to take samples for laboratory analyses.

- Case study 1: The research calculates the required sample size to meet a given level of accuracy, the calculation methods are detailed in Section 9.4. He/she numbers the students in the class from 1 to 45, then generates a random sample of the required size from the numbered list using a random table generator.
- Case study 2: The simple random sampling method is not commonly used for the water sampling in this case, although in principle the x, y and z coordinates of the sampling locations can be randomly generated. Non-probability sampling, such as the quota sampling just discussed, normally yields better results and is more cost-effective.

9.2.2.2 Systematic Sampling: Grid Sampling

In systematic sampling, the sampling units are taken from a population according to a plan, at predetermined intervals in time and/or in space. The systematic sampling design provides a practical and easy method for defining sample locations and ensures the uniform coverage of the area or process under the study. The initial sampling unit is selected at random, but the remaining sampling units are specified so that they follow a regular pattern. Therefore, the results may contain some bias, and a large sample size is required to produce representative results.

- Case study 1: The researcher has the list of students in the class and randomly selects the first student (of any number from 1–45) and then selects the other students at a sampling interval of 5 (every fifth from the first selected) for the survey.
- Case study 2: The researcher can select a starting sampling point upstream of the industrial wastewater discharge point into the river and take samples at downstream locations along the discharge plume, say at every 200 m. Note that in practice, a large number of samples are required, especially near the discharge point when the plume size is still relatively small. To check the temporal variations of water quality, the researcher can take samples or make measurements at a fixed time interval with the initial time of the measurements selected randomly. For example, the measurements are done every hour to determine the diurnal variations of the monitoring parameters.

Grid sampling is a specific type of systematic sampling method, which selects the sampling units using a grid, normally triangular or rectangular. The initial sampling unit is chosen randomly, e.g., by a random number generator. In practice, grid sampling is commonly used for sampling on land/soil surfaces, in conducting forest inventory or agricultural research (USEPA, 2002). The grid sampling method is useful in determining the spatial distribution of the monitoring parameters, for example, to examine the spatial distribution and find the hotspots of chlorinated pesticide contamination in the rice paddy soil. A simple systematic sampling method, on the other hand, is useful for measuring pollutant concentrations at a fixed location of flowing media to reveal the temporal variations in the pollutant concentrations.

9.2.2.3 Stratified Sampling

In stratified sampling, in the first step, the target population is segregated into a number of non-overlapping strata or sub-populations, with each known or thought to be more homogeneous in terms of the characteristics of interest. Further, in the second step, random sampling is applied to take a sample from each stratum. The strata may be chosen based on preexisting information or professional judgment about the population. This design is useful when the target population is heterogeneous, e.g., smoker vs. non-smoker population, or the higher air pollution levels at the roadside vs. lower levels at general ambient sites in a city. The advantage of this sampling design is the lower sampling burden while having the potential for achieving greater accuracy in estimates of the mean and variance than simple random sampling. Further, because random sampling is applied for each stratum, the method allows the computation of statistical estimates for sub-populations of interest.

In this method, it is important to have an optimum segregation of the population into strata. The number of strata is selected in such a way to maximise the difference (maximum variance) between the strata while minimising the heterogeneity (minimum variance) within each stratum. This sampling method is easy to understand and easy to execute and hence is commonly used in social survey studies.

- Case study 1: Like the case of quota sampling, two strata are made for the class of 45 based on the GPAs of the students. However, the researcher will use random sampling in the survey of each stratum which is performed as in the previous case of random sampling.
- Case study 2: Theoretically, the river segment under study may be divided into three strata: upstream, near the discharge point and downstream. Subsequently, simple random sampling may be applied for each stratum. However, for the same reasons as previously mentioned, random sampling is not commonly used in this stage. Instead, a non-probability sampling method is normally applied.

As was seen, the stratified and quota sampling designs are similar in the first step when the population is divided into sub-populations. They are different in the second step when the sampling units are selected to make up a sample. Specifically, in this step, stratified sampling follows the random sampling method, whereas quota sampling uses the availability sampling method. As a result, the stratified sampling can minimise the bias, and the error can be estimated using statistical analyses, which is not the case for quota sampling. Nevertheless, quota sampling is more cost-effective, and with sufficient knowledge of the population, satisfactory results can also be obtained.

9.2.2.4 Cluster Sampling

In the cluster sampling method, in the first step, the population is divided into groups or clusters that are intended to be homogeneous regarding the monitoring parameters. In the second step, researchers select clusters using a simple random or systematic sampling method for data collection. The accuracy of the method increases if there is less variation between the clusters (more homogeneity) but more variation within each cluster (more heterogeneity). This sampling method therefore is usually applied when dealing with a large population that can be split into sub-populations (clusters), and each of them would have almost similar characteristics as the target population. Cluster sampling design may be single-stage or multistage. In a single-stage design, a few clusters are selected by random sampling, and the surveys are performed on the entire selected clusters. Details on multistage sampling designs are further provided in Section 9.3.

Cluster sampling is often used in environmental survey studies. The implementation is straightforward. It is particularly useful when the sampling frame cannot be extended to cover the whole target population; i.e., some parts of the population may not be accessible

TABLE 9.1

Comparison between Probability and Judgmental Sampling Designs

	Probability sampling	Non-probability sampling
Advantages	Having low sampling bias, as every sampling unit has an equal chance of getting selected Allowing calculation of uncertainty associated with estimates Allowing making statistical inference Can handle decision error criteria	Can be less expensive than probability sampling designs Can be efficient when investigators have profound knowledge of the sampling population Easy to implement
Disadvantages	Difficult to define random geographical locations for sampling Requiring accurate conceptual model to produce an optimal design	Sampling bias is not considered as the units are selected based on judgment Errors of estimates cannot be reliably evaluated Data interpretation relative to study objectives depends on personal judgment

Source: Adapted from USEPA (2002).

at the time of the survey. It consumes less time and resources, yet the results may be sufficiently accurate if a successful division of the population into homogeneous clusters could be made so that every cluster contains the characteristics of the target population. Otherwise, this method can lead to unrepresentative sampling because a large part of the population may not be selected for sampling.

Note that both stratified and cluster sampling methods involve a two-step procedure. The key difference between the stratified and the cluster sampling methods is in the first step, i.e., in the way the population is divided into sub-populations. In stratified sampling, it is important to have large differences (large heterogeneity) between the strata, e.g., male vs. female or smokers vs. non-smokers, youth vs. elderly, while maximising the homogeneity within each stratum. For cluster sampling, it is the opposite: the homogeneity between clusters should be maximised, but the heterogeneity within each cluster should be the same as, or close to, the total target population.

A comparison between probability and non-probability sampling approaches is presented in Table 9.1. It is noted that the advantages of one approach are often the disadvantages of the other. Overall, each sampling method has its advantages and disadvantages. Therefore, a combination of sampling methods in the multistage sampling design could provide an optimal solution. The multistage sampling design allows the use of different methods in different sampling stages and is further detailed in Section 9.3.

9.3 Multistage Sampling Design

An example of a two-stage sampling design of eggs at a supermarket for the PCB study was discussed in Section 9.1. In the first stage, the researcher selected 20 boxes (out of 200 boxes of eggs), and in the second stage, they took 3 eggs from each box (containing 10 eggs) for the measurements. That makes the sample size of the first stage of 20 boxes and the final sample size of 60 eggs.

In the multistage sampling design, different sampling methods may be used in different sampling stages. When only the probability sampling methods are used in all the stages, the design is multistage probability sampling. If the methods used in all stages are judgmental, the designs are multistage non-probability sampling. When both probability and non-probability methods are used, the design becomes multistage mixed sampling (Figure 9.1). The sampling units in the first stage are the primary sampling units; those in the second stage are the secondary sampling units, and so on. The smallest sampling units in the final stage are the ultimate sampling units used for data collection, i.e., for PCB measurement in this example.

9.3.1 Multistage Probability Sampling

Multistage random sampling is commonly used, particularly in large-scale population-based surveys. Two types of multistage probability sampling are discussed here: multistage random sampling and multistage cluster sampling. In multistage random sampling, researchers use the different random sampling methods described in Section 9.2.2 to define the sampling units at each stage until they get the final sampling units for data collection. In the multistage cluster sampling, the study population is divided into clusters, which are further segregated into smaller and smaller sub-clusters at each of the following stages. At the last stage, investigators only select a sample containing some members of each sub-cluster for data collection. A multistage sampling design involving a combination of cluster and stratified sampling methods is also useful for large-scale surveys to help ensure that the sampling units are representative of the larger population. Case study 3 below illustrates different multistage probability sampling designs.

Case study 3: Investigators aim to find out the amounts and types of fuels and stoves used for domestic cooking in the rural areas of a developing country. A typical province is selected for the survey. There are 300 rural communes in the province, and a preliminary study shows that these communes are quite similar in social, demographic and economic conditions, i.e., having a common range of income, similar accessibility to cooking fuels etc., and each has about 400 households. Thus these 300 communes may be considered homogeneous clusters.

- Single-stage cluster sampling: The researchers use systematic random sampling to select 20 communes (a sample size of 20) from 300 communes (primary sampling units). They then conduct the survey for the households (secondary sampling units) in the selected communes.
- Multistage cluster sampling: In the first stage, 20 communes (clusters) are selected. Each commune consists of several villages (10–15 villages) which are considered the sampling units in the second stage, and the systematic sampling method is applied to select 3 villages (clusters) from each commune for the survey. In the third stage, each selected village is segregated into a number of household groups (clusters), and random sampling can be used to select the groups of households for data collection. In this design, it is important to ensure that each small cluster in the following stage should be sufficiently diverse internally to represent the characteristics of the target population.
- Multistage random sampling: Stratified or other random sampling methods may be used in certain sampling stages. For example, each village selected in the second sampling stage contains several social groups, e.g., agricultural farmers, aquaculture farmers, fishers and businesspeople. These social groups can be considered strata in the third sampling stage, and simple random sampling is applied to select the households (ultimate sampling units) from the strata. Using the stratified sampling at this stage, the investigators want to ensure that all these social groups are represented in the sample.

9.3.2 Multistage Mixed Sampling Designs

Multistage mixed designs apply either a probability or a non-probability sampling method in a stage based on what is appropriate. The choice of the sampling method for each stage is subjective; hence investigators need clear reasoning for the decision in choosing the most appropriate. Note that the use of probability sampling methods allows for stronger statistical inferences but may be more resource-demanding.

An example of a multistage mixed sampling design is presented in a study by Huy et al. (2020) who reported a survey conducted to collect data on amounts and types of fuel used for domestic cooking in a Metropolitan Region (HMR) of Southeast Asia. At the time of the study, the region had a population of more than 7 million people living in 29 districts (including 10 urban, 5 suburban and 14 rural districts). The districts were further administratively divided into wards (for urban and suburban districts) and communes (for rural districts), and HMR had about 600 wards and communes in total. In the first stage, the quota sampling was applied with the HMR population divided into 6 strata: urban old district (U1), urban new district (U2), suburban with less agricultural activities (S1), suburban with more agricultural activities (S2), rural with more industrial activities (R1) and rural with less industrial activities (R2). Each stratum contained a number of homogeneous wards/communes, which were considered as the sampling units, and a sample including a number of wards/communes was collected from each stratum based on a given quota. In the second stage, simple random sampling was used to select the households (the ultimate sampling units) from each selected ward/commune for data collection by questionnaire.

Multistage mixed design is commonly used for sampling the environmental matrices such as soil, air, water, solid waste and biota, especially when the "environmental population" or the site is highly stratified. For example, when a location in a soil plot is suspected of containing high levels of heavy metals, then the area containing the location is identified as a stratum (boundary is decided subjectively) for which judgmental sampling design is applied to collect the soil samples for analysis.

The adaptive cluster and ranked set sampling designs are examples of multistage mixed sampling designs, which are detailed in USEPA (2002), and are further discussed in this section.

9.3.2.1 Adaptive Cluster Sampling Design

This sampling method allows additional units or sites to be selected depending on the interpretation of observations made during the survey. In the initial stage, the simple random sampling method is used to select the sampling units from a studied environmental matrix, such as a soil plot in Figure 9.2, which is subjected to investigation for heavy metal contamination. In the following stage, additional samples are taken in the surrounding locations where measurements done at the initial stage show considerable pollutant levels, e.g., exceeding a given threshold decided by the investigators. Additional rounds of sampling are done until the adjacent locations of the sampling units of the preceding stage show the pollutant levels below the threshold. In this way, adaptive cluster sampling may be useful in defining the borders of a pollution plume.

Generally, this sampling method is useful for estimating or searching for sporadic but highly aggregated characteristics in a population, e.g. the shaded areas in Figure 9.2. For this pattern of contamination distribution, the traditional statistical approach has problems because the contamination is negligible (non-detectable) over most areas of the soil plot. Random sampling may miss most of the spots of higher concentration. Possible environmental applications of this method include soil pollution studies, hazardous waste site characterisations and determining the extent of occurrence of effects of a source of harmful air pollutants on nearby flora and fauna (USEPA, 2002).

Adaptive cluster sampling concentrates efforts in the areas of greater interest, e.g., the areas with high contamination levels in the hotspot investigation; hence it would require fewer resources than simple random sampling. There are, however, several limitations of this sampling design, for example, the iterative procedure of the process of sampling, testing, resampling and testing may take considerable time. If quick and inexpensive field measurements are not readily available, the total monitoring costs could increase substantially during this process. Because the sampling process stops only when no more units are found to contain the characteristics of interest, the final overall sample size is not known. If the assumption on

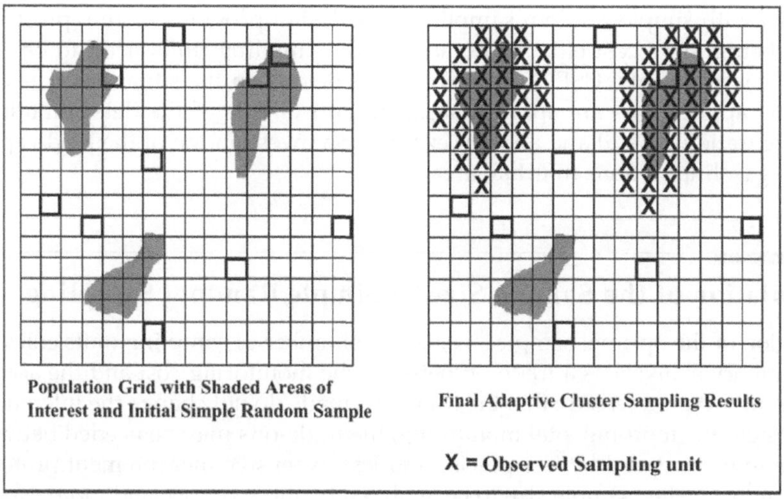

FIGURE 9.2 Illustration of adaptive cluster sampling.

Source: USEPA (2002).

the sporadic presence of the pollutants of interest is not valid and the contamination is more widespread, the follow-up sample size will be large resulting in a much higher total cost. There is also a risk that the contaminated areas are omitted in the process, for example, the third darkened area (contaminated area at the bottom left) in Figure 9.2 could not be detected.

9.3.2.2 Ranked Set Sampling

Ranked set sampling is a two-stage sampling design that combines simple random sampling with professional judgment to identify where in the field a detailed investigation is needed. Ranked-set sampling is a useful and cost-effective way to obtain estimates of the average characteristics of environmental matrices (USEPA, 2002). In the first stage, some sets of field locations are identified using simple random sampling. The locations are then ranked independently within each set using professional judgment and/or field screening measurement results deploying fast, inexpensive or surrogate measurement methods. One sampling unit from each set is then selected (based on the observed ranks) for subsequent measurement for the contaminant of interest using a more accurate method, normally conducted in the laboratory.

Typically, professional judgment is applied to visually assessed locations/sampling units and ranks them from the smallest to largest concerning the characteristic of interest. For example, such visual assessment can be based on the soil colour or vegetation density/height in each plot of a study field. Alternatively, onsite inexpensive measurements, if available, can be used for the ranking. Remote sensing information (aerial photographs, satellite products etc.) with large spatial coverage is especially suitable to identify sampling locations in a large study area.

The use of ranked set sampling increases the representativeness of measurements as they cover the range of low, medium and high values in the population, hence providing better estimates of the mean. Ranked set sampling design is applicable when the cost of locating and ranking locations (sampling units) in the field is low compared to the final laboratory measurements. If surrogate or rapid field measurements are used for ranking, the results should be strongly correlated with those produced by the more expensive laboratory method used to produce the final results.

Compared with simple random sampling, this design provides more representative samples while being more cost-efficient because fewer sampling units need to be collected for laboratory measurements. USEPA (2002) provides details on the situations when the ranked set sampling applications are appropriate, with an example of the determination of heavy metal concentrations in soil and another example on investigation of the Reid vapor pressure (RVP) from gasoline stations in urban areas.

9.4 Calculation of the Sample Size in Simple Random Sampling

Determination of the optimal sample size, i.e., the number of sampling units collected for survey or laboratory analysis, is a trade-off between the monitoring cost and the accuracy of the results. It is important to ensure that the trade-offs made do not change the inferences from the planned design. In environmental monitoring, the trade-offs may be needed between a larger sample size to be analysed by less precise and less expensive measurement protocols, on the one hand, and a smaller sample size to be analysed by more precise and more expensive measurement protocols, on the other. Formulae are available for estimating sample sizes needed for different sampling designs. This section focuses only on the calculation of the sample size in simple random sampling, which is commonly used in social surveys.

In this case, the sample size n is defined as the number of sampling units that are randomly selected to represent the population of size N, i.e., consisting of N sampling units, with a given margin of error. The appropriate sample size depends on the desired level of precision, the level of confidence, and the degree of variability in the attributes being measured in the population (Miaoulis and Michener, 1976). The confidence level or risk level is based on the central limit theorem. When a population is repeatedly sampled, the values obtained by those samples are distributed normally around the true value, and their average is equal to the true value. In a normal distribution, approximately 95% of the sample values are within two standard deviations of the true value. Regarding the degree of variability, a heterogeneous population with a large variance of the monitoring parameters will need a larger sample size to get the desired level of precision as compared to a more homogeneous population.

9.4.1 Sample Size for an Infinite Population

To estimate the proportion of elements that possess a specific characteristic (such as the proportion of individuals exhibiting a particular symptom in an epidemiological study) with a desired level of precision for a large population, the required sample size (such as the number of subjects in an epidemiological study) can be determined using the sample size calculation formula. For sufficiently large or infinite populations that do not require finite corrections, Cochran (1963) developed a formula (Equation [9–1]) to calculate the sample size for the proportions.

$$n_0 = \frac{Z^2 pq}{e^2} \tag{9–1}$$

where n_o is the sample size, Z is the Z score, e is the desired level of precision in portion, i.e., 0.1 for 10%, p is the estimated proportion of the study attribute that is present in the population, and q is $1 - p$.

Z score is the area under the normal curve which can be easily calculated or readily found in statistical tables. Z score increases with the increase in the confidence level: a confidence level of 80% corresponds with $Z = 1.28$; 90% with $Z = 1.65$; 95% with $Z = 1.96$; and 99% with $Z = 2.58$, and so on.

The value of p is unknown in most cases. Since the product of $p*q$ is the largest when $p=q=0.5$, to obtain the conservative result with the largest sample size, Equation (9–2) is used.

$$n_0 = \frac{Z^2}{4e^2} \tag{9–2}$$

Examples of the sample size calculated using Equation (9–2) (for infinite populations) are illustrated next. Note that the sample size is rounded to the nearest integer.

- Precision of 5%, confidence level of 95%: $n_0 = (1.96)^2/(4*0.05^2) = 384$
- Precision of 10% (less precise), confidence level of 95%: $n_0 = (1.96)^2/(4*0.10^2) = 96$
- Precision of 10%, confidence level of 99%: $n_0 = (2.58)^2/(4*0.10^2) = 166$

As seen here, a larger sample size is required to obtain data with a higher precision and a higher confidence level.

9.4.2 Sample Size for a Finite Population

For finite populations, i.e. those have a finite number of elements or events discussed in Section 9.1, finite population corrections are required. For a random sample of size n_o from an infinite population with variance σ^2, the variance of the sample mean $V(\overline{x})$ is σ^2/n_o. However, when samples are drawn from a finite population of size N with variance σ^2, the variance includes a finite population corrections factor expressed by Equation 9–3.

$$V(\overline{x}) = \frac{N - n_o}{N - 1} \times \frac{\sigma^2}{n_o} \tag{9–3}$$

When the population size N is sufficiently large, Cochran (1963) recommends that the denominator of the finite correction factor be N rather than $N - 1$. For the sample size calculation, when the population is finite with a population size N, Israel (2012) proposed Equation (9–4), which determines the adjusted sample size n for a finite population N from the size n_0, estimated using Equation 9–1 or 9–2, for an infinite population.

$$n = \frac{n_o}{1 + \frac{(n_o - 1)}{N}} \tag{9–4}$$

Examples of the sample size calculated for a desired precision of 5% and a confidence level of 95%, but for different N, using Equation (9–4), are provided next. Note that $n_0 = 384$ was obtained by Equation (9–2).

- $N = 5000$: $n = (384)/(1+383/5000) \sim 357$
- $N = 10{,}000$: $n = (384)/(1+383/10000) \sim 370$
- $N = 15{,}000$: $n = (384)/(1+383/15000) \sim 374$

There is only a small increase in the sample size when N increases from 10,000 to 15,000. In general, with N above 10,000, the sample size becomes stable and approaches the result for the infinite population ($n_0 = 384$).

Yamane (1967) provides a simplified formula to calculate the sample size for finite populations in the case of $p=q= 0.5$ and a confidence level of 95% (Equation 9–5).

$$n = \frac{N}{1 + N(e)^2} \tag{9–5}$$

Equation (9–4) is used to construct the sample size tables (Yamane's tables) for given combinations of precision, confidence levels, and degree of variability, which are readily available in statistical textbooks or from internet sources.

For illustration, the results of the (rounded) sample size calculated using Equation (9–5) for a desired precision of 5% and a confidence level of 95%, but for different N are:

- N=5000, n= $(5000)/[(1+5000(0.05)^2] \sim 370$
- N=10,000, n= $(10000)/[(1+10000(0.05)^2] \sim 385$
- N=100,000, n=398

Note that the results obtained by Equation (9–5) are slightly higher than those obtained by Equation (9–4). For $N \geq 10{,}000$, the sample sizes produced by Equation (9–5) are higher than the sample size calculated for an infinite population ($n_0 = 384$).

It is important to emphasise that the sample size in a social survey defines the number of responses obtained after the survey; hence in planning a larger size, i.e., by adding a few percent (5–10%) to the theoretically calculated number, is normally used to compensate for potential non-responses. Further, the calculated sample sizes using all the preceding formulas and those presented in the sample size tables assume that the attributes being measured have a normal or nearly normal distribution. If this assumption cannot be met, then the entire population may need to be surveyed. This is true also for small populations when a high precision ($e \leq 5\%$) and a high confidence level (~99%) are required. In reality, it is difficult to define the margin of error, and the monitoring attributes do not necessarily have a normal distribution. Therefore, it is important to investigate a larger number of sampling units, as long as resources allow, to gain the characteristics of the population. In addition, in a survey study, a pretest of questionnaires is useful before making it operational to obtain the desired information to meet the study objectives.

9.5 Application of Sampling Designs for Environmental Measurements

Environmental data include any measurements or information that describe environmental processes, locations, or conditions; ecological or health effects consequences; or the performance of environmental technologies. Environmental data are produced by recording online, by laboratory measurements and modelling, or by observations. This section focuses on the numerical continuous data generated by environmental measurements: for example, levels of $PM_{2.5}$ in the air, COD in a wastewater stream or pesticides in soil. Other types of numerical data, such as discrete data by counting, and rating scale data from questionnaire surveys are addressed in Part III.

9.5.1 Specific Considerations for Representative Sampling

The sampling design for the environmental measurements is a description of the sample collection plan that specifies the number, type, and positions (spatial) and timings (temporal) of sampling units to be selected for measurements with the aim of obtaining data of required DQOs. For less dynamic matrices such as soil or biota, the sampling units may be designated by longitude and latitude. For more dynamic matrices such as water and air, besides the spatial coordinates, the exact timing of the sampling should also be indicated. In addition, for air pollutants the integrating or averaging period over which the sample collection occurs is important to specify. For example, the sampling protocol for 24-h $PM_{2.5}$ should indicate the start and stop time accordingly. A well-developed sampling design has a critical role in ensuring that representative data are collected with a minimum expenditure of resources while providing sufficient information to draw conclusions needed for effective environmental quality management. The representativeness here means that the obtained data accurately represent a characteristic/attribute of the monitored environmental matrix, its variations in space and time and the variation with environmental conditions, e.g., dry vs. rainy weather.

It is important to note that, in environmental monitoring, the sampling design is not the only important component, and the measurement protocols should also be considered. A measurement protocol is a specific procedure for making observations or performing analyses to determine the characteristics of interest in the study matrix. "The measurement protocol would include the procedures for collecting a physical sample, handling and preparing the physical sample, applying an analytical method (equipment, measurement error) to obtain the measurement data" (USEPA, 2002). Nevertheless, it is important to note that non-representative samples will produce unreliable conclusions on the environmental state regardless of how accurate and expensive methods of sample analysis are applied. Also, the chance of resampling may not exist, for example, if a water sample from a flowing river taken at a given location and at a given time is lost or wrongly handled, there is no chance of repeating it.

A census (100%) sampling is practically impossible in environmental studies, due to the large volume of the environmental matrices under examination, such as the volume of water in a lake, water in a flowing river or continuous emissions of air pollution from an industrial stack. Examining a very large number of sampling units is highly desirable but would induce a very high monitoring cost. Hence the characteristics of an environment are commonly obtained based only on a finite number of sampling units with the aim of generating representative data in the most cost-effective way. A good sampling design is therefore required, which needs multidisciplinary inputs from chemists, experts with profound knowledge of the sampled environmental matrices (air, water, sediment, biota etc.), risk assessment experts, statisticians and data management experts to ensure the representative data are obtained and used effectively.

In environmental sampling design, the use of conceptual models may greatly enhance the selection of appropriate sampling sites and times. A conceptual model describes the expected sources of the contaminant and the area of concern, identifies its fate and transport pathways in relevant environmental matrices and defines the potential exposure pathways (USEPA, 2002). The model can also identify potential factors causing the variability among sampling units that lead to the variability in the measurement data. For example, in the monitoring of dynamic matrices such as water and air, the conceptual model is important to show the dispersion patterns from a source, such as emission from an industrial stack in the ambient air. In industrial wastewater sampling, a detailed flowchart of the production process showing the flow of materials and wastewater is important for selecting sampling points and pollutants of importance for monitoring.

Several considerations in the environmental sampling are listed here:

- ***Population:*** The concept of environmental population is normally not well-defined, unlike in social surveys. For dynamic environmental matrices, such as water and air, the population should be defined with both spatial and temporal dimensions.
- ***Sampled population:*** The sampling frame may be very different from the target population due to the limited accessibility to a large part of the sampling units. For example, not all the air at the breathing level in a city is accessible for sampling or measurements. Therefore, the selection of the right sampling sites and times is crucial to represent the target population.
- ***Sampling unit:*** In some environmental studies, the sampling units may be distinct and well-defined, such as individual trees, animals, or drums of waste material because they are discrete. However, in sampling other solid or fluid matrices such as soil, water and air, such discrete sampling units are not available; hence the units are arbitrarily defined by investigators. It is important to define the units that are appropriate to make a representative sample of the matrix material: for example, an amount of surface soil or water at given sites at specific times to be taken as samples.

- ***Environmental sample:*** An environmental sample normally refers to the amount or volume extracted from the matrix that is subjected to the measurement protocols. Types of environmental samples, grab vs. composite, are discussed in Section 9.5.2. The term "statistical sample size" is not normally used; instead, it is normally referred to as the "number of samples", which may be equivalently interpreted as the "number of sampling units" taken for measurements.
- ***Measurement protocol:*** The environmental samples are collected for subsequent physical, chemical and/or biological analyses; hence analysis methods should be planned together with sampling. For example, if a water sample is used for multiple measurements, it is important to collect a sufficient amount for all intended analyses.
- ***Sample collection technique:*** A manual or automatic technique can be used that varies with sample type, as detailed in Section 9.5.2.
- ***Sample handling and preservation:*** Onsite measurements are preferable but not possible for many parameters. Rapid field measurements are used for unstable parameters that are difficult to preserve, such as temperature, conductivity, DO and pH of water. For those parameters that cannot be measured onsite, suitable preservation methods are used to retard physical, chemical and biological processes to minimise the changes in the composition of environmental samples after collection. Several parameters require specific preservation methods, and the preservatives used for one parameter may interfere with the analytical results of others. Therefore, a sample (water, soil or biota) may be split into different containers and preserved separately.
- ***QC samples:*** These are additional samples used for quality control of the data. Blank samples and spiked samples are used to check the potential errors (loss and/or gain of the analytes) during the sampling, transportation, storage, and analysis. QC samples should be handled in the same way as environmental samples.

9.5.2 Types of Environmental Samples and Collection Techniques

Samples may be classified by the collection and preparation techniques as manual or automatic, grab or composite and passive or active. Different types of samples require different sampling collection techniques, which also depend on the target pollutants to be measured.

9.5.2.1 Broad Categories of Environmental Sampling Techniques

9.5.2.1.1 Active Sampling Technique

In this technique, a pumping device is used to take a sample over a short period and produces the data with an average time between less than 1 h to a few days. Depending on the target pollutants and purpose of sampling, the devices are designed to collect bulk matrix samples or selective samples, i.e., to retain only specific compounds. For example, in air pollution sampling, the bulk air samples are collected in a sampling bag, syringe or canister, which takes the actual air mixture at the site in the sample. On the other hand, selective air sampling retains the pollutants of interest on selected sampling media, such as using filter paper to retain the particulate matter (PM) or an activated carbon sorbent trap to retain volatile organic compounds (VOCs). The amount of pollutants retained is quantified in the laboratory, and the sampled air volume is calculated based on the pumping rate to subsequently determine the concentrations.

9.5.2.1.2 Passive Sampling Technique

A sample is collected when no moving device, such as a pump or evacuated container, is used to move the matrix material. Instead, the pollutants are collected based on physical principles such as diffusion, e.g., of target gases in the diffusive samplers, or settling of particles on a surrogate collecting surface. Passive sampling normally collects a sample over a longer period, e.g., 2–3 weeks to months. In the passive sampling, the volume of air or water passed through the collector is not readily known and hence should be calculated using the matrix conditions.

9.5.2.1.3 Manual and Automatic Sampling

Manual sampling is normally used for relatively static matrices such as soil, biota, sediment etc. For more dynamic fluid matrices such as water or air, either a manual or automatic sampling technique is used depending on the data quality objectives. Manual sampling is simple but labour-intensive when conducted over an extended time. Also, it may not be practical for sampling at a high frequency, for sampling unpredictable events or for detection of illegal discharges. Automatic sampling can provide samples with a high frequency; hence it is preferable when time-series of data are required. However, it is complex and costly and may not be possible to deploy at all sampling locations.

9.5.2.2 Grab vs Composite Sampling

Two common types of environmental samples include grab samples and composite samples. These samples can be collected using either manual or automatic sampling techniques or a combination of both.

9.5.2.2.1 Single Grab Sample

The term "grab" indicates a discrete environmental sample that is grabbed from a point in space (location) at a certain point in time. A grab sample is used to characterise the instant level of an environmental parameter that occurs at a single moment and at a given place. For example, 1 kg of soil from a paddy plot, 1 L of water from a location in a river or 1 m^3 of air taken from a roadside in a city at a given moment in time. For highly dynamic environmental matrices with large variations of monitoring parameters in time and space, a single grab sample is not sufficient to provide representative data, and more samples need to be collected.

A single grab sample is normally collected manually. Here are a few examples of taking grab samples of environmental matrices.

- ***Grab water sampling:*** A bucket is used to collect a grab surface water sample, or a specially designed subsurface grab sampler is used to collect samples at a certain depth.
- ***Grab air sampling:*** The technique is used to take samples for measurements of gaseous pollutants. The samples can be taken using an evacuated container or a large air sampling syringe, which can be momently filled with the sampled air at the sampling site. Sampling for reactive VOCs has specific requirements, which are listed in the USEPA Method TO15 (USEPA, 1999).
- ***Grab sampling for soil and sediment:*** Samples at a certain depth are commonly collected using a core sampler (the area of collection is defined by the core size). To collect a sediment sample over a larger area, a bottom grab sampler may be used.

9.5.2.2.2 Series of Discrete Samples

"Discrete sample" is a term usually applied to a sample that is kept in a separate container but is part of a series of samples collected consistently. Such a series of samples is used to characterise the fluctuations of the monitoring parameters in time or space. For example, a series of grab samples collected manually on an hourly basis from an industrial wastewater stream during the day are used to reveal the diurnal variations of wastewater characteristics.

Discrete samples forming a series may be either grab or composite samples. For example, an automatic sampler is programmed to pump a grab sample from an industrial wastewater stream into a separate container at 1-h intervals. If the sampler is programmed to pump the wastewater over a period, say 20 min, into a container on an hourly basis, a series of composite samples is obtained. If the pumping rate is constant, the discrete samples are time-composite. If the pumping rate is made proportional to the flow-rate of the wastewater stream, the discrete samples are flow-composite.

9.5.2.2.3 Composite Sample and Compositing Protocol

A composite sample may be made up of discrete samples from different locations or at different times. In composite sampling, individual samples are pooled in one container and mixed to form a single homogeneous sample for analysis. For example, a daily composite sample of water from a lake is made up of hourly grab samples collected during the day for analysis of water-soluble inorganic ions. Such composite samples provide average characteristics of the monitoring parameters but not the variability in the levels among the discrete samples.

If composite samples are prepared for analysis, the measurement protocol should also include a composite sampling protocol, which describes the compositing procedures and indicates how many composite samples are to be formed, how many samples, and which samples are used to form each composite. There should be no potential biases (for example, loss of volatile components) generated by the compositing process itself.

For samples taken from more static matrices, such as soil samples for pesticides or dioxin measurements, manual composite sampling is used, whereas for more dynamic fluid matrices, such as water, either manual or automatic composting techniques may be used. Some advantages and disadvantages of composite samples are given here:

- Compositing helps reduce the number of samples subjected to laboratory analyses; hence it is cost-effective when the analysis cost is much higher than the sampling cost. For example, due to the high cost of dioxins analysis, when monitoring dioxins contamination in a large soil plot, a composite of several discrete samples collected from the plot may be prepared and analysed. If dioxins are detected in the composite sample, the discrete samples may be analysed individually.
- Compositing may lead to a false negative error. In the example of dioxins monitoring, if the compounds are present only in a few discrete samples, compositing may dilute it to a level below the detection limits of the analytical method, which leads to a false negative error.
- Compositing may change the sample composition. When individual samples contain mutually destructive compositions and/or organisms, the compositing may change the sample compositions that cause errors in the monitoring data.

In reality, compositing is commonly used in air pollution sampling. For example, the $PM_{2.5}$ sampling for ambient air is, in fact, an automatic time-composting procedure when a pump is operated at a constant flow-rate to draw air through a filter over the averaging period, normally 24h. The retained particles on the filter present a 24h composite sample. For stack gas sampling, USEPA method 5 (USEPA, 2020) recommends the use of isokinetic sampling to collect representative PM samples. The method requires no change in the momentum of collected particles; hence an automatic stack gas sampler is programmed to withdraw gas into the sample with the same velocity as the stack gas. Because the stack gas velocity is not uniformly distributed over the cross-sectional stack area, several traverse points (points to place the probe for sampling) are used to collect a representative sample. At each traverse point, the isokinetic sampling is conducted over a period, say about 15 min, and the sampled PM are pooled together on a single filter for subsequent laboratory analyses of mass and chemical compositions.

In water sampling, composite sampling may be done either manually or automatically. If a series of discrete grab samples are taken periodically, a time-composite sample can be prepared manually by combining the equal volume of aliquots taken from these discrete samples. This is equivalent to pumping the water at a constant rate into a single container.

For industrial wastewater when both compositions and the flow-rate strongly fluctuate in time, the flow-composite sample is recommended (Kim Oanh and Bengtsson, 1996). Manually, the flow-composite sampling can be done by taking the amount of aliquot from each discrete sample proportional to the flow-rate at the sampling time. Automatically, the pumping rate can be programmed to be proportional to the wastewater flow-rate measured online. Alternatively, intermittent pumping may be used, which is done at a constant rate but with the pumping period proportional to the wastewater flow-rate. Overall, the automatic flow-composite sampling requires information on the instant flow-rates. It is therefore more complicated but would provide more representative data of the waste stream (ISO 5667-11, 2009). More details on the use of compositing methods for industrial wastewater are provided in Chapters 10 and 11.

Further, in clinical laboratories, composite samples are useful in screening tests for widely transmitted diseases, such as COVID-19, in a community. In this case, a composite sample collected from every member of a family may be tested, and, if the result is positive, individual samples will be tested.

9.6 Summary

Sampling designs for environmental measurements can rely on non-probability, probability sampling methods or a combination of both. Each sampling approach has its advantages and disadvantages. Multistage mixed sampling designs are commonly used, which deploy both probability and non-probability sampling methods, varying with the sampling stage. Simple random sampling is normally used in one of the stages, e.g., in the first step (ranked set sampling, adaptive cluster sampling) or in the final stage when individual units are selected for survey. A good understanding of the environmental matrices under study is a prerequisite for developing a good sampling design.

In environmental studies, it is important to emphasise that the sampling design is not the only important component. A measurement protocol should be planned together with the sampling design to provide representative data to achieve the monitoring objectives. The use of a concept model such as a process flowchart or a dispersion pathway of contaminants in the environment may greatly help in selecting the right locations and times for collecting representative samples of different environmental matrices.

REFERENCES

Cochran, W.G., 1963, *Sampling techniques*, 2nd edition, John Wiley and Sons Inc., New York, United States, ISBN 978-0471162384

Dodge, Y., 2003, *The Oxford dictionary of statistical terms*, Oxford University Press Inc., New York, United States, ISBN 978-0-19-850994-3

Huy, L.N., Kim Oanh, N.T., Phuc, N.H. and Nhung, C.P., 2020, Survey-based inventory for atmospheric emissions from residential combustion in Vietnam, *Environmental Science and Pollution Research*, 28, 10678–10695, https://doi.org/10.1007/s11356-020-11067-6

International Standard Organization (ISO), 2009, *ISO 5667–11:2009, water quality – sampling – part 11: Guidance on sampling of ground waters*, published by the International Standard Organization. https://standards.globalspec.com/std/14348437/iso-5667-1

Israel, G.D., 2012, *Determining sample size, one of a series of the agricultural education and communication department*, Florida Cooperative Extension Service, Institute of Food and Agricultural Sciences, University of Florida, United States.

Kim Oanh, N.T. and Bengtsson, B-E., 1996, Design of a representative and cost-effective sampling programs for industrial wastewater with examples from bleached Kraft pulp and paper mills, *Resources, Conservation and Recycling*, 18(1–4), 107–124.

Miaoulis, G. and Michener, R.D., 1976, *An introduction to sampling*, Kendall/Hunt Publishing Company, Dubuque, Iowa, ISBN 978-0840316035

U.S. EPA, 1999, *Compendium method TO-15: Determination of volatile organic compounds (VOCs) in air collected in specially-prepared canisters and analyzed by gas chromatography mass spectrometry (GC/MS)*, U.S. Environmental Protection Agency, retrieved on 25 September 2023 from https://www.epa.gov/sites/default/files/2019-11/documents/to-15r.pdf

U.S. EPA, 2002, Guidance for choosing a sampling design for environmental data collection (EPA QA/G-5S), EPA/240/R-02/005, U.S. Environmental Protection Agency, retrieved on 25 September 2023 from https://www.epa.gov/sites/default/files/2015-06/documents/g5s-final.pdf

U.S. EPA, 2020, *Method 5 determination of particulate matter emissions from stationary sources, 40 CFR part 60, Appendix A, methods 1–5*, U.S. Environmental Protection Agency.

Yamane, T., 1967, *Statistics: An introductory analysis*, 2nd edition, Harper and Row, New York, United States.

10

Industrial Pollution Monitoring

Pollution from the industrial sector, if not properly managed, would cause serious harmful effects on the receiving environment. The principles of industrial environmental management are to produce better while polluting less, which necessitates minimising pollutants discharged in all forms: air emissions, solid waste and wastewater, using appropriate preventive measures. To develop and implement appropriate environmental management in an industrial plant or factory, its pollution discharge status and potential effects on the environment should first be understood through a comprehensive monitoring program. This chapter provides details of the design and operation of such programs, illustrated with a case study of cleaner production auditing for a chlorine-bleached Kraft pulp and paper mill in a developing country. The obtained monitoring data were used to develop statistical models for the assessment of the fibre and chemical losses with the discharge effluents. Principal component analysis was used to identify the cluster of wastewater samples associated with the normal operation of the production units and the results were used for the construction of statistical process control charts.

10.1 Industrial Pollution Management

10.1.1 Industrial Pollution

Industrial production generates and discharges different types of pollutants into the environment, which should be properly controlled to safeguard the environmental quality. Industrial wastewater (WW) contains organic and inorganic substances that are harmful to the receiving water bodies. Atmospheric emissions released from fuel combustion and from manufacturing processes of industrial plants include toxic air pollutants of PM and gaseous species, such as NO_x, SO_x, CO and semi-VOCs, along with GHGs. Industrial processes also release a large amount of liquid and solid wastes, some of which belong to hazardous waste categories. This pollution, once discharged into the environment would alter the chemical and biological characteristics of the receiving bodies (air, soil, water, groundwater, coastal areas etc.) and will eventually affect human health. Developing countries face great challenges in managing industrial pollution due to fast and unplanned industrial growth. More problems are often associated with old industrial factories, normally of small and medium sizes that still rely on outdated technologies and have inadequate pollution control devices.

10.1.2 Approaches to Reduction of Industrial Pollution

The strategy for industrial environmental management is to minimise pollution discharge at sources by avoiding the generation of waste in all forms. Several management tools are commonly used for this purpose, for example, environmental impact assessment, life cycle analysis, responsible care (global voluntary initiative of the chemical industry), corporate social responsibility (CSR) etc.

Management tools such as command-and-control (CAC) and market-based instruments (MBI) are commonly applied to reduce industrial pollution discharges. CAC involves setting

DOI: 10.1201/9781003597216-12

and enforcing standards for industrial discharges, fuel quality, energy efficiency, use of BACT (best available control technology) and compliance management. It is a traditional approach that helps to reduce industrial pollution in many countries. However, CAC is normally not cost-effective especially when the standards for environmental quality, such as water quality or ambient air quality, become increasingly more stringent. MBIs are economical tools that provide incentives for the reduction of pollution discharges. These involve the "polluter pays" principle and have aspects of charges, tax incentives, subsidies and trading. An example is the emission trading to control air pollution under the USEPA bubble policies (USEPA, 2023). As compared to CAC, MBI is more cost-effective and flexible, which in principle can encourage industrial factories to reduce emissions, not only to meet the regulatory requirement but also to go lower than that if it is economically feasible.

The term "end-of-pipe" control describes a pollution control approach that aims to reduce polluting substances at the end of the production line before discharge into the environment. These control technologies are physically separated from manufacturing processes. The end-of-pipe control, although it has made significant progress in industrial pollution control, cannot solve many problems especially when the regulations are getting more stringent. The control devices consume large amounts of energy and chemicals, but some of them merely shift pollutants from one environmental component to another, i.e., without eliminating them. For example, an electrostatic precipitator (ESP) or a baghouse is applied to remove fly ash particles from the flue gas of a coal-fired power plant by concentrating and turning them into a dry solid material, which should be properly handled in the next step.

The pollution preventive or cleaner production (CP) approach aims to produce better while polluting less. CP focuses on avoiding the generation of environmental pollutants by source reduction, waste minimisation, and improving efficiency of energy and resource use while maximising product output. CP measures range from better housekeeping, substitution of toxic and hazardous materials, recycling, and reuse of waste products, to more technology-driven strategies with production process modifications and the redesign of products and/or processes toward increasing eco-efficiency. Housekeeping measures, for example, can reduce accidental discharges and spills, whereas radical modifications and/or redesigns of production processes would lower the technology discharge level.

CP presents a holistic and integrated approach to environmental protection that provides a win-win solution to gain both environmental and economic benefits and hence is recommended as a long-term strategy for sustainable development. From a broader perspective, CP is an essential pillar of the circular economy, a model of sustainable production and consumption that has emerged since the 1990s. Through cleaner production and product life cycle management, circular economy (CE) aims at the longer use of the products, materials, equipment and infrastructure; hence it can minimise both the resource inputs and the generation of waste. CE thus aims at creating harmony between economic development and environmental protection (Geissdoerfer et al., 2017).

10.2 Broad Categories of Industrial Pollution Monitoring

To develop the environmental management strategy of an industrial plant, the foremost step is to obtain a comprehensive characterisation of its prevailing pollution discharge status. This in turn requires well-designed pollution monitoring programs to provide data for identifying the main pollution discharge process units, which should be prioritised for control to minimise and/or avoid the pollution discharges. The design and operation of an appropriate environmental pollution monitoring system is an essential component of all the mentioned management tools. Such programs first help the company to understand its environmental performance even if it does not aim to obtain the ISO standard certification.

To provide accurate measurements of industrial pollution, the monitoring program should be based on a well-prepared sampling design and an accurate measurement protocol to generate representative monitoring data in the most time- and resource-effective manner. Industrial pollution monitoring programmes vary with the production type and the monitoring objective; hence there is no "blueprint" that comprehensively covers all cases. Typical objectives of industrial environmental monitoring include:

- Auditing the industrial pollution discharges and/or environmental management system,
- Providing information on production process operations to optimise process control,
- Assessing the regulatory compliance,
- Assessing potential impacts on the receiving environment,
- Leading to research and development for related activities, etc.

The first two monitoring objectives are more directly related to pollution prevention and sometimes are referred to as cleaner production auditing for pollution prevention, which is the focus of this chapter. The last three objectives are more relevant for the assessment of the environmental effects and "cure" measures. Chapter 11 will detail the design of monitoring for assessment of the industrial discharges into the receiving environment.

10.2.1 Monitoring for Industrial Pollution Auditing

Until the early 1990s, environmental auditing was mainly an effort to evaluate the environment or environmental performance of a factory or a company. Following the spread of ISO 14001 (ISO, 2019), the audit target has changed from the environmental performance itself to the environmental management system (EMS). Details of requirements for EMS, a management system for managing the environment, of a company are specified in ISO 14001, which is the essential part of the ISO 14000 family standards published in 1996 by the International Organization for Standardization (ISO). Environmental auditing can be considered a management tool involving an evaluation (systematically documented, periodic and objective) of how well the environmental organisation, management and equipment are performing with the aim of protecting the environment. The audit enables control of environmental practices and assesses compliance with company policies including meeting regulatory requirements.

In the context of this chapter, the following discussions are more related to the environmental performance audit, with a focus on the evaluation of pollution discharges from industrial factories. In this context, an environmental performance audit is a careful examination of the actual environmental status, and the results are used internally for pollution control, although they can also be used externally such as for reporting. Monitoring for auditing aims to establish a basis for pollution abatement measures and improvement of the operation efficiency to reduce pollution discharges. It also provides necessary information on the sources and amounts of material losses and on accidental discharges and spills from the industrial facilities. NIEM (1992) recommended conducting this monitoring in two stages: (1) a preliminary survey and (2) a permanent environmental monitoring program. This concept is useful for industrial environmental management in developing countries, where old-fashioned production facilities still exist but the environmental management capacity is not necessarily sufficient.

10.2.1.1 Preliminary Survey

The survey is done to collect information on the production process flowchart with the material flows, locations of environmental discharge points, such as a sewage map, approximate water use balance, as well as the production units with high potential material losses and high pollution loads. For example, onsite measurements or a series of grab samples can be collected manually from a WW stream for subsequent analyses. Onsite measurements are used for relevant parameters to detect temporal variations of WW characteristics. If (systematic) cyclic variations occur due to the cyclic nature of a production process, the period of sampling should cover the entire cycle. Otherwise, the sampling times should be planned in such a way as to detect the maximum and minimum levels of pollutant concentration and/or amount of the discharges.

10.2.1.2 Permanent Monitoring

This monitoring program is normally established based on the information generated in the preliminary survey. Sampling should include all the important discharge points, i.e., those with a high potential for material loss and pollution load. These sampling points are not necessarily confined to the outlets from the main production units or the final discharge point from the industrial plant. In reality, loss and discharge may occur in many locations, including the evaporative loss of chemicals as well as the spillage from storage silos. Automatic sampling is preferable, especially at the final discharge point to the environment, e.g., a continuous emission monitoring system (CEMS) to be installed to measure the stack emissions or automatic WW monitoring stations placed at the final effluent outlet or at waste streams with high pollution loads. Flow-composite sampling is recommended to collect samples from highly variable WW streams, as detailed in Chapter 9. The composite period may be 24 h to provide a daily average or if the factory is working on two or three shifts per day, the composite period could be the duration of each shift. Concentrations of pollutants are normally measured, but the data have little significance if no information on the discharge rate, e.g., the WW flow-rate, is available to quantify the pollution load. The latter is normally calculated to express the pollution strength of a waste stream, namely in terms of kilograms of a pollutant (PM, SS, COD etc.) discharged per tonne of product or tonne of raw material input. These specific loads are useful in the evaluation of the environmental performance of the monitored industrial factory against international norms or in comparison with other industries.

10.2.2 Pollution Monitoring for Providing Information on the Operation Conditions of Production Process

It is essential to properly understand the various operating conditions of different production units of an industrial factory. The required information includes data on the amounts and concentrations of pollutants generated from key manufacturing units and on the pollutants finally discharged from the factory, e.g., with stack flue gas and WW effluent. The amount or concentration of pollutants that are different from those during the rated "normal" operation suggests that abnormalities may occur in the production process. This information thus also helps maintain equipment, prevent accidents, encourage worker safety, and prevent serious environmental pollution discharges. Figure 10.1 schematically illustrates the "normal" and "abnormal" discharges from an industrial production unit. Note that the "normal" discharge levels presented in Figure 10.1 are considered technology-specific or technology-driven for the given facility. These "normal" levels change when technology changes, i.e., will be lower when cleaner and more advanced production technology is applied.

The use of pollution monitoring to assess the production operation processes implies that the material losses are related to the operation problems of the production process, and these losses can be spotted by analysing the pollution monitoring data. Therefore, rapidly obtainable pollution data are required to timely respond to the problems. The pollution data are used first to determine the origin, nature and causes of the discharges and further can be used to describe the overall discharge status of the factory. These constraints impose two requirements on the pollution monitoring data:

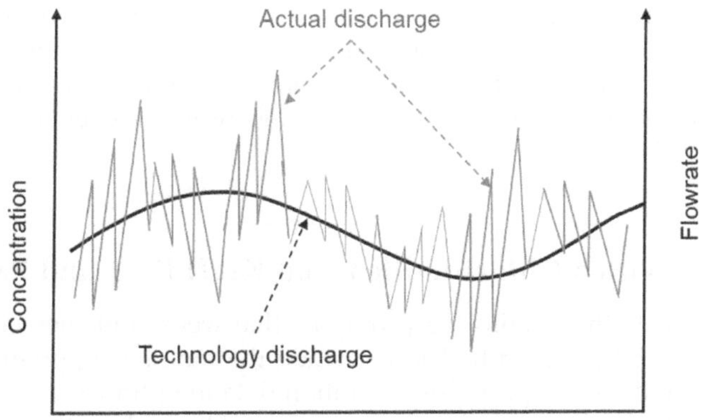

FIGURE 10.1 Normal technology discharge level vs. actual discharges with accidental and spill events.

Source: Modified from Kim Oanh and Bengtsson (1996).

- ***Having strong links with the production processes:*** This in turn requires locating sampling points at the discharge points from each production unit with a priority given to those having a high potential of material loss. Data on the final discharge from the factory is used to provide information on the total discharge and to set up a material balance for cross-checking the data obtained from separate production units.
- ***Providing rapid feedback to the right operational level:*** This requirement entails that the information on the material loss can be correlated with the pollution monitoring parameters, which should be measured rapidly, preferably automatically online, cost-effectively and with acceptable accuracy. Note that a somewhat lower accuracy of the data is acceptable as, in most cases, the relative trend, i.e., higher or lower than the normal discharge levels, is more important than single absolute values. Surrogate parameters that can be quickly measured onsite may be used for the purpose, such as WW conductivity, pH or turbidity.

10.2.3 Regulatory Monitoring

Monitoring to confirm compliance with regulations is carried out to check whether industrial discharges such as the WW, stack emissions and solid waste of a factory meet the standards set by the government. These standards may be set per plant or facility. In some cases, the discharge standards are based on the volume of discharges, while in other cases they are based on the concentration of the discharges. If the discharge levels of a factory exceed these standards, there may be strict measures such as penalties or immediate suspension of operations in some cases, but more commonly a certain grace period is granted for a factory to submit a facility improvement plan.

The regulatory monitoring may also be done to produce the data for the environmental agencies to set the industrial discharge standards. In this case, the assimilation capacity of the receiving environmental bodies is important; hence the existing pollution status in receiving environmental matrices (water, air, soil and biota) should be considered. More stringent standards should be set and enforced for those factories located in populated areas, areas where the environment is already polluted, or areas having highly sensitive ecosystems or national treasures.

In monitoring whether or not the factory discharges comply with regulatory standards, it is necessary to use the methods specified in the regulatory standards and cover all regulated parameters. The monitoring sites should be placed at the final discharge points to the environment, for example at the stack to check air pollutants released into the atmosphere with flue gas or at the WW discharge outlet to the receiving water body. Designing monitoring programs for the assessment of potential harms of industrial discharge to the environment requires a well-planned combination of biological and chemical analyses which is detailed in Chapter 11.

10.3 Case Study with a Chlorine-Bleached Kraft Pulp and Paper Mill

The case study presents the monitoring programs that were implemented in an integrated bleached Kraft pulp and paper mill. These include the monitoring programs designed for (1) environmental audit and (2) providing pollution data to optimise the production process control. Monitoring for the assessment of environmental hazards of the discharges from the mill to the receiving environments, involving both chemical analyses and toxicity testing, will be discussed in Chapter 11.

The mill produced writing and printing paper with an annual production of 55,000 tonnes, using cellulosic raw materials for pulping, 50% bamboo and 50% mixed tropical hardwood, during the study time at the beginning of the 1990s. The chips were digested in a cooking liquor made of sodium hydroxide (NaOH) and sodium sulphide (Na_2S) to dissolve the binding lignin and free fibres. The mill used a bleaching process that consumed about 70–80 kg of total active chlorine which contained 40–50 kg of elemental chlorine per tonne of air-dried pulp (ADP). The bleaching sequence was Chlorine-alkali Extraction-Hypochlorite-Hypochlorite (C-E-H1-H2). The bleached fibres were sent to the paper mill where the final paper products were made with the use of various additives (clay, alum, starch) to produce different paper grades.

The mill represents a typical case of transferred technology and was operating far below optimal conditions. The operation was unstable with a high frequency of stops of some parts of each production unit. The operation time of the mill was only 60–70% of the available time due to the shortage of raw materials and cooking liquor, equipment failure and lack of coordination among the production units. The unstable operation of the equipment caused frequent accidental discharges and spills, which affected the mill revenue and polluted the surrounding environment. The losses of materials, including chemicals (in cooking liquor) and fibres discharged with WW, could be minimised by optimising operating conditions and implementing good management practices. Cooking liquor is expensive; hence maximising the chemical recovery is essential for the Kraft process to be cost-competitive. Large amounts of toxic pollutants may be formed in the manufacturing processes in the mill and discharged into the environment which necessitated a comprehensive study on the consequent environmental effects (Kim Oanh, 1994). These pollutants include also highly toxic and bioaccumulating dioxin compounds (Bonsor et al., 1988) that will be further detailed in Chapter 11.

The main production blocks of the study mill included (1) raw material preparation, (2) pulp mill, (3) paper mill, (4) chemical recovery plants, (5) chemical plant, (6) power plant, (7) raw water treatment plant, (8) WW treatment plant and (9) maintenance plant, as shown in Figure 10.2. The key pollution discharges from the mill to the environment are as follows.

Waste: Solid waste mainly included rejected chips, bamboo, wood bark, lime mud and coal ash from the power plant boiler. Most of the combustible waste components were sold or used as fuel, coal ash was largely recovered, while lime mud was stored in a lagoon. The pollution effects of solid waste from the mill were considered more local and relatively less serious.

Air pollution: Emissions were mainly associated with the stacks from the black liquor recovery boiler, the coal-fired power plant and the shaft lime kiln. Emitted pollutants also included hydrogen sulphide (H_2S) from the green liquor tank and chloroform from the pulp bleaching. The typical odour of Kraft pulping caused by reduced sulphur compounds (H_2S, methyl mercaptan etc.) was also observed. The emissions also include greenhouse gases (GHG), mainly CO_2 from the coal-fired power plant and the lime kiln. Air pollution control

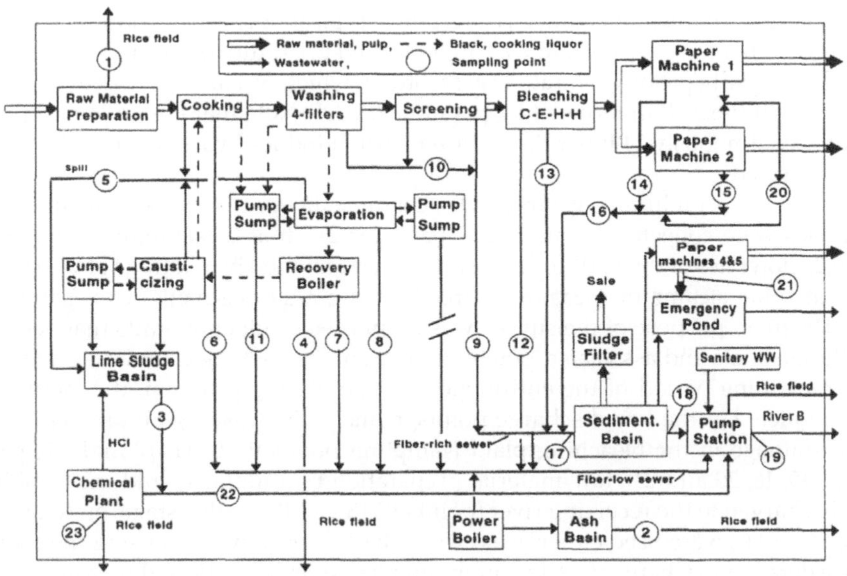

FIGURE 10.2 Flowchart and sampling points (circled numbers) at the mill. (*Note:* the arrows to "rice field" refer to only one field.)

Source: Adapted from Kim Oanh and Bengtsson (1995).

devices were deployed. Accordingly, an ESP was used for the removal of PM, mainly the sulphate particles (Na_2SO_4), from the recovery boiler stack gas. These particles were recovered to be used as cooking chemicals and therefore were not discharged into the environment. The stack emissions from the power plant were controlled using an ESP and a wet flue gas desulfurisation (FGD) unit to PM (fly ash) and SO_2.

Wastewater: high pollution loads of COD, BOD, SS, toxicity, as well as the toxic and persistent organochlorinated compounds (OCl) discharged with the mill effluents had the potential to cause long-term and far-reaching effects (see further details in Chapter 11). The treatment was only for fibre recovery and pH adjustment. Filters were used to recover fibres from the effluents of the pulp screening units and the paper machine. The fibre-rich effluents in the mill were collected and directed to a sedimentation basin for fibre recovery, while other effluents were adjusted for pH and directly discharged to a pump station where the effluent was discharged directly to a receiving river (River B). Some parts of the mill effluents were discharged into a rice field, including an overflow from the pump station and the coal ash basin, WW from the raw material preparation and WW from the chemical plant, as seen in Figure 10.2.

10.3.1 Monitoring for Environmental Performance Auditing

10.3.1.1 Sampling Design

This monitoring program was conducted to get an overall picture of the WW discharge from every production unit and the combined mill effluent. Conventional physicochemical parameters relevant to this industry type were monitored, including, BOD, COD, SS, pH, conductivity, settleable solids (SetS), TKN, OP (ortho-phosphorous), TP (total phosphorus), as well as the water-soluble ions of sodium (Na^+), chloride (Cl^-) and ammonium (NH_4^+). Since the operating conditions of the equipment in this plant were not stable, it was necessary to obtain monitoring data over a sufficient period in order to get the representative values of various operating variables and environmental characteristics under typical operating conditions. For this reason, a one-year monitoring program was implemented to capture the average discharge status of the mill. The daily production rate was recorded, which was used to calculate the specific discharges from the mill in terms of kg pollutant/t ADP. Note that the mill was operating in three shifts, continuously 24 h per day; hence no batch-induced variation of WW characteristics were expected.

A site visit with a preliminary survey was conducted over two weeks, during which grab WW samples were collected on an hourly basis for 2–3 h at 23 sampling points shown in Figure 10.2. Flowrates of the WW streams were either recorded by readings of weirs or plumes available at some steams or measured directly using a simple method of "stopwatch and bucket". The main purpose of the survey was to find the production units that had relatively high pollution loads and associated potential losses of the cooking chemicals and fibres.

In the remaining period of the entire one-year monitoring program, the monitoring was mainly conducted at the key discharge points, namely, the washing and screening effluent, the floor drainage in the bleaching plant (sampling points 9, 10, 11, 12 and 13), paper mill (points 14, 15, 16, 20 and 21), raw material preparation unit (point 1), and the combined mill effluent discharged to the receiving river (points 17, 18 and 19). In this stage, manual composite samples, TC or FC, were used either over every 8 h shift or daily (24 h) using the compositing procedure detailed in Chapter 9. A few grab samples were also collected occasionally at other sampling points than those listed for cross-checking the material balance of different pollutants and the water balance.

The monitoring results were presented as the average and variation of the pollutant concentrations, as well as the pollution loads in terms of kg/day, kg/t ADP or kg/t of related product. The key findings included the water balance, chemical balance, SS balance and the balance for main pollutants of COD, BOD, TKN etc. (Kim Oanh, 1994).

10.3.1.2 Results of Material Balance Analysis

An example of the material balance is presented in Figure 10.3 for water which shows the average intake of 72,000 m^3/day for all usages that corresponded to an average usage of 515 m^3/t ADP. This was a high specific water consumption as compared to other similar industries, and it was due to the low percentage of water recycling and reuse within the mill, i.e., only about 10% overall.

The material balance for the cooking chemical, expressed as sodium sulphate (Na_2SO_4), is presented in Figure 10.4. Note that the balance analysis included only the discharges associated with the potential cooking chemical loss while excluding those associated with the use of sodium in other production units, namely the E-stage bleaching and the chemical plant. With an average make-up chemical of 100 kg/t ADP, the mill had only a chemical recovery rate of 88%, which was much lower than the average level of above 96% for mills in Sweden at the time of the study. The chemical balance thus provides useful information on the potential extent of minimising the chemical loss to gain more revenue for the mill. The most significant loss of the cooking chemical was with effluent from the pulp washing unit and the flue gas from the recovery boiler stack, followed by the causticising unit (where the green liquor is converted to the white cooking liquor). A better performance of the ESP can minimise the loss carried with the recovery boiler flue gas. Loss with accidental spills in various production units contributed about 20%, which can be minimised by good housekeeping measures.

The material balance for SS in Figure 10.5 was used to identify the key effluents associated with the fibre loss. SS70 is the suspended solid retained on a filter wire of 70 µm pore; hence it is more relevant to represent the discharge of good (long) fibres from the pulping and paper-making units. SS is the suspended solids retained on a glass fibre filter of 1.6 µm pore; hence it is more related to the total solids present in the effluent. The SS discharged with the raw material washing (point 1, Figure 10.2) and coal ash basin (point 2, Figure 10.2) are excluded because they were not related to the fibre loss. Based on the SS material flows, it

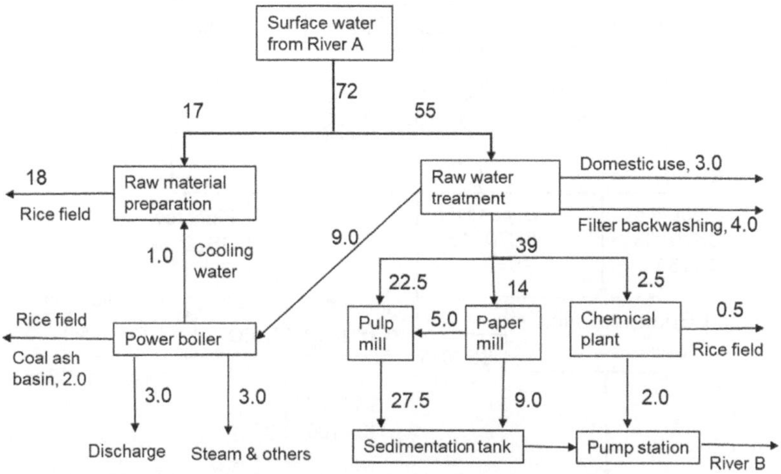

FIGURE 10.3 Water balance for the mill for an average production of 140 t ADP/day, (The numerical values are the water flow in 1000 m^3/day.)

Source: Adapted from Kim Oanh (1994).

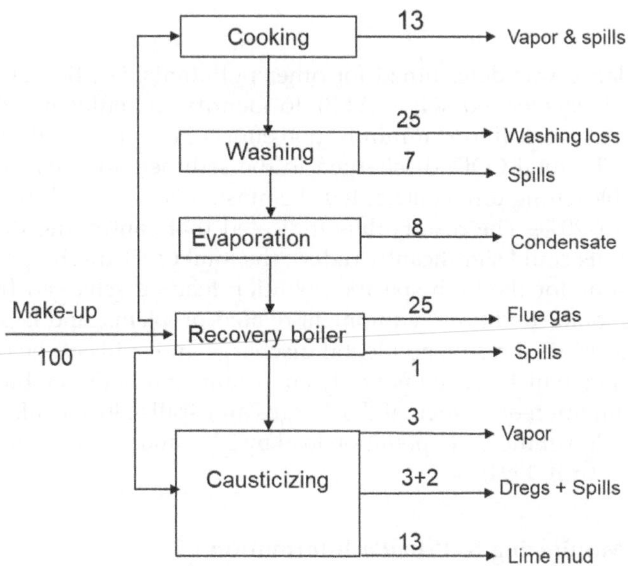

FIGURE 10.4 Chemical balance at the mill, in %, for an average of make-up chemical of 100 kg Na_2SO_4 per tonne of ADP.

Source: Kim Oanh (1994).

was evident that some parts of the fibre-rich effluents had bypassed the sedimentation basin and discharged directly to the pump station. This bypassed amount was estimated using the material balance and is also presented in Figure 10.5.

The fibre loss of SS70 with 73 kg/t product (7.3%) was significant and hence should be minimised by better management practices at the production units, i.e., screening, bleaching and paper-making. Besides, the end-of-pipe treatment (sedimentation basin) can minimise the fibre discharge, but it was not economically sound because of the low quality of the recovered fibre which was only either used to make cardboard products or sold externally at a low price, i.e., less than 10% of the paper product.

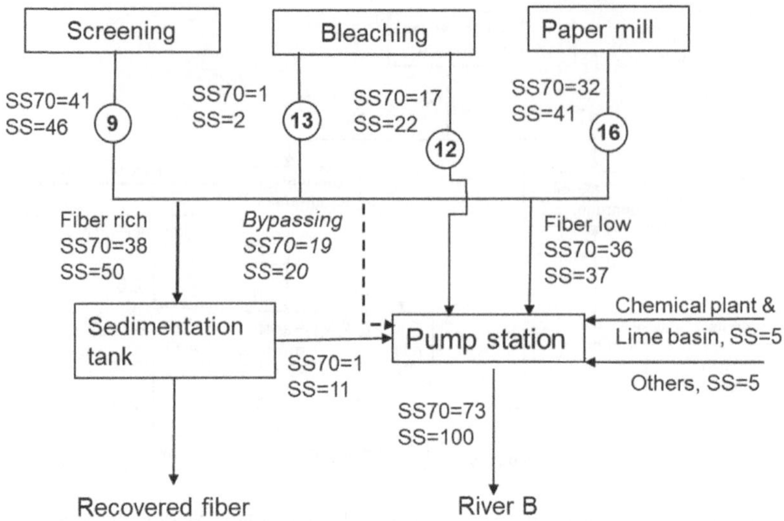

FIGURE 10.5 Material balance for suspended solids in kg/t product. The circled numbers are the sampling points (indicated in Figure 10.2). (*Note:* SS70 – SS on filter wire of 70 μm pore; SS – SS on GF/A [1.6 μm].)

Source: Adapted from Kim Oanh (1994).

The material balance was determined for other pollutants, i.e., BOD, COD, OP, TP, TKN, chloride and volatile suspended solids (VSS), to identify the effluents with high pollution loads in order to focus the pollution minimisation efforts. For example, the balance for filtered COD_f showed that the total COD_f discharged to the sedimentation basin was 100–140 kg/t ADP to which the bleaching unit contributed the most, 50%, followed by accidental spills in the cooking unit (15–20%). The results thus indicated that controlling the discharges from these production units could significantly reduce the total COD_f discharged from the mill.

Overall, the reasons for the high specific pollution loads discharged from the mill were the unstable operational conditions, equipment-related problems, and lack of efficient waste management practices. As a result, accidental discharge and spills of the cooking liquor and fibres happened often, which caused not only environmental pollution but also a significant economic loss. Elimination of accidental discharges and spills alone could reduce the chemical loss by over 20%, lower the COD_f pollution load by 30%, and eliminate most of the fibre loss from the mill (Kim Oanh, 1994).

10.3.2 Pollution Monitoring to Provide Information for Optimising Production Control

This monitoring program was developed to incorporate into the production control in order to minimise potential upsets and associated abnormal discharges. It was an attempt to connect environmental protection work with the mill operation for detecting malfunctioning events in the production process to issue a warning at early stage to prevent accidental spills. The ultimate goal was to find internal solutions such as good housekeeping measures that help improve the mill economy by minimising material loss, along with improving its environmental performance. The monitoring was focusing on the key discharge points associated with large pollution loads and material losses to rapidly assess the material losses using the pollution parameters. For this purpose, the monitoring program was developed in three steps: (1) select the appropriate pollution parameters to express material losses; (2) identify the "normal" or "typical" discharges from the production units; and (3) construct the control charts.

10.3.2.1 Selection of Pollution Monitoring Parameters

The pollution parameters selected for quick feedback to control the material loss from the production units should satisfy the following criteria:

- Be rapidly obtained, preferably by online/real-time monitoring,
- Be quantified by simple but reliable measurement methods, and
- Having durable measurement equipment for the fieldwork.

As mentioned previously, for the Kraft pulping, the sodium content could be used to express the cooking liquor loss, whereas SS70 and/or SS in WW effluents could be used to assess the fibre loss. Determination of Na^+ and SS70/SS in WW is resource- and time-consuming; hence it was essential to find a way to express them through other parameters that should satisfy all the criteria. To identify a set of suitable rapidly assessed pollution parameters (RAPP), simulated/synthetic WW was prepared using diluted cooking liquor to study the chemical loss, while that prepared using diluted pulp was used to study the fibre loss.

To assess the chemical loss, several empirical models were developed to express the Na^+ content in the simulated WW using conductivity and pH, respectively. The models were tested with the actual WW monitoring data for the mill to select the most suitable model for the range of Na^+ concentrations typically found in a particular effluent (detailed in Table 10.1).

For the fibre loss control, the simulated study was conducted separately for the fibres in the pulping line (pulp materials after washing) and in the paper-making line (pulp stock) because of their different quality and size. In principle, the SetS pollution parameter is used to determine the volume of particles that settle out of a WW sample during a 1-h settling test, in millilitres per Liter (mL/L), using an Imhoff cone following the standard method. To express the fibre content in the WW sample, this study used the SetS with shorter settling periods, i.e., 15 min and 30 min, respectively, because they require a shorter measurement time. High correlation coefficients of over 0.998 were found between all three measurements for SetS (15 min, 30 min, 1 h) for the fibres in the effluents from the pulping and paper making. Therefore, SetS15 was selected as RAPP to assess the fibre loss, which is expressed by SS. The regression models were developed between SetS15 and SS separately for the pulping and paper-making line. The turbidity was also tested as a candidate RAPP for the fibre loss, but due to the strong colour and high content of fibres in the effluents, the parameter was considered not suitable.

The selected models applicable to different WW effluents are presented in Table 10.1 along with the coefficients of determination (R^2) and results of ANOVA. The fibres in the pulping line and the combined fibre-rich inlet to the sedimentation basin were larger in size and mostly of fibrous nature; hence they were expressed as SS70. The solids in effluents from the paper-making line consisted of not only fibres (mainly of fine size) but also various additives, e.g., kaolin; hence it would be better to express them as SS. High R^2 values were found for most of the models, indicating the high percentages of the total variation explained by the regression equations. The *null hypothesis* of no linear regression between the dependent and independent variables in each equation was tested at the significance level of $\alpha = 0.01$. The F statistics obtained for the models were higher than the critical F value ($F_{0.99,1,n-2}$)l hence the *null hypothesis* was rejected at 1% significance or, in other words, with 99% confidence.

10.3.2.2 Determination of Typical Normal Discharge Levels of the Production Units

An important task in establishing the process control procedure is to determine the normal condition that the production unit is "typically" operating with and use it as a reference to determine the abnormal operation and associated pollution discharges. For stably operating equipment and facility, operating parameters of the normal condition may be obtained straightforwardly based on a monitoring record over a production cycle or based on the reference level obtained from an appropriate historical dataset. However, for old-fashioned

TABLE 10.1

Selected Models Based on RAPP to Express the Material Losses in the Mill

Chemical loss in pulping line				
Model and applications	Equation	R^2	*F* ratio	$F_{0.99,1,n-2}$
Model 1: Cond<1500 µS/cm (Na<7500 mg/L)	Na = −169 + 0.54 Cond	0.9998	36670	13.74
Model 2: Cond<1000 µS/cm (Na<250 mg/L)	Na = −58 + 0.35 Cond	0.8567	29.9	16.26
Fiber loss in the pulping line				
Model 3	SS70 = 46 + 3.9 SetS15	0.9733	474	9.07
Model 4	SS = 47 + 11 SetS15	0.8369	25.7	16.26
Fiber loss in paper-making line				
Model 5 for whole data	SS = 393 + 2.6 SetS15	0.5797	52	7.35
Model 6 for days with kaolin used	SS = 290 + 3.5 SetS15	0.7028	73	7.53
Fiber loss in mixed fibre-rich WW influent to sedimentation tank				
Model 7	SS70 = 70 + 3.7 SetS15	0.8450	125	7.88
Model 8	SS = 95 + 4.1 SetS15	0.7814	32.2	10.6

Note: SetS15 (settleable solids after 15 min of settling) in mL/L; Cond: conductivity in µS/cm; Na (sodium) in mg/L; SS in mg/L; SS70 in mg/L.
Source: Extracted from Kim Oanh (1994).

factories with unstable operations like the mill in this study, setting the ideal targets is challenging, and meeting those ideal targets is even more challenging. The targets for unstably operating factories could not be entirely borrowed from other state-of-the-art factories because it would be too ideal to be achievable.

Therefore, for the study mill, the "normal" discharges from production units when the equipment was running smoothly were used as the targets for the loss control. Note that these targets were still far from the ideal technology discharges with a perfect performance of the existing equipment. Rather, these represented the "normal" discharges under the typical existing status of the equipment performance. These control targets thus should be continuously revised to reflect the improvement in the equipment and/or operation status of the production units.

It is possible to obtain the normal discharge levels based on long-term monitoring data of the discharges with simultaneous observations of production status, spills and accidental discharges of every production unit. The data should be sorted out to get the periods when the equipment in particular units was operating smoothly and without spills and accidental discharge events. Multivariable analysis of the pollution data can be applied to objectively identify the data points representing the effluent characteristics of these periods.

In this case study, the PCA was applied for the measurement data of effluents from main production units: pulping, paper-making, combined WW influent to the sediment basin, and at the final discharge from the pump station. Discharge status was characterised by conductivity, pH, TDS (total dissolved solids), and T°C (temperature), which were recorded online for each stream at intervals of either 10 min, 30 min, or 1 h, and over a measurement period between 8 and 24 h, varying with the effluent type. Simultaneously, WW samples were taken for measurements of SetS15 and SetS30. A logbook was kept documenting equipment starts, stops, spills, and any other incidents for comparative analysis with the pollution data. The WW flow-rates were measured using the installed weirs. An example of the scores and loadings plots is given in Figure 10.6 for the effluent from the floor drainage of the bleaching plant (point 13). For this discharge point, the dataset of the monitoring parameters consisted of

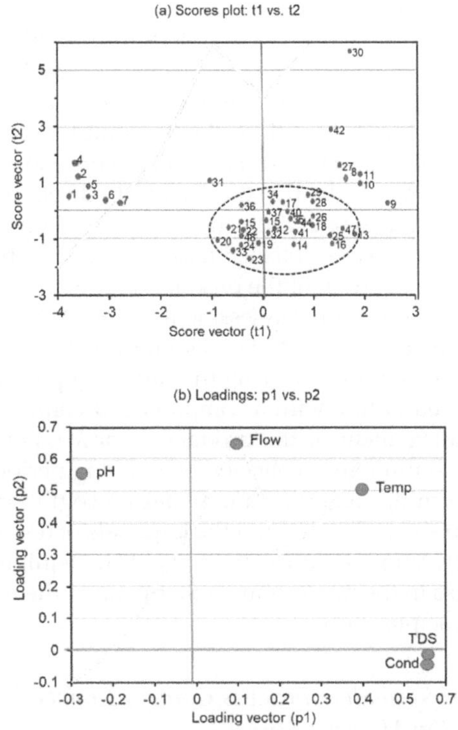

FIGURE 10.6 Results of PCA for measurement data of the wastewater effluent from the bleaching plant floor drainage (sampling point 13) presented for the first two PC: (a) scores plot with numbers indicate the sequence of 10-min sampling over an 8-h shift; (b) loadings plot showing the monitoring parameters.

10-minute measurements (numbered in sequence) taken over an 8-hour shift of the production unit. The PCA retained four principal components (PC), which explained more than 90% of the total variance of the original dataset. Only the first two PCs, explaining 82% of the total variance, are presented in the plots.

The scores plot (Figure 10.6a) shows a large cluster of the data from the samples (circled) collected during the period when the production unit was observed to operate stably without a stop due to the equipment malfunction or any significant spills or accidental discharges. These samples therefore can be considered to show the pattern of the normal discharge of the unit. The average normal discharges and the upper and lower limits of the 95% variation range of the dataset, respectively, can be used as the warning levels. A small group of data (samples 1–7) was also identified on the plot collected during a period when an overflow from the pulp storage tank was observed; hence these were related to spills. The rest of the data points were scattered and were considered outliers, e.g., data points 30, 31, 42 etc.

The loadings plot can be used to interpret the correlation between the parameters. Accordingly, the correlation between two parameters is determined as the cosine of the angle formed by two lines connecting the origin (0,0) and the two points representing the two parameters, respectively. For example, TDS and Cond are highly correlated because the angle is almost zero ($r = 0.999$).

Directions in the loading plot correspond directly to the directions in the score plot. Hence superimposing the two plots helps indicate parameters that were responsible for the outliers in the discharge regime. For example, data point 30 is located high in the first quarter, in the same direction as the flow, suggesting that the high flow may be the reason for this outlier.

Checking the actual measurement results also showed the peak of flow appeared at the 30th sample while the second peak was at the 42nd sample. Similarly, high pH values (caused by the overflow) were responsible for the horizontal deviations of samples 1–7.

The dominant cluster of data points shown in the PCA score plot was used to determine the "normal" discharge from every considered production unit for setting the respective control target. If there were also other clusters in a particular score plot, the analysis of the simultaneous process data was done to identify the "normal" discharge of the production unit. Note that for a plant with unstable operation such as the mill under this study, the "normal discharge" would not be accurately established based merely on the statistical mean without a good understanding of the operation of the production units.

The PCA results thus can be used for assessment of the WW discharge status at a given monitoring site. The resulting plots provide a quick interpretation of the abnormal discharges and the parameters that were abnormal during the sampling period. Such a WW monitoring program can be implemented online with a computerised control system to timely provide discharge data for the management of the production process operation. The use of large WW monitoring datasets, continuously collected over a long period from the main effluents, is highly desirable in performing clustering and pattern recognition. Further, the data can be collected using sensors and smart devices, and the large sets of data can be analysed using Big Data Technologies to timely provide information about the equipment performance to optimise the production process in the modern manufacturing of the new industry Era, Industry 4.0 (Gupta et al., 2021; Suryacipta, 2023).

10.3.2.3 Development of Statistical Process Control Charts for Minimising the Material Losses

Control charts were developed for the effluents with a high potential for material losses. The effluents associated with the chemical loss were identified as those associated with the spills and overflow of the cooking liquor in the pulping line at points 4, 5, 9, 10 and 11 (Figure 10.2). The effluents associated with the fibre loss in the pulping line were from pulp washing, screening, and bleaching floor drainage (points 9, 10 and 13), those from the paper-making machines (sampling points 14, 15 and 16), and the sedimentation basin influent (point 17, Figure 10.2).

The targets for the loss control were the "normal" discharges from production units determined using this method. The selected RAPPs were used to express the losses, namely conductivity for chemical loss and SetS15 for fibre loss. As a way of example, a chart constructed for controlling the chemical loss associated with spills from the cooking and evaporation plant (point 5) is presented in Figure 10.7a, whereas that for the fibre loss from the paper-making is in Figure 10.7b (point 15).

The mill had a continuous operation over 24 h; hence the target was a straight line without any batch-induced variation. The average normal discharge target (solid lines) and the 95% variation interval of the concentrations and the flow-rates during normal operation (dashed lines) are indicated. The targets provided separately for the concentration and flow-rates were necessary to readily reveal the reasons for the abnormal discharges, instead of the combined parameter of the discharge load because a high flow-rate due to overuse of water is normally associated with low pollutant concentrations. The economic value of the loss could also be calculated and indicated directly in the charts. The discharges within 95% of the variation interval are considered "normal" whereas the values beyond this interval are alarming. Too high discharges indicate higher material losses, and too low discharges may also indicate certain abnormal conditions in the production lines, which should be checked accordingly.

Similar charts were established for each effluent with high potential losses of fibre and chemical of the mill, respectively. Monitoring RAPP for the WW effluent thereby can provide necessary information for the timely control of the production units to minimise material losses. For example, the spills of the cooking chemical at point 5 (Figure 10.7a) could be minimised and eventually eliminated with better housekeeping measures. On the other hand,

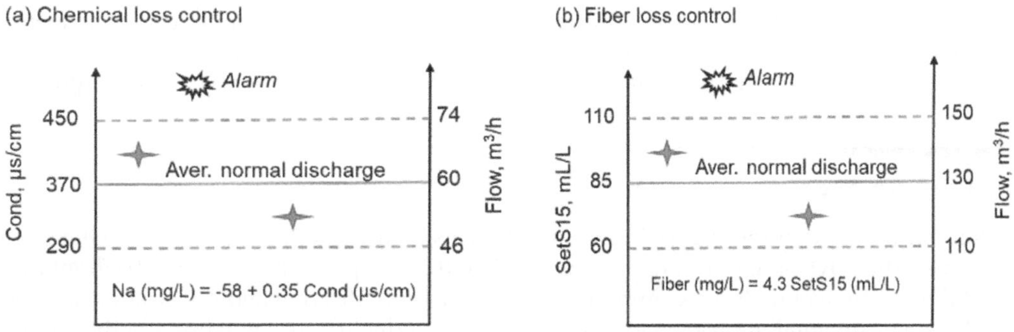

FIGURE 10.7 Suggested statistical graphs for controlling material loss in different production units: (a) chemical loss based on conductivity for the spill stream from cooking and evaporation units (sampling point 5 in Figure 10.2); (b) fibre loss control based on SetS15 for paper machine 2 (sampling point 15). Dashed lines show the 95% interval of the respective normal discharge. The values correspond to the average mill production of 140 t ADP/day.

Source: Adapted from Kim Oanh (1994).

the chemical loss associated with the pulp washing and screening unit (e.g., point 10) was mainly caused by the performance of the washing filters, which should be checked accordingly. Likewise, the fibre loss in the effluent from paper machine 2 (sampling point 15) was the highest in the mill due to a high frequency of overflows from the white water storage tanks, which also could be minimised by good housekeeping measures. This fibre was of a high quality and hence had more economical value.

It is necessary to note that the statistical control charts in Figure 10–7 are not exactly the commonly known Shewhart control chart mentioned in Chapter 1. The basic Shewhart control chart is a common tool used in monitoring the mean of a process to ensure that it remains in control. With the centre line being the in-control mean value and 3σ limits on either side of the centre line, the Shewhart chart signals an out-of-control condition if any observed sample mean falls beyond the 3σ limits (Davis et al., 1990). The statistical control charts in Figure 10.7 developed in this case study for this Kraft pulp and paper mill, also with the purpose of controlling and monitoring the manufacturing process variations, but the main concern was to minimise the excessive material losses with the wastewater effluents. The dashed lines show the 95% interval of the respective normal discharge; the measurement values falling beyond the upper limit indicate the high loss conditions, alerting for immediate actions to control the operation of respective production units. On the other hand, the values falling below the lower limit also indicate abnormal operational conditions that also require attentions.

It is emphasised that the discharge targets presented in Figure 10–7 for the study mill were still far from the ideal technology discharges with a perfect performance of the existing equipment. Rather, these represented the "normal" discharges under the existing status of the equipment performance; hence the targets should be continuously revised to reflect the improvement in the equipment and/or operation status. In principle, the loss should be minimised step by step with a more stringent target for every next step. A control chart could be constructed using a few principal components. However, the multivariate control chart based on latent variables does not directly spotlight the possible causes of the discharge upsets and therefore may not be superior to the charts constructed separately for each monitoring parameter.

The online monitoring results of suitable RAPP for WW could be directly incorporated into a computerised process control system of the industry. In this study, conductivity is a suitable RAPP for chemical loss as it can be easily measured online, providing quick feedback to adjust the production operation to minimise the loss. SetS15, however, takes some time

(15 min) to measure; hence a more appropriate RAPP is desirable. Turbidity can be measured online but was not suitable due to the high fibre content and strong colour of the WW, i.e., the dark colour of pulping line effluent and the white colour of the paper-making line effluent.

10.4 Summary

Industrial production releases pollution in all forms, which should be properly managed to minimise the adverse environmental effects. For this purpose, several industrial environmental management tools are commonly used, such as certification and registration schemes for the environmental management system, cleaner production, or more broadly the circular economy approach, among others. A foremost requirement embedded in these management tools is a well-designed pollution monitoring program, which should be effectively implemented to provide comprehensive information on the status of pollution discharges from the industrial plants. Such information, for example, helps identify, prioritise and implement suitable in-plant pollution control measures when pursuing cleaner production. Industrial pollution monitoring program varies with the production type and the objective of the monitoring program, and no "blueprint" program is comprehensively suitable to all cases.

This chapter focused on the principles of monitoring environmental pollutants discharged with industrial effluents for the purpose of cleaner production auditing. The case study of the monitoring programs implemented at a chlorine-bleached Kraft pulp and paper mill showed that wastewater monitoring data can be used to reveal the production units associated with high potential losses of chemicals and fibre due to accidental discharges and spills. Rapidly assessed pollution parameters can be used to express the material loss with waste streams to provide timely feedback for improving production management practices. The accidental discharges and spills can be minimised using low-cost housekeeping measures to bring in benefits to both the environment and the mill revenue. The examples discussed in the case study are not necessarily applicable to state-of-the-art facilities with ideal operation conditions but rather to industrial factories using equipment discharging high pollution loads and production operation practices involving frequent accidental discharges and spills.

REFERENCES

Bonsor, N., McCubbin, N. and Sprague, J.B., 1988, *Stopping water pollution at its source – Kraft mill effluents in Ontario, Report prepared for the Technical Advisory Committee,* Pulp Paper and Sector of MISA, Ontario Ministry of the Environment, Toronto, Ontario, Canada.

Davis, R.B., Homer, A. and Woodall, W.H., 1990, Performance of the zone control chart, *Commun Stat Methods,* 19(5), 1581–1587.

Geissdoerfer, M., Savaget, P., Bocken, N.M. and Hultink, E.J., 2017, The circular economy: A new sustainability paradigm? *Journal of Cleaner Production*, 143, 757–768, https://doi.org/10.1016/j.jclepro.2016.12.048

Gupta, H., Kumar, A. and Wasan, P., 2021, Industry 4.0, cleaner production and circular economy: An integrative framework for evaluating ethical and sustainable business performance of manufacturing organizations, *Journal of Cleaner Production*, 295, 126–253, https://doi.org/10.1016/j.jclepro.2021.126253

International Standard Organization (ISO), 2019, *Environmental management systems – guidelines for using ISO 14001 to address environmental aspects and conditions within an environmental topic area – part 1: General*, published by International Standard Organization, Edition 1 2019-11, https://www.iso.org/obp/ui/en/#iso:std:iso:14002:-1:ed-1:v1:en.

Kim Oanh, N.T., 1994, *Wastewater monitoring, a tool to optimize process control and reduce contamination from bleached Kraft pulp and paper industry: Vietnamese case studies. AIT dissertation no. EV-02-94,* Asian Institute of Technology, Bangkok, Thailand.

Kim Oanh, N.T. and Bengtsson, B-E., 1995, Toxicity to Microtox, micro-algae and duckweed of effluents from the Bai Bang paper company (BAPACO), a Vietnamese bleached Kraft pulp and paper mill, *Environmental Pollution*, 90(3), 391–399, https://doi.org/10.1016/0269-7491(95)00008-F

Kim Oanh, N.T. and Bengtsson, B-E., 1996, Design of a representative and cost-effective sampling programs for industrial wastewater with examples from bleached Kraft pulp and paper mills, *Resources, Conservation and Recycling*, 18(1–4), 107–124, https://doi.org/10.1016/S0921-3449(96)01172-X

Network for Industrial Environmental Management (NIEM), 1992, *Training package 1, section B, improved mill management*, Swedish International Development Authority (SIDA) and United Nations Environment Programme (UNEP) Regional Office for Asia and the Pacific, Bangkok, Thailand.

Suryacipta, 2023 October, Big data impact: The role of analytics in industrial 4.0, https://suryacipta.com/en/, accessed December 2023.

U.S. EPA, 2023, *USEPA bubble policy*, U.S. Environmental Protection Agency, retrieved on 21 September 2023 from https://www.epa.gov/history/bubble-policy

11

Monitoring to Assess the Impacts of Industrial Discharges on the Environment

A well-designed monitoring program is required to assess the harmful effects of industrial pollution discharges on the receiving environment. The monitoring parameters are selected depending on the industry and can cover a wide range of physical, chemical, and biological variables. This chapter provides principal monitoring approaches applicable to industrial discharges, illustrated with a case study of a chlorine-bleached Kraft pulp and paper mill. The chemical analyses were conducted to quantify the levels of organochlorinated compounds, including dioxins, and heavy metals in the wastewater samples collected inside the plant and at final discharge points, as well as the water, sediment and biota samples in the receiving environmental bodies. Acute toxicity tests were performed to assess the potential biological effects of the mill effluents on the ecosystems.

11.1 Overview of Monitoring for Potential Harms of Industrial Discharges to the Environment

The composition of industrial discharges is often so complex that specific chemical analyses for all substances in the discharges are often technically and economically not feasible. A wide range of physical, chemical, and biological variables may be monitored, and the selection of the parameters for a particular plant is made based on the industrial type, the equipment and raw materials used, along with its operating conditions and environmental management practices. A well-planned combination of biological and chemical analyses may provide a reasonable basis for the assessment of environmental hazards of the discharge at affordable cost. Biological effects reflect the integrated expression of pollution loads, i.e., the sum of harmful effects of the discharges to the aquatic and terrestrial ecosystems. When considering the environmental impact to human health and the biosphere, there are a variety of issues to consider, including, for example, air pollutants such as fine particulate matter and radioactive materials and bioaccumulating pollutants. This chapter particularly focuses on issues related to pollution discharges with the WW effluents and potential harm to the receiving environmental bodies, including surface water, sediments and biota.

Biomonitoring includes both field surveys and laboratory studies. The latter include toxicity tests and bioaccumulation studies. The tests may also be conducted in artificial ecosystems that simulate the "real world". For industrial wastewater (WW), biomonitoring is inevitably an important part of assessing the environmental effects because the ultimate consequences of any form of aquatic pollution are of a biological nature. Parameters such as photosynthetic assimilation of carbon, liberation of oxygen, respiration, biomass, growth, uptake, morphological abnormalities, diseases in animal populations, transformation, and bioaccumulation of pollutants can be considered as possible indicators of both short-term and long-term biological effects. The need for extensive physicochemical analyses may be identified based on the results of biological tests.

A detailed physicochemical characterisation of the discharges helps reveal causative factors of the ecosystem effects. The chemical parameters could be aggregated, i.e., non-specific such

DOI: 10.1201/9781003597216-13

as total Kjeldahl nitrogen (TKN), or specific compounds, such as ammonium, nitrite, nitrate, heavy metals, specific chlorinated organic compounds etc. These analyses are recommended for discharged effluents and the components of the receiving environment including water, bottom fauna, sediment, algae and lipophilic substances in biota and fish. Note that the physicochemical measurements for a pollution stream itself and the receiving environment per se, separately from the biotic systems, cannot give a true picture of discharge effects as well as the history and consequences of the pollution (NIEM, 1993).

Recommended techniques for the characterization of potential environmental effects of industrial effluents include (1) degradability tests, (2) biological effects (toxicity tests), (3) bioaccumulation study, and (4) chemical analyses. According to SEPA (1990), a stage-wise process is recommended, starting with the screening stage based on simpler tests to the final stage with more complex analyses to confirm the effects.

In the screening stage, the degradability tests may include BOD and COD analyses, and the biological effect study may be limited to acute toxicity testing with fish, plant (e.g., duckweed), algae, zooplankton, specific bacteria (e.g., Microtox bacteria) etc. In the bioaccumulation study, the presence of specific heavy metals and lipophilic substances is tested. The recommended physicochemical monitoring parameters include pH, SS, conductivity, total organic carbon (TOC), nitrogen, phosphorus and any known or suspected substances in the waste streams that are considered to be toxic, persistent or bioaccumulating. The monitoring parameters are selected based on the potential toxic substances that may be typical for a particular type of industry, the production processes, the raw materials and the chemicals used in the plant. For example, heavy metals are included in the monitoring of the impacts from a coal-fired boiler or metal processing plant, e.g., iron and steel production or a leather tannery industry. Aggregated parameters for chlorinated compounds, e.g., extractable organochlorines (EOCl) and adsorbable organic halogens (AOX) may be monitored for the discharges from a chlorine-bleached pulping industry.

After the screening tests, specific biological and chemical analyses are recommended with more advanced methods in the intermediate and final assessment stages of potential environmental effects from industrial discharges. For example, chemical analyses are performed to reveal and determine levels of specific substances such as dioxins, heavy metals etc. in industrial effluents. The biological effects of the effluents are evaluated using acute, chronic toxicity and genotoxicity tests. In the final stage, the confirmatory tests of the biological effects are conducted with the ecosystem models. Chemical analyses of sediment and biota in the receiving environmental bodies are recommended to demonstrate the bioaccumulation effects.

11.2 Monitoring for Assessment of Environmental Impacts from Chlorine-Bleached Kraft Pulp and Paper Mill

This case study illustrates a step-by-step approach, from a preliminary screening using basic tests to detailed biochemical analyses in the following stages, for evaluation of environmental impacts from a chlorine-bleached Kraft pulp and paper mill.

11.2.1 Environmental Discharges from the Mill

The mill in this case study has been described in Chapter 10 (Section 10.3), which was an integrated bleached Kraft pulp and paper mill with an annual production of 55,000 tonnes of writing and printing paper in the early 1990s. The mill had the main production blocks of (1) raw material preparation, (2) pulp mill, (3) paper mill, (4) chemical recovery plants, (5) chemical plant, (6) coal-fired power plant, (7) raw water treatment plant, (8) wastewater treatment plant and (9) maintenance plant, which are illustrated in Figure 10.2 (Chapter 10).

The cellulosic raw materials for pulping consisted of 50% bamboo and 50% mixed tropical hardwood with a daily production of 140 tonnes of air-dried pulp (ADP). The pulp bleaching sequence was Chlorine-alkali Extraction-Hypochlorite-Hypochlorite (C-E-H1-H2), which used a large amount of total active chlorine, about 70–80 kg per tonne ADP, of that the elemental chlorine was 40–50 kg. The mill operation was unstable, which caused frequent accidental discharges and spills that affected the mill revenue and polluted the surrounding environment, as already detailed in Chapter 10.

The main discharges from the mill to the environment included solid wastes, air pollution and WW. The latter was considered to have the highest potential harm to the receiving environment due to the high pollution loads, but the WW treatment methods were only limited to fibre recovery and pH adjustment. The water balance analysis presented in Chapter 10 showed a high specific water consumption, about 515 m^3/t ADP, which consequently led to a large volume of effluent discharged. Chemical loss and fibre loss were substantial and eventually ended up in the discharges to the environment.

As a result, high pollution loads of COD, BOD, SS, toxicity, as well as the toxic and persistent OCl compounds were discharged with the mill effluents. The discharges thus had a potential of causing long-term effects, especially in River B, which received most discharges from the mill. In addition, some parts of the mill effluent were discharged directly into a rice field (see Figure 10.2, Chapter 10), which included the effluents from the raw material preparation, the chemical plant and the overflow from the coal ash basin and the pump station. A comprehensive assessment of the environmental effects of the mill discharge was therefore urgently required (Kim Oanh, 1994).

Of special concern was the large amount of active chlorine used in the pulp bleaching process because of the potential formation of OCl compounds. These OCl compounds include dioxins, chlorinated phenolics and chlorinated resin acids that are mainly discharged with WW. Particularly, for this type of chlorine-bleached Kraft pulping process, dioxins, consisting of polychlorinated dibenzo-*p*-dioxins (PCDD) and polychlorinated dibenzofurans (PCDF) or PCDD/F for short are typically found in the liquid effluent, sludge from WW treatment, and pulp itself. These include 2,3,7,8-TCDD (2,3,7,8-tetrachlorinated dibenzo-*p*-dioxins) and 2,3,7,8-TCDF (2,3,7,8-tetrachlorinated dibenzo-furans), the two most important members of the dioxin family from the toxicology point of view. The precursors of PCDD/F are presented in the lignin content of the pulping raw materials; hence the chlorination of these precursors in the bleaching process is responsible for a significant portion of the dioxins formed. If not properly treated, these highly toxic dioxins once discharged into the environment would cause harmful effects (Rappe et al., 1990). Note that chloroform is also formed, but, due to its volatility, the compound is released into the atmosphere causing air pollution (Bonsor et al., 1988).

11.2.2 Assessment of Potential Harms of the Mill Discharges to the Environment

Bleached Kraft mill effluents are extremely complex in composition hence a complete chemical characterisation is not practically possible due to extensive resource requirements. The effluents therefore could be assessed using an appropriate combination of biological tests and chemical analyses to evaluate the potential environmental effects. The biological effects of the effluents can be quantified using toxicity testing, whereas physicochemical analyses are made to reveal the effluent compositions. The latter include conventional parameters, such as BOD, COD, SS, pH etc., along with specific chlorinated organic substances, such as chlorinated resin acids, chlorinated phenolics and PCDD/Fs, along with several aggregated parameters, e.g., EOCl and AOX. In this case study, all these parameters were analysed for the main WW effluents discharged from the mill, as well as for different components (water, sediment and biota) of the receiving environment bodies.

11.2.2.1 Physicochemical Parameters of Effluents

The mill discharged high pollution loads, in terms of BOD, COD and SS. The average specific discharge per tonne ADP into River B during the study period was 140 kg of filtered COD

TABLE 11.1

Average Physicochemical Characteristics for Selected Wastewater Effluents

Effluents	pH	Cond (µS/cm)	COD_f (mg/L)	BOD_f (mg/L)	SS (mg/L)
Combined effluent (point 19)*	6.5 (4–8)	1850 (1600–2100)	400 (350–460)	180 (100–260)	280 (140–400)
Raw material (point 1)	8.0	220	20	30**	250
Sedimentation basin (point 17)	8.0	680	480	370	800
Chemical plant (point 22)	11.0	4100	30	NA	70
Combined bleaching (point 12)	3–4	2000	500	NA	130
C-stage bleaching	2–2.5	4600	390	NA	120
E-stage bleaching	10–11	4300	1200	NA	140
H1-stage bleaching	8–9	3400	470	NA	110
H2-stage bleaching	7–8	2200	100	NA	80
Coal ash overflow (point 2)	7.0	250	30	NA	4000
Combined paper mill (point 16)	4.5–5.5	430	20	60**	900

* The range (lowest and highest value) of 95% confidence limits of the analytical results for the samples taken on different days is given in brackets; sampling points are the same as in Figure 10-2 (Chapter 10).
** BOD of unfiltered wastewater.
Source: Extracted from Kim Oanh (1994).

(COD_f), 62 kg of filtered BOD (BOD_f) and 100 kg of fibrous SS. The specific water consumption in the production line was 390 m^3/t ADP product, and for the entire company, 515 m^3/t ADP product, which was 3–6 times as much as an average Scandinavian mill during the study time (Kim Oanh and Bengtsson, 1995a). An average of 50,000 m^3/day of the combined effluent was discharged into River B. The rice field received about 20,000 m^3/day of different WW, which further fed into a fish pond (Figure 10.3, Chapter 10). The results of physicochemical characteristics of the samples analysed by the toxicity tests are shown in Table 11.1.

11.2.2.2 Toxicity Testing to Assess Potential Biological Effects

11.2.2.2.1 Types of Toxicity Tests

Three types of ecotoxicological tests were used to characterise the potential biological effects of the WW discharges from the mill. These included the acute toxicity tests with Microtox bacteria, green micro-alga (*Selenastrum capricornutum*), and duckweed (*Lemna aequinoctialis*), as detailed in Kim Oanh and Bengtsson (1995b). The Microtox test exposes a luminescent marine bacterium, *Photobaeterium phosphoreum*, to water samples and measures the toxic effect on the organism, usually after 5 or 15 minutes. The result is expressed as the EC50 value, which is the effective concentration of a water sample that causes a 50% reduction in the light output of the bacteria. Since the bacteria are sensitive to pH<6 and pH>8, the pH of the WW samples was adjusted to about 7.0. Microtox bacteria have been shown to classify toxic substances in the same way as higher organisms, e.g., fish or daphnia, hence the test is considered as a suitable pre-screening test for a wide range of effluents (SEPA, 1990). The duckweed test was conducted with a 7-day exposure and the test indicator was the growth over 7 days, i.e., 7d EC50, calculated using the dry biomass production. Note that this study did not analyse the production of new fronds (a leaf-like structure) and *chlorophyll*, which have been previously reported to be useful test parameters (Bengtsson and Triet, 1994). The micro-algae test was conducted with a 3-day exposure and the cell density (counts) was used as the test parameter to calculate the 3d EC50 (Bengtsson and Cerino, 1993). The EC50 results of all the toxicity test types were also expressed in the toxicity unit (TU) which was calculated as 100/EC50 for every tested sample.

11.2.2.2.2 Wastewater Sampling

The tests were also performed for the WW samples collected from different production units in the mill. These include the WW samples collected from the following sampling points (locations are indicated in Figure 10.2, Chapter 10): 1) raw material preparation (point 1), coal ash basin overflow (point 2), combined effluent from the paper mill (point 16), sedimentation basin influent (point 17), mill combined effluent (point 19), bleaching filtrates from each bleaching stage, and the combined bleaching effluent (point 12). The WW samples used for the toxicity tests were both individual (grab) and composite. Individual grab samples were preserved deep-frozen (-20 to -30°C) and the compositing was done just before the tests. Either the time-proportional (TC) or flow-proportional (FC) samples were prepared depending on the variations of flows and concentrations in the effluents, as explained in Chapter 9. To identify the most potentially harmful effluents, the specific toxicity discharge factor (TDF) from each production unit was calculated as the product of the TU and the specific WW flow-rate (m^3/t product) from the unit.

11.2.2.2.3 Sampling of Water in the Receiving Bodies

The potential effects of the discharges on the receiving environmental bodies were assessed using the toxicity tests not only for the combined mill effluent (point 19) but also for water samples collected around the discharge pipe outlet in River B (main receiving water body), from the rice field and the fish pond that received some parts of the mill effluents.

11.2.2.2.4 Toxicity Test Results

The Microtox test results shown in Table 11.2 indicate that the bleaching plant effluents were the most toxic, i.e., with the lowest EC50 (or highest TU), especially for C- and H-stages. The C-stage filtrate had the highest discharge flow-rate of 60 m^3/t ADP whereas other stages had about 20 m^3/t ADP each, and this explained its highest specific TDF value. Except for the bleaching effluents, most of the effluents from the mills were shown to be non-toxic for Microtox bacteria (EC50>100%), as seen in Table 11.2. The toxicity of the filtrates from all the bleaching stages was found to increase considerably from 5-min to 15-min exposure, whereas the results were stable after 30 minutes of exposure as reported in Kim Oanh (1994). The bleaching plant effluent had a 15-min TDF of about 750–860/t ADP, mostly from the C-stage. The average 15-min TDF of the combined effluent discharge (point 19) was 645–825/t APD, i.e., showing a less than additive (antagonistic) mode of toxicity was observed. This means that the sum of the individual TDFs of the incoming WW streams was higher than the TDF measured at the combined effluent, which is a common pattern observed for bleached Kraft pulp effluents (Bonsor et al., 1988).

The Microtox tests conducted for the water samples collected from the receiving water bodies, i.e., the fish pond and the rice field, showed no toxic effects. Samples taken around the discharge outlet in River B were also not toxic during the period of high water (in the rainy season) when the discharge pipe was submerged at about 50 cm below the water surface. However, in the period of low water (dry season) when the discharge pipe was less than 10 cm submerged, the samples were toxic with EC50 in the range of 40–70%. At a distance of 100 m downstream in the river no toxic effects were detected in all water samples taken in both dry and rainy seasons. The high dilution capacity of the river reduced the toxicity in the water samples considerably as compared to that at the pump station.

The Microtox test was performed for all collected WW samples while the duckweed and micro-algae tests were only done for the combined mill effluent (point 19). In total, the Microtox tests were conducted on 11 grab WW samples and 8 TC (time-composite) WW samples collected on 10 monitoring days. The micro-algae tests were conducted for 6 TC samples (8-h and 24-h TC). Two TC samples (one 8-h and another 24-h) of the combined effluent were also tested with duckweed. The Mictorox test results for the grab WW samples showed large variations in EC50, from 7% to over 65%, and even >100% in one sample. A comparative analysis of the results showed that the micro-algae were the most sensitive, followed by Microtox bacteria,

TABLE 11.2

Summary of the Microtox Test Results for Mill Effluents and Water in Receiving Environmental Bodies

Effluents	5min-EC50 (%) and range[a]	TDF	15min-EC50 (%) and range[a]	TDF
Combined bleaching (point 12)	16–24 (9–26)	500–750	14–16 (8–21)	750–860
C-stage bleaching	10–26 (8–33)	230–600	5–13 (4–20)	460–1200
E-stage bleaching	10–26 (8–28)	77–200	11–27 (8–64)	74–180
H1-stage bleaching	46–52 (38–70)	38–43	38–45 (24–60)	44–53
H2-stage bleaching	60–61 (40–92)	33–34	39–83 (35–130)	25–51
Pump station (point 19)	45 (36–53)	695–795	45 (38–53)	645–825
Sedimentation (point 17)	NA	NA	15 (10–22)	—
Paper machine (point 21)	>100	—	>100	—
Paper mill combined (point 16)	>100	—	>100	—
Chemical plant (point 22)	>100	—	>100	—
Raw material washing (point 1)	>100	—	>100	—
Coal ash basin (point 2)	>100	—	>100	—
Fish pond	>100	—	>100	—
Discharge outlet in River B in low water (dry season)	NA	—	40–70[b]	—
Discharge outlet in River B in high water (rainy season)			>100	
100 m downstream from outlet in River B	NA	—	>100	—

[a] The range presents the lowest and the highest value of 95% confidence limits of the toxicity test results for the samples taken on different days.

[b] Lower value (more toxic) when the outlet was near the surface, higher value when the outlet was 10 cm submerged.

NA-: not analysed

Source: Extracted from Kim Oanh (1994).

while the duckweed was not sensitive (EC50>100%). Specifically, the average EC50 for the combined mill effluent (point 19) based on the Microtox test was 45% (Table 11.2), and that for the algae test was 25% with a 95% confidence interval of 10–40% (Kim Oanh and Bengtsson, 1995b).

Overall, the combined effluent was considered only "moderately" acute toxic for the test organisms used in this study, and the results are considered typical for Kraft mill effluents (Kim Oanh and Bengtsson, 1995b). However, the large volume of the WW discharged into the environment could result in adverse sub-lethal effects on the biota, especially in the period of low water in the receiving water bodies. The large variability of EC50 in the Mictorox test results of the grab samples indicated that there was a possibility of discharging acutely toxic WW from the mill intermittently, which was expected for this unstable operating industry. In addition, many organochlorinated compounds discharged from the bleaching plant are bioaccumulating and persistent and hence would cause sub-lethal toxic effects to aquatic life in the long term and eventually would affect the organisms on a higher level in the food chain including humans.

11.2.2.3 Chlorinated Compounds in Effluents and Receiving Environment

The conventional bleaching sequence (C-E-H1-H2) in the mill consumed an average of 40–50 kg of elemental chlorine and 30 kg hypochlorite per tonne ADP to produce the pulp with brightness of 76–80% ISO. This large amount of chlorine compounds used in the process ultimately would generate large amounts of toxic OCls, which would be released into the environment. Most of the bleaching filtrate (120 m^3/t ADP or nearly 17,000 m^3/day) from the mill was discharged to the pump station, directly or via the sedimentation basin, and eventually into River B (Figure 10.2, Chapter 10). The rice field and the fish pond received the WW from the raw material preparation unit, as well as the overflows from the pump station (containing bleaching filtrates), the coal ash basin (located in the power plant) and the emergency pond. The overflow from the coal ash basin was relatively low (2000 m^3/day) but may contain heavy metals.

11.2.2.3.1 Sampling

For the bleaching WW effluents, one sample of 2 L of the bleaching filtrate was collected from the storage tank of every bleaching stage of C, E, H1 and H2. A large volume of filtrate was accumulated over the period while being stored in the tanks; hence these filtrate samples were considered equivalent to the time-composite. A grab sample of well-mixed bleached pulp was taken after the last bleaching stage. The blank QC samples included one water sample from the inlet (from River A) into the raw water treatment plant, one sample of tap water distributed in the mill and one distilled water sample (a trip blank). These QC samples were handled in the same manner as the WW effluent samples.

For the receiving environments, Figure 11.1 shows the sampling points for water (marked as W1), sediments (S1–S4), rice grain (R1–R4), fish (F), crab (C), and mollusc (M1–M2). The biota samples for OCl analyses were taken from the rice field and the fish pond, which received only a fraction of the mill effluents containing bleaching filtrates, but the exposure was unavoidable due to the confined and small size of the receiving bodies. Two reference samples (QC samples) of rice grains (R5 and R6) were collected from the local paddies receiving no mill effluents.

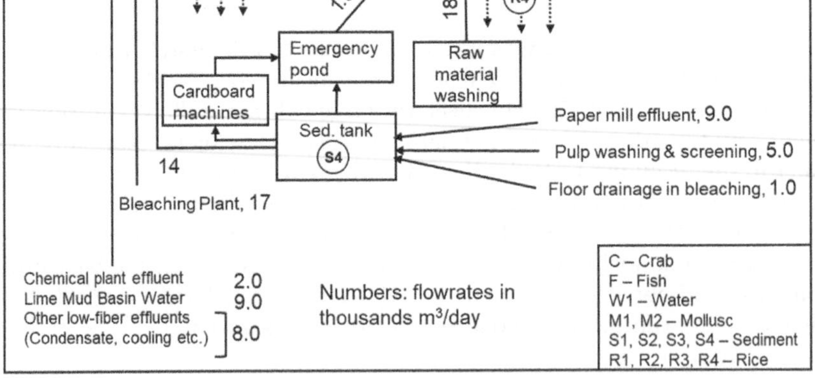

FIGURE 11.1 Sampling locations in the monitoring programme to assess potential effects on the receiving environment. (Numerical values are the effluent flow-rates in 1000 m^3/d, and circled letters are the sampling substances shown in the legend.)

Source: Adapted from Kim Oanh et al. (1995).

11.2.2.3.2 Sample Analyses

The monitoring parameter of adsorbable organic halogens (AOX) was used as a rough non-specific summary expression of the overall organochlorine flux in the mill WW effluents. The AOX analysis is simple and fast; hence it would be a good monitoring parameter for process control and regulatory purposes. In this case study, AOX analyses were done for the TC water samples collected in River B at a distance of 1 km downstream from the discharge outlet, and either TC or FC samples collected from the mill WW effluents (sedimentation basin influent, combined paper mill effluent, combined bleaching effluent, and combined mill effluent).

Extractable OCls (EOCls) were analysed for the sediment samples. These compounds present the hydrophobic part of AOX, i.e., the chlorinated compounds that have the potential to bioaccumulate, as well as a tendency to resist biodegradation, and hence are considered to be a more relevant parameter to assess ecological and environmental effects than AOX. The sample preservation and analytical methods are detailed in Kim Oanh et al. (1995, 1999).

The bioaccumulating and persistent compounds, i.e., chlorinated phenolics and dioxins, in the mill effluents and receiving water bodies were also analysed to better assess the environmental impacts of the mill. The chlorinated dioxins and furans, from tetra (4 Cl) through octa (8 Cl), and non-ortho polychlorinated biphenyls (dioxin-like PCBs: IUPAC#77, 126 and 169) were analysed for the bleaching WW effluents, bleached pulp, sediment samples and biota (fish, mollusc and crab). Note that the non-ortho PCBs or coplanar PCBs are also known as dioxin-like PCBs because of their similar molecular structure to dioxins, which allows them to act alike and contribute to overall dioxin toxicity. The rice samples were analysed for chlorinated phenolic compounds (22 compounds) and heavy metals. Heavy metals were analysed for the TC samples of the coal ash basin overflow (W1). A summary of the samples that were analysed for OCls and heavy metals is provided in Table 11.3.

TABLE 11.3

Environmental Samples for Organochlorinated Compounds and Metal Analyses

Sample identification	Receiving mill effluents	Parameters analysed
Rice 1	Overflow of the combined effluent	OCls; heavy metals
Rice 2	Overflow of the combined effluent Coal ash basin overflow	OCls; heavy metals
Rice 3	Overflow of the combined effluent Coal ash basin overflow	OCls; heavy metals
Rice 4 (sticky)	Same as Rice 3	OCls; heavy metals
Rice 5	No mill effluents[a]	OCls; heavy metals
Rice 6 (sticky)	No mill effluents[a]	OCls; heavy metals
Mollusc, M1	Same as Rice 3	Heavy metals; bioconcentration
Sediment, S1	The combined effluent (sampling in River B)	EOCl; dioxins
Sediment, S2	Overflow of the combined effluent	Dioxins; PCBs
Sediment, S3	Same as Rice 3	OCls; Heavy metals; dioxins, PCBs
Sediment, S4	Fiber-rich effluents (sedimentation tank sludge)	EOCl; Dioxins
Mollusc, M2 Fish, F Crab, C	Same as Rice 3	Dioxins; PCBs
Water, W1	Overflow from coal ash basin	Heavy metals

[a] Reference samples were taken from the paddies close to the target field but did not receive the mill effluents.

11.2.2.3.3 Results for Levels of Chlorinated Compounds

The analytical results for OCls in the samples of effluents and receiving environment components are highlighted next.

AOX: This was detected in all samples, and the highest was in the combined bleaching effluent (27.5 mg/L), followed by the combined mill effluent (pump station) with an average of 13.8 (range: 12.6–14.8) mg/L. Lower levels were found in the paper mill effluent and the influent to the sedimentation basin with 1–1.6 mg/L. Overall, a high specific discharge of AOX from the mill was found, 4.7 kg/ton ADP, which was more than 20 times above that of the mills with relatively more modern technologies in the 1990s (Axegård et al., 1993). AOX was also detected in the water in the receiving river (1 km downstream in the discharge plume) but at a low level of 0.16 mg/L. Further, the fact that the active chlorine was found in the WW influent to the sedimentation basin and in several samples of the mill combined effluent indicated the spills of bleaching chemicals and filtrates.

EOCls: High levels of EOCls (70–100 µg/g dry weight) were found for the sediment samples of the accumulated fibres in the sedimentation basin (S4). The levels were increasing with depth, which may be explained by the increase of the sludge residence time. The levels in River B (S1) were low, 0.4–0.8 µg/g dry weight. The high-water flow in River B, perhaps, washed and transported away the sediment deposit on the riverbed.

Chlorinated phenolics: Chlorophenolics were analysed for the bleaching filtrates, sediment, and rice samples. These included five groups, namely, chlorophenols (five compounds), chloroguaiacols (five compounds), chlorovanilins (four compounds), chlorocatechols (six compounds) and two chlorosyringols (two compounds).

In the mill effluents, the bleaching stages of H1 and H2 had the lowest levels of chlorinated phenolics, with only two compounds detected, each at about 1 µg/L. For the first three groups, the E-stage filtrate had the most compounds detected (14) and with the highest levels (1–87 µg/L), followed by the C-stage (13 compounds, 1–19 µg/L). However, for the last two groups, C-stage filtrate had higher levels, especially for 3,4-dichlorosyringol, which was detected at 320 µg/L as compared to 100 µg/L in the E-stage. The chlorophenolics discharge load was the highest from the C-stage with 23 g/t ADP, followed by the E-stage with 8 g/t ADP, while the two H-stages collectively had only 0.08 g/t ADP. The total discharge of the chlorophenolics from the mill was around 31 g/t ADP or around 4.5 kg/day for the average production of 140 t ADP/day. These chlorophenolics can be toxic (e.g., chlorophenols, guaiacols and catechols) and bioaccumulating (e.g., chlorophenols and chloroguaiacols); hence the discharges with the mill effluents may pose immediate, long-term and far-spreading environmental effects.

In the receiving bodies, the fish pond sediment sample (S3) had nine chlorophenolics detected with the highest level of 3,4,5-trichlorocatechol (120 µg/kg) followed by 3,4-dichlorocatechol (100 µg/kg), and other compounds in the range from 2.1 to 31 µg/kg sediment on the dry weight basis. No distinct differences were seen between the chlorophenolics levels in the four rice grain samples, R1–R4, or between these rice samples and the reference rice samples of R5 and R6. The highest concentrations of the chlorophenolics in the rice grain samples were found for 3,4,6-trichlorocathecol in R3 with 5.7 µg/kg and 3,4-dichlorocatechol in R2 with 2.9 µg/kg grains on the dry weight basis. A reference rice sample, in fact, had the highest level of 4,6-dichloroguaiacol with 4.3 µg/kg. Hence there was no evidence of bioaccumulation of the compounds in the rice grains caused by the exposure to the mill effluents. The low levels of these compounds found in the samples may be explained by the degradation by microorganisms in the soil, low uptake/metabolisation of the compounds by the plants and the association with the sediment/soil making the compounds less bioavailable.

Dioxins: The chlorinated dioxins are highly toxic and persistent compounds that belong to two families of chemicals, the chlorinated dibenzo-p-dioxins (75 congeners) and the chlorinated dibenzofurans (135 congeners). In this case study, the dioxins were analysed for the mill WW effluents, the bleached pulp and sediment samples S2 and S3.

The results of dioxins for bleaching effluents and bleached pulp are summarised in Table 11.4. Note that no dioxins were detected in the trip blank samples. In the water QC samples, only OCDD was found, and it was higher in the raw water than in the tap water, i.e., 529 pg/L vs. 47 pg/L, suggesting that a possible removal of the compound associated with the removal of suspended solids in the raw water treatment plant. The results are presented for individual compounds along with the Nordic Toxicity EQuivalents (N-TEQ) which is an aggregated parameter to express the total level of dioxins in a sample (equivalent to the most toxic dioxin congener/compound of 2,3,7,8-TCDD).

The highest dioxin levels were found in the filtrate from the E-stage, followed by C, H1 and H2-stage with the N-TEQ of 33.5, 1.15, 0.56 and 0.014 pg/L, respectively. The corresponding N-TEQ loads discharged from the four bleaching stages were in the same ranking with 670, 69, 11.2 and 0.28 ng/t ADP. Large numbers of congeners were detected in the C- and E-stage filtrates, and the bleached pulp, with 2,3,7,8-TCDF and 2,3,7,8-TCDD found at relatively high levels. Some penta-, hepta- and octa- congeners were also detected, especially in the E-stage filtrate. Since the bleaching filtrates were partly recycled within the bleaching plant, dioxins were also carried over with the progressive transfer of bleached pulp from one stage to the other. The ranking found therefore may not necessarily reflect the rank in levels of dioxins formed in the bleaching stages, i.e., it is likely that more dioxins were formed first in the C-stage, but the levels in the E-stage were the highest of all. Several factors affect dioxin formation in the bleaching process, e.g., the amount of elemental chlorine used, cooking process, lignin content, use of condensate in pulp washing, possible dioxin precursors in the water supply and in the raw pulping materials etc. (Luthe and Berry, 1996).

The large amount of WW discharged from the mill diluted the compounds and lowered the concentrations of dioxins in the bleaching plant effluent (N-TEQ of 6.3 pg/L) and in the final mill effluent (1.9 pg/L). However, the specific discharge load, as expected, was comparable with other mills using similar technologies (Kim Oanh et al., 1999). The dioxins load discharged with the effluent of this mill into the receiving bodies, however, was remarkably higher than that from the mills, with more advanced bleaching sequences using less chlorine. Rappe et al. (1990) reported a level of dioxins in the bleached pulp of 0.1–0.2 pg N-TEQ/g on the dry weight basis for a sulphate mill using less chlorine (with oxygen pre-bleaching) and good pulp washing, which was much lower than the level of 4.96 pg/g found in this study. The high dioxin loads from the study mill into the environment should be reduced through radical modifications in the bleaching process to reduce the chlorine use, i.e., with elemental chlorine-free (ECF), and eventually to eliminate chlorine use with the total chlorine-free (TCF) bleaching technologies.

In the receiving environment, dioxins including 2,3,7,8-TCDD/F were detected in most sediment samples with the levels in N-TEQ ng/kg of 0.5–2.3 for S4, 2.2 for S2, and much lower (0.3) for S3. Both samples S2 and S4 demonstrated a "pulp bleaching pattern" of dioxins. Sediment samples from River B (S1) had low levels with N-TEQ of 0.03–0.09 ng/kg dry weights, which were close to the blank levels.

In the biota, both 2,3,7,8-TCDF and 2,3,7,8-TCDD were detected in all samples, but the levels of individual dioxins were relatively low, well below 1 ng/kg of body on a wet weight basis. In terms of N-TEQ, the crab had the highest dioxin concentration of 2.5 ng/kg, followed by mollusc with 0.85 ng/kg, and the lowest was in fish with about 0.2 ng/kg. The low levels found in S3 and the biota samples may probably be because only a small amount of bleaching filtrates overflowed into these receiving environmental bodies (Kim Oanh et al., 1995).

Non-ortho PCBs: These compounds are commonly referred to as the dioxin-like PCBs due to their similar toxicological properties to dioxins, and they are often considered together with dioxins in the context of public health.

In the biota samples, all three analysed non-ortho PCBs compounds were detected with PCB#77 and PCB#126 at about 5–10 times higher than PCB#169. The first 2 PCBs had the highest levels in crab (20–35 ng/kg body wet weight), followed by mollusc (4–8 ng/kg wet weight) and fish (about 2 ng/kg wet weight). In the sediment samples, the levels of the first 2 PCBs were 2–24 ng/kg dry weight, but PCB#169 was not detected. Note that in the bleaching filtrates, the first two PCB levels were close to those in the trip QC blank and lower than the raw

TABLE 11.4

Dioxins in Bleaching Effluents and the Bleached Pulp

		Bleaching stage effluent, pg/L					
Dioxin compounds	**Bleached pulp, pg/g ADP**	**C**	**E**	**H1**	**H2**	**Combined bleaching[a]**	**Load in the filtrate, pg/g ADP**
2348/2378-TCDF	12.8	0.64	1.25	nd	nd	0.53	0.06
Total TCDF	153	19.8	138	16.4		35.6	4.3
2378-TCDD	3.65	1	19.7	nd	nd	3.8	0.46
Total TCDD	25.4	52	1440			266	32
12348/12378-PnCDF	nd	nd	nd	nd	nd	0	0
23478-PnCDF	nd	nd	2.7	nd	nd	0.45	0.05
Total PnCDF	17.1		100			16.7	2
12378-PnCDD		nd	nd	nd	nd	0	0
Total PnCDD		16.1	2520			428	51
123479/123478-HxCDF	nd	nd	nd	nd	nd	0	0
123678-HxCDF	nd	nd	nd	4	nd	0.67	0.08
123789-HxCDF	nd	nd	nd	nd	nd	0	0
234678-HxCDF	nd	nd	nd	nd	nd	0	0
Total HxCDF	7.5		8.3			1.4	0.17
123478-HxCDD	nd	nd	12.1	nd	nd	2	0.24
123678-HxCDD	nd	nd	68.1	nd	nd	11.4	1.4
123789-HxCDD	nd	nd	38.3	nd	nd	6.4	0.77
Total HxCDD			796			133	16
1234678-HpCDF	nd	nd	9.6	nd	nd	1.6	0.19
1234789-HpCDF	nd	nd	nd	nd	nd	0	0
Total HpCDF			9.6			1.6	0.19
1234678-HpCDD	2	4.1	35.1	16.3	1.14	10.8	1.3
Total HpCDD	3.4	4.1	63.5			12.6	1.5
OCDF	nd	19.5	nd	nd	nd	9.8	1.2
OCDD	7.5	25.5	34.4	nd	2.5	19	2.3
N-TEQ, pg/L	**4.96** [b]	**1.15**	**33.5**	**0.56**	**0.014**	**6.3**	**0.75**
N-TEQ load, ng/t ADP	**4960**	**69**	**670**	**11.2**	**0.28**	**750**	**750**

Note: nd – not detected; N-TEQ: Nordic Toxicity EQuivalent.

[a] Values for the combined bleaching effluent were the weighted averages against the specific effluent flow-rate of each stage.

[b] Value in pg/g ADP.

Source: Extracted from Kim Oanh et al. (1999).

TABLE 11.5

Levels of Heavy Metals in Samples (aver. ±1SD of 3–4 repeated analyses)

Element	Water (W1), μg/L	Sediment (S3), mg/kg	Mollusc (M1), mg/kg	BCF for Mollusc	Average (R1–R3), μg/kg	Ref. R5, μg/kg	MDQ*, ng
As	727 ± 7	13.2 ± 0.6	16.6 ± 0.33	23	197 ± 14	248 ± 24	0.5
Cd	12.3 ± 0.2	0.36 ± 0.01	3.84 ± 0.01	313	8.5 ± 0.9	22 ± 7	0.07
Co	73.4 ± 0.5	10.4 ± 0.6	10.9 ± 0.14	149	4.5 ± 2.5	nd	0.07
Cr	72.6 ± 1	67.4 ± 0.5	1.4	1304	171 ± 6.2	154 ± 16	2
Cu	251 ± 4	104 ± 4	185 ± 0.1	738	837 ± 33	1060 ± 60	2
Hg	25.9 ± 0.9	0.12 ± 0.04	0.62 ± 0.001	24	113 ± 9	nd	0.5
Ni	201 ± 2	19.3 ± 1.2	30.1 ± 0.44	150	nd	nd	1
Pb	621 ± 45	73.4 ± 1.5	34.5 ± 0.48	56	129 ± 14	91 ± 6	0.5
Zn	1010 ± 3	155 ± 3	409 ± 0.3	405	16,500 ± 16	15,900 ± 100	30

Note: nd: not detected;
* *MDQ:* minimum detectable quantity of the method.
Source: Extracted from Kim Oanh et al. (1995).

water samples, while the third PCB was not detected. The results thus did not show a direct link between the levels of non-ortho PCBs found in the environment with the discharge from the mill (Kim Oanh et al., 1995, 1999).

Heavy metals and bioconcentration: The coal ash basin contained the fly ash discharged from the built-in coal-fired power plant, which had an overflow to the environment (Figure 11.1). This was one of the most important discharge sources of heavy metals from the mill. In total, nine elements were analysed, and the results are presented in Table 11.4.

Among the analysed samples, namely, the coal ash basin overflow water (W1), sediment (S3), mollusc (M1) and rice samples, the higher levels of metals were detected in the sediment and mollusc samples. Although metals contained in sediments are less bioavailable in principle and hence are less toxic to aquatic organisms than those in water, the contaminated sediments are a source of heavy metals to overlaying water to affect the aquatic organisms in the long term. The bioconcentration factor (BCF) was calculated as the ratio between concentration in water and that in the mollusc, and the results indicated significant bioaccumulation with BCF ranging from 23 to 1304. The high bioconcentration factors of heavy metals found in molluscs posed a potential risk for human intoxication through the food chain.

There was no evidence of bioaccumulation of the heavy metals in the rice grains because the levels in all rice samples were lower than the reference rice samples taken from the local paddies receiving no discharge from the mill. Among the rice samples, R1 was from an upper land paddy and was less exposed to the coal ash basin overflow, while Rice 2–4 were exposed to the overflow to a similar extent (Figure 11.1). Nevertheless, similar levels of every element were found in the samples of R1, R2 and R3, which belonged to the same rice cultivar. The reference sample R5 (normal rice) even demonstrated higher levels of As, Cd and Cu than the R1–R3 samples. It is worth mentioning that, in R4, a glutinous rice strain, higher element levels were detected, which were close to those found in the glutinous rice reference R6 (not shown in Table 11.5), suggesting a possible influence of the rice strain (Kim Oanh et al., 1995).

Briefly, in this case study, a combination of biological tests and chemical analyses was carried out to assess the potential environmental impacts of the mill discharge on different components of the receiving bodies (water, sediments, biota). The results provided an important basis for the planning of environmental control measures for this factory.

11.3 Summary

The most important function of an environmental monitoring system of industrial factories is to prevent pollution discharges from causing environmental problems occurring outside the factory. This in turn requires a well-designed monitoring program to characterise potential harmful effects of the industrial pollution discharged to the receiving environment. This chapter highlights the principles of designing such a program illustrated with a case study of monitoring for assessment of the environmental effects of a chlorine-bleached Kraft pulp and paper mill.

The mill used facilities generating large amounts of environmental pollution including highly toxic OCls. Relevant parameters typical for this type of industrial facility were selected and necessary sampling and analytical protocols were implemented. These included the acute toxicity test to assess the potential biological effects, and the physicochemical parameters analysis to reveal the causes of the toxic effects. In particular, the study also investigated specific organochlorinated compounds of dioxins, non-ortho PCBs and chlorinated phenolics, along with heavy metals in the mill effluents and different components of the receiving environment, i.e., biota, sediment and water.

The combined effluent was considered "moderately" acute toxic for the test organisms used in this study, but the large volume of the effluent discharged into the environment could result in adverse sub-lethal effects on the biota. The concentrations of chlorinated phenolics, dioxins and PCBs in rice grain and biota samples were found at relatively low levels. Nevertheless, the high amount of chlorine used in the bleaching process suggested that a substantial amount of these compounds could be formed and discharged into the environment, and their low levels found in the samples may be merely a result of high dilution factors. Radical modifications of the bleaching process toward more advanced technologies, such as the element chlorine-free and total chlorine-free, are required to reduce and eventually eliminate the discharges of these toxic chlorinated compounds.

REFERENCES

Axegård, P., Dahlman, O., Haglind, I., Jacobson, B., Morck, R. and Stromberg, L., 1993, Pulp bleaching and the environment-the situation 1993, *Nordic Pulp & Paper Research Journal*, 8(4), 365–378.

Bengtsson, B-E. and Cerino, G.F., 1993, *Method for toxicity test with micro-algae, Selenastrum capricornutum, project report for the AIT/SIDA cooperative environment programme* (p. 24), Asian Institute of Technology (AIT), INRDM.

Bengtsson, B-E. and Triet, T., 1994, Tapioca starch wastewater toxicity characterized by Microtox and duckweed tests, *Ambio*, 23(8), 473–477.

Bonsor, N., Mccubbin, N. and Sprague, J.B., 1988, *Stopping water pollution at its source – Kraft mill effluents in Ontario, report prepared for the technical advisory committee, pulp paper and sector of MISA*, Ontario Ministry of the Environment, Toronto, Ontario, Canada.

Kim Oanh, N.T., 1994, *Wastewater monitoring, a tool to optimize process control and reduce contamination from bleached Kraft pulp and paper industry: Vietnamese case studies. AIT dissertation no. EV-02-94*, Asian Institute of Technology, Bangkok, Thailand.

Kim Oanh, N.T. and Bengtsson, B-E., 1995a, Development of a wastewater monitoring program incorporated into process control for mitigation of chemical and fiber loss from the Bai Bang paper company (BAPACO), a bleached Kraft pulp and paper mill in Vietnam, *Resources, Conservation and Recycling*, 14(1), 53–66, https://doi.org/10.1016/0921-3449(95)00005-4

Kim Oanh, N.T. and Bengtsson, B-E., 1995b, Toxicity to Microtox, micro-algae and duckweed of effluents from the Bai Bang paper company (BAPACO), a Vietnamese bleached Kraft pulp and paper mill, *Environmental Pollution*, 90(3), 391–399, https://doi.org/10.1016/0269-7491(95)00008-F

Kim Oanh, N.T., Bengtsson, B-E., Baetz Reutergårdh, L., Hoa, D.T., Bergqvist, P.A., Broman, D. and Zebühr, Y., 1999, Persistent organochlorines in the effluents from a chlorine bleached Kraft integrated pulp and paper mill in Southeast Asia, *Archives in Environmental Contamination and Toxicology*, 37, 303–309.

Kim Oanh, N.T., Bengtsson, B-E., Reutergårdh, L-B., Bergqvist, P.A., Hynning, P-X., Remberger, M., 1995, Levels of contaminants in effluent, sediment, and biota from Bai Bang, a bleached Kraft pulp and paper mill in Vietnam, *Archives of Environmental Contamination and Toxicology*, 29, 506–516.

Luthe, C.E. and Berry, R.M., 1996, The role of dibenzo-p-dioxin and dibenzofuran precursors in the formation of tetrachlorinated dibenzo-p-dioxins/-furans during bleaching, *Chemosphere*, 32(5), 881–891, https://doi.org/10.1016/0045-6535(96)00022-7

Network for Industrial Environmental Management (NIEM), 1993, *Training package 2, vol. 1: general introduction: Receiving water quality evaluation and receiving land quality evaluation*, Swedish International Development Authority (SIDA) and United Nations Environment Programme (UNEP) Regional Office for Asia and the Pacific, Bangkok, Thailand.

Rappe, C., Glas, B., Kjeller, L.O., Kulp, S.E., De Wit, C. and Melin, A., 1990, Levels of PCDDs and PCDFs in products and effluent from the Swedish pulp and paper industry and chloralkali process, *Chemosphere*, 20(10–12), 1701–1706.

Swedish Environmental Protection Agency (SEPA), 1990, *Biological chemical characterization of industrial wastewater*, Swedish Environmental Protection Agency, Stockholm, Sweden.

12

Design and Implementation of Ambient Air Pollution Monitoring Programme in Asian Regional Project of AIRPET

Air pollution presents a significant threat to human health and the environment globally. High pollution levels in the Asian developing countries and the associated adverse effects make air quality improvement a priority agenda of many national, regional and urban environmental agencies. To develop effective air quality management strategies for a domain, e.g., an urban area, a region, or a country, the foremost requirement is to have thorough information on the nature of air pollution problems in relation to the local emission sources, long-range transport, and atmospheric transformation processes. This in turn requires a comprehensive analysis of air quality in the domain, which can be obtained by using well-designed and properly implemented air quality monitoring programs. Ambient air quality is a dynamic and complex environmental phenomenon that exhibits large temporal and spatial variations reflecting changes in the source emission and meteorological and topographical conditions. Therefore, it is always a huge challenge to obtain representative data to comprehensively characterise air quality, especially when constrained by resources availability. This chapter discusses the design and operation of ambient air quality management programmes conducted within the Asian Regional Air Pollution Research Network, or AIRPET. One of the main research objectives of AIRPET was to provide a comprehensive assessment of the PM pollution focusing on $PM_{2.5}$ and PM_{10} levels and composition in the participating six countries, harmonised by a consistent QA/QC procedure.

12.1 Introduction: Overview of AIRPET

Developing Asia has high levels of air pollution and a high number of attributable mortalities, which are the consequence of the large amounts of toxic pollutants emitted from human-made and natural sources. Key anthropogenic emission sources include road traffic, industrial facilities, power plants, domestic combustion, solid waste open burning, biomass open burning, construction and fugitive emissions such as road dust etc. Among the air pollutants, of the most concern from the health effect point of view is the fine particulate matter or $PM_{2.5}$, which has been observed at episodic levels in many places during the dry season. According to the State of Global Air report (HEI, 2020), in 2019 air pollution was the fourth leading risk factor for premature death worldwide. It caused about 7 million deaths due to exposure to ambient $PM_{2.5}$, O_3 and household air pollution, sharing 12% of the global total mortality from all causes. About 70% of these air pollution-related death cases occurred in Asia and the Pacific. Air pollution puts a burden on economic development due to the excessive cost associated with mortality, morbidity, damage to crops and properties, and loss of tourism, among others.

There has been growing research interest in the characterisation of levels and compositions of size-segregated PM, especially in developing Asia where such information is only fragmentedly available. Intensive monitoring programs are required to gather such data, which helps in understanding the nature of particulate pollution observed in a particular geographical domain. This kind of information is essential for the development of effective air quality

DOI: 10.1201/9781003597216-14

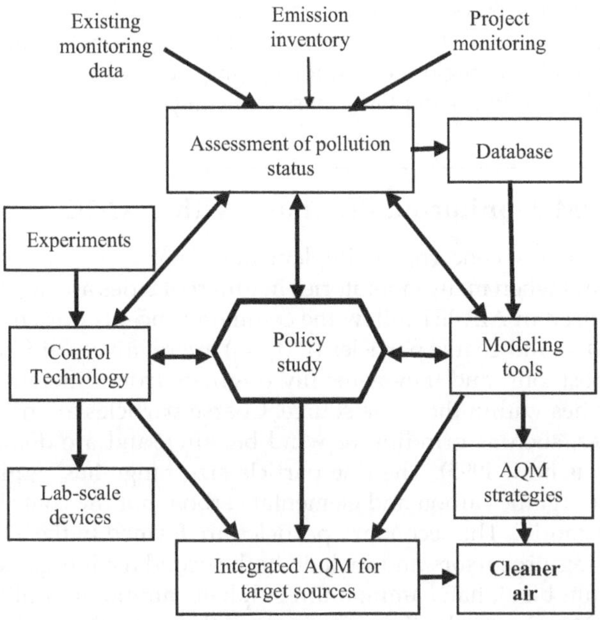

FIGURE 12.1 Overall framework of AIRPET.

Source: Modified from Kim Oanh and Polprasert (2014).

management strategies. However, systematic, regionally compatible, and long-term PM data records were not available in many Asian developing countries during the 2000s. To partly fill up the gap, the Asian Regional Air Pollution Research Network (AIRPET) was designed and implemented in six Asian countries (Figure 12.1). Activities were conducted in the participating countries from 2001 to 2010 focusing on the key elements of integrated air quality management:

- Establishment of a comprehensive assessment of air pollution status in the countries;
- Development of suitable air pollution control technologies;
- Development of emission inventory and application of modelling tools for integrated air quality management, and
- Demonstration of showcase projects of integrated air quality management to reduce air pollution from target sources in the countries.

The air pollution monitoring component was the most extensively implemented within AIRPET. The monitoring program design was done aiming at generating data for a comprehensive assessment of air pollution status in six participating cities/metropolitan regions, namely, the Bangkok Metropolitan Region (BMR, Thailand), Bandung (Indonesia), Beijing (China), Chennai (India), Metro Manila (Philippines) and Hanoi Metropolitan Region (Vietnam). The focus of the AIRPET monitoring was on the levels and composition of $PM_{2.5}$ and PM_{10}. The monitoring was done not only at urban sites but also at selected remote sites in the participating countries to investigate the potential regional and long-range transport pollution to the selected cities. AIRPET produced more than 6000 valid 24-h PM samples (3300 $PM_{2.5}$ and 2500 PM_{10} or $PM_{10-2.5}$) in six cities and more than 400 samples of $PM_{2.5}$ and PM_{10} at remote sites. The PM composition data were analysed by receptor modelling to quantify the source contributions to the PM levels measured at the representative sites in the cities (Kim Oanh et al., 2012). The AIRPET monitoring activity also produced some of the first datasets on carcinogenic pollutants including VOCs (600 samples of the BTEX group, including benzene, toluene, ethylbenzene and xylenes) and semi-VOCs (200 samples of PAH, pesticides and PCBs). This chapter focuses on the design of this regional monitoring program with an emphasis on QA/QC to ensure the data quality for $PM_{2.5}$ and PM_{10} mass and composition concentrations.

12.2 Design of PM Monitoring Programs with QA/QC

Data quality is the foremost concern, and implementation of a harmonised QA/QC procedure is especially important when many laboratories in different cities are involved. The terminologies of QA and QC used in AIRPET follow the common concepts presented in Chapter 8.

Fine particles ($PM_{2.5}$) and coarse particles ($PM_{10-2.5}$) are relatively different in their physical and chemical compositions, and they generally originate from different types of sources or even different activities within the same source. Coarse particles are normally generated by mechanical activities, such as grinding or wind blowing, and are dominated by materials of geological origin (Chow, 1995). The fine particle size range has significant shares of the primary particles of organic carbon and elemental carbon, but the contribution of secondary particles is also substantial. The secondary particles are formed in the atmosphere by chemical reactions involving precursors and include both secondary inorganic aerosols (SIA: sulfuric acid, ammonium bisulphate, ammonium sulphate, ammonium nitrate) and secondary organic aerosols (SOA). Due to the time required for the formation and atmospheric mixing processes; secondary particles tend to have a more uniform spatial distribution in an urban area than primary particles. Some secondary particles, e.g., ammonium nitrate, may be highly volatile and hence can be lost during the sampling and sample handling process. These are some important points to consider when selecting monitoring sites, conducting PM sampling, and doing sample preservation and analysis.

This section emphasises the QC/QC measures implemented in the AIRPET network. The QC of the AIRPET monitoring network was implemented at all possible stages from sampling (siting, sampling method selection, sampler calibration, collocated sampling, selection of filter media, filter preparation, sampling execution, sample preservation), and chemical analyses (filter segregation, weighing, trip blanks, laboratory blanks, blind samples, split samples etc.) and finally to data analysis and data management. The QA measures implemented included, for example, the inter-laboratory comparison of the measurements of black carbon for the AIRPET samples collected in different cities in the six countries. The data from all AIRPET cities were compiled and quality-checked to flag or remove suspicious data points by professional experts. In addition, the use of common monitoring methodologies and/or inter-comparison of data generated by different sampling and analytical techniques in the network was important to obtain a harmonised database generated by all six project teams. For this reason, as uniform methodologies as possible were planned, with an established QA/QC procedure, in order to produce reliable and comparable data within the network (Kim Oanh et al., 2006).

12.2.1 Sampling Design

12.2.1.1 Monitoring Sites

Siting is always a challenging task in crowded and congested urban areas in Asian developing countries, where mixed land use patterns commonly exist. It is even more challenging when only limited resources are available, which constrains the number of sites to be deployed in a domain. Sampling sites with known strong impacts from certain sources such as roadside or industrial facilities are expected to have higher contributions from these nearby sources and hence can be used to characterise the associated impacts on ambient air quality but may not represent the general air quality in a domain. Because of complex urban activities, a large urban area may consist of several different sub-airsheds, hence needing an adequate number

TABLE 12.1

AIRPET Sampling Sites for $PM_{2.5}$ and PM_{10} in Cities and Remote Areas

Country	Urban and remote site	Site characterisation	Period
Thailand	BMR	6 urban sites (upwind, traffic, mix, residence, suburban, agricultural)	Mar. 2002–Feb. 2008
	KhaoYai	Remote mountain background (120 km, NE to Bangkok)	Jan. 2005–Jun. 2005
China	Beijing	5 urban sites (upwind, traffic, residence, industry, sub-urban)	Jul. 2001–Jan. 2007
	Duolun	Remote rural background (180 km N to Beijing)	Jul. 2006 – Jan. 2007
India	Chennai	5 urban sites (upwind, traffic, industry, mixed traffic, urban background)	Nov. 2002–Nov. 2006
	Kavalur	Remote mountain background (250 km, NW to Chennai)	Feb. 2006–Jan. 2007
Indonesia	Bandung	5 urban sites (upwind, traffic, mix, residence, industry)	Jun. 2001–Jul. 2007
	Rancabali	Remote mountain background (35 km, SW to Bandung)	Aug. 2005–May 2005
Philippines	Manila	6 urban sites (urban background, traffic, mix, industry, commercial, urban-mixed)	Aug. 2001–Oct. 2007
	Gabaldon	Remote mountain background (150 km, N to Manila)	Jan. 2006–Jun. 2008
Vietnam	Hanoi	5 urban sites (upwind, traffic, mix, residence, commercial)	Oct. 2002–Mar. 2007
	Tamdao	Remote mountain background (70 km, NW to Hanoi)	Sep. 2005–Jan. 2006

Note: Bangkok Metropolitan Region (BMR) includes Bangkok city and the surrounding provinces.

of monitoring sites for characterisation (Kim Oanh et al., 2009). These include, for example, the upwind, downwind, city centre, commercial, institutional, residential sites etc. Thus careful consideration of the distribution of emission sources and typical meteorological conditions (wind pattern, rain pattern etc.) in a study area during a year is needed in selecting monitoring sites. Sites located upwind and downwind of a city help in understanding the air pollution transport across the city.

The AIRPET monitoring program for PM was designed to provide harmonised datasets of $PM_{2.5}$ and PM_{10} in the participating cities. The six participating cities were different in terms of topography, meteorology, population, primary energy use, industry, vehicle fleet composition, and population density and hence required a large number of monitoring sites to capture these variations. Accordingly, six types of sampling sites were used to characterise an urban area: upwind, traffic, residential, commercial, industrial and urban mixed (downwind) sites, as shown in Table 12.1. The classification of sites was based on the major influencing emission sources in the sub-airsheds and was done to produce as much consistency as possible for all six participating cities. Due to the monsoon effects in the cities, the downwind site in the dry season, which is a more polluted season, was generally the upwind site during the rainy season.

The AIRPET urban monitoring sites were divided into two categories: the intensive sites with a few hundred PM samples collected during the project period and the minor sites with only about 30 valid samples each. One urban area had one or two intensive sites to provide a necessary number of samples for statistical analyses of the data by multivariate receptor modelling. To provide information on the regional transport air pollution, the monitoring was also done at a remote site located 35–250 km upwind (considering the prevalent monsoon in the dry season) of the cities.

12.2.1.2 Sampling Equipment for Receptor Modelling Study of PM

An important requirement of a sampling system is its ability to collect enough PM amount on filters for subsequent chemical and physical analyses to produce the required PM compositional dataset for statistical analysis by receptor modelling. The AIRPET samples for $PM_{2.5}$ and PM_{10} or $PM_{10-2.5}$ were collected using Minivol (Airmetrics), dichotomous (dichot) or other equivalent samplers depending upon availability. A dichot sampler simultaneously collects both coarse ($PM_{10-2.5}$) and $PM_{2.5}$ samples and was used as a reference sampler in the network. Each city was provided with one Sierra Anderson dichot sampler, which was collocated with other samplers used in the city for collecting about 30 samples for performance evaluation (Kim Oanh et al., 2006). The PM mass results produced by the dichot sampler were found consistent with those produced by most other samplers used in the six cities, i.e., yielding a slope of 0.93–1.06 for $PM_{2.5}$ and a slope of 0.97–1.05 for PM_{10} and $PM_{10-2.5}$ (R^2 of 0.86–0.99), as detailed in Kim Oanh et al. (2006). The evaluation thus indicated that the samplers used within the AIRPET network produced comparable mass results and hence can be used for intercity comparison.

Furthermore, the performance of the dichot sampler in collecting $PM_{2.5}$ mass and composition was tested in a collocation study with an US EPA FRM sampler (RASS and Partisol®-Plus) in BMR. The results, seen in Figure 12.2, show that the dichot performance for $PM_{2.5}$ sampling agreed with the US EPA FRM sampler (41 pairs of samples) with a good linear correlation between these two methods (R^2=0.98). The $PM_{2.5}$ mass concentrations measured by the dichot samplers were on average about 4% higher than the FRM. The results of individual ions in $PM_{2.5}$ collected by the two sampling techniques satisfactorily agreed as well as the ion balance (Figure 12.2).

12.2.1.3 Sampling Time and Frequency

Filter-based PM sampling for mass concentration measurements is normally performed over 24 h, which provides a convenient value for checking the compliance with the 24-h ambient air

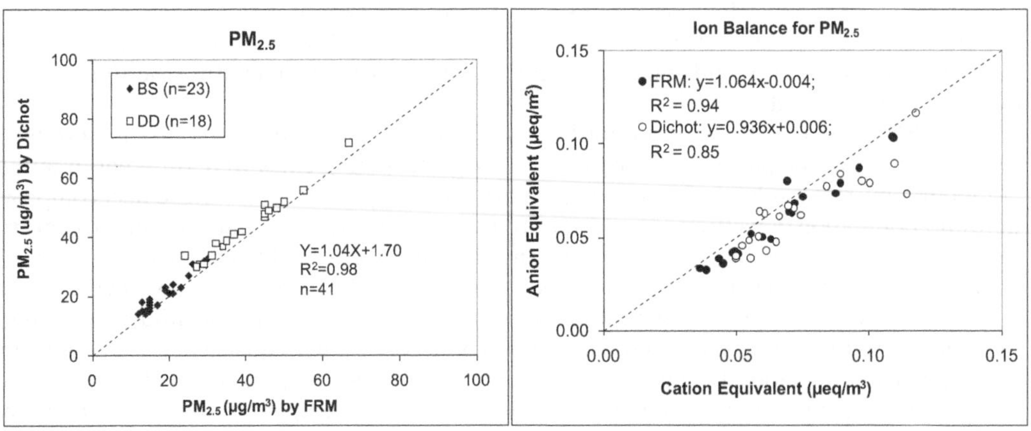

FIGURE 12.2 Comparison between the collocated FRM and the dichot samplers in the Bangkok Metropolitan Region for $PM_{2.5}$ mass and the ion balance in $PM_{2.5}$. (*Note:* BS and DD are the sampling sites with the number of valid samples indicated.)

Source: Kim Oanh et al. (2009).

quality standards. However, shorter sampling intervals provide better insight into the temporal variations of contributing sources, meteorology, as well as the temporal variation in the atmospheric chemistry, and hence are highly desirable. However, due to the large resources required to produce these online or semi-continuous measurements for PM mass and compositions, the AIRPET network relied mainly on 24-h filter-based measurements. Nevertheless, in some instances the samples were also collected separately for daytime and nighttime and over a shorter averaging time, e.g., covering 4 h as reported by Hai and Kim Oanh (2012). The climate of the Asian region is dominated by monsoons with two distinct seasons, dry and wet, which are characterised by different pollution levels; hence the monitoring was conducted in both seasons to capture the annual variation. Accordingly, at the intensive sites, the samples were collected every day or every two–three days to cover at least every month to make a full year sampling in a given city. At the minor sites, the sampling was done every two–three days and mainly to cover the representative months of the dry period and wet period, respectively.

12.2.1.4 Filter Selection, Filter Pretreatment and Weighing

Selection of filters was made to accommodate the intended subsequent analytical parameters and analytical methods. Teflon, quartz fibre and mixed cellulose ester filters were accordingly used in the network. Quartz filters were pre-baked at 550°C for 4–5 h prior to sampling to remove potential organic contaminants.

12.2.1.5 Sampled Filter Retrieval and Visual Check

The exposed filters were retrieved after the end of the sampling period following the standard operation protocol (SOP) of the respective samplers. The sampled filters were visually examined for potential damage or changes in sample mass due to particle loss, passive deposition. Only those samples collected continuously over a sampling period of 24±1.0 h and having no visible damages on filters were considered valid samples and were subjected to the analyses.

12.2.2 QA/QC for Chemical Analyses

12.2.2.1 General QA/QC

The collected samples were refrigerated, and the storage period was kept as short as possible to avoid losses of the PM components. For example, the samples collected in BMR for ionic species were extracted within 24 h after the collection. Laboratory blanks and trip blanks were prepared and treated in the same manner as the actual samples. Repeated and split analyses of selected samples were done to check the consistency of the results produced by different analytical equipment and methods. At AIT, the split samples were prepared using 30 sampled filters of $PM_{2.5}$ and $PM_{10-2.5}$, each was segregated into three–six parts that were analysed separately to quantify the precision of the results obtained for ions and elements (Kim Oanh et al., 2009).

12.2.2.2 Analytical Plan

The PM samples were analysed for mass and compositions. Overall, the analysed compositions included at least 20 elements (Fe, Ni, Pb, Sr, Ti, V, Zn, Si, Al, Ca, Cd, Co, Cr, Cu, Mg, Mn, Mo, Tl, etc.), eight water-soluble ions (Na^+, NH_4^+, K^+, Ca^{2+}, Mg^{2+}, NO_3^-, SO_4^{-2} and Cl^-) and BC (black carbon, by reflectometry) in the participating laboratories. In addition, for most samples collected in Bangkok, Beijing and Hanoi, the components of element carbon (EC, operational definition of BC) and organic carbon (OC) were also quantified using EC/OC analysers. The analytical methods for elements used by the participating laboratories were evaluated using a standard reference material (SRM). The average analytical results by the analytical laboratories used in the six cities agreed reasonably with the corresponding certified values, i.e., with

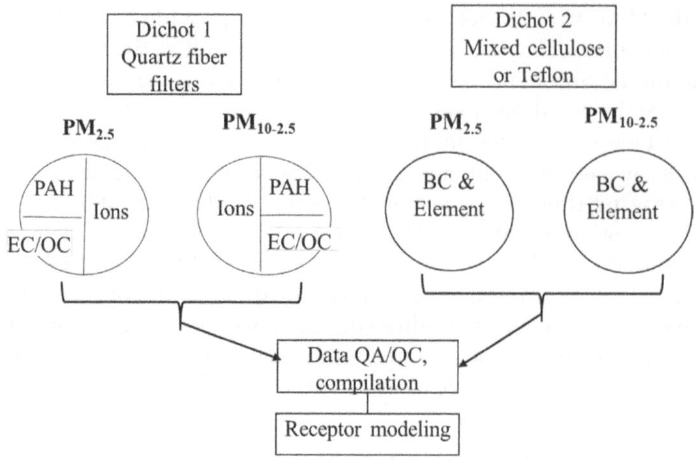

FIGURE 12.3 Segregation of PM filter samples for composition analyses.

Source: Modified from Kim Oanh et al. (2009).

the difference less than 30% for most elements while the ranges were largely overlapped (Kim Oanh et al., 2006).

In way of example, different analyses performed for each PM sample collected in BMR are presented in Figure 12.3. Two co-located dichot samplers were placed about 2 m apart on a rooftop of a selected automatic monitoring station or on a specially prepared platform, and they produced simultaneously the 24-h samples for fine and coarse PM. One dichot sampler used a quartz fibre filter (Advantec, pore size 1.2 μm) and the other used mixed cellulose ester (Advantec Mfs. Inc., pore size 0.8 μm) or Teflon. All filters were used for mass quantification, but the quartz filters normally produced unstable mass results. This was because the quartz filters were fragile, and some parts could not be properly recovered on the highly humid days in the wet season. Therefore, the mass results produced by mixed cellulose or Teflon were selected for the data analysis.

12.2.2.3 Comparison of Elemental Results by Different Analytical Methods

For this purpose, 30 samples collected on Teflon filters in BMR were analysed for elements by XRF (X-ray fluorescence spectrometers), and the results were compared with those produced by PIXE (proton-induced X-ray emission spectroscopy) using the mixed cellulose filters collected by the collocated dichot samplers. Figure 12.4 shows that the results produced by the two methods generally agreed, although the PIXE produced somewhat lower values than XRF for most elements. Note that the results of PIXE analysis were already successfully evaluated against SRM and showed a good agreement with the certified values of SRM.

12.2.2.4 Calibration of BC Monitoring Equipment

BC measured by the optical method, also referred to as the equivalent black carbon (EBC), is a measure of blackness of PM collected on a filter, whereas EC is determined using the thermal oxidation method; hence they may differ due to the nature of the particles being measured. In AIRPET, BC in the PM samples was mainly measured by EELS M43D Digital Smoke Stain Reflectometers. For calibration purposes, the BC measurement results produced by the reflectometer were compared with the EC results produced by EC/OC Sunset analyser using thermo-optical transmittance (TOT), NIOSH 5040 protocol. The PM samples of candle

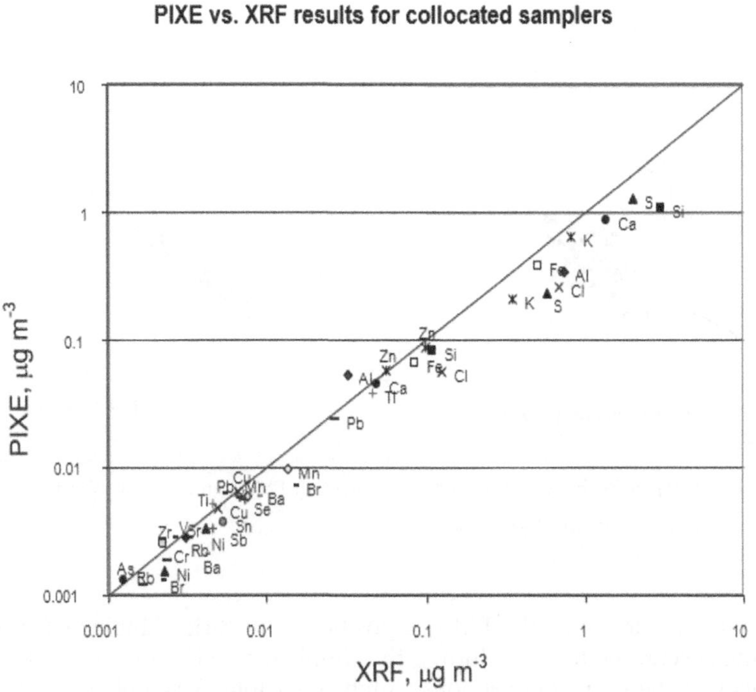

FIGURE 12.4 PIXE vs. XRF results for different elements, averaged for 15 samples with two data points for each element (1 for $PM_{10-2.5}$ and 1 for $PM_{2.5}$).

Source: Kim Oanh et al. (2009).

burning smoke and diesel exhaust collected by AIT were analysed for both BC and EC, and the relationship between the two datasets were investigated. The results showed that this relationship varied with the BC level. The results were divided into three intervals of BC, and each was fit with a linear regression equation (Kim Oanh et al., 2009). These equations were used to convert the reflectometer readings to the BC values (which therefore can be considered equivalent to EC by the TOT method).

In addition, selected samples collected in BMR were analysed by both reflectometer and TOT method, and the results showed that the relationship between the two datasets varied with monitoring site and season, with BC by the reflectometer generally higher than EC by TOT (Kim Oanh et al, 2009). Principally, the BC optical measurement should be used when only a single layer of particles is collected on the filter. Too much loading of PM on the filter results in multiple layers of particles and would increase the uncertainty of BC measurement results. In particular, the lower loading of BC during the wet season may be a reason for a better agreement between the results by the two methods for BMR samples shown in Figure 12.5.

12.2.3 QA/QC for Data Quality

Data quality was evaluated in several steps. The mass results were verified to produce explainable $PM_{2.5}/PM_{10}$ ratio that was shown to vary with the site type. The coarse PM fraction of the dichot samples was corrected for the inherited carried-over fine fraction (Kim Oanh et al., 2009). The PM mass produced by the consecutive samples taken at a shorter interval (daytime and nighttime, or 4 h) during a monitoring day was summed up to produce the mass of a 24-h sample. The PM concentrations of AIRPET were reported at the standard conditions of temperature and pressure (25°C, 1 atm).

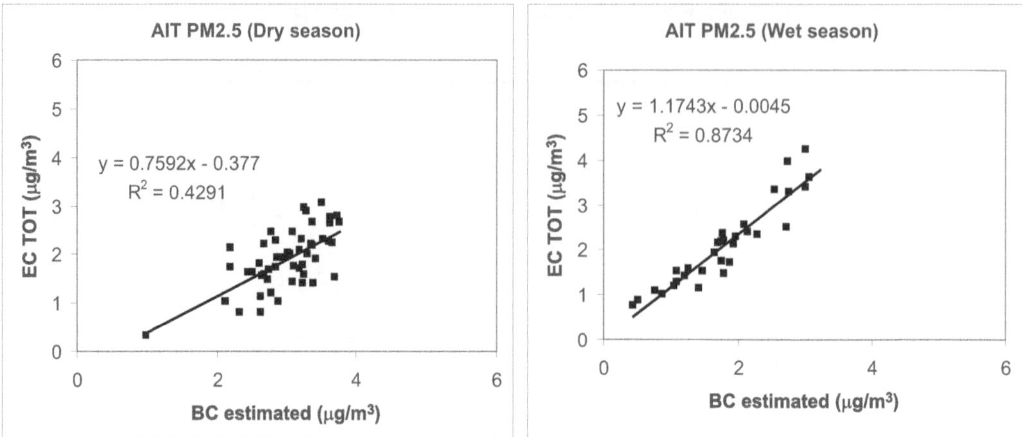

FIGURE 12.5 EC (TOT) vs. BC by reflectometer for ambient air $PM_{2.5}$ samples at a suburb site.

Source: Extracted from Kim Oanh et al. (2009).

The internal consistency of the PM composition data series obtained for each site was checked using several methods, including the simple mass closure, the reconstructed mass, the consistency of chemical compositions (sulphate vs. total S, soluble K^+ vs. total K, soluble Cl^- vs. total Cl), and the anion and cation balance. The latter was calculated using the ratio of the sum of the cation equivalents (NH_4^+, Na^+, K^+, Mg^{2+}, Ca^{2+}) to that of the anion equivalents (Cl^-, NO_3^-, SO_4^{-2}), an example of the ion balance is presented in Figure 12.2.

Simple mass closure for every PM sample was estimated as the ratio between the gravimetric PM mass and the sum of the analysed chemical species (without conversion to their possible compounds commonly found in the air). For example, the simple mass closure includes only the elements and organic carbon, but they commonly present in the ambient air in the forms of certain chemical compounds, e.g., metal oxides and organic compounds (Kim Oanh et al., 2012). Therefore, in principle, this ratio should be considerably lower than 100% because there are non-quantified components of the PM samples. Note that as a QA/QC measure, all the analysed species were scrutinised to avoid double counting, for example soluble Ca^{2+}, Cl^- and K^+ are normally excluded if Ca, Cl and K are already included in the sum of species.

The reconstructed mass, as an effective QC measure, was calculated using different mass groups, including the organic matter (OM), soot, crustal elements, trace elements, sea salt, sulphates, nitrates etc., which also provide some information on the contributing sources. Note that OM is estimated as the product of OC with a factor f (f=1.4–2.1) to convert the organic carbon mass to the associated organic compounds, and for the AIRPET samples, f=1.7 was used (Kim Oanh et al., 2006). The data from the AIRPET network were compiled and quality-checked for consistency and removed suspicious data points before using for characterisation of the status of PM pollution in the cities and to find key contributing sources by receptor modelling.

12.3 Key Findings of the AIRPET Monitoring Programme

The large number of samples generated by the AIRPET monitoring activity provided a consistent database of PM mass and compositions with harmonised QA/QC procedure for the six countries during the monitoring period 2001–2008. The statistics of $PM_{2.5}$ mass concentrations for different sites in the six cities were calculated using the measurements for 3300 of 24-h $PM_{2.5}$ samples and are presented in Figure 12.6, separately for the dry and wet seasons. Higher levels

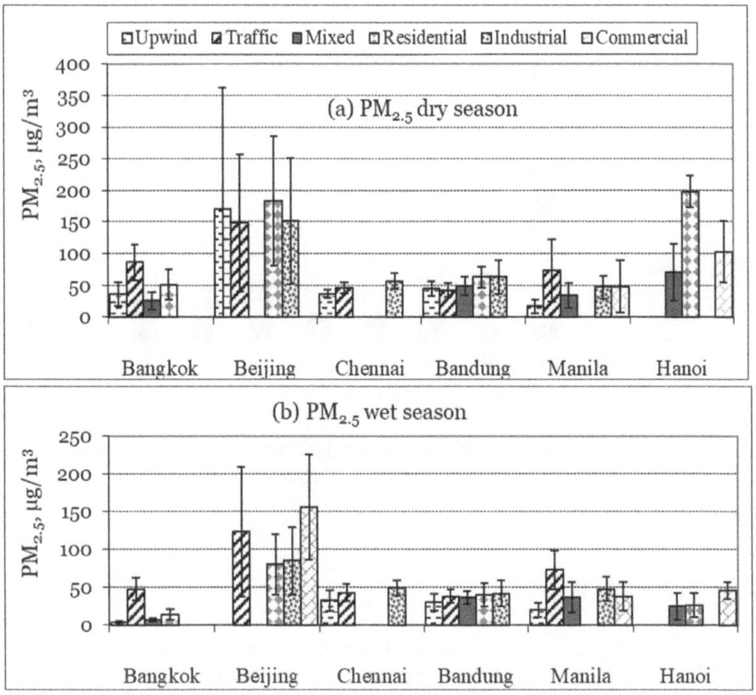

FIGURE 12.6 Seasonal levels of $PM_{2.5}$ (average ±1 SD, shown as whiskers) at different sites in the AIRPET cities. The average and standard deviations were calculated using the 24-h data for respective season during the period of 2001–2008.

Source: Raw data from AIRPET project.

of PM were observed during the dry season. The daily data showed that most measured levels in the six cities exceeded the WHO-2005 guideline value for 24 h $PM_{2.5}$ of 25 µg/m^3. The highest daily PM levels during the period were recorded during winters in Beijing, and the second highest were in Hanoi, also in winters, which can be explained by stagnant meteorological conditions coupled with the increased local emissions (biomass open burning and residential combustion) and regional transport pollution. Based on the more complete monitoring data produced for Bangkok and Manila during the AIRPET period, there was no declining trend observed for the annual average $PM_{2.5}$ levels over the 8-year period (Kim Oanh et al., 2012).

The $PM_{2.5}$ composition data obtained for six participating cities were used to calculate the average reconstructed mass for the dry season and wet seasons, respectively, over the report of 2001–2008, as seen in Figure 12.7. The PM mass consisted of several major mass groups, including OM, sulphates, nitrates and ammonium, and a substantial amount of soot (BC) particles. The "unexplained" mass is the difference between the measured mass and the reconstructed mass, which was still significant, suggesting that additional compositions need to be analysed. Note that the inclusion of OC in the composition data helps to improve the mass closure significantly for Bangkok (BMR), Beijing and Hanoi (dry season).

The statistics of $PM_{2.5}$ in the remote areas, shown in Figure 12.8, were estimated using 400 24-h samples, which show lower $PM_{2.5}$ levels than the urban areas, as expected. Similar seasonal variations were also observed at the remote sites with higher levels in the dry season than the wet season. Among the sites, higher levels were observed in China (Duolun), Vietnam (Tamdao) and Thailand (Khaoyai, dry season). However, the measurements at the remote sites were not fully comparable between the countries because of the difference in the distance from the remote site and the populated urban areas, as seen in Table 12.1. In addition, these

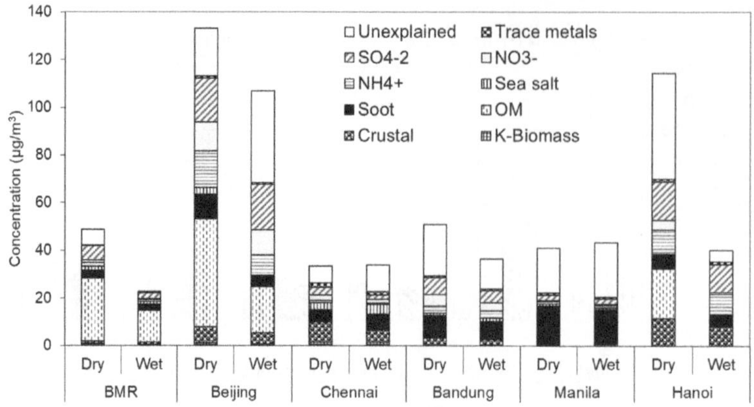

FIGURE 12.7 Average reconstructed mass of $PM_{2.5}$ with major mass groups of the samples taken in different sites in each city, separately for the dry and wet seasons during 2001–2008.

Source: Raw data from AIRPET project.

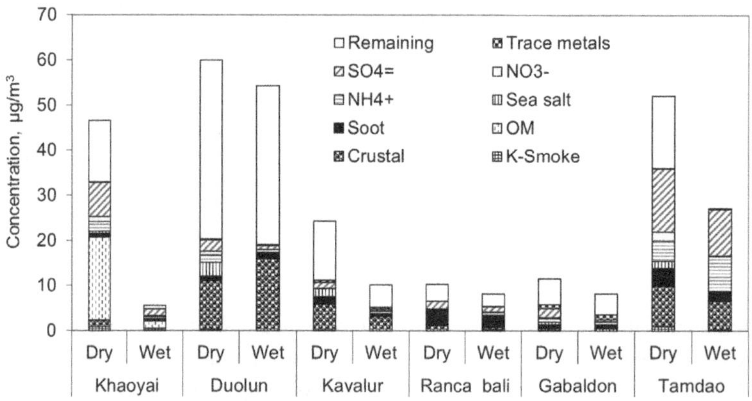

FIGURE 12.8 $PM_{2.5}$ mass and reconstructed mass groups, averaged for 2005–2006, at the remote sites in different countries.

Source: Raw data from AIRPET project.

sites were also different in terms of the height above sea level and the influence of local emissions in the vicinity. Note that the inclusion of OC in the composition data helps to improve the mass closure significantly for Khaoyai (Thailand) during both seasons.

To quantify the contributions of major source categories of PM pollution, the obtained dataset of PM mass and chemical compositions was analysed by several types of receptor models, for example the chemical mass balance (CMB) and the positive matrix factorization (PMF). The results showed that the key source factors contributing to the $PM_{2.5}$ pollution in the cities were diesel vehicles, biomass burning and secondary sulphate and nitrate particles. Collectively, these three source categories contributed about 55–85% of the $PM_{2.5}$ mass. For the coarse particles ($PM_{10-2.5}$), the most important sources included soil/fugitive dust, construction activities and sea spray (sea salt), as detailed in Kim Oanh et al. (2012).

Based on the monitoring and source apportionment results, the important sources causing the $PM_{2.5}$ pollution in each city were identified. Subsequently, the AIRPET demonstration projects were conducted with the aim of reducing emissions from these key sources to improve the air quality in the cities. These include, for example, rice straw open burning in Thailand, phasing out of leaded gasoline in Indonesia, emission control for brick kilns in Vietnam, and VOC emission control in China (Kim Oanh et al., 2014). Several policy recommendations have been put forward, as further detailed in Zhuang et al. (2014).

12.4 Summary

The monitoring programs, designed and implemented in six participating countries of AIRPET, aimed to obtain a harmonised regional database. To ensure the data quality, the foremost concern of these regional monitoring programs, a harmonised QA/QC procedure was developed and implemented by all participating research institutions. Common methodology was applied for sampling and analyses to the greatest possible extent. In the case that different methods were used, comparison between these sampling and analytical techniques was done to harmonise the results across all participating countries. The QA/QC procedure was followed at all possible stages from sampling, chemical analysis to data analysis in order to ensure the data quality.

The monitoring results showed high levels of PM pollution, especially during the dry season, in all the cities, which were well above those in the remote areas. The levels also widely varied among different sites in each city. Analysis of PM compositions using the reconstructed mass and receptor modelling approaches confirmed that the most important sources contributing to $PM_{2.5}$ in the cities included diesel vehicles, biomass burning and secondary sulphate and nitrate particles, which in turn could be associated with the precursors emitted from various fuel combustion activities located both inside and outside the cities. Based on the findings, these important sources were selected for the AIRPET demonstration projects to showcase how various emission control measures can be implemented for particular sources to improve air quality in the areas.

REFERENCES

Chow, J.C., 1995, Measurements methods to determine compliance with ambient air quality standards for suspended particles, *Journal of Air and Waste Management Association*, 45(5), 320–382.

Hai, C.D. and Kim Oanh, N.T., 2012, Effects of local, regional meteorology and emission sources on mass and compositions of particulate matter in Hanoi, *Atmospheric Environment*, 78, 105–112, https://doi.org/10.1016/j.atmosenv.2012.05.006

Health Environmental Institute (HEI), 2020, *State of global air 2020, special report*, Health Effects Institute, Boston, MA, ISSN 2578-6873

Kim Oanh, N.T. and Polprasert, C., 2014, *Overview of AIRPET activities and findings, chapter 1 in the book improving air quality in Asian developing countries: Compilation of research findings*, Vietnam Publishing House of Natural Resources, Environment and Cartography (NARENCA), Hanoi, Vietnam, ISBN 9786049044106

Kim Oanh, N.T., Pongkiatkul, P., Upadhyay, N. and Hopke, P., 2009, Design a monitoring program of ambient particulate matter for source apportionment study by receptor modeling, *Atmospheric Environment*, 43(21), 3334–3344.

Kim Oanh, N.T., Prapat, P., Melliza, T.C., Nghiem, T.D., Ligy, P., Zhuang, G. and Puji, L., 2012, *Monitoring and source apportionment for particulate matter pollution in six Asian cities, chapter 3 in the book integrated air quality management: Asian case studies*, Taylor and Francis Group, CRC Press, ISBN 9781439862254

Kim Oanh, N.T., Upadhyay, N., Zhuang, Y.H., Hao, Z.P., Murthy, D.V.S., Lestari, P., Villarin, J.T., Chengchua, K., Co, H.X., Nghiem, T.D. and Lindgren, E.S., 2006, Particulate air pollution in six Asian cities: Spatial and temporal distributions, and associated sources, *Atmospheric Environment*, 40(18), 3367–3380.

Kim Oanh, N.T., Zhuang, Y., Puji, L. and Co, H.X., 2014, *Development of integrated management strategies for emission reduction from target sources in Asia, chapter 18 in the book improving air quality in Asian developing countries: Compilation of research findings*, Vietnam Publishing House of Natural Resources, Environment and Cartography (NARENCA), Hanoi, Vietnam, ISBN 9786049044106

Zhuang, Y., Kim Oanh, N.T., James, S., Ligy, P., Puji, L., Co, H.X., Achim, M., Wang, H., Zhang, W., Wang, X., Genevieve, L., Melliza, T.C. and Didin, A.P., 2014, AIRPET country specific policy recommendations, chapter 19 in the book improving air quality in Asian developing countries: Compilation of research findings, Vietnam Publishing House of Natural Resources, Environment and Cartography (NARENCA), Hanoi, Vietnam, ISBN 9786049044106

13

Sampling and Analysis for Particulate Matter: Kashima SPM Study

This chapter discusses various issues associated with the measurement of airborne particulate matter. Particulate matter (PM) is not a single pollutant but is composed of many chemical constituents. Particles in the atmosphere differ in size, shape, chemical composition, specific gravity, light absorption and scattering properties, hygroscopicity and so on. These characteristics are also relevant to our health, meteorological phenomena such as precipitation and cloud formation, and global climate. In addition, PM sample collection and analysis present many challenges due to its complex configuration. The receptor model is a model that estimates the contributions of emission sources using the chemical composition data of PM that will be further detailed in Chapter 23. The discussions included in this chapter focus on the data acquisition methods essential for receptor modelling, especially considerations related to sampling of the particulate matter and chemical analysis of its components. The case study of IPCAJ conducted around the Kashima industrial area in Japan is used to illustrate the various challenges involved.

13.1 Overview of the Kashima SPM Study

The Industrial Pollution Control Association of Japan (IPCAJ) conducted a large-scale air quality study around the Kashima industrial area. As explained later in Section 13.2.3, in Japan, airborne particulate matter with a particle size of 10 µm or less is called suspended particulate matter (SPM) under the regulation that is technically equivalent to PM_{10}. The main purpose of this IPCAJ's Kashima SPM study was to achieve the following:

- Examination of SPM (PM_{10}) measurement method
- Confirmation of the applicability of an air quality model (atmospheric dispersion model) to SPM
- Evaluation of the feasibility of a receptor model (still under investigation at that time)

Regarding air quality dispersion modelling research, administrative operations were already carried out for SO_2, so the most important concern to investigate is whether the model could be improved and expanded to include SPM. The main task here was to validate the model using data from air quality monitoring networks operated by local governments and source inventories created by the authorities. On the other hand, in a receptor modelling study, it was necessary to prepare the measurement data by the own efforts. An example of that is discussed in Chapter 23. This chapter addresses implementation of a statistical method in the planning and analysis processes, in accordance with the necessary sampling techniques. Some of the major achievements of the Kashima SPM study (IPCAJ, 1984a, 1986; Okamoto et al., 1986; Okamoto and Kobayashi, 1991), along with results of an additional study (Pongkiatkul et al., 2010), are used for illustration.

DOI: 10.1201/9781003597216-15

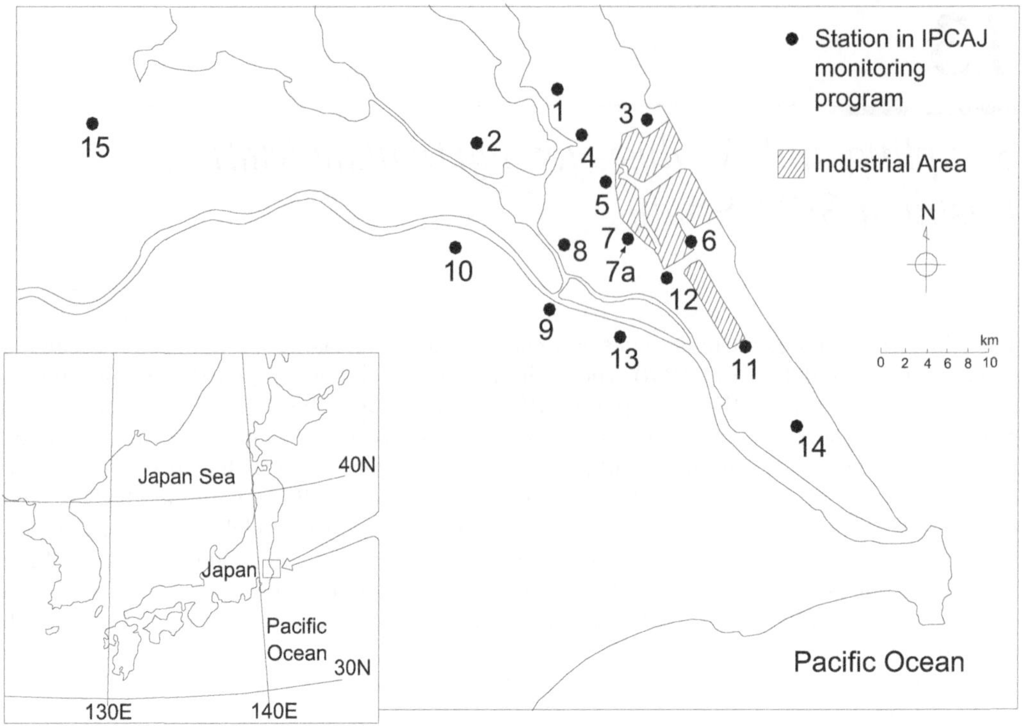

FIGURE 13.1 Sampling sites for the IPCAJ Kashima SPM study.

13.1.1 Sampling Design

13.1.1.1 Monitoring Sites

The Kashima area has one of the largest industrial complexes, which is located about 80 km ENE of Tokyo. This area also has a high-density air quality monitoring network operated by several local governments. The industrial area is divided by a digging harbour, as shown in Figure 13.1. A steel mill was located in the northern area; the southern area has an oil refinery, petrochemical industries, and a 4400-MW oil-fired power plant.

In this field study, the IPCAJ installed the necessary samplers at 15 sites around the Kashima industrial area. The main reason for selecting these locations was that they were already used as the air quality monitoring stations, which were selected as suitable sites for monitoring air pollutant concentrations. As for the seventh site, Kamisu Town Hall monitoring station, an auxiliary site 7a was also set up on the roof of the new government building (six-story building) at a distance of about 30 meters.

13.1.1.2 Sampling Equipment, Sampling Time and Frequency

The Kashima SPM study used an Andersen impactor to examine spatial distribution and particle size distribution and a high-volume air sampler to see daily time-series changes. Table 13.1 displays the collection conditions and sampling durations for each sampler.

A low-volume air sampler was used for comparison with the environmental standard and also to see the differences in concentrations for different weather conditions.

The PM was sampled for about two weeks in each of the four seasons of 1983. All samplers were calibrated for flow-rates according to specified conditions before the four seasons of the survey, and all calibration data were accurately recorded. After that, all samplers were

TABLE 13.1

Sampling Conditions for the Samplers

Sampler	Location (site number)	Season	Collection substrate**	Sampling duration	Species determined
Andersen impactor	1–15, 7a	4 seasons	Polyethylene (I) Teflon(F)	14 days	Elements and 3 anions
Hivol	2, 4, 6, 7	4 seasons	Teflon(F)	1day (24 h)	Elements and 3 anions
Hivol	2, 7	Summer and fall	Quartz(F)	1day (24 h)	Carbons and 4PAHs
Lovol* (wind-selective Lovol)	2, 4, 6, 7	4 seasons	Teflon(F)	14 days	Elements and 3 anions

* The low-volume air sampler was built in parallel so that the drive of the suction motor of the six samplers could be selected according to the wind direction and speed. The six cases here were four wind directions, a calm condition of 0.5 m/s or less, and the entire period.
** I: impaction substrate, F: filtration substrate

deployed at their respective sampling sites. The sampling with high-volume air samplers and low-volume air samplers started at 9:00 AM on the first day of the survey. For the Andersen impactor, due to the large number of units, sampling was started at 9:00 AM using a timer. All sampling sites were patrolled twice daily, once in the morning (about 9:00 AM) and once in the afternoon, and the indicated values of the flow-meters were recorded. In addition, if the flow-rate deviated from the target value, a correction was made, and after confirming that the indicated value was stable, the flow-rate was recorded again.

13.1.2 Analytical Plan and Related Composite Sampling

In this study, Teflon filters were utilised with all high-volume air samplers, low-volume air samplers, and Andersen impactors. These filters were desiccated and weighed both before and after being loaded into the samplers. The filters were cut and analysed for elemental and ionic composition using instrumental neutron activation analysis (INAA), X-ray fluorescence (XRF), and ion chromatography (IC). A cutting plan for Teflon filters is shown in Figure 13.2. This method was employed in the four-season survey. However, as a supplementary study, two high-volume air samplers were used in the summer and fall surveys, one of which was fitted with a quartz filter for carbon and polyaromatic analysis.

In the analysis of trace components contained in particulate matter and the like, the cost of analysing chemical or elemental components is much higher than the cost of collecting the samples. The biggest budget constraint in this type of study is the number of samples subjected to necessary chemical analyses. Therefore, to reduce the analytical cost, composite samples were prepared. As a result, the number of samples to be subjected to expensive chemical analysis can be reduced, and an accurate average value can still be obtained. This method of sample preparation is called composite sampling, which is detailed in Chapter 9 (Section 9.5.2). For samples collected using the Andersen impactor, efforts were made to reduce the total number of samples subjected to chemical analysis by adjusting the contents through composite sampling.

In the field survey of the Kashima SPM study, the particulate matter was collected at 16 locations by eight stages of particle size by using the Andersen impactors. Weight concentration was measured at eight particle sizes for all of the sampling sites. At four intensive monitoring stations, elemental concentrations were analysed for all eight stages. However, for the remaining 12 locations, elemental concentrations were analysed in two categories: coarse particles (stages 1–4) and fine particles (stages 5–7 and backup filter). This can reduce the cost by a

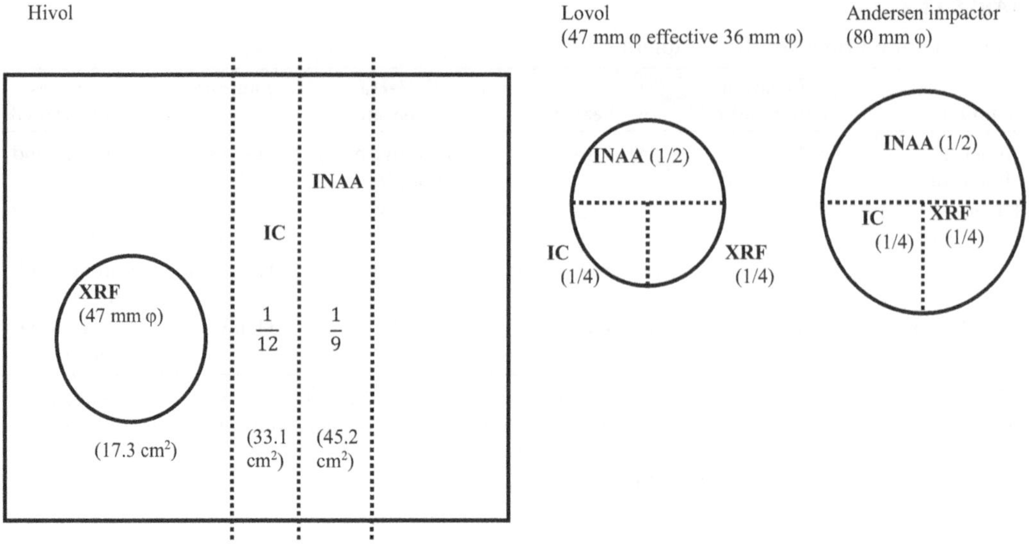

FIGURE 13.2 Cutting design of the filter.

Source: Adapted from Okamoto (1994).

factor of four compared to directly analysing the eight-stage samples for chemical analysis and calculating the concentration of coarse and fine particles by using the individual composition data. The number of samples for each sampling site, sampling device and element is shown in Table 13.2.

13.2 Field Survey Program Designs and Quality Assurance in the Kashima Study

In the IPCAJ's Kashima SPM study, various laboratory-level studies were also conducted before the field survey, and several statistical analysis methods such as analysis of variance for the data were effectively conducted. The efforts of quality assurance for these data are very unique. This study paid close attention to quality assurance of measured values. The issues considered include:

- Particulate matter (PM) sampling positions/points/sites,
- Selection of the PM sampling devices and their calibration and flow volume calculation,
- Filter collection, storage, weighing (effect of humidity) etc., and
- Chemical analysis method selection, calibration etc.

In this case, PM sampling locations were assigned to the rooftops or other suitable locations of each local government air quality monitoring station, so it was not anticipated that there would be any major obstacles to data quality in terms of location selection.

13.2.1 Sizing Efficiency and Instrumental Differences Effects

As for the PM sampling, the selection of sampling equipment is an important issue. In particular, it is necessary to check the performance of the pre-cut device mounted on the sampler inlet, and it should be consistent with the desired size selection of the particles. In this study,

TABLE 13.2

Number of samples for each sampling location, sampling device and elements (Autumn season survey)

	Analysis item	Metals and anion			Ammonium ion	(NH_4^+)	Carbons	Polycyclic aromatic hydrocarbon
Site	Method	INAA, XRF and IC			AA	IC	TCA	HPLC
	Sampler	Hivol	WD–Lovol	A–impactor	Hivol	A–impactor	Hivol	Hivol
	Filter media	Polyflon	Polyflon	Polyflon	Quartz	Polyflon	Quartz	Quartz
1	Kyuchu			2				
2	Itako	15	6	8	15	8	15	15
3	Aohama			2				
4	Shimohanawa	15	6	8				
5	Fukashiba			2				
6	Karunohigashi	15	6	8				
7	Kamisu	15	6	8	15	8	15	15
7a	Kamisu 2			2				
8	Kashima			2				
9	Omigawa			2				
10	Ohkura			2				
11	Ohta			2				
12	Seihannren			2				
13	Sasagawa			2				
14	Uematsu			2				
15	Edozaki			2				
Total		60	24	56	30	16	30	30

Note: INAA: Instrumental neutron activation analysis (30 elements);
XRF: Fluorescence X–ray analysis (Pb, Cd, S and Si); AA: Atomic absorption spectrophotometry analysis; IC: Ion chromatography analysis; HPLC: High-performance liquid chromatography; TCA: Thermal carbon analyser.

high-volume air samplers, low-volume air samplers, and Andersen cascade impactors were used to collect PM samples. In particular, the former two types of samplers were equipped with size-selection inlets to remove particles larger than 10 μm. Therefore, a comparative test was conducted to understand the characteristics of the PM collected by these samplers.

All three types of samplers were needed to compensate one another to provide the required data on the spatial variability and daily variations of the PM levels and composition. The low-volume air sampler, which is the standard official collection method in Japan, cannot provide samples for different particle size ranges to determine the composition. However, the PM amount collected on a filter by this low-volume air sampler during one day is not sufficient for all the chemical analyses. To overcome these challenges, this study used the Andersen impactors at all sampling sites (16 sites).

The Andersen impactors were either the AN-200 model manufactured by Dylec Co., Ltd. or its equivalent. The specifications are shown in Table 13.3. In addition, the high-volume air sampler was placed at each of the four intensive sampling sites to examine daily changes (day-to-day changes). The particle sizing efficiency curve of the Andersen impactor is very sharp, as shown in Table 13.3, there is not much difference between 100% cut-off particle size and 50% particle cut size. On the other hand, as shown in Figure 13.3, the sizing characteristic curve of the cyclone attached to the high-volume air sampler lies close to horizontal.

TABLE 13.3

Specifications of the Anderson Cascade Impactor

Stage	50% cut-off diameter (µm)	Catalogue value (µm)
0	10.0	11.0
1	7.0	7.0
2	4.7	4.7
3	3.3	3.3
4	2.1	2.1
5	1.03	1.1
6	0.62	0.65
7	0.42	0.43

Note: Particles with diameters of 7 to 11 µm are collected on the first-stage filter paper. Particles smaller than 0.43 µm are collected by the backup filter. The airflow rate is 28.3 L/min.

Source: Manufacturer catalogue (Koritsu Kiki. Co. Ltd.) (also, see the following address: https://www.t-dylec.net/service/an-200/).

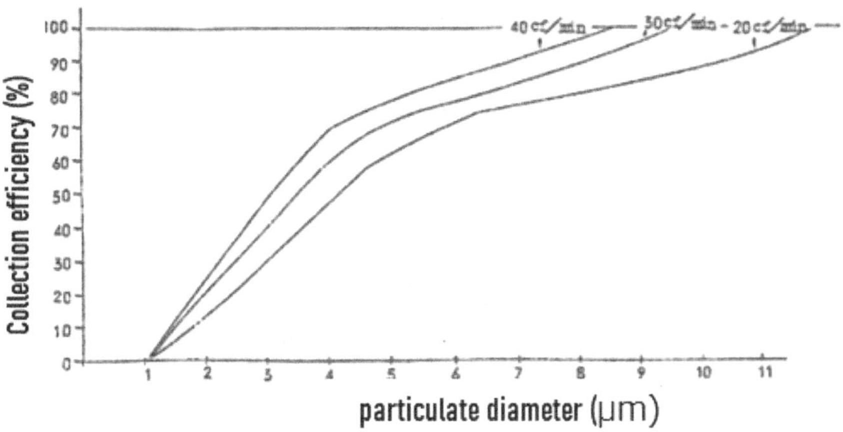

FIGURE 13.3 Efficiency of the cyclone attached to the high-volume air sampler.

Source: Adapted from IPCAJ (1983).

To provide the precise cut-off of each size range, it strictly requires a flow-rate of 800 L/min (28.3 ft^3/min). There was a problem with this setting in the initial mechanical specification, and in the winter survey the flow-rate was set to 1000 L/min (35.3 ft^3/min), but after the remodelling, it could operate at 800 L/min.

The PM concentration by the low-volume samplers was measured during a sampling period of two weeks in each of the four seasons. Therefore, to compare with these measured values of the low-volume sampler, the average values of a light-scattering digital dust meter and the other PM monitors in the corresponding duration were calculated. Table 13.4 shows the comparison of the average concentration of each instrument during these measurement periods.

Here, to investigate the characteristics of the sizing device attached to the air sampler, it was also compared with the measured value by the Andersen impactor. First, the measured mass concentrations by the low-volume air sampler were compared with those by the first and subsequent stages of the Andersen impactor. This result is shown in Figure 13.4. The low-volume air sampler has a 100% cut-off of 10 µm, whereas the Andersen impactor has a 50% cut-off of

TABLE 13.4

Comparison of Average Concentration of Each Instrument (unit: μg/m³)

							Andersen impactor		
Season	Site	β-ray	Piezo-balance	Light-scattering	Hivol.	Lovol.	Under 2.1 μm	Under 11 μm	All stage
Winter	Itako	26.2	26.7	33.0	18.6	34.9	23.2	39.4	43.9
	Shimohanawa	23.5		28.6	18.2	34.3	18.0	31.3	35
	Karuno	26.5		30.6	17.6	37.3	24.3	45.5	55.2
	Kamisu	34.0		36.3	16.2	38.5	24.0	41.4	46.4
Spring	Itako	36.3	15.8	28.5	18.5	35.7*	16.0	30.9	33.9
	Shimohanawa	33.1		26.3	21.7	32.3	16.3	32.8	36.1
	Karuno	36.1		25.1	19.4	32.1	15.6	33.8	37.5
	Kamisu	35.7		27.7	23.7		18.9	34.7	37.5
Summer	Itako	37.7	21.3	29.5	21.1	29.6	15.3	28.8	32.5
	Shimohanawa	40.3		27.6	19.6	29.2	16.9	32.7	37.1
	Karuno	42.9		29.2	19.1	30.0	14.5	30.0	33.3
	Kamisu	40.9		36.8	24.5	31.3	16.5	31.2	36.5
Autumn	Itako	27.3	17.9	27.4	15.6	30.3	15.2	32.8	36.6
	Shimohanawa	26.3		22.8	14.9	28.6	13.4	30.4	33.9
	Karuno	23.3		19.2	11.8	27.3	11.1	28.2	32.3
	Kamisu	27.1		23.2	14.7	30.4	16.3	34.4	38.0

Note: Winter season: January 10–23, 1983, Spring season: April 11–24, 1983, Summer season: July 31–August 13, 1983, Autumn season: October 28–November 10, 1983.

* Due to the failure of the sampling by the normal low-volume sampler, the spring data of the Itako monitoring station was calculated from the segregated data with the wind-direction-selective collection device.

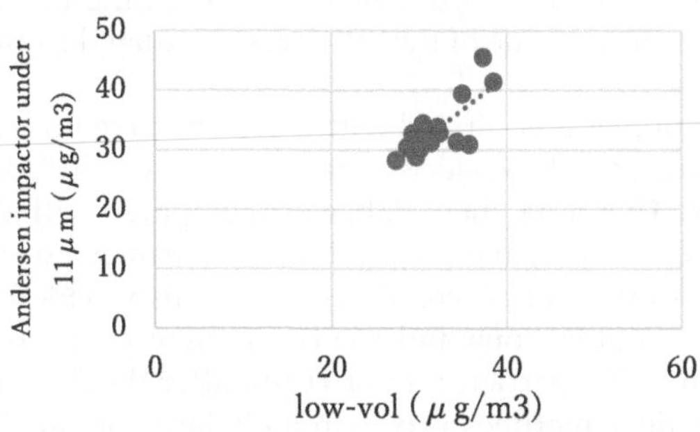

FIGURE 13.4 Scatter diagram of mass concentration by the low-volume air sampler and that by the Andersen impactor for PM ≤ 11 μm ($r = 0.77$, $n = 16$).

10 μm; hence there is a slight difference between them. Next, Figure 13.5 shows a comparison between the measured mass concentrations by the high-volume air sampler and that by the fifth and subsequent stages of the Andersen impactor. The data from the winter season were excluded because they were sampled before the flow-rate settings were adjusted.

Table 13.4 shows the relationship in terms of mass concentrations. The characteristics curve of the particle sizing efficiency of the cyclone attached to the high-volume air sampler is not so sharp; therefore further investigation may be necessary. Since the chemical constituents in the particles have different particle size distributions, it is possible to characterise the sizing properties by investigating their properties using several chemical constituent data.

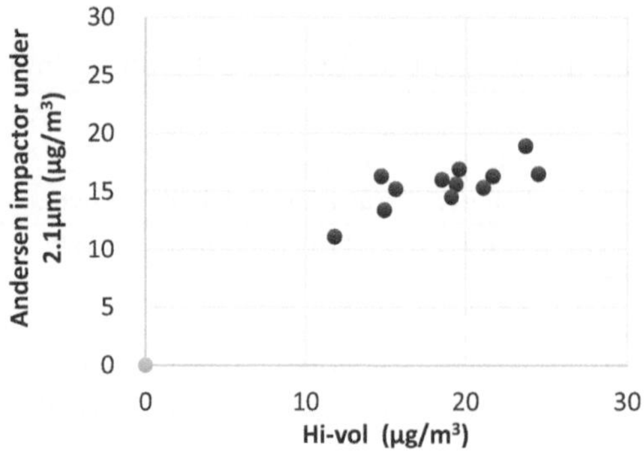

FIGURE 13.5 Scatter diagram of mass concentration by the Andersen impactor for PM ≤ 2.1 µm and that by high-volume air sampler ($r = 0.76$, $n = 12$).

Table 13.5 shows the ratio results between the geometric means of the component concentrations measured by different samplers at four sites and four seasons. From these results, it can be seen that, for many components, it is possible to see some degree of consistency between the elemental concentrations by the high-volume air sampler (with cyclone) and by the Andersen impactor (from the fifth stage to backup filter).

Note that the data of the last day Hivol was excluded due to the longer sampling duration than the data of Lovol and Andersen impactor. Although 15 days of sample data were produced by the Hivol, the last day's sample was excluded to match the sampling duration of other sampling devices. For the WD-selective Lovol, the weighted average value was calculated according to the frequency of appearance of the wind direction during the sampling period. Hivol data were omitted if they were ≥3 data points below the detection limit out of 14 data points during the 14 days. At that time, for the data below the detection limit, 1/10 of the lower limit was set. Finally, the ratio was calculated excluding the data below the detection limit. Then the geometric mean value of the ratio was calculated from a total of 16 data at four sites in four seasons.

The high-volume air sampler is equipped with a cyclone to remove any particles larger than 10 µm, but since 10 µm is set to 100% selective particle size, the 50% particle size is said to correspond to about 2–3 µm. Therefore, these data were compared with the total of the fifth-stage to backup filters (PM size ≤ 2.1 µm) and the total of the fourth-stage to backup filters (less than 3.3 µm) of the Andersen impactor. A comparison was also made between the low-volume air sampler with a 10-µm cut cyclone and the component concentrations measured by the Andersen impactor with a 50% particle size of 11 µm after the first stage. From these results, it can be seen that the ratio sometimes was extremely large or small for components having a small absolute amount or components having a low analysis accuracy. For example, such a tendency was seen for Cd and Pd obtained by fluorescent X-ray analysis.

From these considerations, it seems that the data provided by the high-volume air sampler with the attached particle sizer (separator) can be considered to be the concentration of the fine fraction of about 2.5 µm or less. However, since the efficiency curve of this particle sizer is not sharp, it cannot be denied that some parts of coarse particles are mixed.

13.2.2 Flow Rate and Flow Volume Calculation

When measuring the PM concentration with an air sampler equipped with a particle sizer, incorrect setting of the flow-rate will not only increase the error of airflow volume but also change the size-selecting characteristics of the particles, so that the measured value becomes unreliable. If only the former is the problem, the measurement value can be corrected later. However, the latter problem cannot be solved at a later date, and the data must be discarded.

TABLE 13.5

Average Value (Geometric Mean Value) of the Ratio of the Chemical Compositions and Mass Concentration by the Sampling Devices ([A, B, C, D, E, F])

Component	Ratio <B>/<A>	Ratio <C>/<A>	Ratio <E>/<D>	Ratio <F>/<D>
Ag	—	—	1.22	0.90
Al	1.05	2.68	1.12	1.33
As	1.22	1.40	0.85	0.79
Ba	—	—	1.25	0.50
Br	0.98	1.04	1.18	1.15
Ca	1.15	1.58	0.89	2.18
Ce	—	—	1.19	0.99
Cl	0.46	1.01	1.38	1.68
Co	—	—	1.25	0.66
Cr	0.68	0.93	1.34	1.00
Cs	—	—	1.11	0.60
Cu	0.47	0.60	0.84	0.79
Fe	1.00	1.68	1.04	1.13
Hf	—	—	1.46	0.59
K	—	—	0.67	0.46
La	—	—	1.20	1.04
Lu	—	—	—	—
Mn	1.18	1.63	1.20	1.19
Na	0.52	1.07	1.07	1.31
Ni	—	—	2.88	0.62
Sb	1.21	1.27	0.95	0.92
Sc	1.24	3.04	1.09	1.26
Se	—	—	1.11	0.81
Sm	—	—	0.92	1.19
Th	—	—	0.84	1.11
Ti	—	—	1.07	1.63
V	1.05	1.19	1.07	1.06
W	—	—	1.96	—
Zn	0.85	0.92	1.31	0.86
Cd	0.15	0.17	1.69	1.43
Pb	0.55	0.65	3.03	3.98
S	0.85	0.96	1.02	1.25
Si	0.45	1.51	1.18	1.75
Cl^-	0.68	1.57	1.20	1.54
NO_3^-	0.85	1.34	1.07	1.12
SO_4^{2-}	1.17	1.26	0.87	0.88
mass	0.93	1.15	1.06	1.08

[A] High-volume air sampler with cyclone (Hivol)
[B] Andersen impactor (stage 5 to backup filter) less than 2.1 μm
[C] Andersen impactor (stage 4 to backup filter) less than 3.3 μm
[D] Low-volume air sampler (Lovol)
[E] Total of the wind-direction-selective low-volume air sampler (WD-selective Lovol)
[F] Andersen impactor (stage 1 to backup filter)

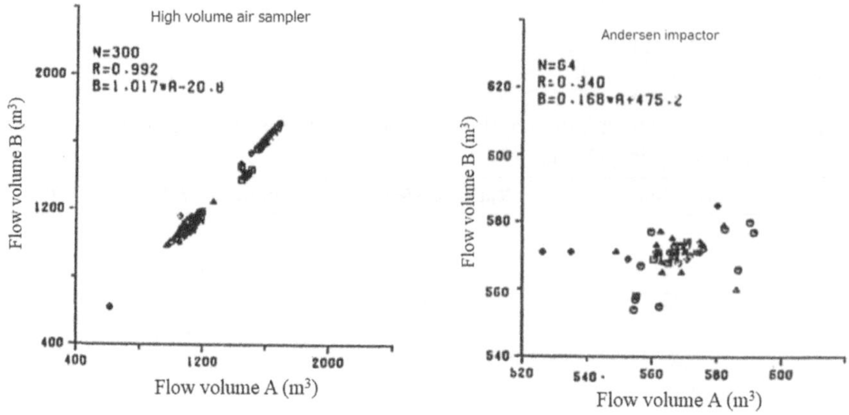

FIGURE 13.6 Comparison of calculated values of air sampler suction air volume: A, measurement contractor; B, calculation contractor (software house).

Source: **IPCAJ (1984a).**

For this study, to guarantee precise flow-rates for the samplers, it was mandatory for the contractor to calibrate the flow-meters on all samplers intended for collecting particulate matter prior to each field survey in every season and to provide the calibration records. For a high-volume air sampler equipped with both a rotameter (float flow-meter) and an integrating flow-meter, a comparison between the two was also performed. In the IPCAJ survey, the contracted measurement service provider responsible for the onsite survey, calibration of samplers in every season, determination of the flow-rates, calculation of PM concentrations reported the results. The flow-rate calculations were derived from a visual interpretation of the calibration curve. To validate these results, IPCAJ initiated a supplementary survey conducted by an independent software firm, which was engaged to compute the air volumes using the calibration reports and patrol inspection records provided by the measurement service contractor. A comparison of the air volumes calculated by each contractor was performed by the IPCAJ. In the calculation by the software company, all the procedures were objectively processed, such as replacing the reading operation of the calibration curve by interpolation using a higher-order regression equation. Details of this procedure, including data quality control, are given in IPCAJ (1984a). Figure 13.6 shows a scatter plot comparing the results of flow-rates and the volume calculations submitted by both companies, which showed a good agreement.

However, the plot for the high-volume air sampler showed a split on the scatter diagram because the flow-rate setting was different in the winter survey part. Therefore, while all the data can be used to consider approximate seasonal changes, a more detailed analysis needs to be carefully considered. Due to this problem, some studies using high-volume air sampler data were conducted for only the summer and autumn seasons. Note that the software company calculated the flow-rate only from the inspection record, calibration record, and the related materials submitted by the measurement company, without any other information from the company. Therefore, it can be concluded that the measurement company provided good measurement data and records with high traceability. In the end, the calculation result of the suction air volume provided by the software contractor was adopted.

13.2.3 Continuous Monitoring Instrument for Particulate Matter

The reference method for the suspended particulate matter (equivalent to PM_{10}) in Japan is a low-volume air sampler with the cyclone inlet to remove particles larger than 10 μm. However, the low-volume air samplers cannot provide hourly PM concentrations; hence the light-scattering type of digital dust monitors were installed in almost all air quality monitoring stations at that time. Furthermore, to formally register as the PM_{10} data, it was necessary to perform calibration based on the measurement of a low-volume air sampler once a month. The correction value in this calibration was called the *F* value. The Environment Agency (predecessor of the Ministry of the Environment) did not want to call the measured values of light-scattering measuring instruments without *F* value correction as suspended particulate matter (PM_{10}), so they called such measuring instruments dust monitors. Subsequently, the Environment Agency began investigating whether this type of measurement was significantly affected by the problem of moisture absorption of particles. Based on the discussions, the Agency had changed the PM_{10} reference measurement method. However, there have long been indications that this decision was somewhat arbitrary. For this reason, IPCAJ conducted research involving light-scattering type dust monitors.

The IPCAJ's Kashima SPM study produced interesting data on the performance comparison tests of measuring instruments, and some of them will be introduced here. Two sets of PM_{10} and dust measuring instruments of the same type by the same manufacturer were used for this comparison test: they were a light-scattering digital dust monitor, β-ray attenuation monitor, and piezo balance monitor. These measuring instruments were co-located on the roof of a laboratory in Kojiya, Tokyo, for 72 hours (10 AM, Dec. 7–10 AM, Dec. 10, 1982), and hourly data were collected.

Statistical evaluation scores of the difference between the indicated values of two sets of the same type of instruments are presented in Table 13.6b. In this comparative test, the true value of the concentrations is unknown. Therefore, although bias cannot be evaluated from this test, the performance of the light-scattering type measuring instrument seems to be the best in terms of precision. This type of instrument seems to be superior to other instruments. If the bias is reduced by proper calibration, it can be the best PM instrument.

13.2.4 Effect of Humidity on Measuring Instruments

Particulate matter includes particles having high hygroscopicity, and humidity is also attracting attention as one of the factors that increase PM concentration. The particles collected by

TABLE 13.6A

Comparison between the 3-Day Average Values (μg/m³) of Two PM Monitoring Instruments of the Same Type

Instrument	Instrument no.1	Instrument no.2	Average
Light-scattering dust monitor	122	126	124
Piezo-balance PM_{10} monitor	149	148	149
β-ray attenuation PM_{10} monitor	148	138	143
Low-volume air sampler	139 (quartz)	141 (Teflon)	140

TABLE 13.6B

Comparative Test Results of Two Instruments of the Same Type

	Correlation coefficient	Root mean square difference (μg/m³)	Mean of difference (μg/m³)
Light-scattering dust monitor	1.000	5.5	4.53
Piezo-balance PM_{10} monitor	0.975	22.1	22.24
β-ray attenuation PM_{10} monitor	0.97	23.6	21.19

Note: Number of data points: 72.

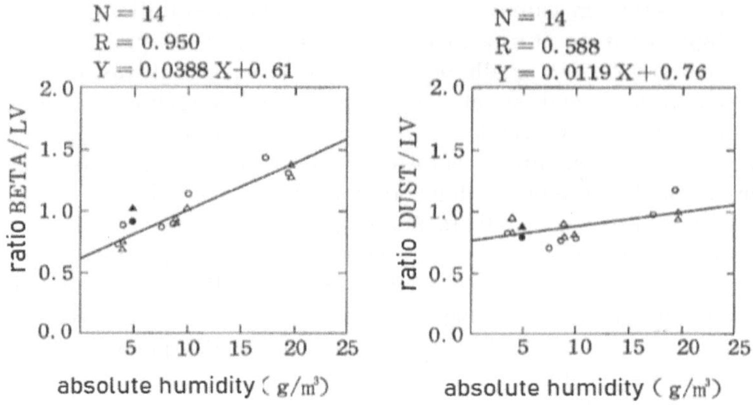

FIGURE 13.7 Comparison of the ratio of two monitors with the humidity (LV, low-volume air sampler; BETA, β-ray attenuation PM_{10} monitor; DUST, light-scattering dust monitor).

Source: IPCAJ (1984a).

a low-volume air sampler were measured after filter conditioning in a humidity-controlled chamber. On the other hand, at that time, they said the light-scattering digital dust meters can be strongly affected by moisture absorption under high humidity conditions. The hygroscopicity of the particles may change the scattering characteristics of the particles, which becomes a significant error factor. During the 1980s, many air quality monitoring stations in Japan were equipped with light-scattering-type measuring instruments. The measurements produced by the light-scattering device therefore were not considered appropriate to compare with the environmental standard for SPM. Accordingly, the light-scattering instrument was called a dust monitor instead of a PM_{10} monitor.

To confirm effects of atmospheric humidity, IPCAJ carried out a test. At the monitoring station equipped with the low-volume air sampler (Lovol) and the β-ray PM_{10} monitor, the ratios of the readings from the light-scattering dust monitor and the β-ray PM_{10} monitor to the Lovol measurements were determined, and their correlation with humidity levels during the sampling period was analysed. The humidity was assessed as absolute humidity, considering the hygroscopic nature of the particles and the filter media. The findings are presented in Figure 13.7.

From these results, it can be seen that the correction value (the concentration ratio) is in the range of 0.6 to 1.4 for the β-ray PM monitor and 0.6 to 1.2 for the light-scattering dust monitor. It appeared that these monitors tend to report higher concentrations (than the reference Lovol) as the humidity increases. Note that the reference values measured by the low-volume sampler already excluded the humidity effects (after desiccating conditioning). Moreover, these experimental results showed that this tendency was even more remarkable for the β-ray PM monitor than for the light-scattering dust monitor. The reason was associated with the effect of the hygroscopicity of the filter media of the β-ray instrument which was more significant than the effect of the hygroscopicity of the particles themselves. This demonstrates that the decision to phase out light-scattering measuring devices, due to their high sensitivity to humidity, in favour of β-ray PM monitor was misguided.

From these results, it is considered that the light-scattering-type measuring instrument is not so much affected by the hygroscopicity of particles as was generally perceived at that time and is a good measuring instrument with extremely high reproducibility. However, the results produced by the light-scattering dust monitor were somewhat lower when compared with the results of the Lovol sampler, which is the official reference measurement method. Therefore, this decision to replace a beta-ray monitor with a light-scattering instrument could help reduce the enormous burden of calibration and did not necessarily seem too bad.

13.2.5 Issues Related to Analysis for PM Chemical Composition

In a study aimed at analysing the composition of particulate matter, it is necessary to determine a sampling method and a component analysis method that correspond to the purpose. Multi-elemental analysis may require the use of multiple samplers with different filter media or splitting a sheet of filter paper and using the parts for different analyses. Here are some examples that might be useful in such cases (IPCAJ, 1984a; Fujimura et al., 1988).

13.2.5.1 Instrumental Differences

Before the field survey of the Kashima SPM study, some preliminary experiments were conducted in the laboratory to examine the performance of the sampling devices and equipment used for chemical analyses. To examine the analytical results of the PM chemical components collected by multiple samplers, first, it is necessary to determine the instrumental differences among the samplers. For this purpose, six sets of the high-volume air samplers were arranged side by side, i.e., collocated a few meters apart, and the PM samples were collected using a quartz filter for 24 hours. The experiment was repeated four times (24 hours each), and the arrangement of the samplers was changed before the second and fourth trials. Therefore, it is not possible to collectively consider the effects of locations, the effects of instruments, and their interactions in a two-way ANOVA, and their effects were analysed by separate one-way ANOVA. In this experiment, it cannot be denied that the interaction is not significant. However, by including the location or instrument effect as one factor in the error term, each can be analysed by a one-way ANOVA. Further, by examining the magnitude of the variance of the error term, some inferences can be made as seen in Tables 13.7a and 13.7b.

The results show that there was no significant difference between the concentrations produced by the six high-volume air samplers, as well as by the installation locations of the samplers. Furthermore, when taking the sampler or the location as a factor, the values of

TABLE 13.7A

ANOVA for Analysing the Differences in PM Concentrations Caused by Instruments

Factor	Sum of squares	Degrees of freedom	Mean square	*F* value
Block (day)	903,035	3	301,012	4240*
Instrument	713	5	143	2.0
Error	1060	15	71	
Total	904,808	23		

* 0.01 significance level

Source: Adapted from Fujimura et al. (1988) and modified.

TABLE 13.7B

ANOVA for Analysing the Differences in PM Concentrations Caused by the Locations of the Collocated Samplers

Factor	Sum of squares	Degrees of freedom	Mean square	*F* value
Block (day)	903,035	3	301,012	3584*
Location	514	5	103	1.2
Error	1259	15	84	
Total	904,808	23		

* 0.01 significance level

Source: Adapted from Fujimura et al. (1988) and modified.

the square root of the error variance are almost equal, which suggests that the influence of both on the measured PM concentration was not so large and hence can be neglected. Thus it can be concluded that both effects can be pooled into the error term, and the mean square error (MSE) was about 88.7 μg/m^3. Therefore, with the coefficient of variation determined using the average measured concentration of 324.8 μg/m^3 and the RMSE (root mean square error) of 9.4 μg/m^3, the measurement accuracy represented by the coefficient of variation is about 3.0%.

13.2.5.2 Effect of Filter Media

In the Kashima SPM study, a Teflon (fluorocarbon polymers) filter was used for analysing elemental compositions, and a quartz filter was used for carbon analysis. The filter paper was weighed at a humidity of 60% or less, based on the results of preliminary experiments. Since it was previously confirmed that the difference in PM concentrations caused by the difference in position between collocated locations and the differences between the air samplers were sufficiently small, it was possible to start the study on the difference due to filtration media of the samples produced by these collocated air samplers. Four types of filter paper including quartz, glass, membrane and Teflon (Polyflon) were selected to use in different samplers, and three trials were carried out. This experiment was carried out on highly polluted days, at Kojiya, Tokyo, a relatively highly polluted area. Table 13.8 shows the comparison test results of measured PM concentrations by the filter media. Table 13.9 shows the results of ANOVA of the results.

The results show that the concentration data based on Teflon filter samples were the lowest, followed by the quartz filter samples, and that obtained with glass fibre filter samples were the highest. The results suggest that the glass fibre filter may have the weight increased due to the adsorption of acid gas by glass fibre, as is commonly believed (Tierney and Conner, 1967).

TABLE 13.8

Result of the Comparative Study on the Effect of Filter Media on PM Concentrations (unit μg/m^3)

Filter media	Trial 1	Trial 2	Trial 3
Quartz	838	138	299
Glass	847	149	326
Teflon	828	123	284
Membrane 1	844	143	309
Membrane 2	844	133	301

TABLE 13.9

ANOVA for the Experiment Results of PM Concentrations Collected on Different Filter Media

Factor	SS	df	MS	*F* value
Block (day)	1,349,480	2	674,740	24,990*
Filter media	1355	4	339	12.6*
Error	214	8	27	
Total	1,351,050	14		

* 0.01 significance level

Source: Adapted and modified from Fujimura et al. (1988).

13.2.5.3 Fluctuations Within the Filter Paper for the Collected Mass

The general procedure for the multi-component analysis of PM is to first divide a filter paper into several parts, which can be used to analyse different chemical constituents. In this case, it is required that the collected particulate matter is evenly distributed over the filter collected area. To examine the uniformity of PM distributed over a filter paper, six points were punched out from the same filter paper as shown in Figure 13.8, and five elements in each punched sample were analysed by atomic absorption spectrometry.

The difference in the concentrations measured at the six locations (A, B, C, D, E, F) was analysed using the ratio of the concentration at a position to the average value for each element. The results for a Hivol using a cyclone inlet are shown in Table 13.10, which reveal certain differences in concentration depending on the position on a filter paper, but for most elements, the error was less than 10%. Note that, in general, a high-volume air sampler using a cyclone inlet may induce lower concentration near the centre, but the results are acceptable.

13.2.5.4 Chemical Analysis and Its Precision

Here, it was considered that there was no serious difference in the amount of particulate matter collected at the multiple positions within the filter paper as previously described. Therefore, four sample pieces were cut out from one filter paper and used for analysis. To compare the analytical results by several analysing methods, four sample pieces of 47 mm φ each were cut out from a membrane filter paper sample collected by the high-volume air sampler as shown in Figure 13.9. Six metal elements in the sample were analysed by four different analysis methods: atomic absorption analysis (AA), inductively coupled plasma emission spectroscopy

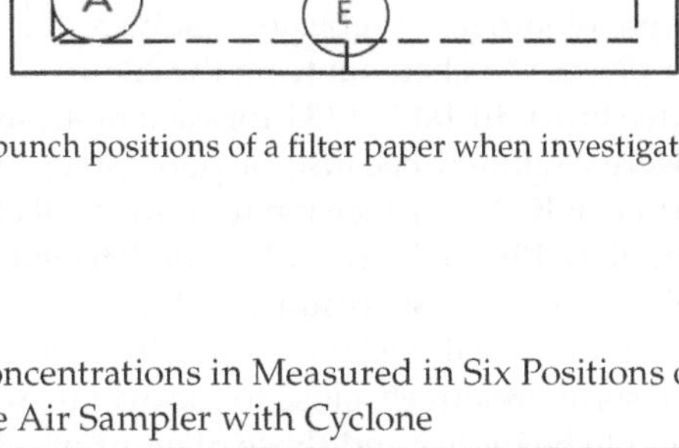

FIGURE 13.8 Example of the punch positions of a filter paper when investigating the uniformity of PM distribution on the filter.

TABLE 13.10

Differences in Elemental Concentrations in Measured in Six Positions on a Filter Paper Sample Collected by a High-Volume Air Sampler with Cyclone

Position/element	Fe	Zn	Mn	Cu	Pb
A	1.03	0.98	0.92	1.1	0.88
B	0.96	1.03	1	1.01	1.03
C	0.96	1.03	1.07	1.01	1.03
D	0.96	0.94	0.92	1.01	0.87
E	1.03	1	1	0.93	1.03
F	1.07	1.03	1.07	0.93	1.2

Note: For each element, the ratio of the concentration at each position is shown, assuming that the average of the concentrations at six positions is 1.0.
Source: Adapted from Fujimura et al. (1988) and modified.

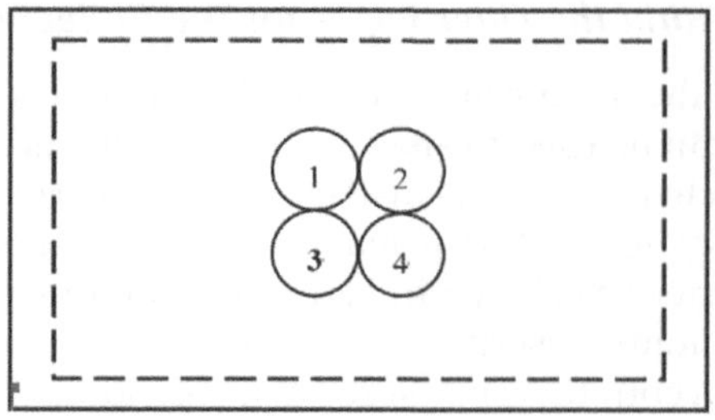

FIGURE 13.9 Punched parts on the filter sample used for examining the repeatability of component analysis by a method and the consistency among the analytical methods.

(ICP), instrumental neutron activation analysis (INAA) and X-ray fluorescence analysis (XRF). Here, the six elements analysed were iron, zinc, manganese, copper, chromium and lead. As for the lead, INAA was excluded. In this comparative analysis, three identical filter papers (sample numbers SM202, SM203 and SM204) were used. Each sample piece was subjected to necessary pretreatment before being introduced into the analytical instrument. The results are shown in Figure 13.10.

Furthermore, based on these data, differences in the measured values by the analytical method were examined by analysis of variance. Here, the differences among analytical methods were examined using three filters as a block factor. No significant difference was found among the analytical values for the elements, except for iron. However, chromium and nickel were found to have some variation among the analytical methods.

It has been explained in Chapter 8 that the quality of the data is represented by the accuracy of the measured values, which can be divided into two parts: precision, which represents the variation around the centre of distribution, and trueness, which represents the deviation from the true value. This study did not use standard samples with known component amounts, so it was not possible to assess the degree of trueness. However, since the same sample was analysed repeatedly, it is possible to understand the precision to some extent. Many factors are related to precision, and this case study tried to evaluate the variability within the same analysis method. This variability can be grasped by repeated measurement as repeatability.

This repeatability is evaluated by the magnitude of the coefficient of variation of the analysis value for the data, which were obtained by repeated analysis of a sample by a given method. In this experiment, the four pieces punched out from the centre of the same filter sample were analysed, and the results are shown in Table 13.11 for each analysis method.

The AA results for Cr were excluded because of poor precision, whereas INAA did not include Pb and Ni. In addition, in ICP, the precision for almost all elements was low compared with AA. INAA could not analyse Pb, but Ni was also not detected due to insufficient sensitivity. Here, the general trends in the analysis results, including the actual sample preparation operations, are grasped. Roughly speaking, the values by atomic absorption spectrometry and X-ray fluorescence analysis are relatively close to the overall average value, whereas those of ICP emission analysis tend to be higher and those of neutron activation analysis tend to be a little lower.

In the weighted least squares receptor model, there is an option to adopt the reciprocal of the measurement error as a weighting coefficient. The examination of the possibility of use of the measurement error in such cases was conducted. In the INAA analysis, gamma-ray emission is a random phenomenon, and the variance can be estimated because the count number of the detector follows the Poisson distribution. Moreover, since the four punched-out samples from the same filter were measured, the unbiased estimated value of the error variance can be calculated from these data. Then the total variance was calculated assuming that they were independent of one another. However, elemental components below the detection limit were excluded. The results shown in Table 13.12 do not contain some components that are expected to be major marker elements in the receptor model, such as potassium.

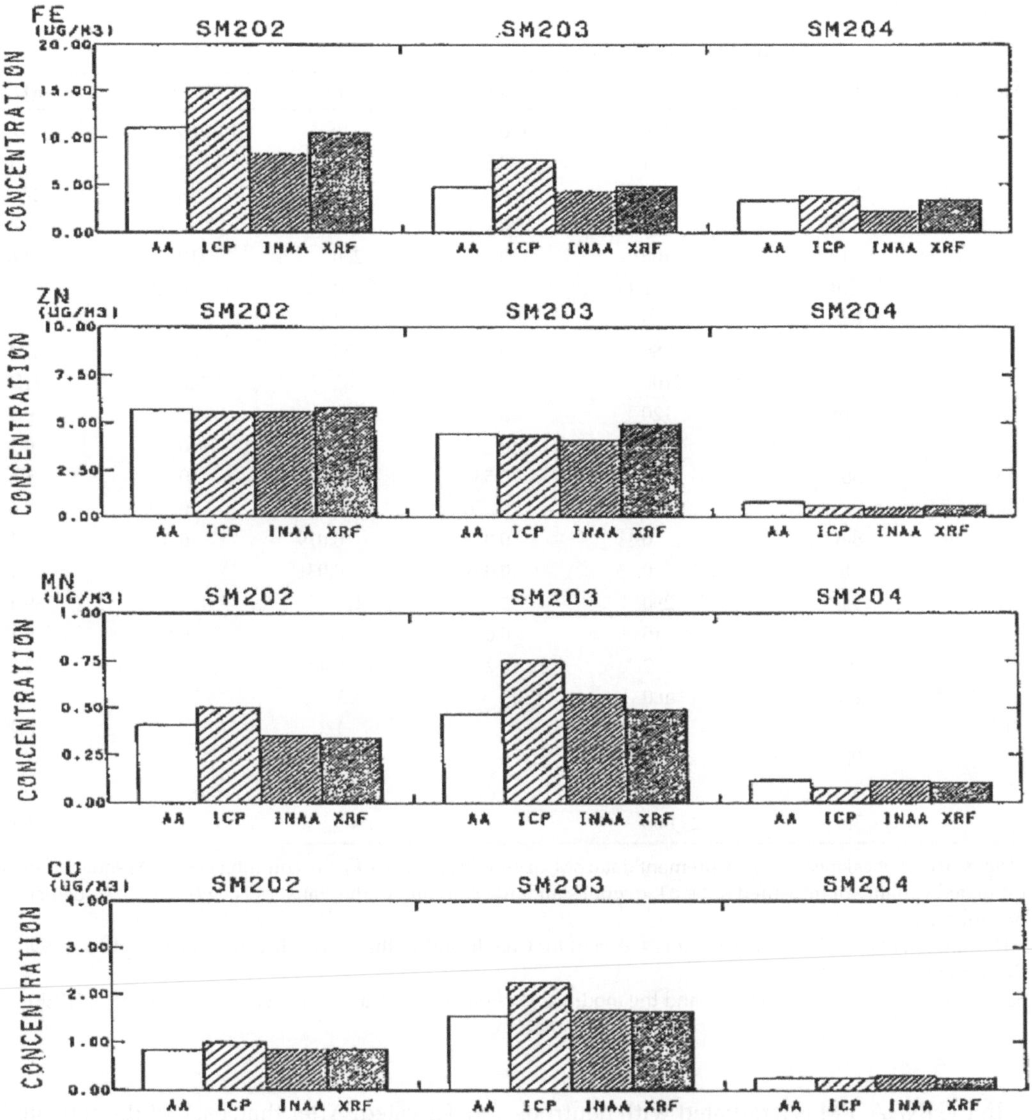

FIGURE 13.10 AA, IPC, INAA and XRF analytical results of chemical composition for PM samples collected by a high-volume air sampler.

Source: Adapted from IPCAJ (1984a).

TABLE 13.11

Repeatability of the Results for Each Element by Four Analytical Methods (ratio of standard deviation/average value), %

Analysis method	Fe	Zn	Mn	Cu	Pb	Cr	Ni
AA	1.1	1.9	1.5	1.7	2.8	—	7.3
IPC	3.8	2.3	6.0	3.0	10.4	11.8	30.3
INAA	4.5	5.2	7.0	19.8	—	7.3	—
XRF	9.6	2.7	5.8	2.0	11	4.7	6.8

TABLE 13.12

Precision of the Measurement for Each Element

Method	Element	$\bar{X}$	σM	σx	σT	σT/$\bar{X}$
INAA	Al	6300	130	340	360	5.8
	As	4	0.82	2	2.1	53.1
	Br	20	3.4	3.9	5.3	27
	Ca	3000	290	390	490	16.6
	Cl	2100	170	260	310	14.6
	Co*	1.1	0.23	0.25	0.32	28.6
	Cr	29	2	2.1	2.9	10.1
	Cu	380	30	75	81	21.4
	Fe	2100	230	82	240	11.4
	Mn	120	3.7	12	12	10.1
	Na	3200	140	290	320	10.1
	Sb	15	0.58	0.58	0.82	5.6
	Sc	0.3	0.026	0.012	0.029	9.6
	Sm	0.11	0.023	0.019	0.03	27.1
	Th*	0.25	0.081	0.031	0.076	30.1
	Ti	290	32	110	110	39.4
	V	15	0.66	0.5	0.86	5.8
	W*	2.7	1.2	0.49	1.2	42.2
	Zn	480	25	25	36	7.4
XRF	Cd	21		1.7		8.4
	Pb	320		35		11
	S	7800		0		0
	Si	29,000		1500		5.1

Note: $\bar{X}$, arithmetical mean of measurement data; σM, arithmetical mean of γ-ray counting error; σX, square root of estimated variance of measured data; σT, overall standard error; unit of the statistics is ng/m^3, σT/ $\bar{X}$: coefficient of variance (%),

* The amount of data for each element is 4, except for Co, Th, and W that have 3 because one sample was below the detection limit.

Source: Adapted from IPCAJ (1986), and the modified version of this data can be seen in Fujimura et al. (1988)

In the INAA, PM is irradiated with neutrons and activated. After that, each of the activated elements emits gamma rays of a specific wavelength and changes to more stable substances. After activation, the elements that can be analysed depending on when the gamma-ray spectrum is measured. In this survey, two measurements of short-period elements and long-period elements were performed as shown in the appendix of this chapter. As a result, the sensitivity of the analysis for some elements such as potassium falling in between, deteriorated. If there is no budget constraint, it is ideal to analyse the gamma-ray spectrum in three periods.

13.3 Key Findings of the Kashima Study

In this Kashima SPM study, various PM samples from 15 locations and 4 seasons were obtained. Although it is limited to 15 days each season, the study could grasp the daily changes to some extent. It was also possible to clarify the geographical distribution of each chemical component around the industrial area and the particle size distribution of each chemical component. Furthermore, since the samplers were installed at the air quality monitoring stations, it was possible to consider the relationship between the concentrations of many gaseous air

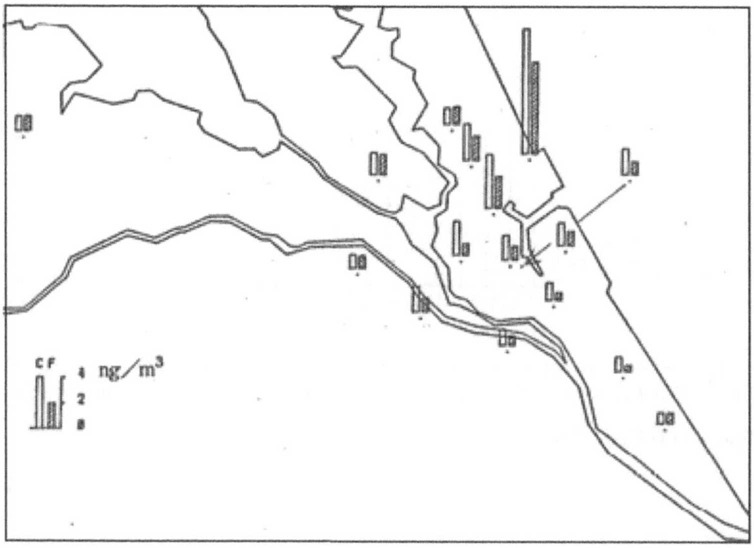

FIGURE 13.11 Spatial distribution of chromium concentrations obtained by the combined sampling scheme. (*Note:* C, coarse fraction; F, fine fraction.)

Source: Adapted from IPCAJ (1984b).

pollutants and the concentrations of chemical components in PM. Here are some interesting examples of these characteristic analysis results.

At each of the four sampling sites where elemental analysis was performed for eight particle sizes, the individual values were used to calculate elemental concentrations for coarse and fine particles. For the remaining 12 sites, samples of coarse particles and fine particles were prepared in advance by means of composite sampling, and their analysis values were used. Based on the data obtained in this way, a regional distribution map of each element was created. Figure 13.11 shows an example of the regional distribution of chromium in coarse and fine fractions at four-season average concentrations. This spatial distribution is very similar to the distribution of manganese shown in Okamoto and Kobayashi (1991), and it can be seen that the concentration is higher near the steel mill.

As for the four intensive sampling sites, elemental analysis was conducted on all of the eight-stage samples of Andersen impactors installed. The particle size distributions were created using the average value of a total of 16 data for these four sites and four seasons. The size distribution for each element was obtained from four sites' data. Using the average concentrations of each of the four intensive measurement stations, the mass median diameter (MMD) of each season and yearly average was calculated. These results are shown in Figure 13.12. The abscissa of this figure shows the component names by arranging from the smallest to the largest MMD of the chemical components, and the vertical axis shows the value of the MMD. Looking at these results, the element with the largest particle size is silicon, and the MMD is large with elements such as Al, Sc and Ti that are abundant in the soil in all four seasons. On the other hand, elements having a small MMD are Ni, Ag, Sb, Cs, and the like.

Next, the relationship between secondary particles generated by chemical reactions in the atmosphere and their precursors SO_2 and NO_2 was investigated. The ionic components in these secondary particles, especially nitrates, are strongly affected by temperature and by change between the solid phase and the gas phase. Although not shown here, it can be observed that the size distribution of anions related to secondary particles varies greatly with the season (see details in Pongkiatkul et al., 2010). To examine the relationship between the particulate and gaseous pollutants, the relationship between the chemical composition in PM collected

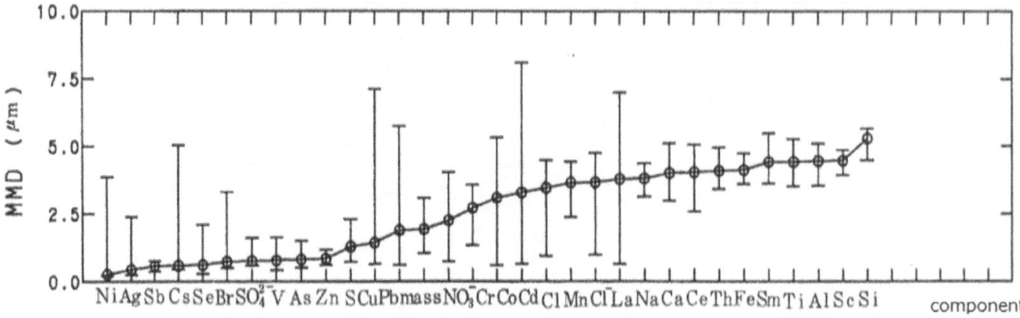

FIGURE 13.12 Mass median diameter (MMD) of chemical components in particulate matter, the average value was calculated from 16 data at 4 locations and 4 seasons. Vertical bars indicate maximum and minimum.

Source: Adapted from IPCAJ (1986).

TABLE 13.13

Correlation Coefficients between the Gaseous and Particulate Pollutants

Gas	Particulate	Winter	Spring	Summer	Autumn	Total
SO_2	SO_4^-	0.295 (60)	0.436 (60)	0.319 (60)	0.144 (50)	0.163 (230)
SO_2	V	0.697 (60)	0.548 (60)	0.587 (60)	0.175 (50)	0.346 (230)
SO_2	Ni	0.337 (5)	0.703 (12)	0.615 (24)	0.356 (8)	0.525 (49)
NO_2	NO_3^-	0.642 (59)	0.395 (60)	0.492 (48)	0.589 (45)	0.375 (212)
NO_x	NO_3^-	0.691 (59)	0.401 (60)	0.335 (48)	0.573 (45)	0.31 (212)
NO_x	Cl^-	0.576 (59)	0.531 (60)	0.025 (54)	0.529 (53)	0.633 (226)
NO_x	Pb	0.351 (47)	−0.257 (40)	−0.119 (50)	−0.205 (44)	−0.202 (181)
NO_x	Br	0.712 (52)	0.417 (58)	0.403 (53)	0.641 (60)	0.385 (223)
NMHC	SO_4^-	0.614 (15)	−0.098 (18)	0.203 (30)	0.739 (30)	0.314 (93)
NMHC	BAP			0.577 (15)	0.082 (15)	0.091 (30)
NMHC	BGHIP			0.338 (15)	0.374 (15)	0.283 (30)
Ox	SO_4^-	−0.432 (26)	0.194 (60)	0.36 (60)	−0.627 (60)	0.056 (206)
Cl(G)	Cl(P)	0.211 (15)	−0.411 (15)	0.585 (15)	0.152 (15)	0.347 (60)
Cl(G)	Cl^-	0.269 (15)	−0.468 (15)	0.542 (13)	0.111 (14)	0.379 (57)
Cl(G)	NO_3^-	−0.021 (15)	−0.498 (15)	0.31 (13)	0.191 (10)	0.056 (53)

Note: NMHC is a non-methane hydrocarbon, OX is an oxidant, Cl(G) is gaseous chlorine, and Cl(P) is chlorine in particulate matter, BAP is benzo[a]pyrene, BGHIP is benzo[ghi]perylene.
The concentration of gaseous pollutants is the average value of 24 h, which has been adjusted to the sampling duration of the air sampler. The number in parentheses is the amount of data.
Source: Adapted from IPCAJ (1986).

by the high-volume air sampler and the corresponding gaseous pollutant concentration was analysed. As mentioned earlier, the high-volume air sampler is equipped with a size-selective inlet that removes larger particles, resulting in data primarily representing fine particles. The average values of the hourly concentrations of gaseous pollutants from 9:00 AM to 9:00 AM of the next day were calculated to match the sampling time of the high-volume samples. Data above the detection limit were considered valid, and the correlation coefficient was calculated when there were four or more valid data points. The results are shown in Table 13.13. The numbers in parentheses in the table indicate the number of valid data points.

The correlations between SO_2 concentration and sulphate ion and between NO_2 concentration and nitrate ion were both positive but not very high. Vanadium and nickel, which are considered to be marker elements for oil combustion in receptor modelling, showed high correlations with SO_2 and NO_x concentrations. Although most of the large combustion facilities in this industrial area are equipped with desulfurisation, denitrification and electrostatic precipitators, this result seemed to suggest that they were not necessarily sufficiently effective in removing these fine particulates. However, PM concentrations around this industrial area are quite low considering the production scale of the companies located there.

The correlation between oxidant concentrations and the concentrations of secondary particle-related components varied greatly depending on the season. In spring and summer, there was a weak positive correlation, but in autumn and winter, there was a clear negative correlation. In autumn and winter, the oxidant concentration was lower when the air was stagnant. Therefore, there may be a negative correlation between the components whose concentrations are higher under such weather conditions. In the Kashima area, the concentration of gaseous chlorine was also monitored at one site as an air quality against leakage from a chemical plant. The particulate chlorine obtained here is not a sea salt particle because it is a fine particle sampled by a Hivol equipped with a sizing device. The correlation between gaseous and particulate chlorine varied with the season, and it was not possible to determine a clear relationship.

13.4 Summary

This chapter outlined key findings from IPCAJ's study on particulate matter which, at that time, was considered the most critical air pollutant and its measurement methodologies were under investigation. It also discussed the potential of utilising the gathered data as input for a receptor model, which was in development then. The chapter detailed the plan for sampling and analysing particulate matter and highlights the principal outcomes of these analyses. An example of a receptor model analysis based on this data is provided in Chapter 23.

REFERENCES

Fujimura, M., Minakawa, N., Kobayashi, H., Ohtoshi, T., Okamoto, S. and Kobayashi, K., 1988, Sampling and analytical errors of airborne particulate matters by high-volume air sampler, *Journal of Japan Society of Air Pollution*, 23(5), 299–310 (In Japanese with English captions), https://doi.org/10.11298/taiki1978.23.299

Industrial Pollution Control Association of Japan (IPCAJ), 1983, *The study on the behaviour of particulate matter, vol. 1: Report on the fiscal year 1982 field survey* (p. 419) (In Japanese), Japan Environmental Management Association for Industry, Tokyo, Japan.

Industrial Pollution Control Association of Japan (IPCAJ), 1984a, *The study on the behaviour of particulate matter, vol. 5: Report on the analysis of field survey data* (p. 560) (In Japanese), Japan Environmental Management Association for Industry, Tokyo, Japan.

Industrial Pollution Control Association of Japan (IPCAJ), 1984b, *The study on the behaviour of particulate matter, vol. 6: Report on the analysis of field survey data II* (p. 340) (In Japanese), Japan Environmental Management Association for Industry, Tokyo, Japan.

Industrial Pollution Control Association of Japan (IPCAJ), 1986, *The study on the behaviour of particulate matter, vol. 9: Final report* (p. 163) (In Japanese), Japan Environmental Management Association for Industry, Tokyo, Japan.

Okamoto, S., 1994, *Report on the study on air quality prediction, source apportionment methods for particulate matter II* (p. 70), Tokyo University of Information Sciences (In Japanese).

Okamoto, S. and Kobayashi, K., 1991, Characterization of aerosols in the Kashima area – a summary of IPCAJ Kashima SPM study, *Atmospheric Environment*, 25B(1), 133–142, https://doi.org/10.1016/0957-1272(91)90047-I

Okamoto, S., Kobayashi, K. and Yamada, T., 1986 August 25–29, Characterization of aerosols in the Kashima area and a source apportionment study, *Proceedings 7th world clean air congress*, vol. 4, 253–262, Sydney, Australia.

Pongkiatkul, P., Kim Oanh, N.T., Wangkiat, A. and Okamoto, S., 2010, Further analysis and PMF application to the chemical composition data of IPCAJ's Kashima SPM study, *International Journal of Environment and Pollution*, 40(4), 337–350, https://doi.org/10.1504/IJEP.2010.031754
Tierney, G.P. and Conner, W.D., 1967, Hygroscopic effects on weight determinations of particulates collected on glass-fibre filters, *American Industrial Hygiene Association Journal*, 28(4), 363–365, https://doi.org/10.1080/00028896709342537

Appendix: Instrumental Neutron Activation Analysis (INAA) Analysis Condition in this Study

Let us briefly explain the specifications of INAA analysis.

- Reactor: Musashi Institute of Technology Nuclear Research Institute, TRIGA-II type (thermal neutron flux density, $1.5 \times 10^{12} cm^{-2} s^{-1}$).
- Irradiation measurement conditions were divided into three element groups. The cooling time and irradiation time for the short group were 3–4 minutes and 1 minute, respectively. For the long-1 group and long-2 group, these times were (2–3 days, 5 h), (about 20 days, 5 h), respectively.
- Detector: Ge (Li) semiconductor detector
- Calibrator: A multi-element standard close to the airborne dust composition was used, in which a fixed amount of the standard solution was removed on 5A filter paper.

This analysis is not compatible with Pb, Bi, Po etc. Elements that are not altered by neutron irradiation in the reactor or whose nuclides do not emit γ-rays cannot be analysed. An example of gamma-ray spectrum analysis for the long cooling time group is shown in Figure 13A.1.

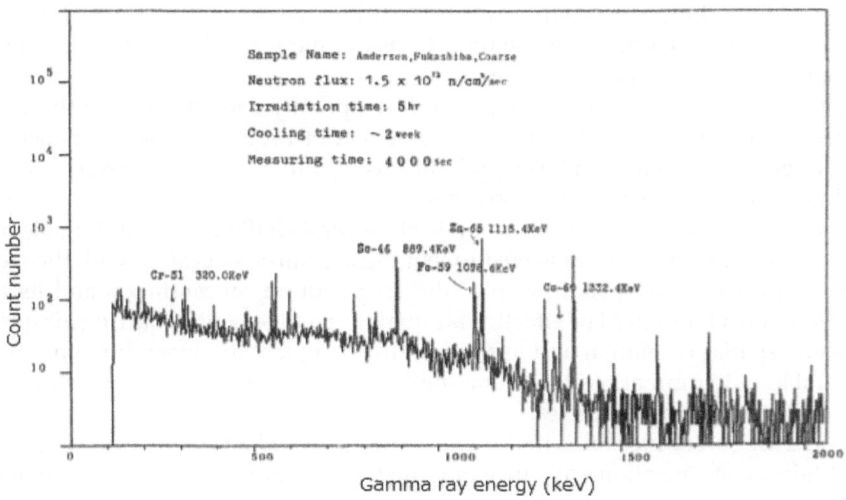

FIGURE 13A.1 An example of a gamma-ray spectrum in instrumental neutron activation analysis.

14

Temporal Resolution and Averaging Time of Concentrations

The concentration of contaminants in the environmental matrices such as air or water changes depending on the location and time. When discussing global problems like climate change and ozone layer depletion, we should be discussing the average concentration of the entire globe. On the other hand, for the quality of air in daily life, the average concentration in a limited space such as inside a house is important. A correct understanding of these different spatial and temporal scale resolutions is very important in discussions of air pollutant concentrations. In air quality management, it is important to know the pollutant concentration at the averaging time corresponding to the degree of its influence. The averaging time (or sampling time) is one of the most important issues in air quality measurement. The issues related to temporal variation and time resolution of the measurements are primarily described in this chapter. The issue of spatial representativeness of measurement, another important consideration in environmental monitoring, is discussed in Chapter 15.

14.1 Spatiotemporal Scope of Air Pollution Problems

Consider what air pollution problems there are at a given time and space scale. The spatiotemporal scale of air pollution problems is schematically shown in Figure 14.1; the horizontal axis is the space, and the vertical axis is the time. Consider the scale of 1 second, 1 minute, 1 hour, 1 day, 1 month, 1 year, 10 years, 100 years.

In addition, the concentration fluctuates both temporally and spatially, so there are many difficulties in measuring it. It is related to the effect of health and other impacts. Because it relates to environmental standards and various administrative objectives, it is naturally required that the measurement also correspond to it.

Measuring air pollution concentration presents several challenges, including technical and resources constraints. It is crucial to ensure that the measurements align with the monitoring program's objectives. The spatial mean of the air pollution concentration and its timescale value are key considerations. For effective air quality management, acquiring air quality data with suitable spatial resolution and averaging time is essential, depending on the target of analysis. Table 14.1 summarises the averaging times needed for air quality analysis, simulation and decision-making processes.

In considering safety measures in the event of leakage of hazardous materials, it is necessary to obtain a concentration with an averaging time of several seconds to several minutes. Therefore, when examining the concentration of such substances, it is not very meaningful to calculate an hourly average concentration. On the other hand, if one tries to calculate the concentration of chlorofluorocarbons (CFCs) and carbon dioxide (CO_2), which are the cause of global environmental pollution, i.e., ozone layer depletion and climate change respectively, annual or monthly averages are more appropriate than hourly average concentrations.

DOI: 10.1201/9781003597216-16

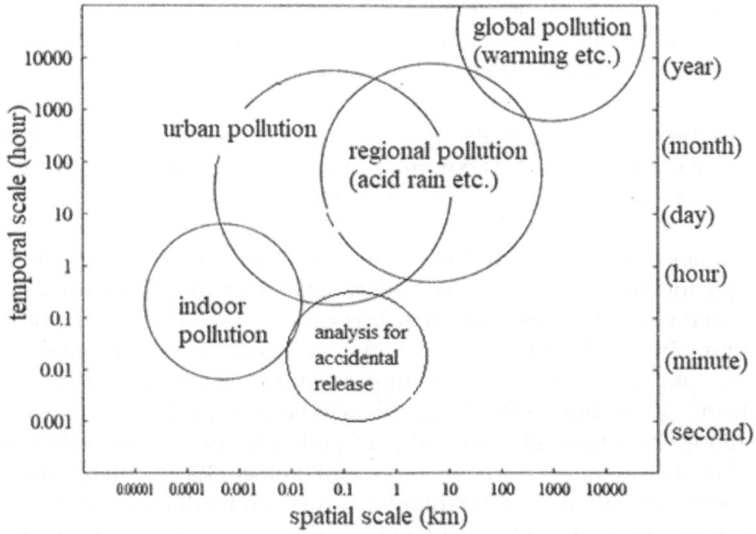

FIGURE 14.1 Position of characteristic air pollution problems on a time and space scale.
Source: Adapted and modified from Okamoto (2001).

TABLE 14.1
Temporal Resolution of Concentrations for Various Environmental Issues

Environmental issues	Timescale
Leakage of highly poisonous substances, flammable or explosive gas	Less than a few seconds
Responsible care for chemicals, odour countermeasure	Several minutes
Measures for air pollutants such as sulphur oxides and nitrogen oxides	1 hour to annual
Global environmental issues such as carbon dioxide and chlorofluorocarbon	More than 1 year

14.2 Averaging Time of Concentrations in Air Quality Management

Statistical analysis of the extensive data gathered by the air quality monitoring network can yield various insights. These include the frequency distribution of pollutant concentrations and their relationship with averaging times. Such information is also pertinent to numerous aspects of environmental management. Even in the case of pollutants for which environmental standards have been set for the hourly average concentration or daily average concentration, annual average values may be calculated using an air quality model, and the results of such predictions may be used to consider various pollution countermeasures. Some of these issues are considered here. While this method is currently not permitted in the United States, it is widely used in Japan. The issues associated with this approach, as well as potential problems if it were implemented, will be examined.

Larsen (1969, 1973) analysed several years' worth of air quality data in several major US cities, including Washington, DC, Chicago, and Los Angeles, and showed the following characteristics for the averaging time data.

- First, the cumulative frequency distributions for these averaging times tend to fit log-normality and thus plot as an approximately straight line on log-normal graph paper (see Figure 2.3). This fit is particularly good in the case of the high half of the concentration distribution, which is important in air quality management.

- Second, the median concentrations for the various averaging times tend to plot as a straight line on logarithmic graph paper (50th percentile line of Figure 14.2, created with a similar concept). Median concentrations are thus proportional to averaging time raised to an exponent.
- Third, maximum concentrations also tend to plot as a straight line (not the maximum but see the 99% value in Figure 14.2). Maximum concentrations are thus approximately inversely proportional to averaging time raised to an exponent.

Although the data is for a limited period of three months instead of one year, it seems to be a useful example for comparison when considering whether the air quality data has the previously mentioned characteristics. Note that this case is taken from a part of the validation study of the air quality model. The result is shown in Figure 14.2. Such studies are also useful in validating air quality models as well as in percentile-concentration curves and Q-Q plots (please see Chapter 19, Section 19.5). The details will be explained in Chapter 19.

By incorporating these three characteristics of pollutant concentration, an empirical, averaging-time mathematical model has been developed (Larsen, 1973). If the geometric mean and standard geometric deviation for a pollutant at a site can be determined for one averaging time, then the model can be used to calculate the geometric mean, standard geometric deviation and annual maximum concentration for any other averaging time.

This analysis was performed based on data from monitoring stations installed in large cities. On the other hand, analytical results around a large emission source such as a thermal power plant in a rural area showed a different tendency, as will be explained in Chapter 18, Section 18.4. Specifically, in the vicinity of a large industrial facility with tall stacks situated in a rural area, the annual average concentration of pollutants is quite low, yet the short-term peak concentrations, that occur at the tail end of the distribution, are significantly high. Such a problem, for example, has been reported for SO_2 concentrations in the surroundings of a large thermal power plant. For this reason, it was considered inappropriate to apply a method that had been developed relying on the statistical relationship between the annual average

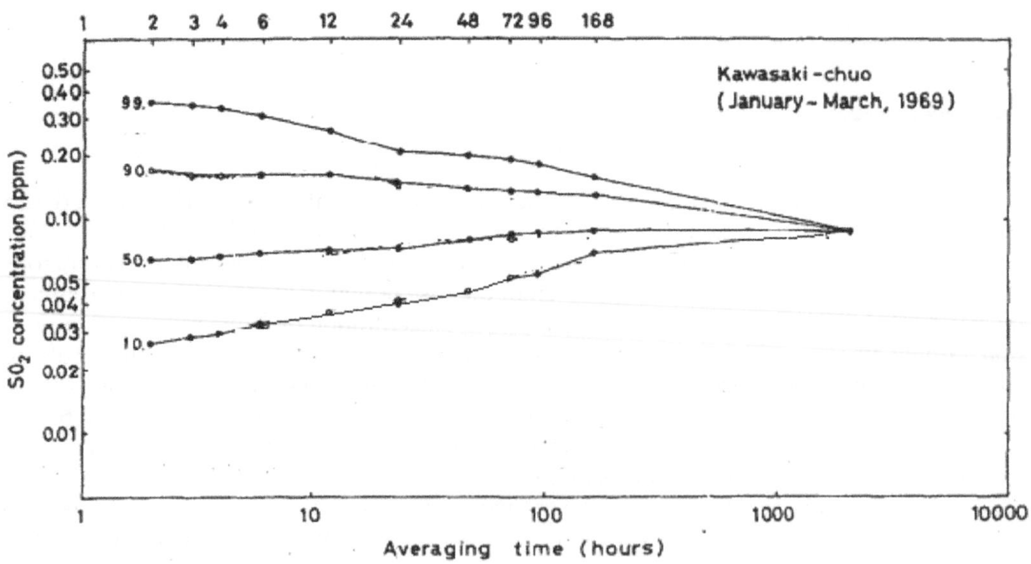

FIGURE 14.2 Results of analysis of hourly SO_2 concentration data at the Kawasaki Chuo monitoring station for three months. It shows the 10-, 50-, 90- and 99th percentile values for each averaging time. (*Note:* Data were retrieved from Okamoto and Shiozawa (1978), and this was originally created to compare the statistical properties of the air quality model output and actual measurements.)

value and the high percentile value, which was estimated based on the data obtained from many air quality monitoring stations in urban areas. The method of Larsen and other similar statistical methods was not suitable for these cases, because it failed to predict the appearance of high concentrations. Subsequently, the revised guidelines of US EPA (2003) stipulate that the conversion of concentrations at different average times by any transformation technique is not allowed in regulatory applications.

In Japan, however, the method of statistically estimating the annual average target value, which corresponds to 98% of the daily average value specified in the environmental standard, is widely used in the application of air quality management. After calculating the annual average value with an air quality model, the 98% daily average value corresponding to that annual average value is statistically estimated. The results are then compared to the prescribed environmental standard. The annual average target value corresponding to the NO_2 environmental standard (0.06 ppm at the 98% value of the daily average concentration) was said to be about 0.03 ppm. However, according to the analysis of Kan'nari et al. (1989), the annual average concentration, which was equivalent to 98% of the daily average concentration of 0.06 ppm, was in the range of 0.025–0.035 ppm, i.e., not exactly 0.03 ppm. The statistical relationship between the annual average concentration and the daily average 98th percentile value varies greatly from year to year (Shiozawa and Okamoto, 1990).

After reading the previous paragraph, you may wonder: there is an air quality model (a dispersion model) that can predict the annual average value; why not use it to predict the 98% daily average value? The answer to this question is explained in Chapter 18. This is because the model formula and its parameters do not correspond to all conditions, particularly the extreme values of less occurrence frequency on the high concentration side. For details, see Chapter 18.

Here, consider how to create a scatter plot between the 98th percentile of the daily average and the annual average and determine the target value by regression analysis. Figure 14.3

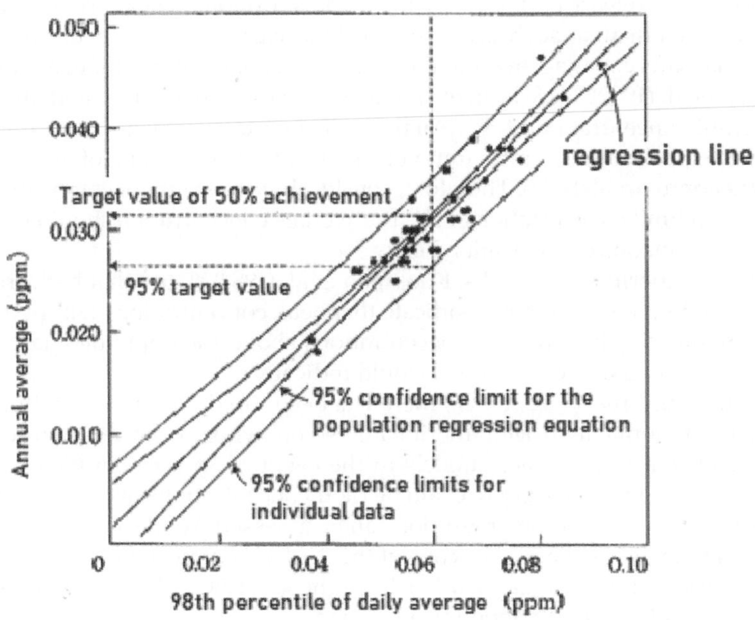

FIGURE 14.3 Regression analysis for estimating the annual average target value corresponding to the 98th percentile of NO_2 daily average specified in the environmental standard. (*Note:* Analysis by data of the fiscal year 1991 [April 1991 to March 1992] at the ambient air monitoring stations in Tokyo.)

Source: Adapted and modified from Okamoto (2004).

shows the 95% confidence limit of the population regression equation and the 95% confidence interval of each data in addition to the regression equation. Here, in the regression equation $y = b_0 + b_1 x$, the prediction interval of the reliability of 100(1 – α)% of the value of the objective variable that will be realised in the future when $x = x_0$ is specified as follows:

$$(b_0 + b_1 x_0) \pm t(n - 2, \alpha)\ [\{1 + 1/n + (x_0 - \bar{x})^2/Sxx\}\ Ve]^{1/2} \quad (14\text{–}1)$$

where b_0 and b_1 are the regression coefficients, $t(\varphi, \alpha)$ is the value of the t distribution table with degrees of freedom φ, n is the number of data, $\bar{x}$ is the average value of data x, Sxx is the sum of squared deviation, and Ve is the unbiased variance of the difference.

To ensure that the environmental standards for each year (98th percentile of the daily average concentration) are achieved with 95% confidence, the bottom line (the 95% limit of the confidence interval of each dataset) in Figure 14.3 must be used to set the target value of the annual average concentration, instead of using the centre line (regression line). In other words, the target value obtained by the central line (regression line) is based on the assumption that the probability of achieving the environmental standard is only 50%.

14.3 Atmospheric Concentration with an Averaging Time Shorter Than One Hour

Recently, attention has been focused on risk assessment and responsive care of chemical substances, which are aimed at an emergency response when harmful substances are released due to accidents. For companies that handle dangerous chemical substances and fuels, it is an essential requirement to build and operate an environmental management system that includes such a risk management plan. Against this background, research is being conducted in developed countries in Europe and the United States to develop dispersion models that can respond to accidental release and to verify the models (Koopman et al., 1989). Such dispersion models have evolved in areas such as the chemical industry dealing with hazardous materials and occupational safety management in the workplace. For example, there are materials such as in Koopman et al. (1982). Here is an example of an experiment using methane with a lower combustion limit concentration of 5%. Analysis of the measurements from this experiment revealed that the peak-to-mean ratio (ratio of the instantaneous value of 3–5 Hz to the average value) was approximately 5.0. Therefore, considering the safety of combustible gas with a lower combustion limit concentration of 5%, the average concentration should be set below 1% to keep the instantaneous concentration below 5%.

Based on these experimental results, Koopman et al. (1982) stated that high-frequency (3–5 Hz) gas-concentration measurements indicate that peak concentrations within the flammable limits are common with 10-s-average concentrations above 1%, implying a larger flammable extent than averaged data or calculations would indicate.

In addition to liquid fuels (e.g., LNG), there has been rapid research growth on other substances such as ammonia. hydrogen fluoride and so on, whose concentrations cannot be calculated by the ordinary dispersion models in the event of an accident (Spicer and Havens, 1989). For such gases, it is necessary to evaluate their concentrations over very short averaging times so that the risk of explosion or ignition can be assessed. While a detailed description of such prediction models is beyond the scope of this text, it is important to understand that the averaging time issues for these gases differ from those for SO_2 PM, and other air pollutants measured in a typical air quality monitoring station.

Many air quality monitoring stations collect data every hour, so a huge amount of data is accumulated every year. For example, in Japan, there are about 2000 monitoring stations even within the range known by the Ministry of the Environment; so annual data of about 2000 stations × 8000 hours × 10 monitoring items means that 160×10^6 data points would be generated. For this reason, the statistical characteristics of temporal scales longer than one hour are

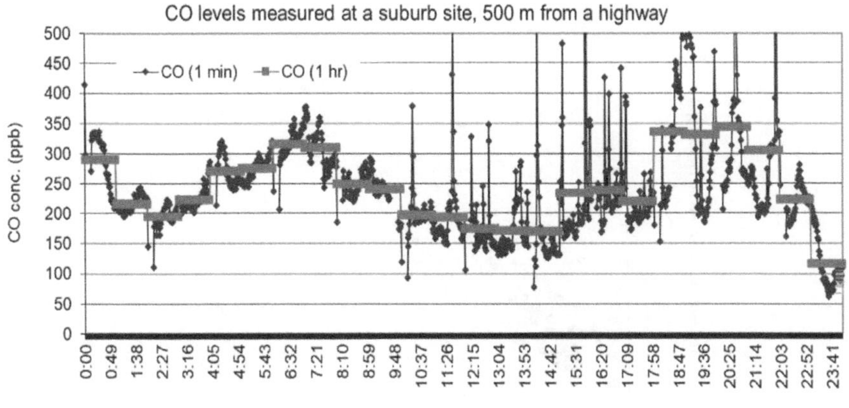

FIGURE 14.4 Highly fluctuating 1 min average vs. an hourly average of CO concentrations.

Source: Raw data provided by a research project of AIT.

well analysed. On the other hand, shorter-scale phenomena have not been captured by hourly monitoring results and hence could not be properly analysed and understood.

There are not many studies reporting the analysis of ambient concentrations with shorter averaging times. This is an important issue when assessing the safety measures for highly toxic chemicals and flammable gases. In such analyses, the threshold value to prevent health damage due to short-term exposure and the peak-to-mean ratio of the certain averaging-time concentration that is easy to measure are also important indicators. The following two examples are used to illustrate the observed concentrations at an average time shorter than 1 hour.

- The CO concentrations that were measured at a place slightly away from a highway, and
- The result of an air tracer dispersion experiment from the roadway.

The example displayed in Figure 14.4 is a measurement of CO concentrations at a suburban site located 500 m away from a busy highway in a Southeast Asian city. It shows the relationship between the 1-minute interval and the hourly average value. The maximum 1-min average value was well above the highest hourly value during the measurement period.

In the IPCAJ dispersion experiment in the Shin-Yokohama area, the concentration of the air tracer SF_6 was measured every 3 min directly by GC, and this dataset is used to examine the concentration with an averaging time shorter than 1 h. To examine the relationship between the average concentration over 1 h and the variation in concentration within 1 h, the IPCAJ data were reanalysed by Hayashi et al. and presented in Yamane and Koizumi (1984) and Tsukioka and Ito (1985).

In this experiment, the average value was calculated for 20 data points every 3 min, and the standard deviation was calculated to see the dispersion of these 20 data points. From Figure 14.5, it can be seen that when the concentration was high, the dispersion (deviation) of the concentrations was also large. It is also shown in Figure 17.4 (see Chapter 17) that this data is in good agreement with the analytical values of the atmosphere collected by the collocated air sampler that was conducted at the same time.

14.4 Other Statistical Analyses on Concentration Data

In environmental science, it is well-recognised that pollutant concentrations in air, soil and water exhibit a log-normal distribution. Blackwood (1992) mentioned that the log-normal distribution has become a common choice to represent intrinsically positive and often highly

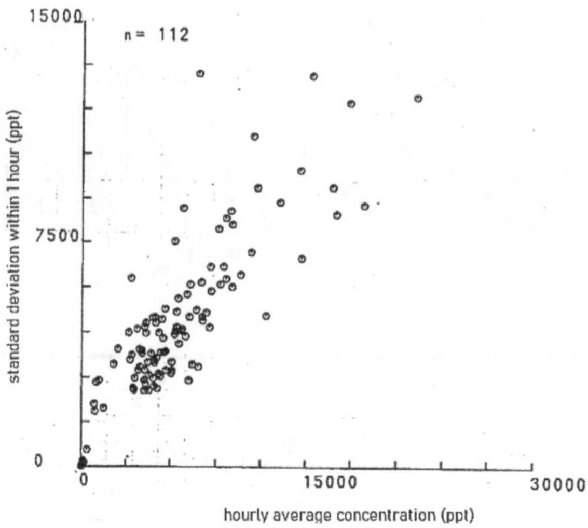

FIGURE 14.5 Relationship between the 1-hr average concentration of SF_6 and the standard deviation of the 3-min concentration within 1 h at the sampling point G-2 (road shoulder) in the air tracer dispersion experiment in the Shin-Yokohama area. (*Note:* For air sampling and analysis in the air tracer dispersion experiment, refer to Section 17.1.3 of Chapter 17.)

Source: Adapted and modified from Tsukioka and Ito (1985).

skewed environmental data in statistical analysis; however, the implications of its use are often not carefully considered. Ott (1990) noted that the widespread occurrence of random dilution processes in the environment contributes to the frequent observation of log-normally distributed pollutant concentrations. In this context, we have conducted an analysis on this matter, and a few findings are presented herein.

This section considers statistical analysis of distributions and time-series trends of air quality concentration data.

14.4.1 Goodness-of-Fit Test

Let us consider the characteristics of the concentration distribution. It is generally said that the concentration of air pollutants follows a log-normal distribution. A statistical analysis method that can be used when discussing such a sampling distribution is the goodness-of-fit test. This is an analysis method related to the chi-square test of the data summarised in the contingency table, which will be described in Chapter 20.

14.4.1.1 Concept and Analysis Method

The goodness-of-fit test is a method of examining the frequency distribution table of a sample survey to determine whether the proportion of each item in the population is a designated arbitrary proportion. Now, let's say that a sample survey has been performed on a certain population and obtained the following frequency distribution table (see Table 14.2). Assuming that the number of bins is m, the population ratio of each bin is $p_1, p_2. . ., p_m$. Let $n_1, n_2. . ., n_m$ be the frequency of each class in the frequency distribution, and n be the total frequency. The value obtained by multiplying the total frequency n by $p_1, p_2. . ., p_m$ is called the expected frequency, and $n_1, n_2. . ., n_m$ is called the measured frequency.

A goodness-of-fit test is to check whether the frequency distribution table obtained from a sample survey is from the population under consideration. That is, check whether $n_1, n_2, . . .$ and $np_1, np_2, . . .$ is consistent with each other. Now calculate the following statistic.

TABLE 14.2

Frequency Distribution Table for the Goodness-of-Fit Test

		1	2	3	...	m	total
Ratio of population		p_1	p_2	p_3	...	p_m	**1**
Sampling Survey	Observed	n_1	n_2	n_3	...	n_m	n
	Expected	np_1	np_2	np_3	...	np_m	n

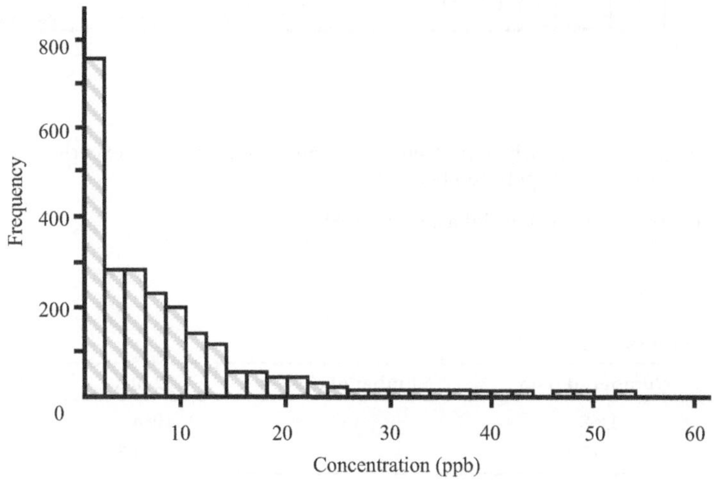

FIGURE 14.6 Histogram of the 3-min SF_6 concentrations (Shin-Yokohama area experiment, G-2 sampling point) mean = 6.48 ppb, standard deviation = 6.65 ppb, number of data = 2240.

Source: Data were retrieved from Tsukioka and Ito (1985).

$$\sum \frac{(n_j - np_j)^2}{np_j} = \frac{(n_1 - np_1)^2}{np_1} + \frac{(n_2 - np_2)^2}{np_2} + \ldots + \frac{(n_m - np_m)^2}{np_m} \tag{14–2}$$

Under the hypothesis that the measured frequency and the expected frequency are equal (in other words, the sample ratio and the population ratio are equal), it is known that the statistics calculated by Equation (14–2) follow the χ^2 distribution when the statistics are obtained by repeating the sample survey many times. Taking advantage of this, the goodness-of-fit test examines whether the proportion of each item in the population is the same as any ratio. In the goodness-of-fit test, the null hypothesis assumes that there is no significant difference between the observed and the expected value. When enough data are obtained, divide the data into m classes. Next the observed frequency of the data entering each class, and the expected frequency in the assumption that data follows the predetermined distribution are summarised as shown in Table 14.2. And if the value calculated by Equation (14–2) is large enough, the hypothesis is rejected.

14.4.1.2 Application

The appearance frequency distribution of the 3-min value (strictly speaking, the instantaneous value every 3 min) and the 1-h value of the air tracer concentration are shown in Figure 14–6 and Figure 14.7.

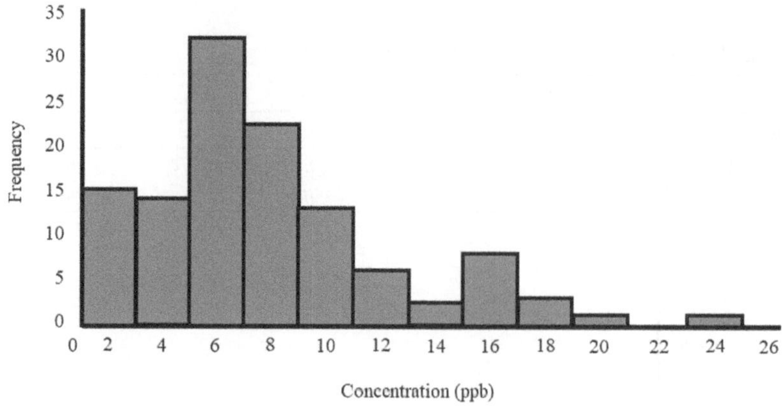

FIGURE 14.7 Histogram of the hourly SF_6 concentrations (Shin-Yokohama area experiment, G-2 point), mean = 6.41 ppb, standard deviation = 4.21 ppb, number of data = 112.

Source: Data were retrieved from Tsukioka and Ito (1985).

TABLE 14.3

Result of the Goodness-of-Fit Test

	Number of data	Number of classes	$\chi 2$ obs	df	$\chi 2$ (0.950)
3-min value	2368	18	1084	15	25.0
1-h average	112	12	101.5	9	16.9

Note: Since the mean and variance of the normal distribution are estimated from the data, the number of degrees of freedom is not the number of classes minus 1, but minus 3.

Looking at these figures, both have a distorted shape with a tail towards higher concentrations and also look like a log-normal distribution. Let us confirm this with the goodness-of-fit test. After the logarithmic transformation of the concentration data, the distribution of the transformed data was examined by the goodness-of-fit test to see whether it followed a normal distribution. The hypothesis to be tested was that the data follow a normal distribution. If it could not be rejected, then the distribution could be assumed to be normal. In this case, one can therefore consider that the dataset before the logarithmic transformation follows a log-normal distribution.

Here, it was examined whether or not the concentration data shown in Figures 14.6 and 14.7 follow a log-normal distribution. First, after logarithmically transforming the concentration data, a frequency distribution table was created. The numbers of the class k in the frequency table were 18 and 12, respectively. From this, the value of χ^2 was calculated, as shown in Table 14.3. From this result, it could be said that the logarithmic transformed value of the concentration was significantly different from the normal distribution. However, near the tail of the distribution, the condition $np_i > 5$ was not satisfied, so it seemed that this test itself might not be appropriate.

Note that, as was explained in Part I, Hypothesis testing, in many cases, the probability of a type II error is not specified. Keep in mind, therefore, that if it fails to reject the null hypothesis, one cannot positively assert that this is the case.

Data on air pollution concentrations are said to follow a log-normal distribution. Here, it was attempted to confirm this by the goodness-of-fit test. However, we could not confirm this within the significance level of the statistical test. Further analysis may be required by devising a better classification of the underlying data or adding more data. Thus only an introduction to the method of goodness-of-fit testing is provided here.

14.4.2 Time-Series Analysis of Concentration Data and Its Relationship to the Law of Large Numbers

In statistics, many metric data are often analysed without any special attention, assuming a normal distribution. Many statistics textbooks state that, in exercises, the following analysis should be performed on the independently obtained data. Almost all readers unconsciously skip this phrase. But here is a hidden important meaning. This is because calculating the mean based on independently acquired data allows us to approximate the mean to the normal distribution. If not, researchers must always check whether the mean of the data can be approximated by the normal distribution.

So what about the concentration data shown in Figure 14.6 and Figure 14.7? Why does the average value not approach the normal distribution even if the 1-h value is calculated by summing up as many as 20 three-min value data shown in Figure 14.7? That is because the concentration data are not independent. If we look at the concentration data in chronological order, we will notice that, the higher the concentration, the more likely that the next data will be higher. If the concentration data are independent, the previous data and the next data are irrelevant, and the correlation coefficient between them should be 0.

In such time-series data, the correlation can be analysed by calculating the correlation coefficient for the shifted data, such as one before, two before and further. The method of statistically analysing data that has been collected sequentially over time is referred to as time-series analysis. Correlogram analysis is one of them. The correlation between the original data and the data shifted in time is called autocorrelation. The graph that shows the lag and the autocorrelation is called a correlogram. If the data show a periodic or persistent trend, this chart will reveal their characteristics.

The IPCAJ's Shin–Yokohama air tracer experiment provided valuable data suitable for analysis in such cases. It is the air tracer concentration data every 3 min. Let us introduce the result of statistical time-series analysis for this data. The autocorrelation coefficient for this dataset was calculated to create a correlogram. This is shown in Figure 14.8. Here a time lag was set up to about 30 min. The results indicate that the autocorrelation coefficient for the concentration after a few minutes exceeds 0.5, demonstrating significant persistence. In other words, for such data, even if multiple consecutive data points are aggregated to obtain the mean value,

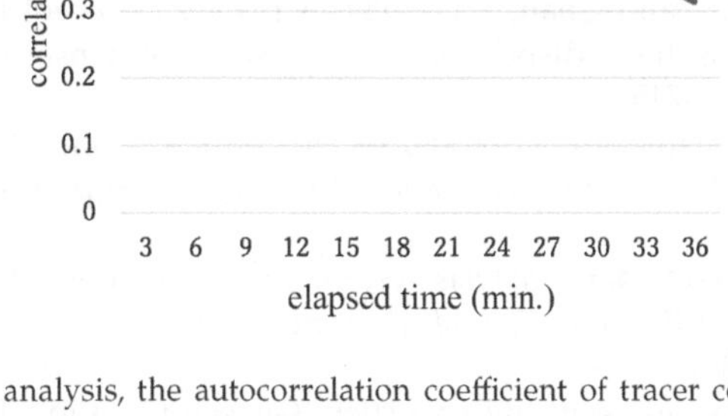

FIGURE 14.8 Correlogram analysis, the autocorrelation coefficient of tracer concentration measured every 3 min in air tracer dispersion experiment. Sampling site G-2 is the roadside, and 4.1 m in height; G-4 is about 50 m apart from the road shoulder and 1.5 m in height.

Source: Data were retrieved from Tsukioka and Ito (1985).

the distribution of the mean values does not easily approach a normal distribution. It can be seen that the correlation was lower at G-2 at the edge of the road than that at G-4, which was a little farther away, due to the effects of air turbulence induced by moving automobiles. This result shows that these data did not satisfy the law of large numbers.

Correlogram analysis is also useful for analysing whether time-series data fluctuates cyclically. By incrementally shifting time-series data and calculating the correlation coefficient, periodicity can be uncovered; typically, the coefficient increases when the data is shifted by one complete period. From these characteristics, it is possible to understand the periodicity of time-series data.

14.5 Summary

The development of monitoring equipment and the data transmission network system have been promoted, and its continuous operation has made it possible to accumulate high-quality air quality data. Nevertheless, it is important to emphasise that, for air pollution data, the averaging period is very important to be reported with the concentration levels. Shorter average time data are relevant to risk management such as the health effects and disaster prevention and help capture temporal fluctuations that reflect changes in the physicochemical transformation of pollutants. Long-term averaging data are important to detect the trend and also for health effects studies such as chronic diseases. The conversion from short- to long-term averaging data for a pollutant is not straightforward, and it depends on the pollutant type and the emission sources. Also, if monitoring data is lacking, it may be possible to convert short-term average data to long-term average data using the modelling results, albeit under limited conditions.

If it is known that the concentration distribution follows a probability distribution such as a log-normal distribution, this property can be used to evaluate the rate at which high concentrations appear, for example. Furthermore, it is expected that this property can be used in the analysis of percentile values specified in environmental standards. However, there are many restrictions on this type of analysis, and a judgment must be made as to whether it is feasible to use.

REFERENCES

Blackwood, L.G., 1992, The lognormal distribution, environmental data, and radiological monitoring, *Environmental Monitoring and Assessment*, 21, 193–210, https://doi.org/10.1007/BF00399687

Kan'nari, Y., Ohara, T. and Ueda, K., 1989, NO_2 long-term average targets and climate change, *Proceedings of the 30th annual meeting of the Japan society of air pollution*, 252 (In Japanese).

Koopman, R.P., Cederwall, R.T., Ermak, D.L., Goldwire, H.C., Jr., Hogan, W.J., McClure, J.W., McRae, T.G., Morgan, D.L., Rodean, H.C. and Shinn, J.H., 1982, Analysis of Burro series 40-m^3 lng spill experiments, *J Hazardous Materials*, 6(1–2), 43–83.

Koopman, R.P., Ermak, D.L. and Chan, S.T., 1989, A review of recent field tests and mathematical modeling of atmospheric dispersion of large spills of denser-than-air gases, *Atmospheric Environment*, 23(4), 731–745.

Larsen, R.I., 1969, A new mathematical model of air pollutant concentration averaging time and frequency, *J Air Pollution Control Association*, 19(1), 24–30, https://doi.org/10.1080/00022470.1969.10466453

Larsen, R.I., 1973, An air quality data analysis system for interrelating effects, standards and needed source reductions, *J Air Pollution Control Association*, 23(11), 933–940, https://doi.org/10.1080/00022470.1973.10469861

Okamoto, S., 2001, *Lectures on air quality prediction*, Gyosei, Co. Ltd., Tokyo (In Japanese), ISBN 4-324-06490-3

Okamoto, S., 2004 November 1–2, *Prediction and evaluation technology of atmospheric environment, chapter 4 of the ministry of environment workshop text on environmental impact assessment*, Japan Association of Environmental Assessment, Tokyo, Japan (In Japanese), https://jeas.org/english/index.html

Okamoto, S. and Shiozawa, K., 1978, Validation of an air pollution model for the Keihin area, *Atmospheric Environment*, 12(11), 2139–2149, https://doi.org/10.1016/0004-6981(78)90169-5

Ott, W.R., 1990, A physical explanation of the lognormality of pollutant concentrations, *J Air & Waste Management Association*, 40(10) 1378–1383, https://doi.org/10.1080/10473289.1990.10466789

Shiozawa, K. and Okamoto, S., 1990 June 24–29, A consideration on preventive measures against nitrogen oxides in Japan, *Preprint of 83rd annual meeting air & waste management association*, 90-153.9, Pittsburgh, USA.

Spicer, T. and Havens, J., 1989, *User's guide for the DEGADIS 2.1 dense gas dispersion model, EPA-450/4-89-019*, U.S. Environmental Protection Agency, https://gaftp.epa.gov/Air/aqmg/SCRAM/models/other/degadis/degadis2.pdf

Tsukioka, M. and Ito, S., 1985, *A consideration on the air tracer experiment*, Graduation thesis under the direction of Professor Hayashi, M., Shibaura Institute of Technology (In Japanese).

U.S. EPA, 2003, Revision to the guideline on air quality models: Adoption of a preferred long-range transport model and other revisions, *Federal Register*, 68(72), 18440–18482.

Yamane, T. and Koizumi, M., 1984, *Statistical analysis on air tracer experimental data*, Graduation thesis under the direction of Professor Hayashi, M., Shibaura Institute of Technology (In Japanese).

15

Spatial Resolution of Concentration Data and Optimal Placement of Monitoring Stations

This chapter considers the spatial variability of air pollution concentrations, another challenge for representative monitoring in addition to the time-scale variability and averaging time discussed in Chapter 14. In particular, when one wants to know the pollution situation in an area, it is first to think about the locations where it is most appropriate to install the measuring instrument. This can be done using a large number of monitors in a saturation study. The monitors, however, should be first tested for their consistency in the concentrations of pollutants of different ranges. Simulated air quality data by a dispersion model and satellite data can greatly help in identifying the monitoring locations to successfully capture the spatial distributions of air pollution in a domain. These issues are discussed and illustrated with examples of real-life case studies in this chapter. In addition, issues such as concentration differences due to slight location differences around the monitoring station are addressed, and the location of the inlet of the measuring device is also considered here.

15.1 Concentration Measurements at Air Quality Monitoring Stations

The concentrations of many air pollutants such as SO_2, NO_x, and PM are often measured at air quality monitoring stations of the governmental networks to fulfil the needs of environmental policy or objectives. The representativeness of the monitored data in terms of the spatial distribution of the pollutant is a great issue of concern, and many problems arise. Some of these insights have been summarised in US EPA documents (Ludwig and Kealoha, 1975; Ball and Andersen, 1977; Ludwig and Shelar, 1978). Questions related to these issues include:

- Are air pollutant concentrations measured at appropriate locations?
- Is the monitoring station strongly unaffected by nearby emission sources, despite the need for representative concentrations throughout the target area?
- Is the air inlet of the monitoring instrument in the right place, at the height above the ground, at the distance from the wall etc.?

The Industrial Pollution Control Association of Japan (IPCAJ) conducted a study to examine such issues, and some of them are introduced here. In this study, it was examined whether measured values at the air quality monitoring station were adequately representative of the air pollution concentrations in the area concerned. In particular, they examined how much the difference in concentration might occur due to the difference in the installation positions of the measuring instruments. In addition, the differences in the measuring instruments used for such a comparative test were also examined. The IPCAJ (1982) study was carried out at the following four sites:

1. Monitoring points within the same building (Kanagawa staff hall building, Site 1),
2. Monitoring sites in the residential area (around Mitsuike Park, Yokohama, Site 2),

DOI: 10.1201/9781003597216-17

3. Monitoring points as an automobile emission monitoring station (Shin-Yokohama, this program included the air tracer experiment), and
4. Same as item 3 but a street surrounded by tall buildings (Aoyama).

15.1.1 Analysis of Differences due to Measuring Instruments

IPCAJ conducted a study examining the spatial distribution of concentrations. First, the study involved conducting calibration and comparison tests on ten sets of Salzman NO_x monitors. This was to confirm that the monitors did not exhibit significant differences that could affect the investigation into whether concentration variances were due to different measurement locations. The Salzman-type NO_x measuring instrument was a standard NO_x measurement method in Japan. Currently, most nitrogen oxide measuring devices installed at air pollution monitoring stations use the chemiluminescence method; however, at that time, most utilised the absorbance method with Salzmann's reagent. The instrument uses a Salzmann reagent, which changes colour by absorbing NO_2 and analyses the amount of light transmitted through the reagent to obtain the NO_2 concentration reading. Regarding NO concentration, NO in the air after NO_2 absorption is oxidised, and NO_2 is measured accordingly. All ten sets of the Salzman NO_x monitors were identical in type, produced by the same manufacturer. Before the experiments, all the NO_x monitors used were calibrated by using the equivalent solution (standard solutions coloured to correspond to given transmitted light doses for calibration). Next, calibrations by the standard gases were carried out, and the relation between the standard gases and indicated concentrations determined by the calibration is shown in Table 15.1. This table shows that the indicated values obtained using the ten sets of NO_x monitors have a small bias; hence the correction must be made based on the calibration results.

Further tests were conducted to ensure that the calibration was successful and that the corrected values for all instruments were within the range of sufficiently small variations. This test was called the same ambient air simultaneous measurement test. Before deploying the calibrated instruments in a designated place, all the equipment was gathered in one place and connected to the distribution branch pipe from the suction pump to measure the concentration in the same ambient air. The result for the site 1 area is shown in Figure 15.1. The standard

TABLE 15.1

Calibration Results Using the Standard Gas for Two Specified Concentrations

Period	Before the Kanagawa area survey		Before the Mitsuike area survey	
Standard gas NO_2 (ppb)	0	85	0	80
Instrument No. 1	2	77	2	72
Instrument No. 2	2	79	1	76
Instrument No. 3	1	80	1	72
Instrument No. 4	2	82	2	72
Instrument No. 5	1	78	2	71
Instrument No. 6	1	73	1	70
Instrument No. 7	1	78	2	75
Instrument No. 8	1	82	2	80
Instrument No. 9	1	77	2	72
Instrument No. 10	1	77	2	75

Note: In the standard gas test, zero gas and a mixture of NO and NO_2 gas were used. Here, only the results of NO_2 concentrations in tests with the introduction of zero gas and mixed gases are shown. Unit is ppb.
Source: Raw data courtesy of IPCAJ (IPCAJ, 1980).

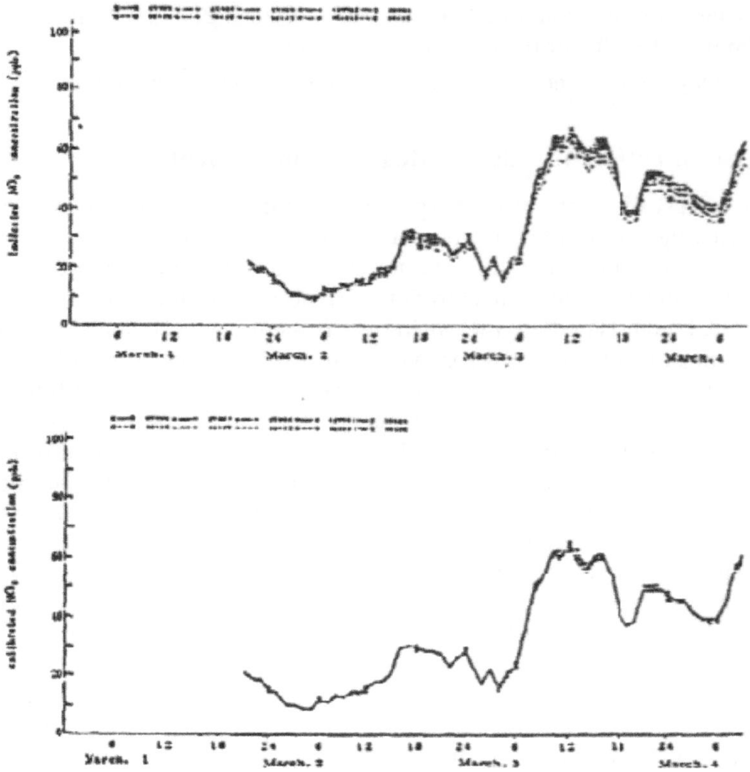

FIGURE 15.1 Results of the same ambient air simultaneous measurement test conducted before the field survey at the Kanagawa Staff Hall (site 1 area). (Top: indicated value before calibration; bottom: calibrated value.)

Source: IPCAJ (1980).

deviation of the estimated measurement error was 1.36 ppb for the site 1 area and 1.18 ppb for the site 2 area. The accuracy of the monitors has been confirmed as adequate for conducting surveys that assess minor concentration variances across locations.

15.1.2 Analysis of Differences due to Measuring Sites

A field survey program was conducted to clarify the variation of measured concentrations, i.e., the concentration difference around a building (see Figure 15.2) and the spatial distribution of ambient concentrations in a residential area (see Figure 15.4). The instrumental differences were assumed to be negligible based on the same ambient air simultaneous measurement test just mentioned. Summary statistics of the NO_2 concentrations in two areas are shown in Table 15.2. The mean concentrations of each area were 39.8 ppb at the site 1 area and 23.3 ppb at the site 2 area.

Kainuma et al. (1987) reanalysed the data to quantify the concentration fluctuations in both areas using the analysis of variance (ANOVA). To highlight differences between locations, the effects of diurnal fluctuation, the biggest factor for the hourly variability, were eliminated by analysing the daily average values. Accordingly, the daily variation of concentration fluctuation and the fluctuation due to the location were analysed. In this case, the daily variation is considered as a blocking factor, and the randomised block design ANOVA can be applied. The results are shown in Table 15.3 and Table 15.4.

TABLE 15.2

Summary Statistics for the Daily Average NO_2 Concentrations Calculated from 24 Hourly Data at the 10 Monitoring Sites (unit: ppb)

Area	Period	Number of data	Mean	Standard deviation	Range
Site 1 area	March 5–11, 1980	70	39.8	8.7	34.7
Site 2 area	April 1–6, 1980	60	23.3	9.7	37.7

TABLE 15.3

ANOVA Table for the NO_2 Data of Site 1 Area

Source	Sum of squares	Degrees of freedom	Mean square	*F* value
Between days (D)	4398.4	6	733.1	277. 7*
Between measuring sites (S)	627.7	9	69.8	26.4*
Interaction (error)	142.4	54	2.6	
Total	5168.5	69		

* Significant at the 1% level,
Source: Adapted from Kainuma et al. (1987) and modified.

TABLE 15.4

ANOVA Table for the NO_2 Data of Site 2 Area

Source	Sum of squares	Degrees of freedom	Mean square	*F* value
Between days (D)	5243.8	5	1048.8	635.6*
Between measuring sites (S)	200.7	9	22.3	13.5*
Interaction (error)	77.1	45	1.7	
Total	5518.6	59		

* Significant at the 1% level.
Source: Adapted from Kainuma et al. (1987) and modified.

These results showed that the effect of location is highly significant. Daily fluctuation is a blocking factor, and, as expected, it turns out that the effect is very large while the day-to-day variation is unavoidable. In principle, the interaction between the two factors of location and day-to-day variation may exist, but in this randomised block design, it is included in the error term and cannot be separately extracted. Nevertheless, the mean square of this term is small enough hence can be ignored and no further discussion may be necessary.

Comparing Tables 15.3 and 15.4, it can be seen that the concentration variation between the sites located in the same building near the main road is larger than the spatial variation of concentrations within about 1 km^2 in a residential area. The former is about three times as large as that of the latter in terms of the mean square. The result suggests that special caution should be given to the position of the measuring instrument's inlet, which is installed in a monitoring station near the main road.

15.1.3 Statistical Explanation of the Additive Nature of Variance

To evaluate a small difference in concentrations measured between adjacent places, special consideration must be given to the accuracy of individual measuring instruments. The statistical knowledge required when considering such issues is described here. As described in Part I of the statistical test of the population mean difference, when comparing the case of estimating the mean for one population and the case of estimating the difference between the two population means, there is a difference in the required accuracy of the data extracted from the population. When there are two individual data, the variance of the sum of them is the sum of their respective variances. This is called the additive nature of variance. When there are two random variables X and Y, and they are independent of each other, the variance of the sum of X and Y, or $X - Y$ is:

$$V(X \pm Y) = V(X) + V(Y) \tag{15–1}$$

By the way, when X and Y are not independent, it becomes:

$$\begin{aligned} V(X+Y) &= V(X) + V(Y) + 2\text{Cov}(X,Y) \\ V(X-Y) &= V(X) + V(Y) - 2\text{Cov}(X,Y) \end{aligned} \tag{15–2}$$

where Cov (X, Y) is the covariance of X and Y.

The notation "When X and Y are independent" means that the covariance is zero. If the difference between the readings of two measuring instruments with the same level of measurement accuracy (variance) is calculated, its variance will be twice the variance for each individual measuring instrument. Therefore, to determine the concentration difference between two adjacent points, it is necessary to use a measuring instrument with sufficient measurement accuracy. For example, looking at the results of the simultaneous ambient air measurement tests before the survey at the site 1 area, the standard deviation of the error of each measuring instrument is 1.36 ppb. Consider using these instruments to determine the concentration difference between location A, which is in contact with the wall, and location B, which is slightly away from the wall. In this case, the readings of both instruments are not considered to affect each other and can be considered independent of each other. Therefore, the standard deviation of the concentration difference between the two points A and B is:

$$\hat{\sigma}_{\text{A-B}} = 1.36 \times \sqrt{2} = 1.92$$

This result can be thought of as follows. If the variance of A and B is 1.36^2, then the variance of (A-B) is the sum of two 1.36^2, due to the "additive nature of variance". Therefore, its standard deviation is 1.92, which is the square root of twice 1.36^2.

15.1.4 Concentration Variability around the Same Building

In places where no major emission sources are nearby, one may not need to worry too much about locating the sampling inlet of the measuring device. However, where there is a source of contamination in the immediate vicinity, such as along a highway or near a chimney or a vent, the measurement data will be unreliable if the location of the instrument intake is not appropriate. Table 15.5 shows the IPCAJ survey results. This survey used three buildings in the Aoyama area of Tokyo, in addition to the Kanagawa prefecture staff hall monitoring (site 1 area) program previously described. The measurement period was about one week.

As for the Kanagawa staff hall experiment layout (Figure 15.2), comparing the levels at the sampling point No. 9 located inside the balcony and at point No. 10 located outside, the levels were higher for NO_2 by about 1 ppb and for NO by about 8 ppb. This was because of the influence of NO emitted directly from automobile exhaust. As for NO_2, it takes some time to form from the oxidation of NO (to NO_2) in the atmosphere, which explains the smaller concentration difference. Figure 15.2 shows that the NO concentrations at the site on floor 1 facing the road were about twice as high as the value at the back of the same building (point 8).

TABLE 15.5

NO and NO_2 Concentrations within the Same Building during the Experiment (unit: ppb)

Name of building	No. of floors	No. of monitoring sites	NO concentration			NO_2 concentration		
			Average	Max.	Ratio (max/min)	Average	Max.	Ratio (max/min)
Kanagawa staff hall	5	9	55.4	83.6	2.0	40.1	44.2	1.2
Building C (Aoyama)	7	7	131.9	197.0	2.9	47.8	55.6	1.3
Building D (Aoyama)	12	3	126.9	208.2	2.6	50.8	61.0	1.4
Building E (Aoyama)	5	4	80.7	88.5	1.2	45.4	48.2	1.1

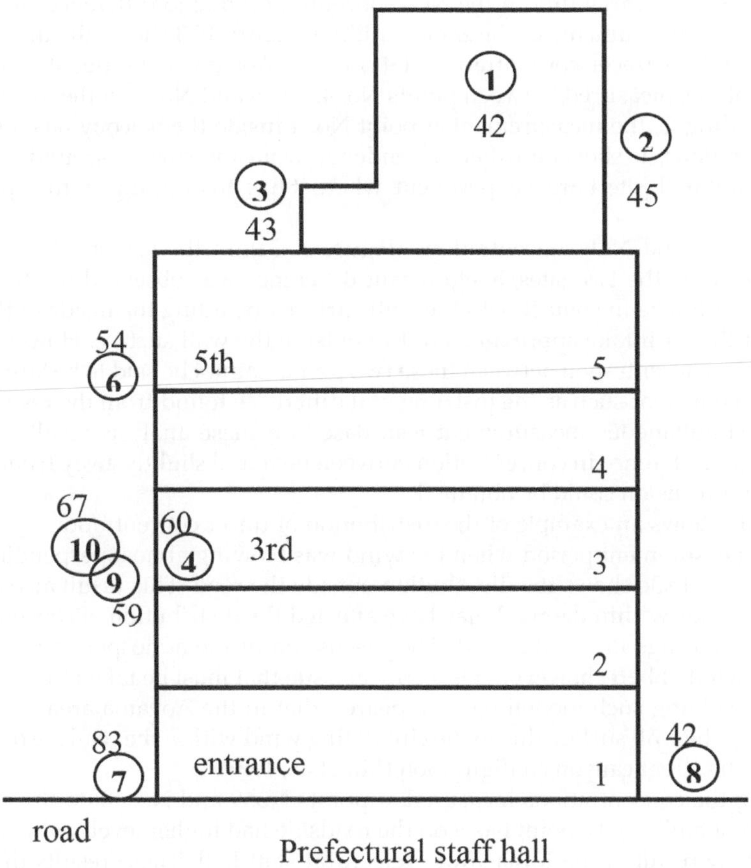

FIGURE 15.2 NO concentrations around the Kanagawa prefecture staff hall building (unit: ppb). Average concentrations from 20:00 on March 4 to 10:00 on March 12, 1980.

TABLE 15.6

NO and NO_2 Concentrations at the Fourth Floor of Building C of Aoyama Area Survey, December 3–11, 1980 (unit: ppb)

Name of building	Floors	Inlet position	Position no. in Figure 15.3A	NO concentration		NO_2 concentration	
				Average	Maximum	Average	Maximum
Building C (Aoyama)	Fourth floor	Place in contact with the wall	4	148.3	467	52.9	107
Building C (Aoyama)	Fourth floor	1 m away from the wall	5A	148.1	476	50.0	130
Building C (Aoyama)	Fourth floor	1 m away from the wall (old type*)	5B	140.5	480	49.6	101
Building C (Aoyama)	Fourth floor	2m away from the wall	6	145.2	430	51.0	108

* The principle of measurement is the same, but it is an old-fashioned measuring instrument with twice the measurement span of the scale.

Some of the experimental results in the Aoyama area are introduced in this section. Table 15.6 shows the variability in the NO_x concentration due to differences in the locations of the measuring instruments in the same building. Figure 15.3 shows the measuring points for the NO_x and air tracer concentrations (SF_6) in the Aoyama experiment. Compared with the concentrations measured between points No. 4, No. 5 and No. 6 on the fourth floor of the roadside building C, the measurement at point No. 4 inside the balcony has higher NO and NO_2 concentrations. It shows a different tendency from the survey around the Kanagawa prefectural staff hall (site 1 area) experiment, where it has no building on the opposite side of the road.

As for the NO and NO_2 concentrations (Figure 15.3a), in the statistical test of the mean difference between the two sites, a significant difference was observed in all combinations between the air intake in contact with the wall surface (including the inside of the balcony or the like) and the air intake approximately 1 m outside the wall surface. However, the difference in average concentration between these two points cannot be said to be large considering the measurement error such as the instrumental difference found from the result of the same ambient air simultaneous measurement test. Based on these analysis results, it seems that there is a slight difference in concentration between near and slightly away from the wall, but no definitive conclusion could be obtained.

Figure 15.3b shows an example of the distribution of tracer concentrations within the road during this measurement period when the wind was blowing almost perpendicularly to the road, and Figure 15.3c shows the distribution outside the road. This result also suggests that the complex winds within the road may have affected the distribution of concentrations near the walls of buildings facing the road. The discussion of the atmospheric modelling study itself is not included here; however, it is a critical issue that must be taken into account in any field study involving such modelling. It appeared that in the Aoyama area, the polluted air may rise along the wall surface due to the circulating wind within the road surrounded by tall buildings or the street canyon configuration (Lin et al., 1991).

Comparing the concentrations measured at points No. 9 and No. 10 on the rooftop of the opposite side of building D, point No. 9 on the roadside had higher levels, which is consistent with the survey result at the Kanagawa prefectural staff hall. These results thus suggested that special consideration is required to locate the monitoring equipment inside the so-called street canyon where the concentrations vary significantly between the leeward and the windward sides which also depend on the wind direction above the surrounding buildings.

In the Aoyama area, IPCAJ conducted an air tracer gas (SF_6) experiment. The air was sampled by the bag sampler, and the data were comparably analysed with the NO_x concentrations

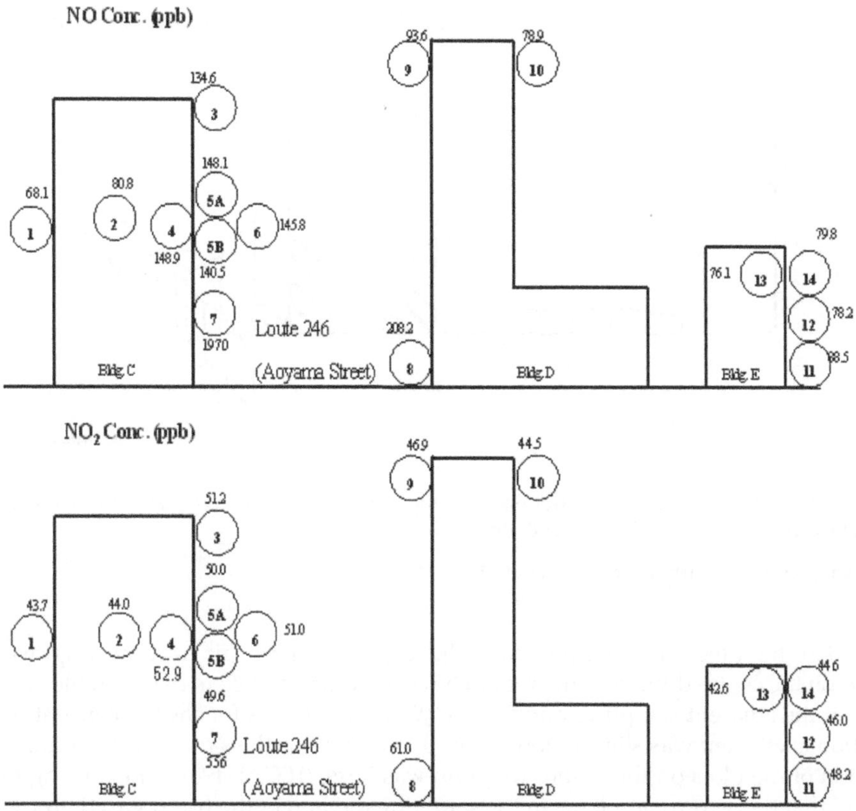

FIGURE 15.3A Average NO and NO_2 concentrations during the field survey at Aoyama area (ppb). Average concentrations from 17:00 on December 3 to 12:00 on December 11, 1980.

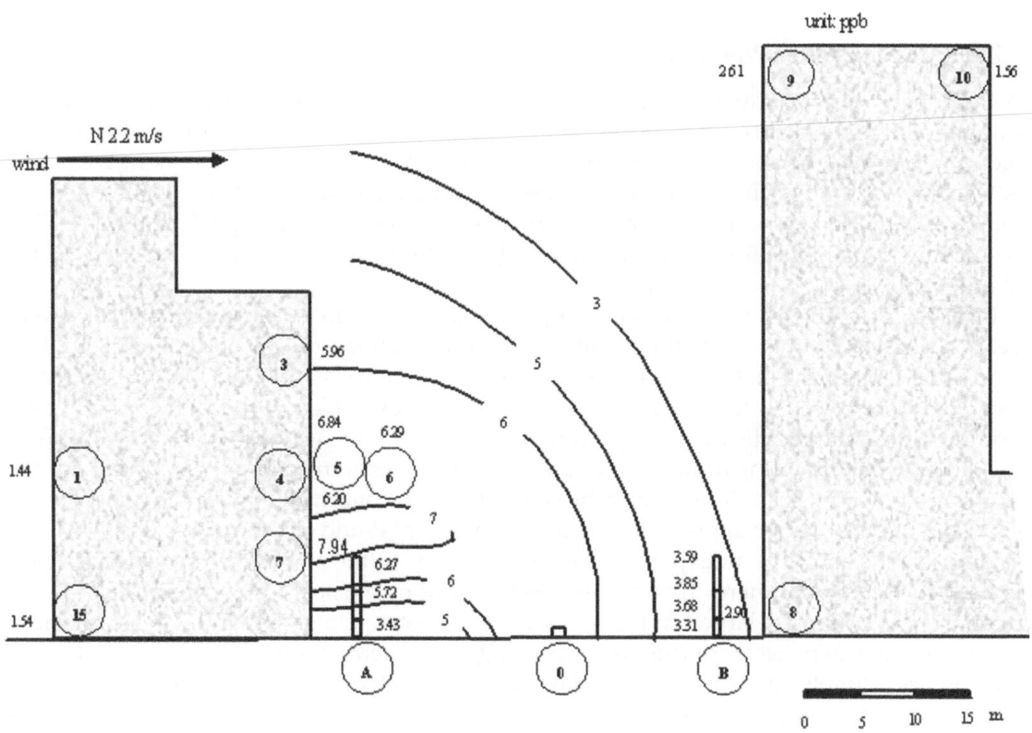

FIGURE 15.3B Air tracer (SF_6) concentration distribution within the street canyon of Route 246 (Aoyama Street) at 8:00–9:00 AM, December 8, 1980. The tracer was discharged from the centre divider.

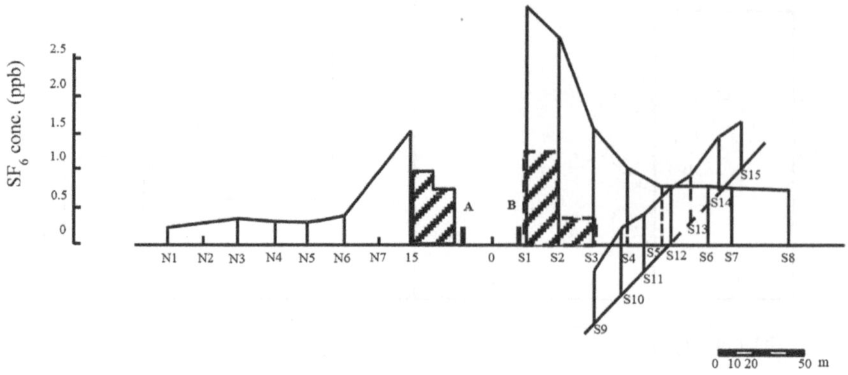

FIGURE 15.3C Air tracer (SF$_6$) concentration distribution for the outside of the street canyon of Route 246 (Aoyama Street) at 8:00–9:00 AM, December 8, 1980.

Source: Adapted and modified from Lin et al. (1991).

measured at the same sampling points. In this experiment, the air tracer was sprayed from a pipe laid on the centre divider of the road. The correlation coefficient between the NO_2 concentrations at the adjacent sampling points was 0.99 or higher. As for the SF_6 concentrations, the correlation coefficient was slightly lower, i.e., 0.95 to 0.99 for the same sampling pairs, but the dispersion of the plots on the scatter diagram was large (IPCAJ, 1981a). Regarding this point, it seemed that there were problems in the sampling method of the tracer gas, therefore further analyses using the tracer gas concentration were not conducted (IPCAJ, 1981a). The sampler for collecting air tracers will be discussed again in Chapter 17, including issues of the data quality such as the reliability of the sampler.

15.1.5 Concentration Variability in a Residential Area

A field survey was conducted in the suburb of Yokohama City to clarify spatial variations in air pollution concentrations in residential areas. Figure 15.4 shows the distribution of NO_2 concentration at ten measurement points installed around Mitsuike Park in Yokohama (site 2 area). Although the concentration was slightly higher at point No. 6 located near the main road, the NO_2 concentration at other points was about 22 to 24 ppb. From the ANOVA result shown in Table 15.4, the mean square due to location is 22.3 ppb^2, and the RMSE is about 5 ppb.

In Tokyo, a complementary survey using a simple sampler has been conducted for several years since 1985 to investigate the detailed spatial distribution of concentrations around the air quality monitoring stations (Tokyo Met. Bureau Env. Conservation, 1985). In these surveys, samplers were placed at tens to hundreds of points, for example around the monitoring stations of higher automobile exhaust gas concentration and intersections with complaints from neighbours. The sampling duration for each sample was 48 hours, and it was carried out several times for each area. This sampler measured the NO_2 concentration in the atmosphere by exposing the filter paper impregnated with the absorbent liquid to the atmosphere.

Here is an example of an experimental result similar to the above-mentioned survey that is used for illustration here. To get a detailed picture of the spatial distribution of air pollutant concentrations, a large number of measuring instruments need to be densely placed. However, many expensive continuous measuring instruments cannot be used. In such cases, one can use a simple passive sampler. Let us introduce such a case here (Okamoto et al., 2000). A field experiment was carried out in a residential area of Kawasaki city. Near Route 1 in the Komukai area, Kawasaki city has a traffic volume of about 50,000 vehicles/day. The survey was conducted in two weeks in 1999. The site was a residential area without major emission

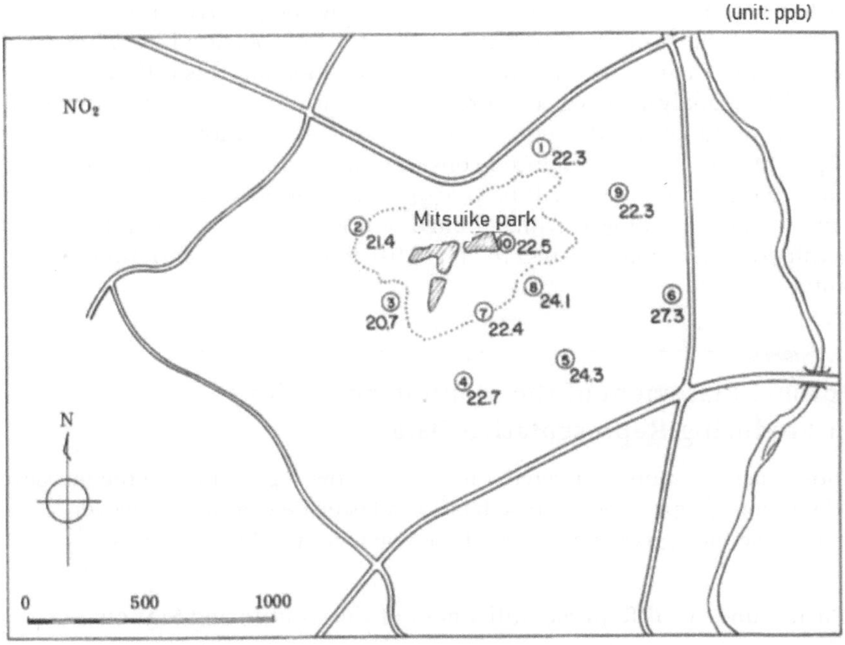

FIGURE 15.4 Spatial distribution of NO_2 concentrations around Mitsuike park, 6-day average values, April 2–7, 1980.

Source: IPCAJ (1981a).

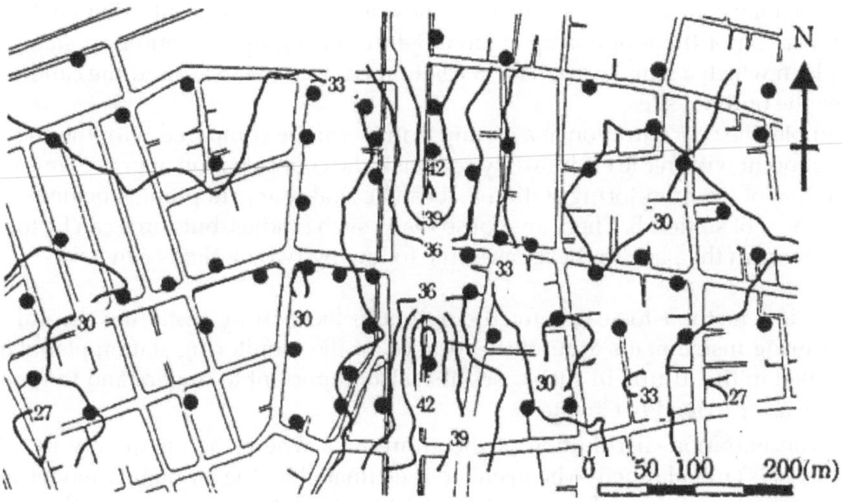

FIGURE 15.5 NO_2 concentration distribution in Komukai area near Route 1 during 10:00, Feb. 8– 10:00, Feb. 10, 1999 (unit: ppb). The dot is the position where the passive sampler was placed.

Source: Okamoto et al. (2000).

sources such as factories. In addition to the ten sets of Salzmann NO_x monitors, several passive samplers were also used in this survey. The survey area (250 m × 250 m) was divided into a 25 m grid, and 70 passive samplers (Filter Badge, Advantech Toyo Co. Ltd.) were deployed on the grid points to obtain the spatial distribution of NO_2 concentration around the road. These passive samplers were exposed to the air for 48 hours. The result is shown in Figure 15.5.

The relationship between the NO_2 concentrations by the passive sampler and continuous NO_x monitor was analysed. Although the concentrations measured by the passive sampler were slightly lower than those measured by the NO_x monitor, this difference was relatively small. A statistical analysis using linear regression was also carried out for the paired data, and its result showed 1.01 for the regression coefficient and 0.91 for the correlation coefficient, indicating that both methods produced consistent measurement results. Although the time resolution of the obtained data will be coarser than that of continuous monitoring instruments that can only be deployed with a limited number due to the cost issue, the simple NO_x monitors can provide a more detailed regional distribution of concentrations with acceptable data quality.

15.2 Optimal Placement of the Monitoring Stations for Producing Representative Data

This section considers more concretely the issue of how representative the measured values are. The definition of representativeness has not yet been determined; however, let us consider how to determine the representative area based on concentration statistics.

15.2.1 Basic Concept of Representativeness of the Monitoring Station

When a certain standard or criterion of the representative area of the station is set by a statistical measure such as correlation coefficient, root mean square error, and so on, one can evaluate the area for each station by using the criteria. However, when one wants to conduct these analyses, it is a must to know the concentration not only at the monitoring stations but also at all locations within the entire study region. Generally speaking, it may be impossible to do this, and one may choose an alternative way how to guess the concentration field around the intended location of the monitoring station. Before setting up a monitoring station, there is an example in which a simple auxiliary measurement was made around the candidate site to determine the optimal site.

For example, the concentration at a certain station will be compared with the concentration at each grid point within the entire study region. If the concentration at grid point A is similar enough to that of the monitoring station B, it can be said that grid point A belongs to the representative area of station B. There are not so many such studies, but some can be found in the research papers in the past. In these cases, the following two methods can help.

1. There is a method to search for the optimum location by installing several simple measuring instruments around the location of the monitoring station planned to be installed in the future. In such cases, it is also important to understand the sampling theory as described in Chapter 9.
2. The concentration distribution of the entire area where the air quality monitoring station(s) is (are) planned to be installed is estimated by the air quality model, and the location of the most suitable candidate site(s) can be determined. This method can also be applied in the following cases. If it is desirable to evaluate a candidate site in an area where the existing monitoring network is not enough or a new monitoring station is to be added, an evaluation score for the candidate site can be obtained based on the calculated concentration with an air quality model. Using this score, it is possible to estimate the most appropriate site for the target monitoring station.

The second method (2) can also be used to evaluate existing monitoring network systems. Interesting case studies are presented here. One is the case of $PM_{2.5}$ in the SEA urban area. and the other is part of an earlier IPCAJ's comprehensive study on the proper placement of monitoring stations.

15.2.2 Estimating the Concentration Spatial Distribution in the Target Area

When considering the installation site of the monitoring station, even if it aims to know the concentration of the place, it is better to grasp the concentration distribution of the entire target area to some extent in advance, so that a more suitable site can be selected. The challenges of preliminary sampling that can be applied in such cases have also been discussed in Chapters 8 and 9. A more effective method is the use of the simulation results of air quality models. Note that the air quality model is a tool for calculating the ambient concentrations using the input data of the emissions and meteorological conditions as further detailed in Chapter 18. By using the dispersion model tool, it may be possible to estimate the concentration distribution with the desired spatial resolution for the entire target area.

In addition, satellite data with large spatial coverage as discussed in Chapter 8 are increasingly available for air pollution research. The satellite products provide the indicators of pollution levels over a coarse grid, a few to a hundred square kilometers. Besides, the satellite data are columnar by nature; hence they may not be directly related to the actual pollution data measured at the ground level. Thus, although supplemental analyses may be required to estimate surface concentrations from the satellite data, they can be used to preliminarily determine the optimal number and location of stations in a domain.

An example of the model-simulated spatial distribution of the annual average $PM_{2.5}$ distribution in an urban area of Southeast Asia is shown in Figure 15.6. The air quality model used is a chemistry transport model of the Eulerian type, which provided the gridded $PM_{2.5}$ concentration values at 2 × 2 km^2 for this urban domain. Thus, although the model did not provide concentration information at the exact location (x, y, z) of the monitoring station, the simulated results are useful to identify the areas of high, medium and low concentration levels. Further, depending on the type and configuration of the models, finer spatial resolutions may be used. Commonly, the chemistry transport models provide outputs on hourly basis and the outputs can be used to calculate daily, monthly and annual average concentrations, which can help in identifying the spatial distributions at a corresponding timescale.

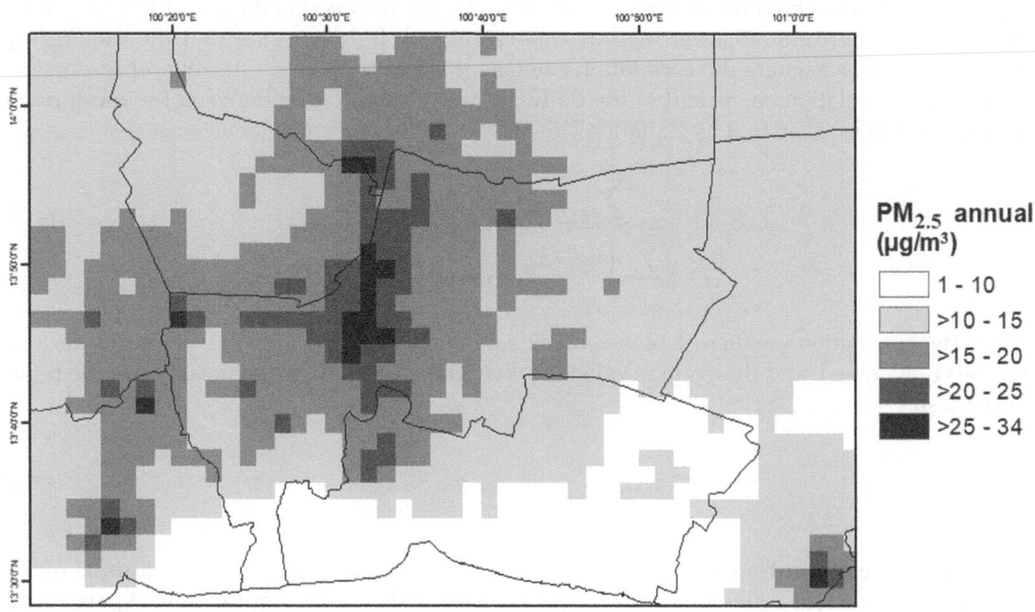

FIGURE 15.6 Annual average $PM_{2.5}$ on a grid net of 2 × 2 km^2 simulated by a chemistry transport model (WRF-CAMx) for an urban area in Southeast Asia.

Source: Ha Chi and Kim Oanh (2020).

As seen in the figure, the Eulerian (grid-type) air quality model calculates the average concentration in each grid area. For this reason, the grid-type models on an urban or regional scale with a grid size above 1 km are not suitable for understanding microscale phenomena such as automobile exhaust pollution on roadsides. However, it is useful in selecting locations of monitoring station candidates for monitoring air quality to some spatial extent such as photochemical air pollution. Note that many grid-type numerical analysis models have also been proposed for calculating the concentration distribution in narrower areas such as roadsides, for example, a model used in the research of Okamoto et al. (1998) and Huertas and Prato (2020), which is not detailed here.

Another example is the study of IPCAJ (1981b), which conducted a re-evaluation of the air quality monitoring network in the Himeji region, which had a good monitoring network system and a good database of emission inventory. In this study, the hourly pollutant concentrations were calculated for both the station location and for the corresponding 1 km grid area (exactly, 0.925 km by 1.143 km) for the entire region in the model year. Concentrations at all grid points throughout the region were simulated using the air quality model, which was very similar to the model used in the Kashima simulation described in Chapter 18 (Section 18.3.1). In this chapter, the analysis following Section 15.2.3 has been done using the simulated concentrations of this air quality model.

15.2.3 Representativeness of a Monitoring Station

When discussing the degree of spatial representativeness of one monitoring station, there is a difficult problem in that the definition of the term "representativeness" is not always clear. Here, when the difference between the concentration measured at a certain monitoring station B and the concentration at a certain place A around the monitoring station B is within the determined reference range, the concentration at place A can be represented by that of station B. In the IPCAJ study, the following statistics regarding the setting of this standard were considered (IPCAJ, 1981b; Okamoto et al., 1984; Okamoto, 1987). The environmental standard for NO_2 in Japan is set for the daily average concentrations, so the statistics used for the analysis were also set to be the daily average value. First, the correlation coefficient of the NO_2 daily average value between adjacent meshes was calculated. It is expected that the monitoring station in the place where the correlation coefficient is low will have a narrow representative range. The correlation coefficient of the daily average concentration between the mesh point (i, j) and $(i, j+1)$ is calculated by Equation (15–3):

$$R_{ij+1} = \frac{\sum_k \left(x_{ijk} - \bar{x}_{ij}\right)\left(x_{ij+1k} - \bar{x}_{ij+1}\right)}{\sqrt{\sum_k \left(x_{ijk} - \bar{x}_{ij}\right)^2} \sqrt{\sum_k \left(x_{ij+1k} - \bar{x}_{ij+1}\right)^2}} \tag{15–3}$$

Next, the correlation coefficient between the reference mesh i, j, and the surrounding four meshes is averaged, and the average value is taken as the correlation coefficient R_{ij} of the reference mesh.

$$R_{ij} = \frac{1}{4}\left(R_{ij+1} + R_{ij-1} + R_{i+1j} + R_{i-1j}\right) \tag{15-4}$$

where x_{ijk} is the concentration at the grid point i,j, of the k-th day data. The distribution of the correlation coefficient is shown in Figure 15.7. Looking at this figure, for the central part with a large population and the main roads passing, the correlation coefficient is low. This tendency is the same as the result described in the previous section.

Next, statistics such as the root mean square of the concentration difference between the neighbouring meshes were calculated. In this study, statistics were calculated between the

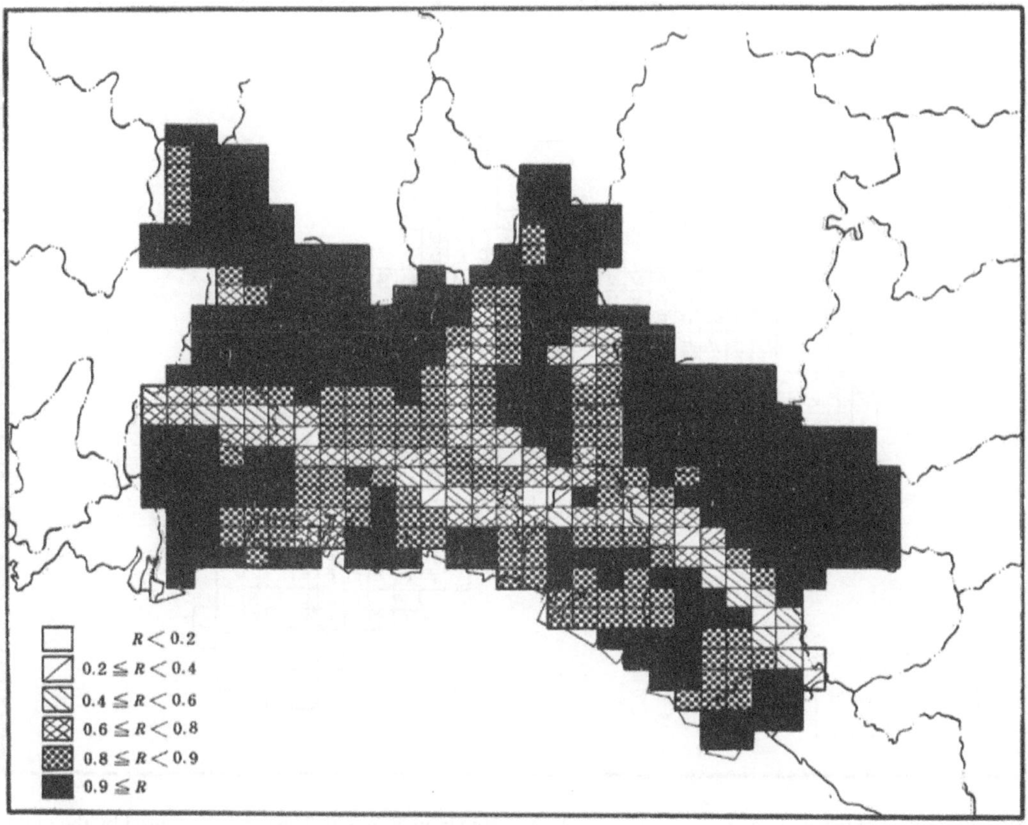

FIGURE 15.7 Distribution of correlation coefficient between the adjacent mesh concentrations.

Source: Okamoto et al. (1984).

TABLE 15.7

Criteria for Representativeness of the NO_2 Monitoring Station

Simulating condition	Correlation coefficient R	Absolute difference of annual mean ΔC	RMS difference σe
Case 1	$R \geq 0.7$	$\Delta C \leqq 5$ ppb	$\sigma e \leqq 5$ ppb
Case 2	$R \geq 0.9$	$\Delta C \leqq 5$ ppb	$\sigma e \leqq 5$ ppb
Case 3	$R \geq 0.5$		$\sigma e \leqq 10$ ppb

concentration at each station and those at all grid points, thereby determining the representative area of the stations. Table 15.7 shows the statistics for the criteria under these conditions. Here, a typical example is case 1. The strict condition is case 2, and the loose condition is case 3. Figure 15.8 shows a part of the calculation results of the representative areas of the monitoring stations when the criteria of Case 1 were applied. Given these criteria, it is possible to look at how many stations could cover most of the concerned region.

Table 15.8 shows the results of the analysis for each criterion-setting condition of the representative area. If a loose criterion is set, the area covered by all monitoring stations and the population ratio included in that area will be large. However, the area covered, duplicated in two or more stations, is extremely large. Conversely, if a strict criterion is set, the area that

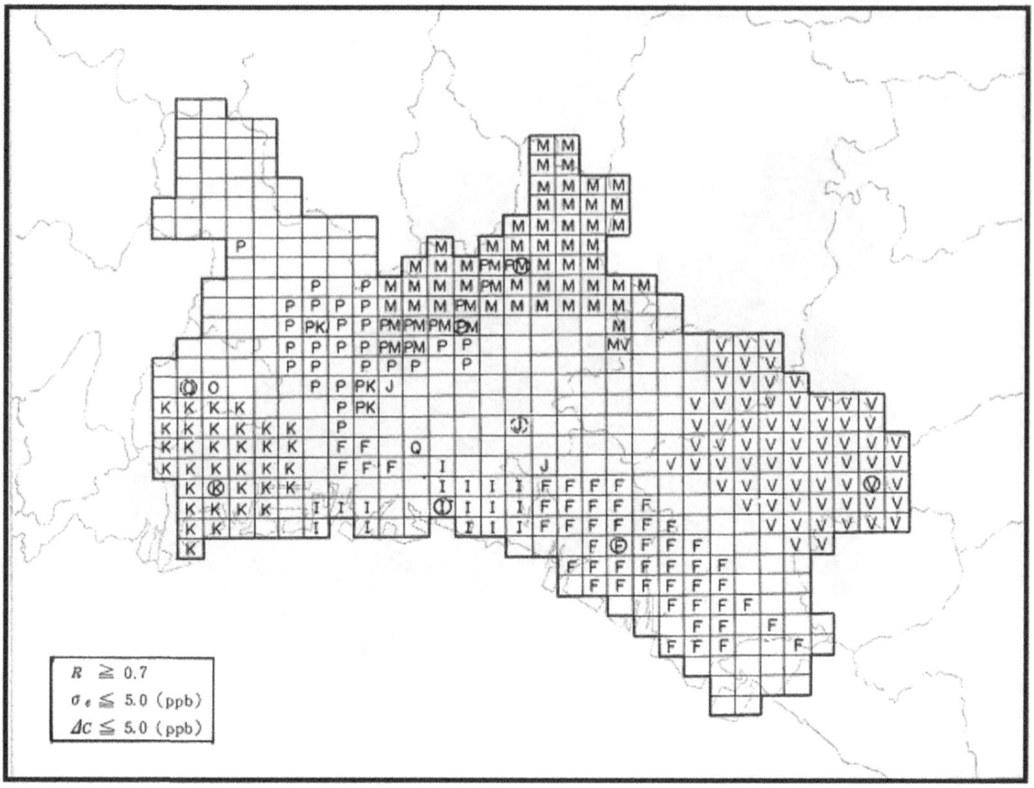

FIGURE 15.8 Representative areas of several stations in the Himeji region. The meshes covered by the monitoring stations are indicated by the station codes. The circled monitoring station code indicates the location of the monitoring station.

Source: IPCAJ (1981b).

TABLE 15.8

Results of Analysis for Each Criterion Setting Condition of the Representative Area

Simulating condition	Percentage of geographical area included in any of the monitoring station coverage areas	Percentage of areas that are duplicated in the representative area of multiple monitoring stations	Percentage of population within the area covered by the representative area of any of the monitoring stations
Case 1	81%	52%	86.1%
Case 2	40%	10%	50.3%
Case 3	95%	90%	97.5%

can be represented and the proportion of the population included in that area will become smaller.

The method explained so far is to prioritise the placement of monitoring stations with a wide representative area. The advantage of this method is that it is suitable when the purpose is to know the average concentration in the target region (for example, in the city) by a smaller number of monitoring stations.

However, the biggest flaw is the lack of understanding of highly polluted areas where the people most suffering from air pollution. In other words, as described in Section 15.1, it may be impossible to grasp the status of pollution along roadsides or near large emission sources. The reason is that in such a high concentration area, the representative area evaluated statistically becomes extremely narrow.

IPCAJ's study in the Himeji region attempted to prioritise not only the area covered by the monitoring stations but also the population within that area as an evaluation scale. Nevertheless, along the major national highways, several meshes appeared that were not included in the representative area of any monitoring station. The criteria for determining the highest concentration and the criteria for covering a broader area seem incompatible.

15.2.4 Evaluation of a Monitoring Network

Now let us consider the evaluation of the existing air quality monitoring network. It is also possible that low-priority monitoring stations in the existing network can be candidates for deletion for some reason or another. In Japan, a large number of monitoring stations had been set up from the viewpoint of pollution prevention during the high economic growth period. In particular, many monitoring stations were set up with national subsidies in line with the installation of annoying facilities such as thermal power plants. However, it became difficult for local governments with weak financial bases to maintain the station's operation. The reduction of the number of monitoring stations was also being considered. Let us also consider the problem in such cases.

In the study of the Himeji region, 22 existing stations during the time of the study were evaluated according to their importance based on the dimensions of a representative area. When another station is added to the monitoring network in the order of its importance, the relation between the number of stations and dimensions of the area to be covered by the whole monitoring network is shown in Figure 15.9.

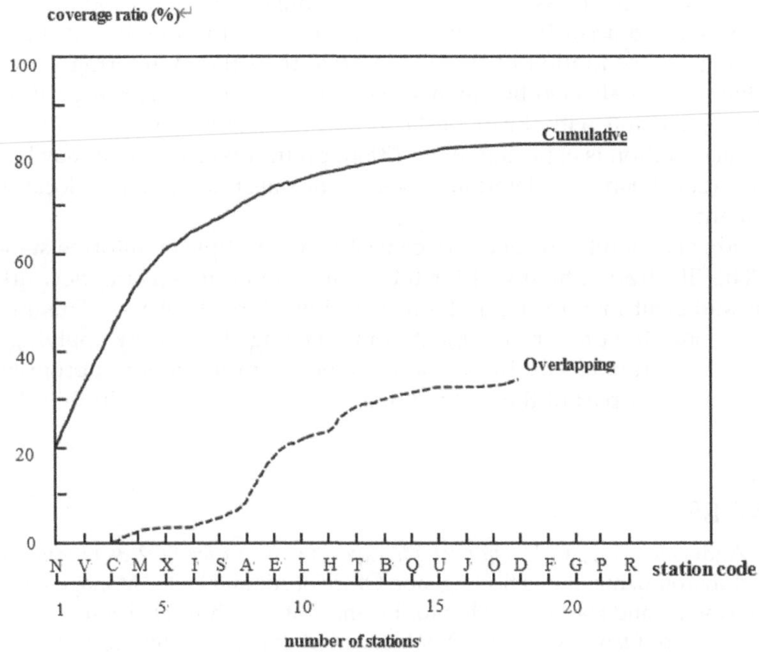

FIGURE 15.9 Relationship between the number of stations and coverage ratio. Ordering by the importance of the area covered by the stations.

Source: Adapted from Okamoto et al. (1984) and modified.

In Figure 15.9, the evaluation criterion is based on the coverage ratio, the cumulative ratio of the area to be covered by the monitoring network. When the dimensions of the representative area are applied as the importance level of the station, the station with the highest priority in this region is station N and the next is station V, as shown in Figure 15.9. Some meshes are covered by more than two stations; that is, some meshes overlap. This section briefly explains the procedure for creating this figure. These meshes are determined to be within the representative area of the station with the higher priority. In this case, evaluation is carried out for each station, and the representative area of the highest-priority station is reserved.

Using this process of determination, that part of the representative area overlapped by more than two stations is rejected from the representative area of the lower-priority stations. The determination of the representative area of the lower-priority stations and ordering of these stations are considered within the remaining area, from which the area covered by the higher-priority station was subtracted.

Here, the case study examined an arrangement that would grasp the concentration distribution of the entire target region to be ascertained with fewer monitoring stations. However, this evaluation score may not be the best. If finding the highest concentration site in the objective region is prioritised, the priority of the monitoring stations will change completely. In general, the closer to the emission source, the higher the concentration, the less representative the concentration measured there. Therefore, it is important to balance both objectives when designing the actual monitoring network (Okamoto et al., 1984; Okamoto, 1987).

Further, Kainuma et al. (1990) proposed an evaluation scheme for the air quality monitoring network. They showed the usefulness of a multi-attribute utility function to analyse the network by using the same basic air quality monitoring network data.

15.3 Summary

This chapter examined some issues related to the siting of air quality monitoring equipment. The location is selected according to the purpose, i.e., what to monitor. If the purpose is to know the average concentration of the whole area, then the monitoring station should be selected so that it is not affected by any nearby emission sources. However, if the purpose is to investigate the hotspots with the highest concentration, then the locations should be where the highest concentration is expected. In addition, preliminary research may be necessary to confirm the location. Such considerations also include questions about the location of the inlet of the instrument.

The evaluation of a monitoring network consisting of multiple monitoring stations was also considered. The illustration here was for the evaluation of an existing network in a model area that has sufficient monitoring and emission inventory databases. This method can be extended to consider the proper placement of monitoring stations. By applying this method to all the gridded candidate sites in the model region, a proposal for appropriate placement can also be obtained. A part of the methodology has been presented in this chapter for this purpose.

REFERENCES

Ball, R.J. and Andersen, G.E., 1977, *Optimal site exposure criteria for SO2 monitoring, EPA-450/3-77-013*, U.S. Environmental Protection Agency, Office of Air and Waste Management, Office of Air Quality Planning and Standards, Monitoring and Data Analysis Division.

Ha Chi, N.N. and Kim Oanh, N.T., 2020, Photochemical smog modeling of PM for assessment of associated health impacts in crowded urban area of Southeast Asia, *Environmental Technology & Innovation*, 21, 101241, https://doi.org/10.1016/j.eti.2020.101241

Huertas, J.I. and Prato, D., 2020, CFD modeling of near-roadway air pollution, *Environmental Modeling & Assessment*, 25, https://doi.org/10.1007/s10666-019-09666-w

Industrial Pollution Control Association of Japan (IPCAJ), 1980, *Comprehensive study on problems related to ambient air monitoring stations, vol. 1: Report on the fiscal year 1979 field surveys, NPMS-1* (p. 505) (In Japanese), Japan Environmental Management Association for Industry, Tokyo, Japan.

Industrial Pollution Control Association of Japan (IPCAJ), 1981a, *Comprehensive study on problems related to ambient air monitoring stations, vol. 5: Report on analytical results of field survey data, NPMS-5* (p. 319) (In Japanese), Japan Environmental Management Association for Industry, Tokyo, Japan.

Industrial Pollution Control Association of Japan (IPCAJ), 1981b, *Comprehensive study on problems related to ambient air monitoring stations, vol. 7: Report on representativeness of air quality monitoring station and its optimal placement, NPMS-7* (p. 252) (In Japanese), Japan Environmental Management Association for Industry, Tokyo, Japan.

Industrial Pollution Control Association of Japan (IPCAJ), 1982, *Comprehensive study on problems related to ambient air monitoring stations, vol. 8: Final report, NPMS-8* (p. 204) (In Japanese), Japan Environmental Management Association for Industry, Tokyo, Japan.

Kainuma, Y., Shiozawa, K. and Okamoto, S., 1987 October 20–23, Study of quality assurance on air pollution measurements, *Proceedings of international conference on quality control, ICQC'87, presentation C-2-01*, 339–344, Tokyo, Japan.

Kainuma, Y., Shiozawa, K. and Okamoto, S., 1990, Study of the optimal allocation of ambient air monitoring stations, *Atmospheric Environment*, 24B(3), 395–406, https://doi.org/10.1016/0957-1272(90)90047-X

Lin, F.C., Okamoto, S., Yamada, H., Kobayashi, K., Kitabayashi, K. and Shiozawa, K., 1991, An air tracer experiment in street canyon and validation of SRI street canyon model, *Journal of Japan Society of Air Pollution*, 26(4), 235–245 (In Japanese), https://doi.org/10.11298/taiki1978.26.4_235

Ludwig, F.L. and Kealoha, J.H.S., 1975, *Electing site for carbon monoxide monitoring, EPA-450/3-75-077*, U.S. Environmental Protection Agency, Office of Air and Waste Management, Office of Air Quality Planning and Standards.

Ludwig, F.L. and Shelar, E., 1978, *Site selection for the monitoring of photochemical air pollutants, EPA-450/3-78-013*, U.S. Environmental Protection Agency, Office of Air and Waste Management, Office of Air Quality Planning and Standards.

Okamoto, S., 1987 November 30–December 2, Representativeness of air quality data and optimal allocation of monitoring network, *Proceedings 3rd joint conf. air pollution studies in Asian areas*, 66–86, UENP Regional Office for Asia and the Pacific, Tokyo, Japan.

Okamoto, S., Kobayashi, K., Yamada, T., Kitabayashi, K., Kuriyama, S. and Sugiyama, H., 1984, A study on representativeness and optimal allocation of ambient air monitoring stations, *Journal of Japan Society of Air Pollution*, 19(2), 129–140 (In Japanese), https://doi.org/10.11298/taiki1978.19.129

Okamoto, S., Sakai, K., Matsumoto, K., Horiuchi, K. and Kobayashi, K., 1998, Development and application of a three-dimensional Taylor-Galerkin numerical model for air quality simulation near roadway tunnel portals, *Journal of Applied Meteorology*, 37(10), 1010–1025.

Okamoto, S., Takahashi, K. and Minakawa, N., 2000 June 5–8, A study on precise distribution of pollutants around roadway, *Proceedings 9th international symposium "transport and air pollution"*, 289–294, Avignon, France.

Tokyo Metropolitan Bureau of Environmental Conservation, Atmospheric Conservation Division, 1985, *Report on actual conditions of high-concentration air pollution areas (Supplemental survey on continuous measurement of air pollution), environmental conservation-related material 2–1-Taiki (Air) 26* (In Japanese), Tokyo Metropolitan Government Office, Tokyo, Japan.

16

Quality Assurance of the IPCAJ's Acid Rain Study

Chapter 16 introduces the importance of data quality assurance and is illustrated with a case of acid rain monitoring. The accuracy of the data obtained from experiments and observations, not just in the field of environmental science, is an important issue in determining the success or failure of research results. Such issues related to the measurement of particulate matter have already been discussed in Chapters 12 and 13. For any research project, the quality of the measured data should be considered at the planning stage, not after the data have been collected. It is too late to notice that the quality of the measurements is poor while doing the data analysis. In many such cases, the corrective actions may not be possible for further data analysis. Quality assurance of such data also requires some knowledge of statistics.

16.1 Overview of the IPCAJ's Acid Rain Study

The Industrial Pollution Control Association of Japan (IPCAJ) conducted one of the largest-scale acid rain studies in Japan (IPCAJ, 1987, 1989). The purpose of the project was to measure the acid deposition, to understand its characteristics, and to develop and validate a regional acid deposition model. In this project, before the field survey, important issues related to data quality such as sampling methods, sample storage methods and analytical methods had been confirmed in advance in the laboratory.

In this project, samples were collected at 15 sampling sites as shown in Figure 16.1. Eleven sites among the 15 sites were annexed to the existing air quality monitoring stations operated by local governments and had the air pollutant data of SO_2 and NO_x, among others. The other three sites were located within several kilometers of an air quality monitoring station, and one site was located in a remote area of a mountainous range.

Each sampling station had a DRS-150 wet/dry precipitation collector, a bucket sampler for bulk deposition and a low-volume air sampler. The low-volume air sampler was equipped with quartz filter paper. The sampled filters were used for the analysis of ionic components and carbons. In Dodaira and Mitaka (Tokyo) sampling sites, two low-volume air samplers were used, one of which was equipped with membrane filter paper and used for the analysis of elements.

In this study, dry and wet depositions were measured for every month in one year. In the Kanto area (eastern Japan including the Tokyo metropolitan area) except the northern edge of Gunma prefecture, winter is the dry season and June–July and September have much rain. However, in 1988, several typhoons and tropical depressions brought much rain in August, and some stations showed records of more than 500 mm/month. The sampling duration for the precipitation and aerosols was usually one month. Duration for the first sample was between October 12 and November 10, 1987, and the second sample from November 10 to December 9, and so on. Therefore, the sample was named after the month, e.g., October and November, etc. However, due to heavy rain brought by typhoons, the collection of the rain samples was carried out twice in August.

DOI: 10.1201/9781003597216-18

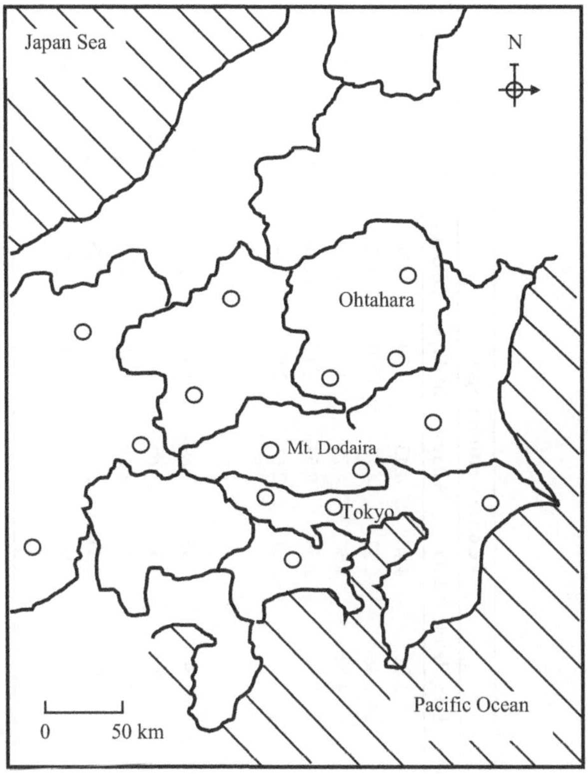

FIGURE 16.1 Sampling network for the IPCAJ acid rain study.

16.2 Measures for Quality Assurance

This section describes quality assurance related to the sampling, storage and analysis of rainwater samples. First, the variability among rainwater sampling devices used was examined. Second, the effects of rainwater storage conditions on changes in rainwater constituents were examined. Next, with the help of the analytical laboratories that participated in the committee for this study, the differences in analytical methods were investigated. In addition, some basic considerations such as the ion balance of the analytical data were also considered, and the data screening was conducted.

16.2.1 Storage Condition of Rain Samples

Generally, rainwater collected outdoors is transported to a laboratory where component analysis is performed, which often takes several days to operate the analyser to produce the results. For this reason, the alteration of the rainwater samples with time during an extended storage period was investigated. For this preliminary survey, rainwater samples were collected in summer and winter at two locations in Kawasaki, near the industrial area, and at Tsukuba, in the suburbs. These samples were analysed for the changes over time in pH, conductivity, sulphate, nitrate, and ammonium ions for up to 60 days. Some of the results are shown in Figure 16.2 and Table 16.1. As for the Tsukuba rainwater, a decrease in ammonium ions and an increase in nitrate ions were remarkable after 60 days. No noticeable change was observed for sulphate ions.

TABLE 16.1

Temporal Change of Components in Rainwater

				Concentration (mg/L)							
Composition	**Sampling site**	**Sampling date**	**Storage condition**	**Sampling time**	**1 day later**	**2 days**	**5 days**	**10 days**	**20 days**	**30 days**	**60 days**
SO_4^{2-}	Kawasaki	Aug. 19, 1986	4°C dark place		4.52	4.40	4.45	4.50	4.56	4.68	4.73
SO_4^{2-}	Kawasaki	Aug. 19, 1986	40°C room	4.51	4.48	4.53	4.72	4.70	4.71	4.81	5.36
SO_4^{2-}	Kawasaki	Aug. 19, 1986	40°C dark place		4.52	4.50	4.75	4.73	4.70	4.78	5.30
SO_4^{2-}	Tsukuba	Dec. 8, 1986	4°C dark place		2.57	2.58	2.58	2.47	2.56	2.93	2.67
SO_4^{2-}	Tsukuba	Dec. 8, 1986	Leaving indoors	2.56	2.57	2.54	2.61	2.53	2.56	3.02	2.84
SO_4^{2-}	Tsukuba	Dec. 8, 1986	Freezing					2.47			2.67
NO_3^-	Kawasaki	Aug. 19, 1986	4°C dark place		1.80	1.85	1.89	1.86	1.82	1.74	1.81
NO_3^-	Kawasaki	Aug. 19, 1986	40°C room	1.89	1.80	1.86	1.85	1.81	1.81	1.73	1.97
NO_3^-	Kawasaki	Aug. 19, 1986	40°C dark place		1.80	1.81	1.89	1.81	1.81	1.73	1.91
NO_3^-	Tsukuba	Dec. 8, 1986	4°C dark place		2.99	2.85	3.23	2.71	2.78	2.66	3.15
NO_3^-	Tsukuba	Dec. 8, 1986	Leaving indoors	2.90	2.93	2.86	3.18	2.71	2.56	2.87	9.01
NO_3^-	Tsukuba	Dec. 8, 1986	Freezing					2.71			3.10

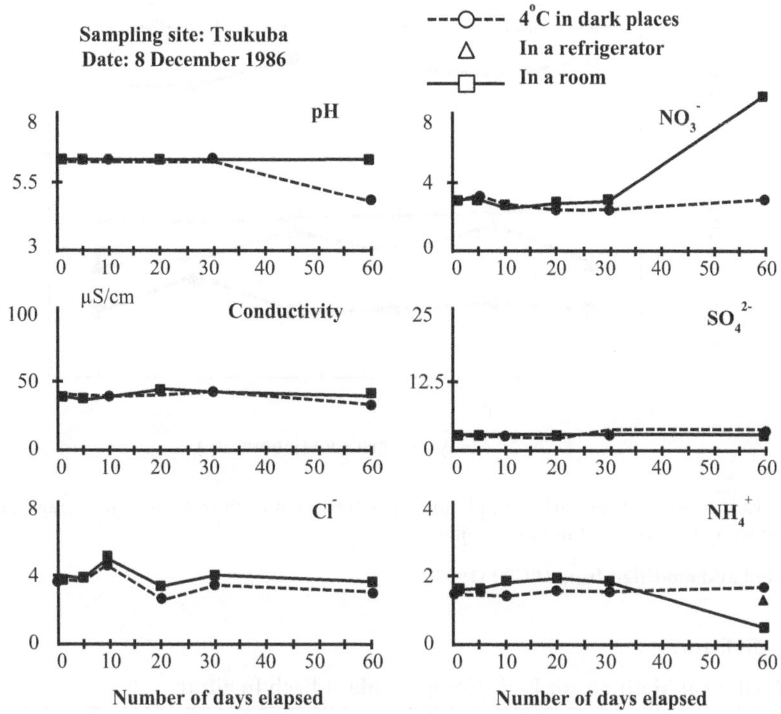

FIGURE 16.2 Degradation of rainwater samples, temporal change of components in rainwater.

Source: Adapted and modified from IPCAJ (1987).

16.2.2 Inter-Laboratory Comparison of the Analytical Results

In this project, the same samples were measured and compared to assess the measurement errors and discrepancies caused by the differences in analytical methods. An inter-laboratory comparison was conducted for six participating institutes consisting of local government laboratories and environmental centres, including the contract research institute, conducted comparative tests itself. Three rainwater samples and one standard sample were sent to each institute. To prevent deterioration during the transport of the sample, the samples were filtered in two stages using a 0.45 µm and 0.15 µm Millipore filter and were stored at room temperature for about 1 week, before bringing them to the participating labs. All samples sent were stored in a refrigerator until the analysis at each lab. In addition to the rainwater samples, standard adjustment materials (blind samples to the participants) with known sulphate ion concentrations were also prepared. This standard sample contained only SO_4^{2-} and NO_3^-. Each lab used its routine method for the analysis of rainwater samples to analyse these test samples (see Table 16.2). At each analytical institute, measurement was carried out five times for each of the four samples (three rainwater samples and one standard sample), and the results are shown in Figure 16.3.

The average of the five analysis values of the five repeated measurements on every sample by each lab is shown in Figure 16.3. For sample A, which has the lowest concentration, the measured value of institute V, which is being analysed by turbidimetry, was slightly higher. In the case of the highly concentrated sample B, the difference among the analytical institutes was small. These analytical data were analysed by analysis of variance. In this case, the sample is considered a blocking factor and analysed by randomised block design, as shown in Table 16.3.

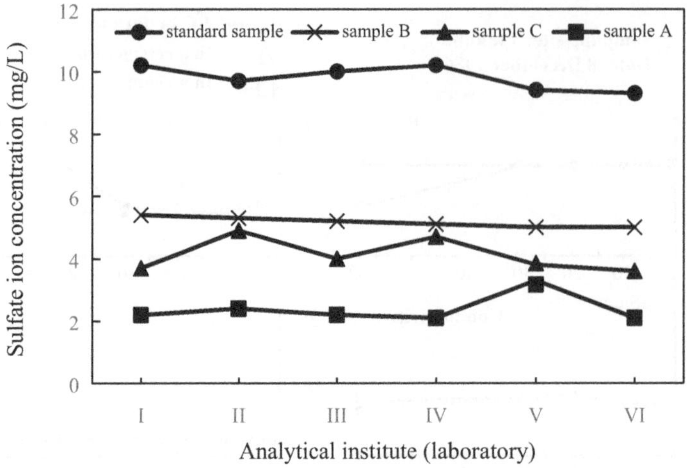

FIGURE 16.3 Comparison of measured sulphate concentration at each institute, an average of five runs for three rainwater samples and one standard sample.

Source: Adapted and modified from IPCAJ (1987).

TABLE 16.2

Analytical Methods for Ionic Components at Each Institute

Analytical institute	I	II	III	IV	V	VI
Composition						
Sulphate ion	IC	IC	IC	IC	T	IC
Nitrate ion	IC	IC	IC	IC	SA	IC
Chloride ion	IC	IC	IC	IC	SA	IC
Sodium ion	IC	AA	FP	AA	FP	IC
Ammonium ion	IC	SA	SA	SA	SA	IC

Note: IC, ion chromatographic analysis; FP, flame photometry; AA, atomic absorption spectrophotometry; SA, spectrophotometric analysis; T, turbidimetry

TABLE 16.3

Results of ANOVA for the Measured Sulphate Ion Concentration at Each Institute

Factor	Sum of squares	Degrees of freedom	Mean square	*F* value	Significance
Sample (block)	949.1	3	316.38	75759	*
Institute	4.2	5	0.82	197	*
Interaction	14.6	15	0.97	233	*
Error	0.4	96	0.004		
Total	968.2	119			

* 1% significance

In such an analysis, the precision of repeated measurements of the same sample, that is, reproducibility, is extremely high. For this reason, the variance of the error term becomes small, and even a slight difference between the factors is often determined to be significant. If the calibration curve is recreated each time during repeated measurements, the error term due to repetition becomes larger, which may be able to mitigate the effects of undesirable too high sensitivity. It should be noted that, in such a test, even if the number of repeated

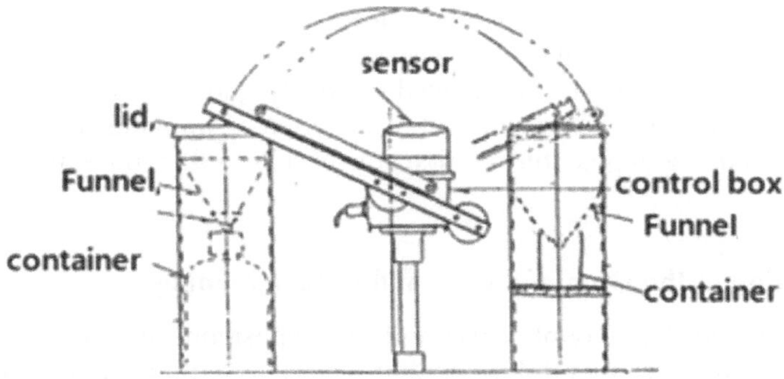

FIGURE 16.4 Koshin Electric Co. DRS-150 wet/dry precipitation collector. (This figure shows the top lid of the precipitation collection section closed when it is not raining.)

Source: IPCAJ (1987).

measurements is increased, the power of testing becomes too high, and the effect of the factor to be investigated may not be evaluated well by ANOVA.

16.2.3 Precipitation Sampler and Its Instrumental Error

The sampling device used in this study is a Koshin Electric Co. DRS-150W type of precipitation collector. A schematic diagram of the device is shown in Figure 16.4. All sampling stations were equipped with this type of sampling device. The device is an automatic precipitation sampling device and has a structure that operates a rain sensor in rainy weather, automatically opens the top lid of the precipitation sampling section to collect rainwater and closes the top lid to prevent fallout from entering when there is no rainfall.

Simultaneously, a bulk sampler, which is an open sampling container that collects all of the fallout, was also installed at each station. This sampler had the same diameter (154 mm) of the opening as the DRS-150 W. Both samplers also had the same height of the intake. A funnel was attached to the collection section, and a fine bottle was used for the storage of rainwater to make less contact with the outside air. Bulk sampler measurements are used to provide information on the total deposition, which can be supplementary especially when either wet or dry data are missing. It is also used as a quality assurance measure by comparison with the total wet and dry deposition collected by the automatic precipitation sampler at the same station.

Before deploying the sampling device at each station, a test was conducted to confirm that there was no significant difference among the samplers used in the project. This test was conducted in Kawasaki with five sampling events, each of which was about one–three weeks in duration during the period from September to November 1986. Six DRS-150W collectors were installed at the same location, and the chemical components in the deposition were analysed and compared. The test for the rain-sensitive operation status – i.e., the rain sensor operation to automatically open the lid when it rains for wet sampling – of the device was also conducted. It was done because there was a possibility of malfunction during drizzle or heavy fog. Accordingly, the operating status of the automatic sampling device was recorded every hour during this comparative test and confirmed that there was no difference between the devices.

Variability in the amount of precipitation water collected by the samplers was also checked and confirmed to be negligible. The results showed that the coefficient of variation of the precipitation amounts collected by the samplers was 0.97–2.82%, with an average of 1.86%. However, the coefficient of variation of the dry deposition amount collected was larger, i.e., from 2.49% to 12.7%, and the overall average was 7.89%. This large fluctuation was explained by the differences in the amount of very large particles (up to about 100 μm), which by nature are unevenly distributed.

Fluctuations in the pH value of the precipitation samples collected in this comparative test were analysed. The results showed that the coefficient of variation of the pH value of precipitation was 0.21–1.09%, and the average was 0.5%, which may be considered negligible. Variabilities in the concentrations of each ion in the collected wet and dry depositions were also analysed, and the results showed relatively good consistency for both the wet and dry depositions of SO_4. The coefficient of variation of sulphate and nitrate (NO_3^-) ions in precipitation and dry deposition simultaneously collected by the same type of samplers was about 3–4%.

16.2.4 Evaluation of the Obtained Data and Data Screening

The project measured the amount of wet and dry depositions at 15 locations in the Kanto region, eastern Japan every month for one year. The collected rain samples were stored in a cool and dark place (not in a refrigerator), and chemical analysis was carried out within one month, i.e., satisfying the conditions to minimise the sample alterations previously detailed (Section 16.2.1).

First, the bulk sampler measurements and the automatic precipitation sampler measurements were compared. Note that the amounts of water collected by the bulk samples would not be compatible with those of the dry/wet collector because of the possibility of evaporation loss and the sampler recovery from the bulk sampler (washing with distilled water). As for the deposition results, the data of bulk samplers tended to show slightly higher values for almost all the components, i.e., ammonium, chloride, calcium and magnesium ions, compared with the automatic precipitation sampler at every station. The monthly data showed that the months having the noticeable deposition amounts collected by the bulk sampler were March, July, August and September, which had relatively high rainfall. In addition, the amount of precipitation collected was compared with the data of the nearest Japan Meteorological Agency observatory stations. As the results of this screening procedure, the doubtful data points due to equipment troubles, quality assurance activities etc. were excluded from the subsequent data analysis. Of the total of 180 data at 15 sites for 12 months, 23 data points were discarded, and 157 data points were used for subsequent statistical analysis. The ion balance for the valid data was analysed, as shown in Figure 16.5 (Shirai et al., 1993). The anions analysed included were SO_4^-, NO_3^- and Cl^-. The cations were H, NH_4, K, Ca and Mg.

There were five cases in winter in which the total anion was larger than 1.5 times the total cation equivalent. On the other hand, there were 25 cases in August and September in which

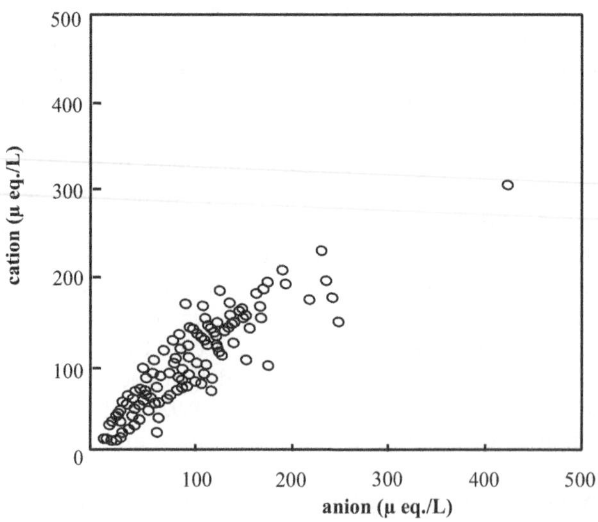

FIGURE 16.5 Ion balance (number of data = 157, corr. coef. = 0.89).

the total cations in equivalence were more than 1.5 times anions. It may be due to a typhoon approaching the sampling domain during this period when the rain caused by tropical air masses may be influenced by CO_3 and other factors. These data were used in a study to understand regional characteristics using principal component analysis but are omitted here. For more details, refer to Okamoto et al. (1992).

16.3 Key Findings of the Acid Rain Study

For most compositions in wet deposition, concentrations were higher in the winter and spring seasons than in the other two seasons of the year. At almost all sampling sites in Kanto Plain, many ion concentrations in precipitation samples were the highest in February because the amount of rainfall was the smallest. Since most of the rain in August was brought on by the typhoon, its rain was not polluted by local anthropogenic sources coupled with the high dilution rate, the concentrations of SO_4^{2-}, NO_3^- NH_4^+ and other ions were very low. As a result, although there was a lot of rain in August, the monthly total and wet depositions of these ions were not as much as the heavy rainfall. An example of the time-series of rainfall and SO_4^{2-} deposition at Mt. Dodaira station is shown in Figure 16.6.

The results of the time-series shown in Figure 16.6 indicate the following: the monthly precipitation from April to July was 100–200 mm, and the monthly precipitation in August was about 700 mm, about five times higher than the monthly average, but the deposition of SO_4^{2-} was about twice as high.

The relation between SO_2 concentrations measured at the collocated or nearby air quality monitoring stations, and SO_4^{2-} in aerosols and SO_4^{2-} total deposition were analysed (Figure 16.7). The correlation between SO_2 concentration and SO_4^{2-} in aerosols was higher (r = 0.62, n = 96) in cooler months (October–May) but lower (r = 0.12, n = 34) in hot months (July–September). The correlation between SO_4^{2-} in aerosols and SO_4^{2-} total deposition was also higher in the cooler season (r = 0.62) and lower in the hot season (r = 0.49). The characteristics in June differed from those in other months. SO_4^{2-} in aerosols was much higher in June, and this tendency can be seen in most monitoring sites. Monthly average suspended particulate matter concentrations were the highest in December, and generally higher in winter than in spring or summer. Carbon concentration in the aerosols was also higher in winter, whereas the SO_4^{2-} and NH_4^+ concentrations peaked during June and had their minimum values in winter.

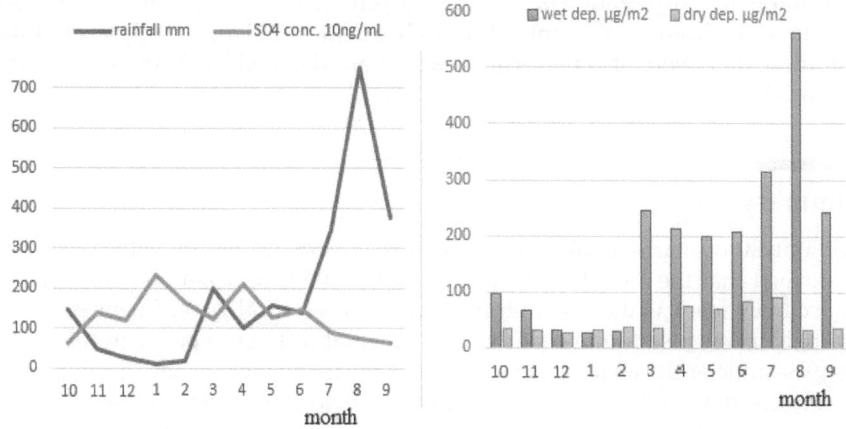

FIGURE 16.6 Seasonal variation of the amount of rainfall, sulphate concentration of rainwater and sulphate deposition flux at Mt. Dodaira station.

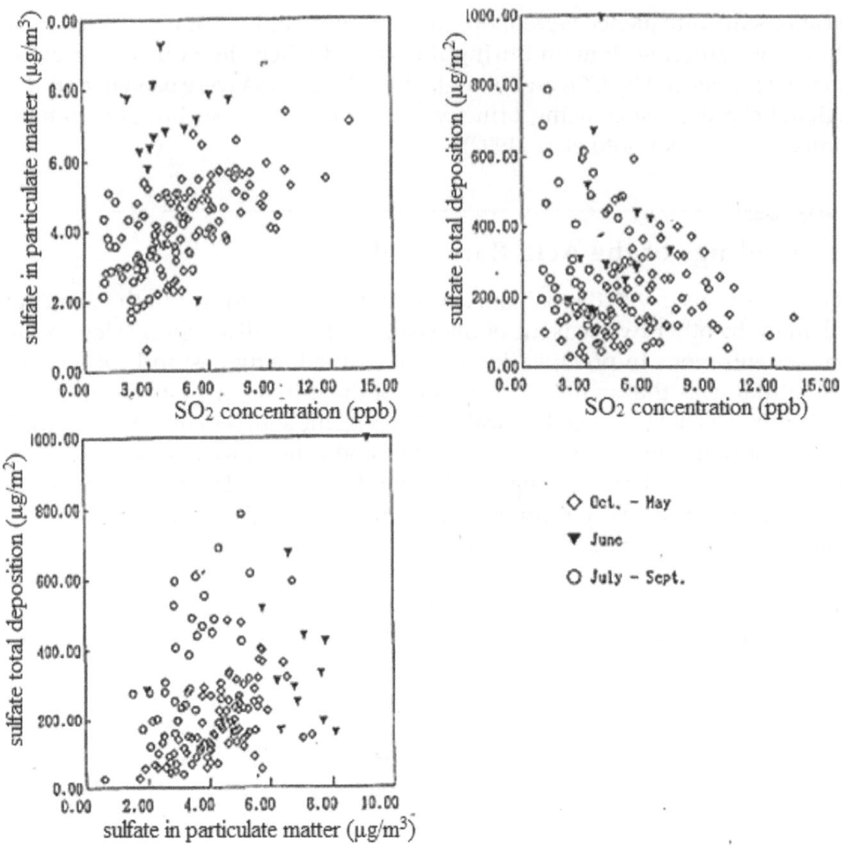

FIGURE 16.7 Relations between SO_2 concentration (ppb), sulphate in particulate matter (μg/m³) and sulphate total deposition (μg/m²) based on the monthly data.

Source: Adapted and modified from Okamoto et al. (1992).

Seasonal variations and spatial distributions for acid rain were investigated based on the monthly data at 15 sampling sites. Annual average pH values for rain samples were between 4.2 and 5.0. Sulphate concentration of rainwater was higher in spring, but total deposition was high in summer due to high amounts of rainfall. Further statistical analyses and modelling studies on these data were also performed, which are detailed in Okamoto et al. (1992) and Katatani et al. (1991).

16.4 Summary

In this acid rain study, various efforts were made to ensure the quality of the data. As a result, it was confirmed that the analytical methods used at the contracted analytical institute had a similar accuracy level as the research institutes of each local government participating in the committee. It was also found that the instrumental difference between the precipitation collectors was sufficiently small in the collocated measurements. Confirming the quality of the measurement data is the foremost requirement before the data can be used for various analyses. These quality assurance issues need to be resolved as much as possible before the actual monitoring activities are conducted.

REFERENCES

Industrial Pollution Control Association of Japan (IPCAJ), 1987, *Study on acid deposition and secondary particulate prediction model development, report on sampling and analysing methods* (p. 179) (In Japanese), Japan Environmental Management Association for Industry, Tokyo, Japan.

Industrial Pollution Control Association of Japan (IPCAJ), 1989, *Study on acid deposition and secondary particulate prediction model development, report on measurement results for dry and wet depositions* (p. 318) (In Japanese), Japan Environmental Management Association for Industry, Tokyo, Japan.

Katatani, N., Murao, N., Okamoto, S. and Kobayashi, K., 1991 September 1–6, Modelling study on acid deposition in eastern Asia, *Proceedings of the 2nd IUAPPA regional conference on air pollution "emerging issues in Asia"*, vol. 2, 59–64, Seoul, Korea.

Okamoto, S., Kobayashi, K., Katatani, N., Nezu, T. and Shiozawa, K., 1992 August 30–September 4, Characterization on acid deposition and secondary aerosols in Kanto area Japan, *Papers from the 9th world clean air congress, critical issues in the global environment*, vol. 2, IU10.05, Montreal, Canada.

Shirai, N., Suzuki, T. and Mochizuki, K., 1993, *On the correlation among the constituents of acid deposition*, Graduation thesis under the direction of Assoc. Prof. Okamoto, S., Tokyo University of Information Sciences (In Japanese).

17

Quality Assurance of Air Tracer Dispersion Experiments for Roadways

This chapter outlines the design and implementation of atmospheric tracer dispersion experiments. The air tracer dispersion experiment involves releasing a tracer substance from a target emission source or a simulated source and subsequently measuring the concentrations in many locations around the source to reveal the dispersion pattern. The results thus provide a clue about the behaviour and the impact of simulated pollution sources or the consequence of a future accidental release of toxic substances. The data obtained from well-designed tracer experiments are useful in the validation of air quality models. This study focused on the dispersion of air pollutants in the vicinity of roadways related to vehicle exhaust pollution, which are less commonly conducted as compared to the numerous experiments for typical industrial chimneys. The data obtained from the dispersion experiments presented here meet the required quality standards and have been used to validate the line source air quality model.

17.1 Methodology of Air Tracer Experiments for Roadways

17.1.1 Overview of Air Tracer Experiments

An air tracer dispersion experiment is a resource-consuming research activity in which a tracer substance is released from a target emission source or a simulated source, and its concentration is measured at many points around the source. The results thus provide a clue about the behaviour and the impact of simulated pollution sources or the consequence of a future accidental release of toxic substances. The tracer substance selected for the experiments should be harmless to humans and other environments. For the air tracer, a substance that is chemically non-reactive, non-toxic, and easy to measure is selected. In the past, sulphur hexafluoride (SF_6) was commonly used but due to the greenhouse effect of the gas, its use is now limited.

If the conditions of an emission source are similar to those of a future or another designated source, then tracer experiment results can be directly utilised for impact predictions in the environmental impact assessment (EIA), namely for assessing proposed new facilities. However, it is practically impossible to guarantee that the meteorological and emission conditions during the experiment will mirror those of the planned emission sources. For this reason, the tracer experiments are rarely used for predictions in EIA or other applications. Rather, they are normally used to generate data for the validation of an air quality model in its development stage. From this point of view, the database accumulating the data obtained by the air tracer dispersion experiments has come to be considered a toolbox for model validation (Olesen, 2001; Irwin, 2017). Measurement of tracer concentration in dispersion experiments requires accuracy equal to or higher than that of ordinary atmospheric concentration measurement because the data quality is of the foremost importance for model validation. In addition, it is necessary to properly address the issue of whether the tracer disperses from the simulated emission source is reproduced correctly. Therefore, it is necessary to understand

DOI: 10.1201/9781003597216-19

not only the atmospheric dispersion processes but also the measuring method of tracers and the meaning of statistics on the spatiotemporal scale of data.

The air tracer dispersion experimental data for many stationary sources have been collected and compiled within the Harmonisation Initiative (Olesen, 2001). However, not many datasets for roadways were published. A study on General Motors test courses is one of the largest-scale experiments up to now for roadways, but the experimental conditions were somewhat different from ordinary road conditions (Cadle et al., 1977). In Japan, many experiments have been conducted that generated meaningful datasets but have not yet been widely published (Okamoto et al., 1999a). This chapter introduces the tracer experiments conducted by IPCAJ (1981) and others. Section 17.1 deals with the data from the air tracer dispersion experiments (1) to (4) in the following list, and Section 17.2 covers the data in (5).

1. Experiment in Shin-Yokohama area (flat area)
2. Experiment in Aoyama area (street canyon)
3. Experiment in Ujiie area (highway on embankment)
4. Experiment in Kanda Jimbocho area (street canyon)
5. Experiments conducted by the Ministry of Construction at eight locations in Japan (various road structures)

The focus of Section 17.1 is on the experimental designs involving the tracer release methods and air sampling methods. Next, Section 17.2 details a case study focusing on the experiment data compilation and analysis.

It is a great challenge to conduct air tracer dispersion experiments on roadways, i.e., how to release the tracer and how to conduct the sampling and analysing of the ambient tracer concentrations. This section describes these methodological issues that have a significant impact on the quality of data collected. Rappolt and Kerrin (2010) detail various technical challenges of dispersion experiments with air tracers. They also discussed efforts to ensure the quality of experimental data and data management techniques for tracer studies. Momoi et al. (1995) also mentioned the issue of data quality in air tracer dispersion experiments but for general dispersion experiments, not specific to issues related to the experimental technique for roads as targets. Experiments concerning roadways necessitate a distinct approach compared to those focusing on smoke from conventional stacks. IPCAJ and other organisations mentioned here strive to guarantee data quality in road dispersion experiments, which are discussed herein.

17.1.2 Air Tracer Releasing (or Disseminating) Plan for Roadway Experiments

In an air tracer experiment, it is important that the discharging condition of the actual effluents from the target emission source and the releasing (or disseminating) condition of the tracer, which is a simulator of the effluent, are not significantly different. There are two methods to reproduce the dispersion of exhaust gas from a roadway:

1. A vehicle equipped with the sprayer of the tracer runs along the designated lane and discharges the tracer during the designated period.
2. The tracer is discharged from the spray pipe laid on the road.

The first method has the advantage of properly reproducing atmospheric turbulence caused by a running vehicle. However, if the number of spraying vehicles is not sufficient, the characteristics as a continuous source cannot be reproduced. It also has the disadvantage of high costs, including arranging a large number of test vehicles and training test drivers. Oppositely, the second method is suitable for long-term continuous experiments without much expense.

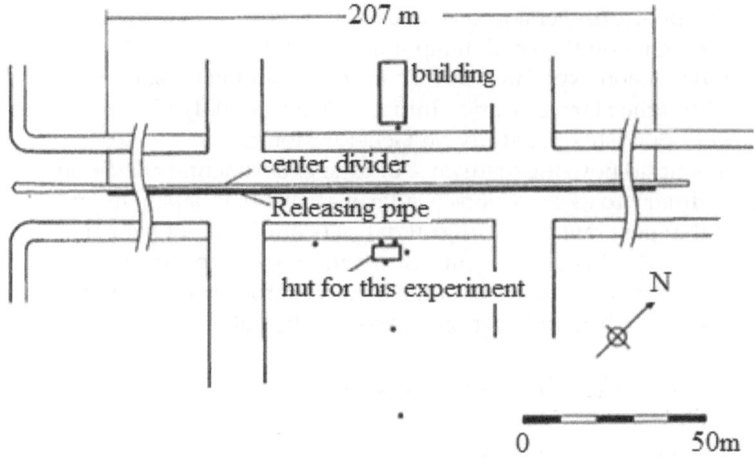

FIGURE 17.1 Location of NO_x monitors (dots) and tracer releasing facility at Shin-Yokohama experiment.

However, in cases where traffic volume is low, there is a difficulty in reproducing the initial dispersion spread which is needed to reflect the atmospheric turbulence caused by a running vehicle. Therefore, one of the challenges is which side of the road to place the pipes on. Another important issue is the lack of two-dimensionality, which is caused by the insufficient length of the pipe. Therefore, it is not possible to simulate the roadway as a line source of infinite length unless these issues are solved.

In the IPCAJ dispersion experiment of the Shin-Yokohama area, the comparison of these two tracer releasing methods was carried out. The experiment was carried out along the Yokohama City Loop Street No. 2. This roadway has three lanes in each direction separated by a centre divider and has a traffic volume of about 40,000 vehicles/day. The surrounding terrain is mostly flat, but at the time there were only a few small buildings and huts along the streets, as shown in Figure 17.1.

The releasing method of the tracer gas was developed for the continuous experiment. This method used the pipe laid down on the centre divider of the roadway. The pipe was 207 m long and had 68 pinholes, and the distance between the pinholes was 3 m. The flow-rate of SF_6 was regulated by an electronic flow controller and thermal mass flow-meter. The regulated flow-rate of SF_6 was 0.5 L/min. A summary of this experiment is available elsewhere, i.e., Okamoto et al. (1990).

Two types of air tracers (SF_6 and $CBrF_3$) were used to understand the differences between the two tracer releasing methods just described. NO_x measurements were also conducted to understand the dispersion behaviour of vehicle emissions. The main tracer SF_6 was released continuously from a spray pipe placed on the road centre divider. The supplementary tracer gas $CBrF_3$ was also used and was released from near the exhaust pipe of a vehicle that was shuttled along the pipe-laid section of the road.

The release duration of $CBrF_3$ was 1 h and added several minutes, which was a dummy release to attain the steady condition, usually 5 min. The number of shuttles was 30–50 cycles. Dual tracer experiments were conducted eight times during the continuous SF_6 release period. Tracer gases were continuously collected by sequential samplers at the same points of NO_x analysers during the SF_6 releasing period. Four sets of SF_6 continuous monitors (reformed from ECD gas-chromatography) were also used. In addition, extensive tracer sampling was carried out at about 40 points of vertical sampling array 23 times during this period. See IPCAJ (1981) for details on the placement and operation of the air sampler. Sequential samplers and grab samplers were installed adjacent to all NO_x monitor and GC locations, as well as on sampling poles installed perpendicularly to the road.

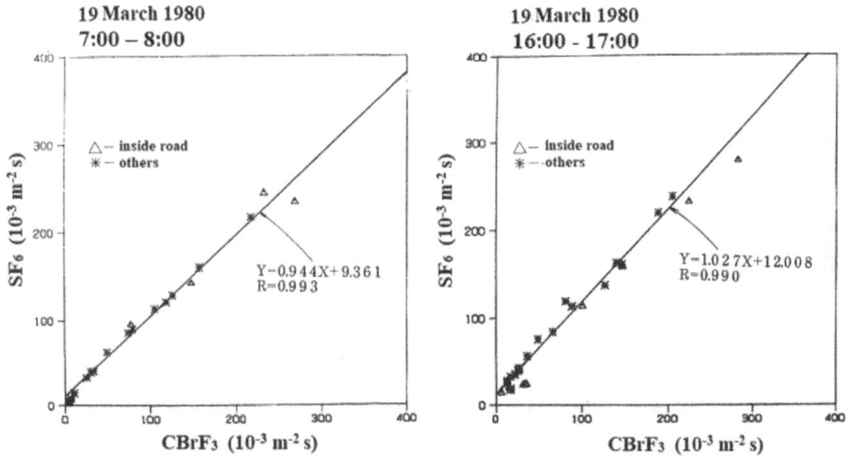

FIGURE 17.2 Comparison between two tracer releasing methods in the Shin-Yokohama experiment, the unit of normalised concentrations is 10^{-3} m^{-2} s.

Source: IPCAJ (1981).

17.1.2.1 Analysis of When the Auxiliary Tracer Is Released from Both Sides of the Road

The vehicle releasing the auxiliary tracer $CBrF_3$ was travelling back and forth on both sides of the road where the pipe releasing the main tracer SF_6 is laid. The $CBrF_3$ was continuously released during the vehicle travel. The measurement results of the two types of tracer concentrations were compared to reveal the effect of differences in tracer releasing methods on the dispersion situation. Figure 17.2 shows a comparison between the normalised concentration per unit emission quantity (C/Q) of SF_6 and $CBrF_3$. From this result, it can be determined that there is no significant difference in the dispersion condition between the tracer releasing methods.

17.1.2.2 Analysis of When the Auxiliary Tracer Is Released from One Side of the Road

In the Shin-Yokohama experiment, the experiments were also conducted to examine the difference between the impact from vehicle emissions on the nearby side lane and that from the opposite side lane. Several times during this experimental period, the release of the auxiliary tracer $CBrF_3$ was carried out only for traveling on one side every 30 minutes. Specifically, the following measures were taken when operating the $CBrF_3$ spray car that reciprocated around the spray pipe laying section of the road. For every 30 minutes, the spraying device was out of operation when traveling on one side of the road. The result is shown in Figure 17.3, which shows that the effect of releasing from the near side lane was large.

17.1.3 Sampling Plan for Roadway Experiments

Key sampling issues to consider are the challenges with bag samplers and the timing of sampling, including the length of dummy releases to guarantee the sampling at the steady-state condition.

- The bag sampler operates on an automatic timer to start and end sampling, but there were concerns about the errors associated with this timer. In addition, the operational stability of the suction pump was also a concern.

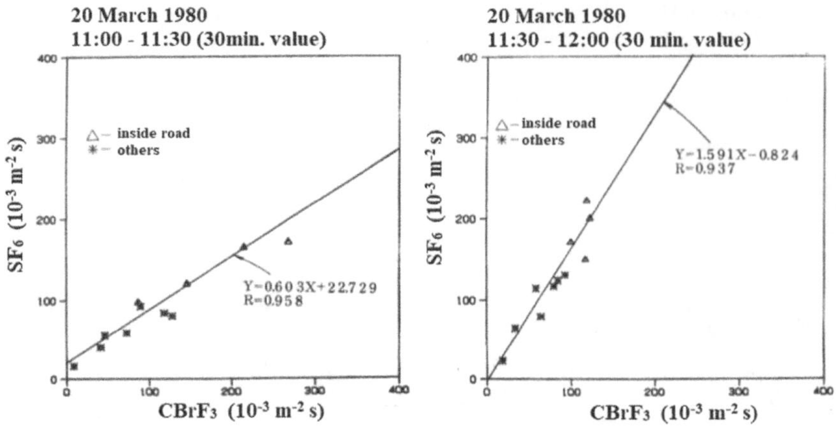

FIGURE 17.3 Comparison between two tracer concentrations, the first half (11:00–11:30) released the auxiliary tracer from the near side lane and the latter half (11:30–12:00) from the opposite lane, the unit of normalised concentrations is 10^{-3} m^{-2} s.

Source: IPCAJ (1981).

- The sampling plan was based on some estimate of the time from the start of tracer releases until the tracer plume reaches a steady state, but verification of this estimate was an issue.

Comparative parallel measurements were conducted to confirm the performance of the air sampler (with timer) and its bag used for sampling the tracer. In other words, gas chromatography equipment (GC), which provided real-time measurements of tracer concentration onsite every 3 min was collocated with the airbag sampler. The latter collected samples over 1 h for later analysis. The simultaneous measurement results by the two methods were subsequently compared. The onsite monitoring data of GC also provides useful information on the concentration fluctuations within 1 h.

During the experimental period of about 1 week, the continuous operation of four GCs was carried out except for the time of periodic inspection, and the concentration data of SF_6 for over 100 h was obtained. This measurement was conducted by aspirating the ambient air every 3 min, namely, intake of the air to GC every 3 min. The hourly concentration data was calculated using 20 data points collected at 3-min intervals. The air sampling with the bag sampler was done only when the wind direction was perpendicular to the road and a total of more than 20 samples were collected during the experiment period. Figure 17.4 shows the data of the G-2 sampling point at the observation hut near the roadside and the G-4 point about 50 m apart from the road. Figure 17.4 shows the results of stratifying and analysing the data based on the wind direction fluctuation σ_θ at the time of measurement, but it seems that a very clear relationship cannot be grasped. From the results shown in Figure 17.4, it was confirmed that there was no significant difference between the GC readings and the analysis results of samples collected by the collocated bag sampler. There were no significant malfunctions of the timer as initially suspected, and the measurements appeared generally reasonable.

The desirable temporal resolution of the tracer concentrations depends on the objectives of the study. In a dispersion experiment for hazardous substances released from an accident, it may be necessary to know the instantaneous ambient concentration, i.e., at every second. However, the purpose of the study on the dispersion of smoke from a stack or on-road automobile exhaust is to know the steady state of atmospheric dispersion. In such a dispersion experiment, the sampling of air should be started after the dispersion condition of the tracer becomes the steady state. The release of the tracer during this waiting time is called a dummy release. In a high-stack release experiment, a dummy release of more than one hour may be required. Generally, the required dummy release time can be calculated by dividing the

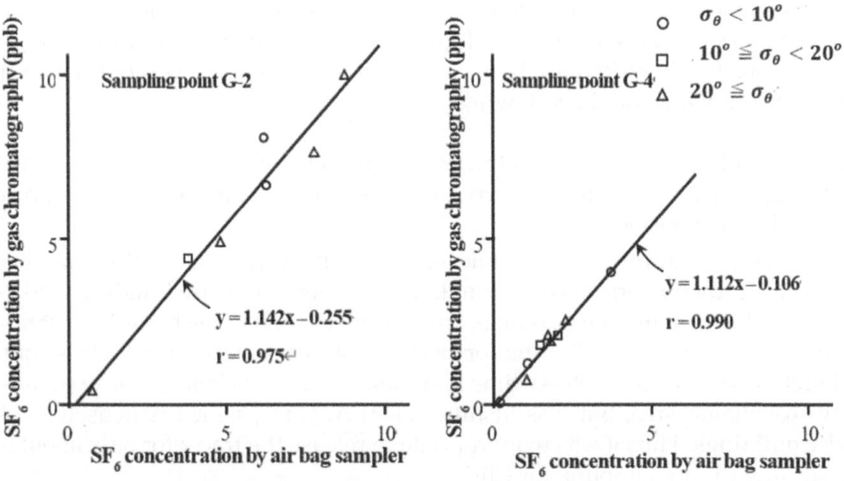

FIGURE 17.4 Comparison of air tracer concentrations between different analytical methods using the data from G-2 sampling points in the observation hut near the roadside and G-4 points about 50 m away from the road. The data were segregated based on the turbulence intensity, measured as the standard deviation of wind direction fluctuations; however, no distinct differences emerged.

Source: Adapted from IPCAJ (1981) and modified.

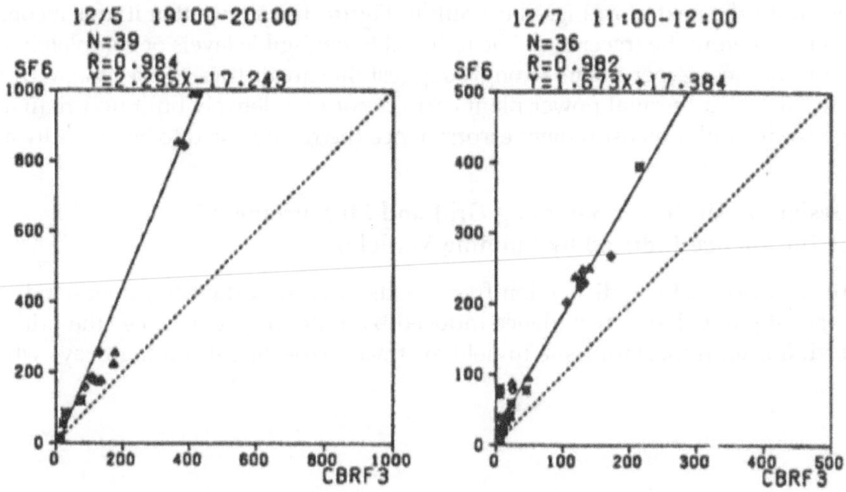

FIGURE 17.5 Comparison between two air tracer concentrations in the Aoyama experiment, the unit of normalised concentrations is 10^{-3} m^{-2} s.

Source: IPCAJ (1981).

distance from the source to the sampling point by the average wind speed and adding a certain amount of leeway (margin). In a dispersion experiment from a roadway, this study initially considered that a few minutes would be enough because the downwind distance was at most a few tens or 100 meters.

For the study of the Aoyama area, the targeting of a street was surrounded by high-rise buildings, and a dual tracer dispersion experiment was conducted. Two types of tracer gas concentrations with different releasing methods were compared, which was the same as the Shin-Yokohama experiment. The comparison of the SF_6 normalised concentration with the $CBrF_3$ normalised concentration is shown in Figure 17.5, which exhibited a strong correlation

(correlation coefficients above 0.95). However, in terms of the magnitude, the normalised concentration of SF_6 was observed to be two–four times higher than that of $CBrF_3$.

Possible reasons for the difference in normalised concentration levels of the two tracers, as shown in Figure 17.5, include the following:

- In a densely built-up area, the main tracer gas (SF_6) released continuously for several hours was drifting between buildings; hence it may be equally considered as the presence of a background concentration.
- The 5 minutes of the dummy release time for the auxiliary tracer gas ($CBrF_3$) may be too short to become a steady dispersion condition. In areas where high-rise buildings are located on both sides of the road and the buildings are densely located in the hinterland, it takes a long time until the concentration distribution becomes steady state after the release of the tracer gas. Therefore, the dummy release time (5 minutes) was sufficient for the experiment in the flat Shin-Yokohama area, but was insufficient in Aoyama, which is densely packed with high-rise buildings. Thus it was inappropriate to release the tracer for only about 5 minutes before starting the air sampling for $CBrF_3$.

The long time it takes to reach a steady state after starting the tracer release means that the time it takes to return to a clean state after stopping the release (i.e., the time to return to the background concentration) is also long. This fact is confirmed by another dispersion experiment conducted in a street canyon in Kanda Jimbocho (Ishii et al., 1995). In this dispersion experiment, an air tracer was continuously released from a pipe laid on the street centre divider. Then, after 21:00 when the release of the tracer was stopped, the air sampling was continued every hour until the next morning. The result in Figure 17.6 shows that it took a considerable amount of time before the tracer could be reduced to negligible levels or completely removed.

The results of these experiments strongly suggest that an air tracer experiment for a stationary source, such as a thermal power plant carried out in a densely built-up urban area, may also suffer from similar measurement errors; hence the results need to be carefully assessed.

17.1.4 Design of Air Tracer Sampling Grid and Measurement of Turbulence Induced by Running Vehicles

In the study of exhaust gas dispersion from roads, it is important to understand the diffusion of exhaust gases due to turbulence induced by a moving vehicle, i.e., the initial dispersion spread. It is an important issue in field air tracer experiments on highways where more

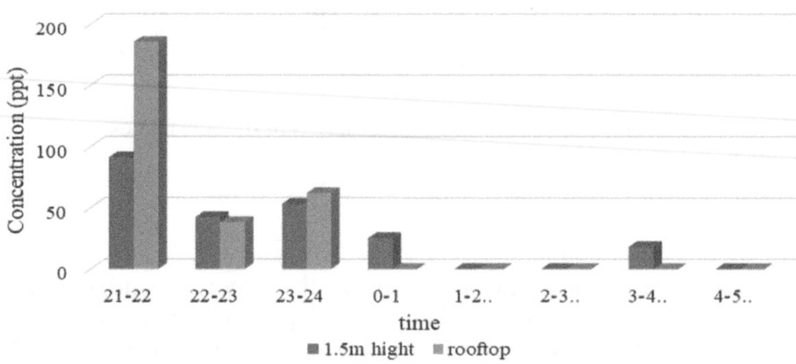

FIGURE 17.6 Time-series variability of the SF_6 concentration within the street. The air sampling was conducted on the ground level and rooftop within the street canyon. The finish time of the SF_6 release was 21:00 on January 18, 1991.

Source: Data were retrieved from Ishii et al. (1995).

vehicles are traveling at high speeds than on urban roads where travel speeds are relatively slow (Okamoto et al., 1998). Here, to illustrate the importance of this issue by presenting a case study of direct measurement of turbulence on the side of a highway is presented. Furthermore, it is important to consider how to design a measurement network (or an air tracer sampling grid) for such cases.

In the air tracer experiment conducted in the Ujiie area of the Tohoku Expressway (Okamoto et al., 1985a, 1985b), the tracer gas was released from the pipe laid in the centre divider of the expressway. The experiments were conducted continuously for 72 hours from 19:00 on December 1, 1981. During this period, the vertical concentration distribution up to 18 m in height was measured at distances of 14 m, 24 m, 40 m, 70 m, and 100 m from the centre divider of the expressway. The sampling pole for vertical measurement located on the road shoulder was at the downwind distance of 14 m, and the 24 m point was just below the embankment. This experimental section of the Tohoku Expressway is almost straight and has a 6-m-high embankment structure. The daily traffic volume in the experimental section was about 18,000 vehicles/day, the large vehicle mixture rate was about 45%, and the average vehicle speed was about 90 km/h. Figure 17.7 shows an example of the SF_6 vertical distribution obtained at different sampling points.

For this experiment, they designed the air tracer sampling grid up to 10 m elevation at the highway shoulder and up to 18 m elevation at other locations. This allowed us quite often to capture the upper end of the concentration distribution. In many experiments, the plume spread is calculated by fitting a normal distribution curve, as shown in Figure 17.7. Therefore, if both ends of the concentration distribution (only the upper end in the case of a vertical distribution) cannot be properly ascertained, an accurate plume spread cannot be calculated. In the actual design of the sampling grid, it is necessary to refer to past cases and literature values. Another item that must be taken care of is the issue of safety for traveling vehicles. Particularly during the installation of various measuring devices at roadside sampling towers, it is essential to implement adequate precautions to prevent the poles from collapsing or the devices from falling, which could pose a risk to passing vehicles.

Determining the plume spread under various conditions is one of the most important purposes of the air tracer dispersion experiment. It is also one of the key parameters of the simulation model for calculating air pollutant concentrations. This will be discussed in more detail in Chapter 18.

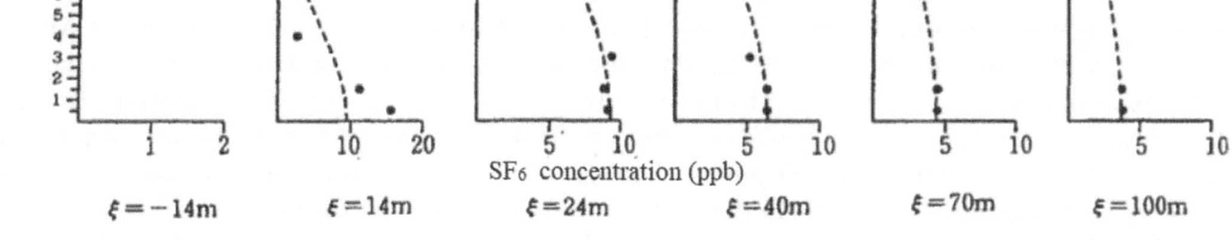

FIGURE 17.7 Vertical distribution of tracer gas (SF_6) concentration and its fitting to the Gaussian (normal) distribution formulas on December 3, 1981, 5:00–6:00 AM. The notation ξ is the downwind distance from the centre divider of the expressway.

Source: Okamoto et al. (1985a).

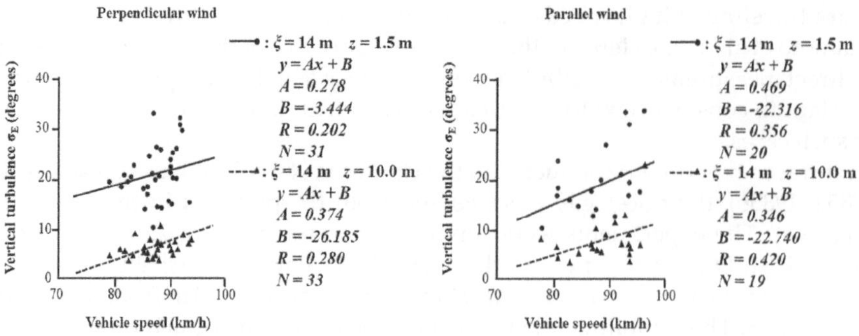

FIGURE 17.8 Relationship between vehicle speed and wind turbulence (vertical wind direction fluctuation).

Source: Okamoto et al. (1985a).

It is noted that the concentration distribution of substances released into the atmosphere can be approximated by the normal distribution, as shown in Figure 17.7. The spread (or dispersion) of quantities that is described by the normal (or Gaussian) distribution can be expressed by the standard deviation. When the spread of the pollutant concentration is expressed by the standard deviation of the concentration distribution, this amount is called dispersion coefficient, diffusion width, plume spread and so on. This text primarily uses the term "plume spread".

On urban roads, it has been confirmed that the tracer concentrations at the roadside showed no significant difference whether released from a moving vehicle or a pipe on the centre divider. However, on expressways with high-speed traffic, the atmospheric turbulence caused by moving vehicles must be carefully considered. Therefore, it is important to conduct turbulence measurements at the roadside. The relationship between the magnitude of atmospheric turbulence measured at the edge of a highway and vehicle speed was investigated. The results are shown in Figure 17.8.

This figure shows the vertical atmospheric turbulence measured at a height of 1.5 m and 10 m at the shoulder of the expressway. These scatter plots indicate the increasing trend of turbulence disturbance with the vehicle's traveling speed. When studying tracer releasing methods, it is important to adopt a method that can reproduce the spread of vehicle emissions due to atmospheric turbulence induced by a moving vehicle. From the results of the dual tracer experiment in the Shin-Yokohama, it is seen that the release from the pipe installed in the centre divider can reproduce it to some extent. However, on highways with high vehicle speeds, it is more important to understand the initial dispersion characteristics at the road edge.

17.2 Database Development for Highway Line Source Modelling

Developing a new air quality model is an expensive task and requires special expertise. Much of this expenditure may be used to get data for model validation, and the air tracer dispersion experiments are one of its central issues. In Europe, the initiative for harmonisation of the air quality models used administratively started in 1991. One of the most important tasks of this initiative is database development. The air tracer experimental data should be used to develop generalised air quality models rather than predicting individual problems. A tool developed under this concept is the Model Validation Kit by Olesen (2001). This database mainly consisted of European experiments, but later it also included Irwin's database, which compiled a vast amount of experimental material produced in the United States since the 1950s. However, as noted previously, the database does not contain sufficient data from line source experiments on roads. Therefore, this section presents a series of air tracer dispersion experiments on highways.

Similar to many other commonly used air quality models, highway air quality models were also validated by comparison with experimental data, and these databases were compiled from many field monitoring programs and air tracer experiments. However, most early studies were conducted only on flat highways. Field experiments for highways with complex structures have been very scarce up to now. To develop and evaluate air quality models applicable to highways with a wide range of road structures, a sufficient database to cover them is required. Therefore, an extensive experimental program was carried out for various traffic and meteorological conditions by the Ministry of Construction, Japan. The main objectives of the study were to analyse the relation between the dispersion of exhaust gas from highway traffic and specific type of highway structures, traffic and meteorological conditions and to find a good indicator for estimating the plume spreads.

A dataset based on the experimental study on atmospheric dispersion from a highway conducted by the Ministry of Construction (MOC) may be one of the most valuable databases for developing or validating a highway line source air quality model. It was intended to be used for the development of the air quality model described in the current Ministry of Land, Infrastructure, Transport and Tourism (formerly Ministry of Construction) technical guidelines for EIA. Not all of them are open to the public, but the limited parts can be introduced by quoting reports from several academic conferences (Okamoto et al., 1999a). This series of experimental studies also included an air tracer experiment to investigate the dispersion in calm conditions as detailed in a published paper (Okamoto et al., 1999b).

17.2.1 Experimental Method and Design

The experimental sites for the chosen highway were flat (at-grade), cut (cut-off), embankment (bank), viaduct (elevated), and highways with noise barriers. The information on the experimental sites and periods is shown in Table 17.1. A straight section of the selected highway was chosen as the experimental site for every experiment.

The SF_6 air tracer was continuously released from the release pipe installed in the centre divider. The flow-rate of the air tracer was regulated by a mass flow control system and diluted by compressed air. The diluted air tracer was transferred into a pipe of 200 m in length and 22 mm in diameter. This main pipe had several fine branch tubes, with a length of 1 m and interval of 3 or 5 m, in a vertical direction to set a source height of 1 m. The release duration of tracer gas was nominally 240 h for the measurement of vertical distribution, 120 h for horizontal measurement, and 168 h for a highway with a noise barrier. The length of the simulated line source was about 200 m, except for the cut-off section and the bank section with a noise barrier. The length of the line source for the cut-off section was about 500 m to confirm the homogeneity of the concentration distribution in the roadway direction. In the case of measuring the horizontal concentration distribution to obtain the horizontal plume spread (σ_y), the tracer was released from one point in the centre divider from a location that varied depending on the wind direction. Here, the position of the release position was changed so that the tracer could be collected without changing the position of the sampler.

A propeller-type breeze anemometer was used to record wind at 10 m above the ground. The hourly average wind data was calculated from the recording of instantaneous values at 1-min intervals. At the same location, wind speeds of 7 m, 5 m, 3.5 m, and 1.5 m height were measured with a 3-cup type anemometer. Basic information is in Table 17.1a, and supplementary information is in Table 17.1b.

In the road sections with noise barrier walls, it was difficult to install the poles in a predetermined place due to various technical restrictions. Therefore, poles were installed at the positions shown in Table 17.1b. The poles on the road edge were installed directly above the noise barrier wall, and the top of the wall was set to 0 m height.

The sampled air was collected in 20-L bags with a flow-rate of 0.2 L/min. The sampling duration for each sample was 1 h. SF_6 in the sampled air was analysed by GC/ECD or an FPD (flame photometric detector) sulphur analyser. Samples collected during unsuitable meteorological conditions, – e.g., opposite wind directions to the monitoring site – were eliminated

TABLE 17.1A

Experimental Site and Period of Highway Air Tracer Experiments

Location	Structure	Noise barrier	Traffic volume (cars/day)	Date	Length of release pipe	Releasing intensity	No. of SF_6 data	No. of NO_x data	No. of perpendicular wind data
Ebetu, Hokkaido	Flat	None	24000	Nov. 20–Dec. 12*	200 m	0.66 L/min	180 h	264 h	96 h
Okayama city	Viaduct	None	80,000	Feb. 3–18	200	1.0	210	240	18
Tsukuba, Ibaraki	Cut	None	67,000	Mar. 6–23	500	7.5	216	268	50
Yamaguti city	Bank	None	35,000	Nov. 16–13	200	0.83	240	240	56
Ohbu, Aichi	Flat	3 m/ 4 m	41,000	May14–20	200	1.5	168	168	35
Mizunami, Gifu	Flat	5 m/ 3 m	29,000	May 28–Jun. 3	200	1.5	168	168	5
Ichinoseki, Iwate	Bank	3 m	26000	Jul. 21–Aug. 6	400	8.0	390	390	7
Chino, Nagano	Bank	5 m	25000	Jun. 23–29	400	7.4	168	168	10

Note: Vertical sampling poles were installed at the road edge and 10 m and 20 m from the edge. However, in Tsukuba, the cut-off section, the position of the second pole was 12 m, i.e., not 10m. Their positions are basically as shown in Figure 17.10.

* As for when the experiments were conducte, all experiments were in 1998, except for Ebetsu in 1997.

TABLE 17.1B

Additional Information for Sections with Noise Barrier

Site	Bank height (m)	Barrier height of measuring side (m)	Distance from the road edge (m)		
			1st sampling pole	2nd sampling pole	3rd sampling pole
Ohbu, Aichi	None	3.0	0	10.3	28.6
Mizunami, Gifu	None	5.0	0	7.3	25.8
Ichinoseki, Iwate	3.6	3.0	0	15.1	24.9
Chino, Nagano	5.4	5.0*	0	11.5	39.5

* The upper end of the wall is curved inward.

and were not analysed. Nitrogen oxides in the air collected in the same sampled bag were also analysed by a chemiluminescence NO_x analyser.

The number of sampling points for each experimental site is shown in Table 17.2. The horizontal plume spread σ_y is commonly used in a Gaussian plume model even for a highway line source model. However, there was no explicit evidence for assisting any σ_y estimating methods. Therefore, some additional air tracer experiments for deriving σ_y were also carried out. In the experiments to determine the horizontal plume spread σ_y, the tracer was released from one point on the centre divider of the highway, and the air sampler was placed along the leeward side of the highway. Regarding the placement of the samplers, in principle, measurement lines were set at distances of 10 m and 30 m from the road. However, in the Tsukuba area, which is a cut section, the distances were set at 15 m and 40 m. Figure 17.9 shows the horizontal air sampling array at this time. Note that the elevated road (viaduct) and noise

TABLE 17.2
Number of Sampling Points and Nominal Sampling Duration

Measurement purpose	Highway structure	Noise barrier	Number of sampling sites							Sampling duration **(h)
			SF_6	NO_x	PM	C_6H_6	WD	WV	Insolation*	
Vertical distribution	Flat	None	27	27	9	5	1	2	1	240
	Embankment	None	27	27	9	5	1	2	1	240
	Cut	None	38	38	9	5	1	2	1	240
	Viaduct	None	31	31	9	5	1	2	1	240
Measurement Horizontal distribution	Flat	None	36	—	—	—	1	2	1	120
	Embankment	None	36	—	—	—	1	2	1	120
	Cut	None	36	—	—	—	1	2	1	120
	Viaduct	None	18	—	—	—	1	2	1	120
Highways with noise barrier	Flat	3 m	23	23	5	5	2	10	1	168
		5 m	23	23	5	5	2	10	1	168
	Embankment	3 m	25	25	5	5	2	10	1	168
		5 m	25	25	5	5	2	10	1	168

Note: * Net radiation. ** This is a set value at the planning stage and was changed depending on weather conditions and the like.

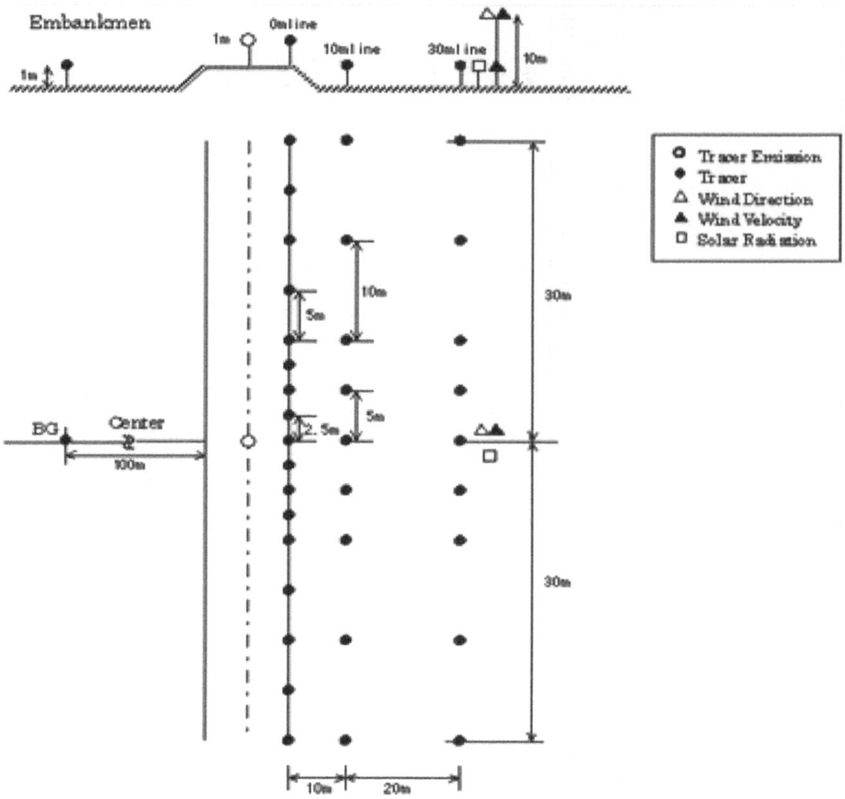

FIGURE 17.9 Layout of sampling points for the horizontal measurement in the case of embankment (Yamaguchi site).

barrier installation sections were not surveyed for estimation. Further, Figure 17.10 shows the vertical air sampling array used in the experiment to determine the vertical plume spread σ_z.

At first, all the hourly NO_x and SF_6 concentrations obtained at all sampling poles and at all surface sampling lines were compiled in the main database. The data for the perpendicular wind conditions were then selected for the analysis. An example of the vertical and horizontal concentration distribution is shown in Figure 17.11.

The available data that can be used for the subsequent analysis were selected from vertical distribution or horizontal distribution of SF_6 and NO_x concentrations. The raw experimental data contained difficult cases in the position of peak concentration was unclear, either because there was a maximum value at the end of the measurement point or because there was an outlier in the measurement value. Such data were excluded from the refined database for estimating the vertical and horizontal plume spreads, i.e., σ_z and σ_y, respectively.

17.2.2 Quality Assurance by Tracer Flux Calculation

As seen from the case studies presented earlier in this chapter, in roadway air tracer experiments, there are many challenges to be addressed, such as setting tracer releasing intensity, setting air sampler timing and calibrating analytical equipment. Further, there are problems such as the uniformity of the releasing intensity of the line source and the sufficiency of its length. Data quality assurance is therefore the utmost importance before the data could be used in model development and validation.

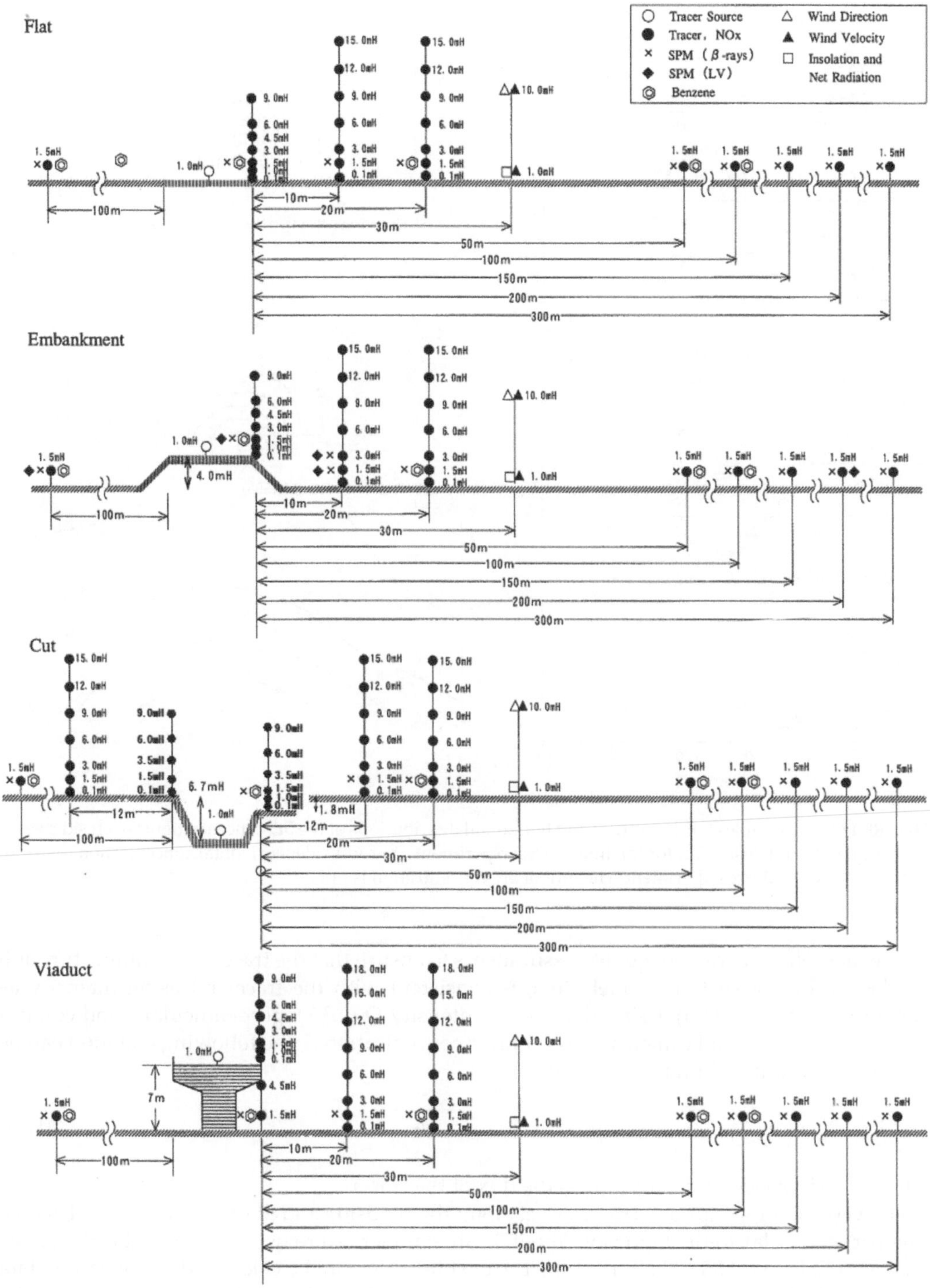

FIGURE 17.10 Layout of sampling points for the vertical measurement.

Source: Adapted and modified from Okamoto et al. (1999a).

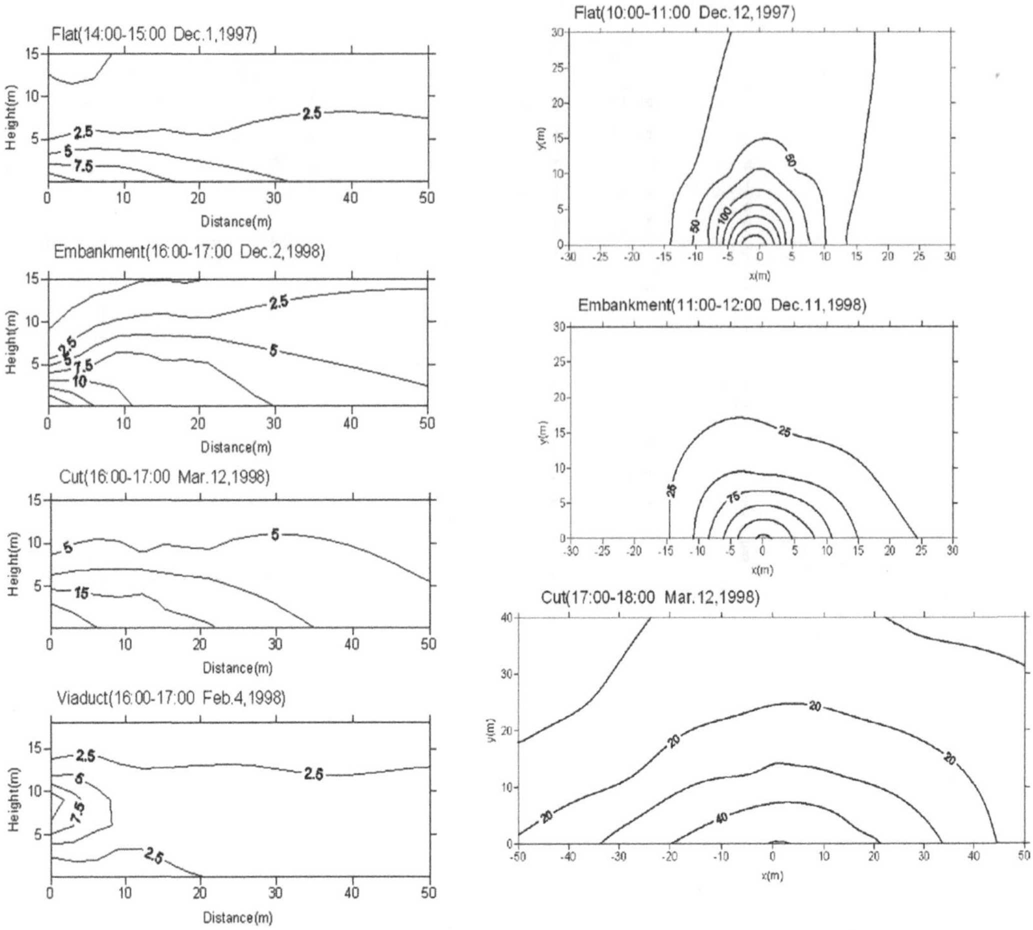

FIGURE 17.11 Examples for the vertical and horizontal distributions of the SF_6 concentration: (a) left side: vertical concentration distribution for the line source experiment; (b) right side: horizontal concentration distribution for the point source experiment. The unit of SF_6 concentration is ppb.

The most effective overall quality assurance is to ensure that the tracer flow volume through the leeward cross-section, namely flux, is consistent with the tracer releasing intensity as shown in Equation (17–1). Using line source intensity Q (m^2/s), perpendicular wind component speed u (m/s), and dimensionless volume concentration C, the following equation can be derived as described in Bullin et al. (1980),

$$Q = \int u\,C\,dz \tag{17–1}$$

where the right side of the formula is the flux of the tracer.

The vertical air sampling array was set along the roadside and at two downwind sites on a line perpendicular to the highway. About 30 air samplers were used to collect the air for the analysis of SF_6 and NO_x. The flux of the tracer substances can be calculated by using the data of the vertical distribution of tracer concentrations and wind velocity components near the road. The tracer fluxes were calculated at three sampling pole locations (at the road edge on the downwind side, 10 m from the road and 20 m from the road). The results are shown in Table 17.3. Since the calculated flux was roughly in agreement with the SF_6 releasing intensity, it could be considered that the overall quality of the tracer dispersion experiment data was

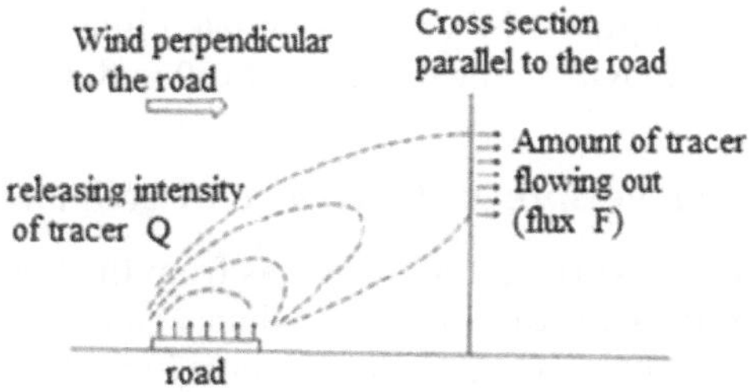

FIGURE 17.12 Relation between the flux at the cross-section and the releasing intensity.

TABLE 17.3

Calculated Tracer Flux at 0, 10 and 20 m Downwind Distance (10^{-6} m^2/s)

Road structure	Place	Flux (x=0 m)		Flux (x=10 m)*		Flux (x=20 m)		Releasing intensity
		Aver.	No. of data	Aver.	No. of data	Aver.	No. of data	
Flat	Ebetu, Hokkaido	0.051	96	0.052	96	0.044	96	0.055
Viaduct	Okayama city	0.044	18	0.094	18	0.088	18	0.069
Cut	Tsukuba, Ibaraki	0.260	50	0.267	50	0.251	50	0.252
Bank	Yamaguchi city	0.155	55	0.217	55	0.196	55	0.155

*As for Tsukuba experiment, x=12 m.

fairly good. Also, the problem of the finite length of the simulated line source did not seem to significantly affect the data quality up to a downwind distance of 20 m from the road.

17.2.3 Key Findings of the Dispersion Experiments

Various analyses were performed on the data obtained from air tracer dispersion experiments conducted by the MOC on eight roads and highways. The key findings are presented here.

17.2.3.1 Concentration Attenuation Tendency in the Direction Perpendicular to the Road

In these experiments, it is important to confirm that the air tracer adequately reproduces the dispersion characteristics of automotive exhaust gases. One of the major objectives of the experiments was to estimate the plume width. Therefore, before calculating the plume width, an examination was conducted to determine whether the trends of concentration decreasing in the downwind direction of the major automotive exhaust gases, NO_x and the tracer SF_6, were consistent. The concentration of NO_x and SF_6 had a similar decreasing tendency in the downwind direction. The results of the analysis of the SF_6 concentration attenuation tendency for each road structure can be seen in Okamoto et al. (1999a). This result showed that, regardless of the atmospheric stability, for the perpendicular wind condition, the average concentration tendency among the four structures was not so much different. However, the initial stage from the viaduct was skewed, due to the distance between the source height (8 m) and the receptor height (1.5 m). The decreasing trends of NO_x and SF_6 concentrations were almost the same up to about 50 m from the road. However, at the distance ≥50 m, NO_x concentrations were slightly higher than that of SF_6. The reason may be the finite length of the simulated line source. These data thus also suggested that this simulated line source (200 m) can be used to describe an infinite line source within the region up to 50 m downwind.

17.2.3.2 Comparison of Plume Spreads Calculated by NO_x and SF_6 Concentrations

For a straight roadway on a flat surface, the tracer gas from the line source of the pipe laid in the centre divider can reproduce the dispersion of exhaust gas emitted from vehicles traveling on that road, as clarified by previous IPCAJ experiments in Section 17.1. However, it was still unclear whether the dispersion could be reproduced on different road structures such as cut or elevated sections and on roads with noise barrier walls. The experimental results therefore provided valuable data to fill in the gaps.

Since there were no significant differences in the overall characteristics of the downwind concentration profiles of SF_6 and NO_X, the vertical concentration profile data for both were then used to calculate their respective plume spreads. The degree of agreement was then evaluated by comparing them. Measured concentrations at the vertical sampling array were fitted to the normal distribution, and the optimal value for vertical spread σ_z was estimated by the least squares fitting. For the analysis, the effective data comprised 96 cases for flat highway structures, 55 for embankments, 50 for cuts, and 18 for viaducts. Almost vertical concentration distributions obtained at the roadside can be approximated by the normal distribution. For the horizontal distribution of concentration, similarly, by fitting the measured concentration to the normal distribution, σ_y was estimated.

The distribution of NO_x concentration around the highway represents the distribution of automobile exhaust gas. Therefore, if the distribution of the air tracer concentration is the same, it can be considered that the behaviour of the air tracer reproduces the dispersion condition of the automobile exhaust gas. This is not a sufficient condition, but it is considered a necessary condition. Now let us compare the plume spread calculated from the NO_x concentration and the plume spread calculated from the air tracer concentration. The results are shown in Figure 17.13. Note that for the experiments of roads with noise barriers, the conditions of perpendicular winds appeared less frequently; hence only a few data points (with required quality) were generated. During the experiments for the sections with noise barriers, the weather conditions were unfavourable. A certain amount of data was obtained only for the Ohbu site (flat, barrier height 3 m). Refer to Table 17.1 for the detailed number of valid data. Nevertheless, to some extent, the dispersion plume spread calculated from both SF_6 and NO_x concentrations did not show so much difference, and it can be interpreted that SF_6 seems to reproduce the dispersion characteristics of an automobile exhaust gas. Within the road of the cut section, it was expected that the frequency of parallel winds inside the road would increase regardless of the wind direction of the upper layers. For this reason, the length of the tracer releasing line source was taken to 2.5 times of other sites, namely 500 m. Still, as seen in Figure 17.13, this result suggested that this length may not have been sufficient. The fact that the plume spread calculated from the tracer concentration did not match the plume spread calculated from the NO_x concentration is likely evidence that the tracer gas flow did not sufficiently reproduce the flow of automobile emissions.

17.2.3.3 Initial Plume Spreads

One of the important parameters in air quality models for automobile emissions is the initial plume spread just off the road due to atmospheric turbulence induced by moving vehicles. The plume spreads were calculated based on the air tracer concentrations. Plume spreads for different road structures and atmospheric conditions are shown in Table A17.1 in the appendix of this chapter. In this study, the plume spread at the leeward side of the road or the road shoulder (for the viaduct) is assumed to be the initial spread, which is respectively called the initial vertical spread σ_{z0} and initial horizontal spread σ_{y0}.

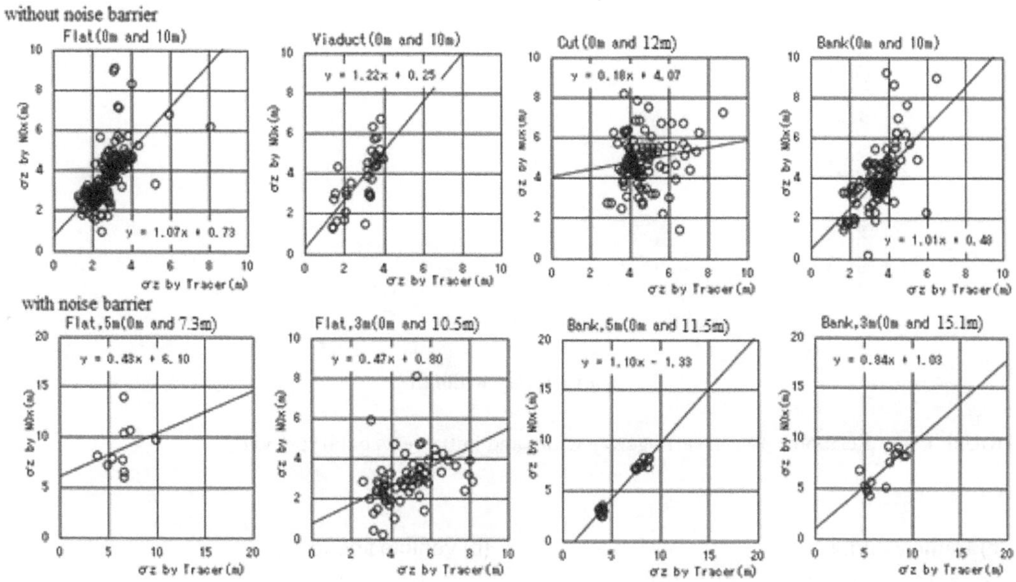

FIGURE 17.13 Comparison of the vertical plume spread (dispersion coefficient of the plume) σ_Z calculated from the SF_6 and NO_x concentrations measured at the roadside. (*Note:* For the sections with noise barriers, data from the first sampling pole [road edge] and the second sampling pole [see Table 17.1b for exact distance from the road] were used.)

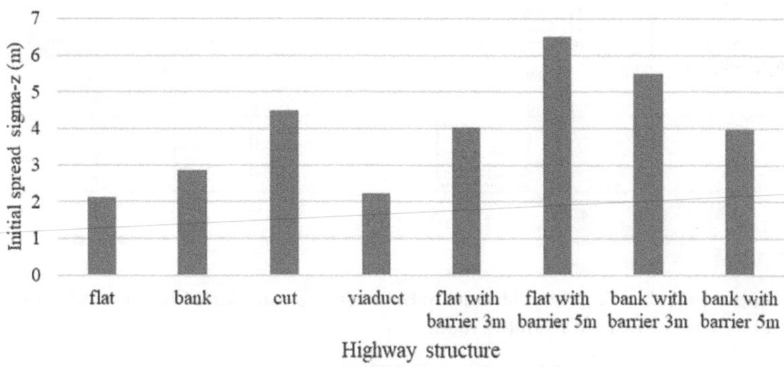

FIGURE 17.14 Initial spread σ_{z0} of highways with different structures.

Figure 17.14 shows the comparison of the initial spread σ_{z0} between a structure with a noise barrier and one without a noise barrier. It was found that the initial spread σ_{z0} for a highway structure with a noise barrier was larger compared with a highway without a noise barrier. The noise barrier enhances the turbulence near the highway and also widens the initial spread. The effect of atmospheric stability on the initial spread seems to be small. Within a downwind distance up to 20 m from a highway, the atmospheric stability can be negligible in estimating the plume spreads. This tendency can be confirmed by the experimental results shown in Table A17.1 of the appendix.

The relationship between the horizontal plume spread at the edge of the road and the carriageway width is shown in Figure 17.15. Since roads with different road widths have different road structures, the possibility that both factors are intertwined cannot be dismissed. However, it can be suggested that wider roads tend to have greater initial spread.

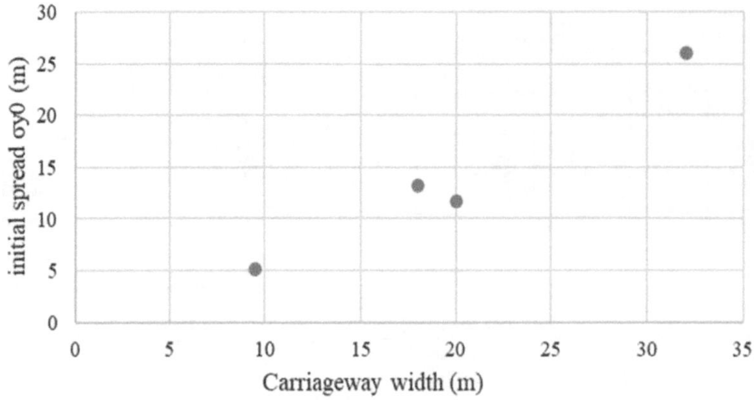

FIGURE 17.15 Relation between carriageway width and initial horizontal spread $\sigma_{y0.}$

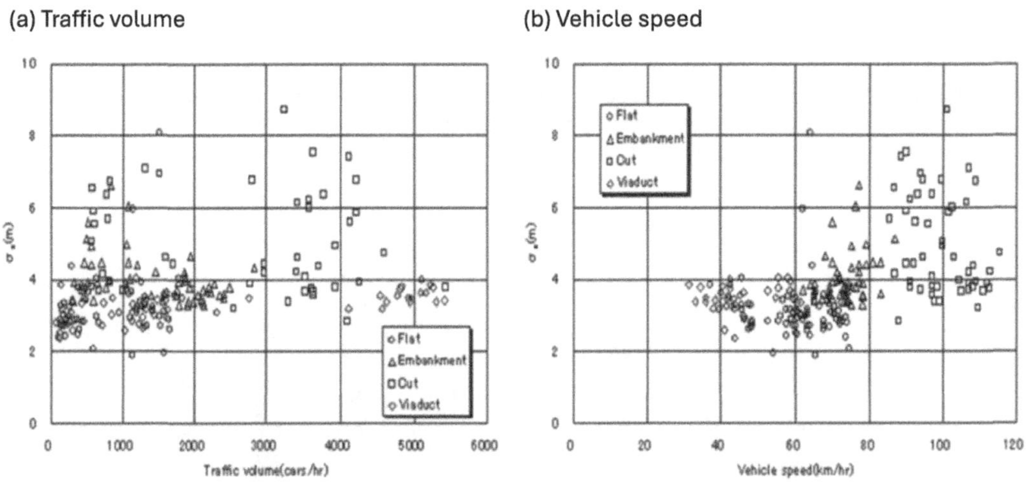

FIGURE 17.16 Scatter plot showing the relationship between the function of traffic conditions and the vertical plume spread σ_z at a distance of 10 m from the road.

Source: Adapted and modified from Okamoto et al. (1999a).

17.2.3.4 Effects of Traffic Conditions on Plume Spread near Roads

The atmospheric turbulence induced by traveling vehicles promotes the dispersion of automobile emissions. Let us consider the extent to which this is the case. Scatter diagrams between the vertical spreads σ_z and the traffic conditions are shown in Figure 17.16. Only a very weak but positive correlation between the vertical spreads σ_z and vehicle speed can be seen. The traffic volume appeared to contribute very little to the vertical spreads σ_z. Thus the effect of the traffic condition is not as important for a plume spread as the road structure.

The atmospheric turbulence along a highway is known to correlate well with the running speed of the vehicles passing by. This was already confirmed by the experimental result in the Ujiie area of the Tohoku Expressway shown in Figure 17.8. However, when the aggregated data for different types of highway structures were analysed together in this case study, the effect of the traffic conditions may be too small compared with that of highway structures and hence was difficult to detect. In an analysis of variance, the effect of the factor under investigation sometimes cannot be detected when the unexpected blocking factor is too large. In other words, the power of testing has decreased because of large blocking factors. It is reasonable to interpret that a similar situation may occur frequently in statistical analyses of various experimental results.

17.3 Summary

This chapter considered the use of data from the air tracer dispersion experiments to understand the dispersion of automotive emissions around roads and to estimate parameters such as those required by air quality models for roadways. The first half of the chapter discussed methodological issues such as air tracer releasing methods, sampling methods, and sampling grid designs. The second half of the chapter described air tracer dispersion experiments conducted at eight locations in Japan, including on highways of different structures.

The air tracer dispersion experiments presented in the latter half of this chapter were conducted to determine the parameters (e.g., plume spreads) in the air quality modelling for roadways. As the quality of the data obtained was the most important concern, several ancillary studies were conducted to verify that the flux of the tracer through the vertical cross-section was consistent with the tracer releasing intensity. This comprehensive method of quality assurance for tracer experimental data will be useful for conducting similar experiments in the future or other locations. Furthermore, this database was used for the development and validation of the line source air quality model in the technical guidelines used for the environmental impact assessment of highways. This technical guideline has been revised several times, and the latest version can be found in NILIM (2013).

REFERENCES

Bullin, J.A., Green, N.J. and Polasek, J.C., 1980, Determination of vehicle emission rates from roadways by mass balance techniques, *Environmental Science & Technology*, 14(6), 700–705.

Cadle, S.H., Chock, D.P., Monson, P.R. and Heuss, J.M., 1977, General motors sulfate dispersion experiment: Experimental procedures and results, *Journal of Air Pollution Control Association*, 27(1), 33–38, https://doi.org/10.1080/00022470.1977.10470389

Industrial Pollution Control Association of Japan (IPCAJ), 1981, *Comprehensive study on problems related to ambient air monitoring stations, vol. 5: Report on analytical results of field survey data, NPMS-5* (p. 319) (In Japanese), Japan Environmental Management Association for Industry, Tokyo, Japan.

Irwin, J.S., 2017, *Atmospheric transport and diffusion data archive, modelling tools in HARMO initiative on harmonisation within atmospheric dispersion modelling for regulatory purposes*, retrieved on 5 October 2023 from http://www.harmo.org/jsirwin/

Ishii, K., Yoshino, N. and Okamoto, S., 1995, Study on atmospheric diffusion in a street-canyon by air tracer dispersion experiment (I), *Journal of Japan Society for Atmospheric Environment*, 30(1), 38–52 (In Japanese), https://doi.org/10.11298/taiki1995.30.38

Momoi, K., Nakano, K., Takahashi, K., Omoto, A., Minakawa, Y., Shiraki, H. and Okamoto, S., 1995 May 25–26, Improving efficiency and accuracy in atmospheric dispersion experiments, *Proceedings of Sapporo quality control conference "creativity and TQC" organized by JUSE*, 171–176, Sapporo, Japan (In Japanese).

National Institute for Land and Infrastructure Management (NILIM), 2013, *Environment impact assessment technique for road project (edition of FY 2012), technical note of NILIM no. 714, landscape and ecology division, road environment division*, National Institute for Land and Infrastructure Management (In Japanese), retrieved on 05 October 2023 from http://www.nilim.go.jp/lab/bcg/siryou/tnn/tnn0714pdf/ks0714.pdf

Okamoto, S., Kobayashi, K., Ide, Y., Ueyama, S. and Nagaseki, H., 1985a, Automobile exhaust gas dispersion experiment on an expressway (II), *J Sangyo Kogai (J Industrial Pollution Control)*, 21(3), 39–51 (In Japanese).

Okamoto, S., Kobayashi, K., Ono, N., Kitabayashi, K. and Katatani, N., 1990, Comparative study on estimation methods for NOx emissions from a roadway, *Atmospheric Environment*, 24A(6), 1535–1544, https://doi.org/10.1016/0960-1686(90)90062-R

Okamoto, S., Kobayashi, K., Ueyama, S. and Nagaseki, H., 1985b, Automobile exhaust gas dispersion experiment on an expressway (I), *J Sangyo Kogai (J Industrial Pollution Control)*, 21(2), 55–61 (In Japanese).

Okamoto, S., Ohnishi, H., Yamada, T. and Itohiya, T., 1999a May 31–June 2, A database for highway air quality modeling, *8th international symposium on transport and air pollution*, 541–548, Graz, Austria, and vol. 76: Report of the institute for internal combustion engines and thermodynamics.

Okamoto, S., Ohnishi, H., Yamada, T., Mikami, S. and Itohiya, T., 1999b October 10–14, A model for simulating atmospheric dispersion in a low-wind condition, *6th international conference on harmonisation within atmospheric dispersion modelling for regulatory purposes*, Rouen, France, also available *International J Environment and Pollution* (2001), 16(1–6), 69–79.

Okamoto, S., Ohta, Y. and Okano, H., 1998 September 14–18, An experimental study on initial spread caused by traffic-induced turbulence, *Papers of the 11th world clean air and environment congress*, vol. 5, paper 14D-5, Durban, South Africa.

Olesen, H.R., 2001 May 28–31, A platform for model evaluation, *7th international conference on Harmonisation within atmospheric dispersion modelling for regulatory purposes*, Belgirate, Italy.

Rappolt, T.J. and Kerrin, S.L., 2010, *Measurement of atmospheric dispersion using gaseous tracers, chapter 25 in the book air quality modeling – theories, methodologies, computational techniques, and available databases and software, vol. IV: Advances and updates*, EnviroComp Institute and Air and Waste Management Association, retrieved on 5 October 2023 from https://www.apsi.tech/material/monitoring/RappoltKerrin2010.pdf

Appendix

At this moment, only the reference from Okamoto et al. (1999a) is available to the public regarding this air tracer experiment; so some important data are provided here in the appendix.

TABLE A17.1A

Estimated σ_y Values from the Air Tracer Experimental Data (m)

Structure	Carriageway width	Atmospheric Stability	Site (distance from the road)			Number of data
			0m*	10m**	30m***	
Flat (Ebetu)	9.5 m	B~C-D	5.5	8.6	12.3	5
		D	5.5	9.3	11.4	10
		E~G	4.6	7.4	9.4	8
		Avg.	5.15	8.27	10.86	23
Embankment (Yamaguchi)	20 m	A-B~C-D	10.0	17.7	23.9	10
		D	12.3	19.9	24.0	10
		E~G	13.4	21.2	23.7	6
		Avg.	11.67	19.38	23.88	26
Cut (Tsukuba)	32 m	A-B~B	28.6	31.8	36.3	5
		D	25.6	28.7	27.9	3
		E~G	24.5	25.9	27.6	7
		Avg.	26.08	29.13	32.33	15
Viaduct (Okayama)	18 m	B-C~C	10.9	—	—	4
		F~G	14.3	—	—	10
		Avg.	13.21	—	—	14

* Road edge. **15 m for cut section (Tsukuba site). ***40 m for cut section (Tsukuba site)

TABLE A17.1B

Estimated σ_z Values from the Air Tracer Experimental Data (m)

Structure	Noise barrier	Atmospheric stability	Site (distance from the road)			Number of data
			0 m*	10 m**	20 m	
Flat (Ebetu)	None	A-B~C-D	3.1	4.8	4.7	5
		D	2.1	3.1	3.7	62
		E~F	2.1	3.1	3.6	29
		Avg.	2.12	3.18	3.67	96
Embankment (Yamaguchi)	None	A-B~C-D	2.7	3.8	4.4	11
		D	2.5	3.9	4.4	15
		E~G	3.0	4.1	4.8	29
		Avg.	2.85	3.96	4.61	55
Cut (Tsukuba)	None	A-B~C-D	4.7	5.5	6.5	16
		D	4.2	4.3	5.6	14
		E~G	4.5	5.0	6.1	20
		Avg.	4.48	4.96	6.13	50
Viaduct (Okayama)	None	A-B~C	2.1	3.5	4.8	8
		D	2.1	3.6	4.9	3
		F~G	2.4	3.7	5.0	7
		Avg.	2.22	3.56	4.90	18

Structure	Noise barrier	Atmospheric Stability	1st sampling pole (0 m)	2nd sampling pole***	3rd sampling pole***	Number of data
Flat (Ohbu)	3 m high	A~C-D	4.5	6.7	10.4	15
		D	3.9	5.6	8.0	7
		E~G	3.5	5.1	6.8	13
		Avg.	4.02	5.87	8.58	35
Flat (Mizunami)	5 m high	B	6.5	6.3	9.4	4
		G	6.5	6.5	7.0	1
		Avg.	6.51	6.36	8.90	5
Embankment (Ichinoseki)	3 m high	A~A-B	5.3	9.3	12.4	2
		D	5.6	8.1	10.5	5
		Avg.	5.50	8.42	11.05	7
Embankment (Chino)	5 m high	A-B~C	4.0	8.2	—	8
		D	3.9	7.9	—	2
		Avg.	3.96	8.05	—	10

* Road edge, **As for the cut section (Tsukuba site), x = 12m. *** See the detail distance in Table 17.1b.
Source: Adapted and modified from Okamoto et al. (1999a).

18

Characteristics of Concentrations Calculated by Air Quality Models

This chapter provides an overview of air quality models and discusses the characteristics of the air pollutant concentrations that are the output generated by the air quality model. The key issue is to examine the consistency of the characteristics of the output concentration values with the intended uses of the model results. The air quality models considered here are mainly used for administrative purposes in air quality management, including comparison with environmental standards and regulations for emission sources. These models are usually called regulatory air quality models.

18.1 Air Quality Model

Air pollutants released from the emission source into the atmosphere are transported by wind and diffused by atmospheric turbulence. The atmospheric turbulence, created by the constantly changing wind speed and direction, scatters and mixes the pollutants with the surrounding air. Then it may be transformed into other pollutants by the chemical reactions and physical processes in the atmosphere. In addition, pollutants are deposited back to the Earth's surface through the dry and wet deposition processes. For example, large particles can be settled out from the air by gravity while water-soluble substances can be removed by rainout and washout processes. The air quality model is used to simulate this behaviour of pollutants in the atmosphere, as schematically shown in Figure 18.1.

Developing an air quality model requires knowledge of mathematics, physics and chemistry. Numerous models have been developed by researchers and are used not only to elucidate the dispersion phenomenon in the atmosphere but also for administrative purposes, such as air quality management. Before using a model for a specific purpose, it is important to ensure that the model is suitable for the intended application. Furthermore, for administrative purposes such as comparing with environmental standards or ensuring compliance with emission limits, it is necessary to know which statistics (e.g., maximum, average or appropriate

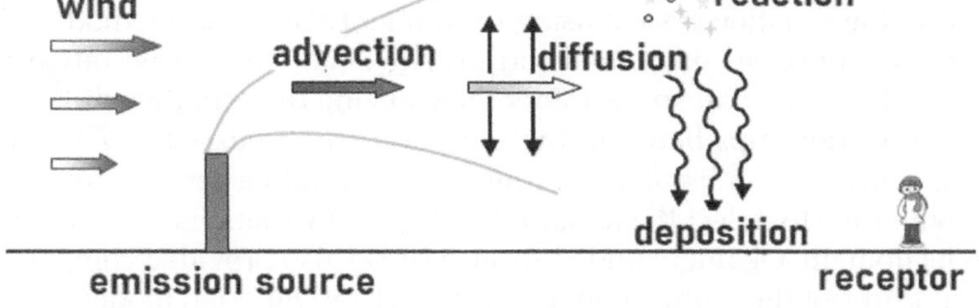

FIGURE 18.1 Behaviour of the pollutants released from a point source in the atmosphere.

Source: Adapted and modified from Okamoto (2004).

DOI: 10.1201/9781003597216-20

percentile values) are required for specific averaging time concentrations. Despite the fundamental statistical nature of this issue, it is frequently overlooked in environmental statistics textbooks. Similarly, atmospheric science textbooks discussing atmospheric dispersion may not sufficiently cover the statistical aspects of concentration data. To fill in the gaps, this chapter focuses on these often neglected issues between atmospheric science and statistics.

Technically, the air quality model is a tool used in air quality management to establish the quantitative causal relationship between the emissions and the resulting air pollutant concentrations and deposition and further to its consequent effects on other environmental factors such as health impact. Particularly, air quality models can be used to analyse past and future emission scenarios to assess the efficacy of various abatement strategies for improving the air quality in a domain.

These air quality models can be broadly divided into two types, the Eulerian and Lagrangian models, depending on how the coordinate system is set up. Each type, the Eulerian or Lagrangian model has different advantages and disadvantages in treating atmospheric processes. Often, the advantages of one model type are the disadvantages of the other. For example, an advantage of the Eulerian models is their capability to handle sophisticated 3D physical treatment and non-linear chemistry, while their high demand for computation time and cost is a disadvantage. The opposite is true for Lagrangian models, for which the extension to 3D treatment is not straightforward and there is difficulty in treating non-linear chemistry, but these models require less computation time and cost (Kim Oanh and Permadi, 2009). It is therefore important to understand the advantages and disadvantages of each model type to select the right model(s) for a particular application. With the development of computing facilities and progress in numerical methods, the Eulerian models now become increasingly popular. Nevertheless, applications of Lagrangian models are also successful in many situations.

Air quality models can be used for historical simulations for investigation of past air pollution episodes and for assessment of associated health, environmental and climate effects. The models can be used to simulate the present (near real-time) and future air quality forecasting using the corresponding emissions and meteorological input data. The simulation results also provide comprehensive spatial coverage for air pollution data that are especially useful for the places where the monitoring data is insufficient, as already discussed in Chapter 15.

In the mathematical model formulation, the phenomena of advection and diffusion of substances in the atmosphere can be described by differential equations. Solving the equation analytically under simple conditions yields the so-called Gaussian-type diffusion equation. A model that uses the Gaussian equation is called a Gaussian model. The Gaussian plume equation is derived when the steady-state ambient condition is applied, and the Gaussian puff equation is derived when the instantaneous steady-state condition is applied. The movement of the puffs can also be expressed in Lagrangian.

As mentioned previously, with the progress of computers and analytical methods, the numerical simulation process has been enhanced and has become widespread in the field of atmospheric science. These numerical models are of the Eulerian type, also called grid models, that solve the equation system using the numerical analysis method such as the finite difference method, finite element method and so on to obtain the concentration at a grid point or averaged value for every grid space, rather than solving the primitive differential equation analytically. The relationships between these models are discussed in Zannetti (1990). The puff modelling approach can break a continuous plume into a series of independent puffs, which evolve with time to reflect the temporal and spatial variations of meteorological conditions in a model domain. Okamoto and Shiozawa (1987) also provided comparisons between the calculation results of the plume model, puff model and Eulerian model.

It is noted that the names such as "puff model", "Eulerian model", etc. are the common terms for the broad categories of the dispersion models. Thus they refer to the entirety of mathematical model categories that use specific basic equations. On the other hand, models like the Industrial Source Complex (ISC) model and the AERMOD model are proper nouns, developed by specific individuals or organisations, namely in this case the US EPA. Software for these US EPA models is provided, and the programs are publicly accessible. In such a model, not only the basic equations for calculating concentrations but also important calculation conditions such as dispersion parameter settings are specified, and arbitrary modifications are strictly prohibited.

In this chapter, the term "air quality model" is considered to be roughly equivalent to the dispersion model. Receptor models can also be considered part of the air quality model but are not discussed in this chapter. Instead, the receptor models are discussed in Chapter 23, which are the statistical analyses of the monitoring data to quantify the contributing sources to air pollution levels monitored at individual sites. Further, predictive models of air pollutant concentrations utilising statistical methods such as regression equations are not included in this chapter but are discussed in Chapter 22.

18.2 Statistical Characteristics of the Concentration Values as the Output of the Air Quality Model

As discussed in Chapters 14 and 15, air pollutant concentrations widely change both temporally and spatially. Therefore, considering what kind of averaging time the concentration should be calculated, and what kind of geographical change should be expressed are the most important concerns in model development or selection. A more important question is at what position on the concentration distribution is the information of concern: for example, the mean, median (50% value), 98% value, second highest value, maximum value etc. Furthermore, it is essential to operate the model whether or not meteorological and emission data corresponding to the purpose of use can be obtained. For example, in a pollution source where the operating conditions fluctuate greatly, it is impossible to predict the concentration every hour when only the emission intensity under the rated operating conditions can be grasped. No matter how sophisticated the model is, if the proper input data cannot be obtained, the prediction results will not be of the expected quality. Also, when using the air quality model administratively, it is necessary to know what characteristics the output of the model has.

To carry out environmental implementation planning or to make a strategy for environmental management, tools for predicting environmental quality are necessary, and the quality of the tools also become an important concern. Air quality models are used for many kinds of environmental decision-making. For many countries, the use of the air quality model has become an unavoidable requirement for determining acceptable emission levels to ensure that the associated ambient concentrations do not exceed the atmospheric environmental standards. The air quality model that is used in environmental planning, for example in environmental impact assessment and permitting the emission, is called a regulatory model. This chapter describes the fundamental components of the air quality models for regulatory purposes. The averaging time of concentrations predicted by air quality models should be consistent with the environmental standard or any regulatory goals. Examples of the national ambient air quality standards (NAAQS) for SO_2 in the United States and Japan are shown in Table 18.1. Because the NAAQS are not the same for different countries, there should be some difference in the air quality models specified for the regulatory purpose of the countries.

Many earlier urban air quality models were developed to predict annual average concentrations. Examples of these categories are the air quality display model (AQDM) (U.S. EPA, 2021, viewed) and the climatological dispersion model (CDM) (Busse and Zimmerman, 1973) of the United States. Even today, almost all of the regulatory models used in Japan belong to this type, i.e., long-term models for predicting annual average concentrations. The concept of these models differs from US models such as CRSTER (U.S. EPA, 1977) and AERMOD (U.S. EPA, 2004), which can directly calculate the maximum concentrations required by the environmental standards shown in Table 18.1.

TABLE 18.1

SO_2 National Ambient Air Quality Standards (NAAQS) in the United States and Japan

Pollutant	Type	Level	Averaging time	Form
US sulphur dioxide (SO_2)	Primary	75 ppb	1 h	99th percentile of 1-h daily maximum concentrations averaged over 3 years
	Secondary	0.5 ppm	3 h	Not to be exceeded more than once per year
Japan sulphur dioxide (SO_2)		0.04 ppm	24 h	Concentration not to be exceeded more than 2% per year (98th percentile value)
		0.10 ppm	1 h	Concentration not to be exceeded

Source: U.S. EPA (2023) and Japan Ministry of the Environment (2023).

TABLE 18.2

Criteria of the Qualified Air Quality Models Used for the Total Pollution Load Control (Implementation Planning Program in Japan)

(1) Regression coefficient b (obs. = $a + b \times$ cal.)	$0.8 \leq b \leq 1.2$ (desirable)
(2) Correlation coefficient r	$r \geq 0.71$ (necessary)
	$r \geq 0.8$ (desirable)
(3) Spatial distribution	The concentration field obtained by the air quality model should be consistent with the observed one.

Note: Statistics are calculated by observed (obs) and predicted (cal) annual average concentrations.

Next, it should be emphasised that the differences in the way of thinking about the air quality model from country to country are not only in the meaning of the output concentration value but also in the concept of quality assurance of the model. In Japan, air quality conservation plans were carried out to regulate sulphur oxides pollution in the 1970s, which is called the total pollution load control (TPLC). Although it has not received as much attention as before due to the improvement in air quality, this system is still effective. An air quality model, called the TPLC model is used in this program, and it is the conceptual model (a manual that describes only texts and basic formulas rather than software).

The total pollution load control (TPLC) in Japan is nearly equivalent to the State Implementation Planning (SIP) in the United States (Japan Environmental Agency, 1982). Its role is to attain the air quality standard by implementing preventive measures. To carry out the TPLC, the relationship between the ambient concentration and emission should be explained by sound and scientific means. The model itself is similar to US EPA CDM, but the model operations differ. The TPLC model system conducts simulations using the historical data of selected past reference years before the prediction to evaluate the model performance. In this evaluation, the calculated and observed concentrations of the reference year must be consistent with one another with reasonable accuracy. The evaluation score and its verification criteria can be found in the relevant reference (Japan Environmental Agency, 1982), which are shown in Table 18.2. A model is considered to be qualified for use if it satisfies all the criteria.

If the model performance is not satisfactory, not only the emission inventory and/or meteorological data must be reviewed, but the dispersion model and dispersion parameters must also be re-examined. Air quality models that cannot conform to the criteria are prohibited to use for TPLC regulations. Strict specifications, as in the "Guideline on Air Quality Models" by the US EPA, are therefore not set for the air quality models used for the TPLC in Japan. The TPLC model evaluation system may be considered to be supported by an extensive air quality monitoring network consisting of approximately 2000 stations situated in nearly every urban and industrial region across Japan.

This chapter discusses the characteristics of the calculated concentration values produced by the air quality model. In developing an air quality model for regulatory use, care should be taken to ensure that the output is consistent with its regulatory purpose. The input conditions should also be consistent with the purpose of its application. However, due to many constraints, it is often not possible to align the input data as expected; even under such circumstances, exploring the reasons for discrepancies by comparing the concentrations produced by air quality models with the observed data may be feasible.

In general, it is difficult to predict short-term concentration values at a specific location and time by using the air quality model. This is because in the prediction for example of a one-hour value at a particular location, a stochastic variation enters a calculation error. Another serious problem is the inadequate time resolution of the emissions input data, which are normally available as the annual average or rated condition of pollutant emissions amount rather than accurate hourly values. However, in the prediction of the long-term average concentration, the error with the positive bias and that with the negative bias offset each other, and this compensation increases the prediction accuracy. However, the reproducibility at a particular position apart from the average (arithmetic mean) value deteriorates. Therefore, even when utilising an air quality model that is designed to predict average dispersion conditions effectively, the prediction of hourly concentrations over a certain period (such as one year) and the creation of their frequency distribution histogram results in poor reproducibility in areas that deviate from the centre of the distribution, as illustrated in Figure 18.2.

18.3 Case Study for the Kashima Area

The case study aims to discuss the characteristics of the air quality model outputs obtained using the emission inventory and meteorological conditions for Kashima. While the features of the air quality model used are briefly introduced, the focus is on the statistical status of its outputs rather than on the basic formulation or parameter settings for the model.

18.3.1 Air Quality Model Used in the Case Study

This section discusses the characteristics of the calculated concentrations that are the output of the air quality model. The model used to make the calculated concentration dataset in the case study is one of the most commonly used air quality models in Japan, the Gaussian plume equation with the Pasquill–Gifford plume spread estimation algorithms. The focus is not on the prediction of the highest hourly concentration but on the prediction of the annual average concentration.

The purpose of this text is to explain not the atmospheric dispersion of pollutants or air quality modelling but the statistical properties of observed and calculated concentration values.

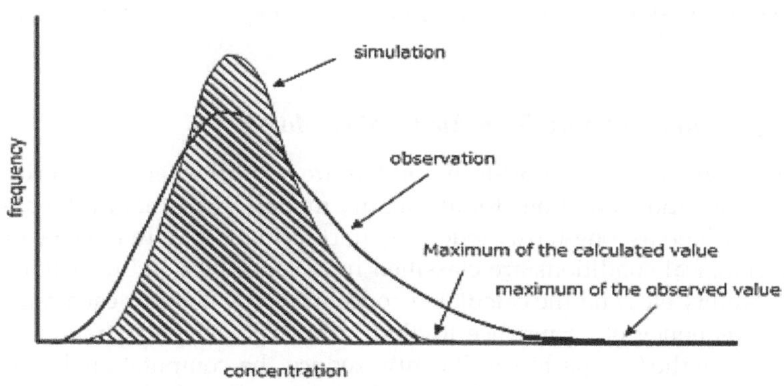

FIGURE 18.2 Relationship of the frequency distributions for the observed and calculated concentrations.

TABLE 18.3

Atmospheric Stability Settings Corresponding to the Pasquill–Gifford Dispersion Coefficient Diagram Used in the Kashima Study

	Daytime			Night				
Original class	**A~B**	**BC, C**	**CD, D**	**D**	**E**	**F**	**Initial spread σ_{z0}**	**Initial spread σ_{y0}**
Modified for factory	C	C	CD	D	DE	DE	5 ~ 30m*	5 m
Modified for automobile	B	BC	D	DE	E	EF	Depending on surrounding building height	5 m
Same but cold season	C	CD	D	DE	E	EF		
Modified for ship	BC	C	CD	CD	DE	DE	3 m	25 m

*$\sigma_{z0} = 0.3\Delta H$ (neutral or unstable condition), $\sigma_{z0} = 0.25\Delta H$ (stable condition), $5m \le \sigma_{z0} \le 30m$ (or 25 m), where ΔH is the height of the plume rise.

The model used in the Kashima study is one of the typical models in Japan. It is a Gaussian model but with the following modifications. In windy conditions, the basic Gaussian plume equation was used. In calm conditions, the integral Gaussian puff equation in which wind speed is zero was used. To estimate the plume spreads, originally the Pasquill stability class was used; however, the correspondence to the Pasquill–Gifford dispersion coefficient (plume spread) diagram was modified as shown in Table 18.3. For other details, see IPCAJ (1985, 1986) and Tada et al. (1989).

18.3.2 Issues Related to Categorical Data as Input Conditions

The input data for the air quality model can be considered essentially as numerical values. However, the atmospheric stability of Pasquill is categorical data. Furthermore, wind direction can also be measured as a numerical value, but, by aggregating to 16 directions, it can be considered to be converted to categorical data. The problem with data measured on such a coarse-grained measurement scale relates to the issue of whether the data have sufficient resolution to serve as the input data required by the air quality model. Let us consider the challenges of inputting such categorical data into an air quality model.

The distribution of output from many air quality models tends to be narrower than the distribution of observed values and tends to cluster around a mean value with less variability, as shown in Figure 18.2. As for the regulatory air quality models, there may be two possible ways to deal with this problem. One is to compensate for the deficiencies of the dispersion model by using a statistical model, and this approach is adopted in Japan. The other is a method adopted in the United States and other similar countries, in which only the positions on the probability distribution need to be matched by dispersion calculation. In other words, the idea is that only the position on the probability distribution needs to be predicted, regardless of the one-to-one correspondence of what concentration appears under what meteorological conditions.

18.3.2.1 Long-Term and Short-Term Air Quality Models

The concepts of long-term and short-term models are helpful when considering the issue of categorical data as input conditions for air quality models. Note that in the past, air quality models were classified as long-term models or short-term models. In the so-called long-term model, meteorological conditions are classified in advance by wind direction, wind speed, atmospheric stability etc., and the calculated concentration values for each category are multiplied by the frequency of occurrence to obtain the average concentration for the period. This calculation method helps to significantly reduce the computational time required to calculate the long-term average concentration (annual mean value). On the other hand, the short-term model calculates concentrations by inputting hourly wind speed, wind direction, atmospheric stability and other meteorological conditions. However, the concept of the long-term models is a residue from an era when computer performance was the key constraint. If atmospheric stability could be measured with a more elaborate boundary layer meteorological parameter and wind direction could be measured in angles (degrees) rather than 16 directions, it would be possible to input hourly meteorological data into an air quality model with a much smaller loss of information. Therefore, the significance of adopting a long-term model in which meteorological conditions are classified beforehand is considered to have been lost. However, grasping this concept may be useful to comprehend the impact of utilising coarse resolution data, such as 16 wind direction categories or a limited number of atmospheric stability classes.

When calculating the annual average concentration with an air quality model, it is also possible to input hourly weather and emission source conditions for 8760 hours per year. However, the following methods can be considered: first, the hourly ambient concentration is calculated by using hourly input conditions, both emissions and meteorology, and then obtain annual average values based on the calculated hourly values. However, if meteorological observatories provide only the wind direction observations in 16 directions (22.5° intervals), the resolution is too coarse as compared to the ordinary plume spread (about 10°) to calculate the hourly concentration accurately. In a sense, the wind direction data in 16 directions can be regarded as categorical data. When such category data are used as an input condition, it is necessary to consider variations within the category. This issue is illustrated using the example shown in Table 18.4.

The values presented in Table 18.4 can be considered the intermediate, which can be used to calculate the annual average concentration. In other words, these are the calculated values of concentrations by condition category subdivided by wind direction, wind speed, atmospheric stability, day and night. In addition, the average of the measured concentrations for each condition category is also shown. Some of the condition categories have more occurrences, while others have less. There are also condition categories that do not appear at all, which are indicated by asterisks in this table.

The value calculated by an ordinary dispersion model is an average concentration under typical meteorological and emission conditions. In reality, the resulting concentrations may vary greatly even under the same meteorological conditions and the same emission conditions. Thus if the regulatory target is set so that the average value of the given averaging time should not exceed the designated concentration level, the maximum concentration with the stochastic fluctuation must meet the target. Thus, from a probability point of view, the fact that the concentration calculated by an air quality model (dispersion model) is exactly consistent with the target value means that the probability of achieving this target is only about 50%.

18.3.2.2 Parameter Setting in the Model for Estimating Annual Average and Hourly Maximum Concentrations

It is not possible to model every situation, including infrequent weather conditions. For example, there is a large variation in the plume rise estimates. The plume rise equation used to predict the annual average concentration would be most appropriate if it was well-matched near the centre of this distribution. However, to predict the maximum one-hour value in a year, a plume rise equation corresponding to the lower limit of the variation of the plume rising height is required. The same can be said for the dispersion coefficient (plume spread), the height of the mixed layer and any other parameters in the model equation. The following two approaches are required when estimating the annual average and hourly maximum concentrations, respectively.

1. ***For estimating annual averages:*** As shown in Table 18.4, when meteorological conditions are classified according to wind direction, wind speed and atmospheric stability,

TABLE 18.4

Comparison of the Observed and Calculated Average SO_2 Concentrations for Each Meteorological Condition at the No. 108 Monitoring Station in the Kashima Area (unit: ppb)

		Station No. 108																	
Wind		(A,A-B)		(B,B-C)		(C,C-D)		(D day)		(D night)		(E)		(F)		(G)		average	
speed	Stability	Obs.	Cal.	Obs.	Cal.	Obs.	Cal.	Obs.	Cal.	Obs.	Cal.	Obs.	Cal.	Obs.	Cal.	Obs.	Cal.	Obs.	Cal.
0.0 – 0.3	CALM	************		5.8	0.6	************		7.7	2.0	6.3	1.0	************		************		7.2	1.2	6.9	1.2
0.4 – 0.9	N	5.0	0.2	4.2	0.2	************		7.0	0.4	5.7	0.1	************		************		5.1	0.7	5.6	0.4
	NNE	12.0	1.0	6.7	0.2	************		6.1	0.7	6.0	1.4	************		************		6.1	0.5	6.4	0.7
	NE	7.0	0.9	8.5	1.1	************		4.4	0.9	6.2	1.0	************		************		4.7	1.4	5.5	1.1
	ENE	6.0	1.8	************		************		8.0	5.3	7.8	4.1	************		************		5.8	4.3	7.2	4.4
	E	6.0	6.8	6.0	2.5	************		9.5	2.1	5.9	5.2	************		************		7.0	2.9	6.6	4.3
	ESE	************		************		************		6.3	6.6	7.1	7.0	************		************		9.4	9.5	7.6	7.6
	SE	8.0	5.0	************		************		7.5	6.6	9.3	6.4	************		************		9.7	9.2	8.6	7.1
	SSE	************		6.7	6.8	************		6.7	3.5	9.3	4.3	************		************		9.7	5.2	8.7	5.0
	S	************		7.7	2.7	************		7.1	3.4	8.9	4.6	************		************		7.8	3.5	8.0	3.8
	SSW	5.7	2.8	8.4	3.7	************		7.7	2.5	9.1	2.3	************		************		10.8	2.9	9.0	2.7
	SW	6.0	3.6	6.2	2.6	************		5.5	3.2	7.2	1.6	************		************		6.6	2.0	6.4	2.3
	WSW	************		6.0	5.2	************		7.6	5.5	5.8	2.5	************		************		6.5	4.9	6.7	4.8
	W	10.0	3.6	5.6	4.2	************		8.5	8.0	5.1	4.2	************		************		5.8	6.7	6.1	5.7
	WNW	************		6.4	2.1	************		7.1	3.6	6.6	2.0	************		************		6.2	2.7	6.6	2.7
	NW	6.5	1.0	4.7	0.5	************		5.8	0.8	5.5	0.5	************		************		6.1	0.4	5.8	0.6
	NNW	6.0	0.2	4.2	0.1	************		4.1	0.4	4.4	0.2	************		************		5.1	0.5	4.6	0.3
1.0 – 1.9	N	7.2	0.1	5.5	0.1	************		4.9	0.1	5.0	0.1	************		************		6.5	0.1	5.8	0.1
	NNE	10.0	0.2	4.8	0.1	************		5.3	0.3	5.9	0.2	************		************		6.2	0.2	5.8	0.2
	NE	6.1	0.9	5.8	0.3	************		6.2	0.6	6.8	0.3	************		************		5.9	0.6	6.3	0.5
	ENE	8.6	2.3	8.5	0.9	************		7.6	1.8	7.1	1.1	************		************		5.6	2.4	7.1	1.6
	E	8.9	4.5	5.7	7.6	************		9.7	5.0	7.6	3.9	************		************		7.4	5.2	8.1	4.9
	ESE	7.4	7.4	13.3	14.3	************		12.9	6.7	7.5	7.5	************		************		10.9	11.9	10.5	9.6
	SE	4.0	4.1	6.5	12.9	************		9.7	12.1	8.8	7.5	************		************		10.0	8.4	9.4	9.3

	SSE	14.0	13.1	11.7	9.0	************		7.6	4.0	10.4	2.9	************		************		9.7	4.8	9.8	4.8
	S	8.2	7.1	6.9	4.2	************		6.7	3.4	9.4	1.0	************		************		8.6	1.4	8.4	2.0
	SSW	10.6	3.6	6.4	3.1	************		6.9	1.3	6.8	0.8	************		************		7.4	0.9	7.4	1.4
	SW	6.4	4.5	6.2	2.0	************		9.4	1.7	4.7	1.0	************		************		6.1	0.8	6.8	1.6
	WSW	8.2	2.9	6.9	1.1	************		9.4	3.2	6.0	2.0	************		************		5.6	2.7	6.8	2.6
	W	6.6	2.5	5.4	2.3	************		5.9	3.0	4.2	2.8	************		************		4.9	3.1	5.3	2.9
	WNW	5.5	1.7	4.6	1.0	************		5.8	2.1	4.0	1.0	************		************		5.1	1.2	5.1	1.5
	NW	5.7	1.3	4.6	0.4	************		5.6	0.3	4.2	0.2	************		************		6.0	0.3	5.4	0.4
	NNW	3.5	0.1	5.1	0.1	************		5.2	0.2	4.9	0.1	************		************		5.4	0.2	5.2	0.1
2.0 – 2.9	N	5.0	0.0	4.0	0.1	3.9	0.1	5.5	0.1	4.1	0.0	5.9	0.1	6.2	0.1	************		5.3	0.1
	NNE	8.0	0.1	6.0	0.1	5.4	0.1	5.7	0.1	6.3	0.0	5.6	0.1	5.8	0.1	************		5.9	0.1
	NE	10.0	1.4	6.9	0.3	6.4	0.3	6.6	0.3	6.4	0.2	5.7	0.2	6.4	0.2	************		6.6	0.3
	ENE	8.2	1.9	6.4	0.7	7.3	0.6	6.7	1.0	6.1	0.5	3.7	0.6	4.6	1.4	************		6.2	0.9
	E	12.1	4.9	7.7	6.3	6.7	5.3	9.0	5.0	6.2	3.1	6.6	3.2	6.0	3.4	************		7.7	4.5
	ENE	16.8	7.1	11.9	8.5	9.7	12.4	12.3	9.7	6.0	5.5	6.0	9.3	9.7	8.0	************		10.4	8.4
	SE	17.0	22.1	23.0	8.6	23.2	17.2	9.3	12.1	5.8	6.0	13.2	10.8	11.5	6.8	************		11.7	9.9
	SSE	17.5	7.9	6.0	23.6	11.7	10.5	10.1	6.9	9.6	2.3	10.0	8.2	10.7	2.6	************		10.3	4.9
	S	10.7	1.3	5.5	0.7	8.0	1.0	9.5	1.0	7.1	0.7	6.9	0.8	7.9	1.1	************		8.1	0.9
	SSW	10.2	1.7	9.8	0.7	6.7	0.6	6.8	0.6	5.6	0.5	5.1	0.7	5.2	0.4	************		6.6	0.7
	SW	11.2	0.7	10.0	1.5	5.2	0.6	7.7	1.0	6.3	0.7	5.5	0.6	5.6	0.8	************		7.1	0.9
	WSW	12.5	0.8	6.3	1.0	5.7	1.3	7.8	1.6	3.0	0.8	3.0	0.9	4.7	1.7	************		5.8	1.5
	W	12.0	0.8	6.6	1.3	5.6	1.7	6.1	2.7	5.3	0.9	4.0	1.2	4.8	1.8	************		5.8	1.8
	WNW	10.3	0.7	4.3	0.8	5.7	1.1	5.5	1.4	4.6	1.2	3.0	0.6	4.5	0.7	************		5.0	0.9
	NW	8.0	0.2	4.5	0.2	4.7	0.2	4.8	0.2	3.7	0.1	************		3.7	0.2	************		4.4	0.2
	NNW	5.0	0.1	4.2	0.1	4.9	0.6	4.5	0.2	3.9	0.0	5.3	0.1	4.6	0.2	************		4.5	0.2

Source: IPCAJ (1985).

Here, only a portion of the calculation results is shown, omitting those for wind speeds of 3 m/s or more.

the parameters in the model must be set so that the average value of each category matches well for the conditions characterising the category. These parameters include the plume spreads (more exactly, the class of atmospheric stability), effective stack height, mixing layer height, and so on.

2. ***For estimating the hourly maximum concentration:*** When the conditions are subdivided according to wind direction, wind speed, atmospheric stability etc., the parameters should be set so that they match well only to the highest concentration for each category. These parameters include the atmospheric stability class, effective stack height and mixing layer height. Furthermore, it may become necessary to consider using the alternative equation of atmospheric dispersion to account for these needs. For example, when considering the height of plume rise, it is necessary to set the lowest plume rise height that occurs about once a year (or with a specified probability) rather than the plume rise height under average conditions.

Thus, due to the discarding by "modelling", it can be assumed that the optimal model differs depending on the change in the prediction target and that all elements in the model, including the dispersion parameters, need to be changed accordingly. For example, if the purpose is to find the maximum value of one-hour concentration in a year (8760 hours), it is necessary to use the parameter corresponding to the highest concentration within a given meteorological category, i.e., not to use the dispersion parameter corresponding to the average concentration for the category. Since data shown in Table 18.4 is a part of the calculated values for each condition with the air quality model that was designed for calculating the annual average value, and the average calculated values under each meteorological condition are in good agreement with the actual measured values. This model is considered to be highly satisfactory in predicting annual average concentration. However, it cannot satisfactorily predict the highest one-hour value during a year.

Because the horizontal plume spread of the Pasquill chart corresponds to the concentration of about several minutes of the averaging time, in the calculation of the one-hour concentration, time correction for the σ_y is done in almost all Japanese regulatory air quality models. This increase in σ_y with time is due to fluctuations of the long-cycle (large-eddy) wind turbulence. In estimating the average concentration, the result can be improved by considering this meandering effect. However, this fluctuation is a stochastic phenomenon and does not always occur. Therefore, there is a high risk of underestimation when predicting the highest hourly concentration in a year. Therefore, as a safety side setting, it may be better in such cases to use directly the σ_y obtained from the Pasquill–Gifford diagram, which is said to correspond to an averaging time of three–ten minutes.

18.3.3 Issues Regarding the Prediction Target, Average vs. Maximum Values

The US regulatory model seeks to more precisely replicate the highest and second-highest concentrations that are expected to comply with US environmental standards. In Japan, the environmental standard for SO_2 and NO_2 is set at 98% of the daily average value, and various environmental administrative decisions are made by calculating the annual average using a recommended regulatory air quality model and statistically analysing the relationship between the annual average and the 98% daily average value corresponding to the environmental standard. Consequently, the air quality model in Japan is a tool for predicting the annual average concentration, not for directly predicting the highest value. The Japanese models, designed to estimate annual average concentrations, demonstrate superior predictive performance for these averages, as illustrated in Figure 18.3. However, their accuracy in predicting the second-highest hourly values is less reliable, as depicted in Figure 18.4.

A single model that focuses on predicting either the highest concentration or the average annual concentration alone would not be able to reproduce the full range of concentration frequency distribution shown in Figure 18.2. In this sense, the model evaluation study by Liu

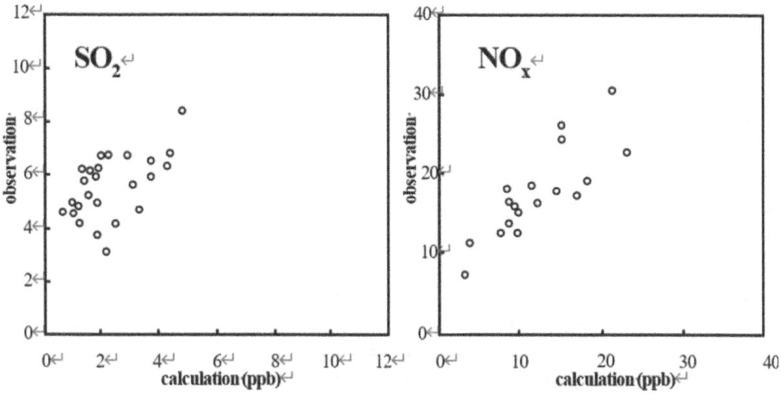

FIGURE 18.3 Scatter diagram for the annual average SO_2 and NO_x concentrations during one year (April 1980–March 1981, Kashima area).

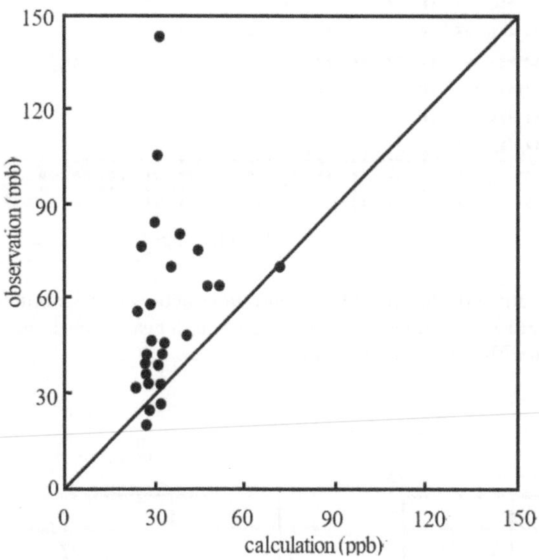

FIGURE 18.4 Scatter diagram for the second highest values of one-hour SO_2 concentrations during one year (April 1980–March 1981, Kashima area).

and Moore (1984) seems to be asking for too much. As far as the annual average predictive performance is concerned, these US models may be inferior to the Japanese system (Okamoto, 1995). On the other hand, it is also clear that a model dedicated to predicting annual average concentrations is incapable of predicting the maximum one-hour value that will appear in a year.

As just mentioned, many of the Japanese models are designed for the prediction of annual average values and not for the highest values. Looking at the correspondence between calculated and observed values for hourly concentrations, as shown in Figure 18.5, models that aim for better reproducibility with average concentrations seem to produce better results.

Further, a comparison is made between the frequency distribution of hourly observed and calculated concentrations at the No. 108 monitoring station in the Kashima area over one year. The results are shown in Figure 18.6. Recall the cumulative concentration distribution

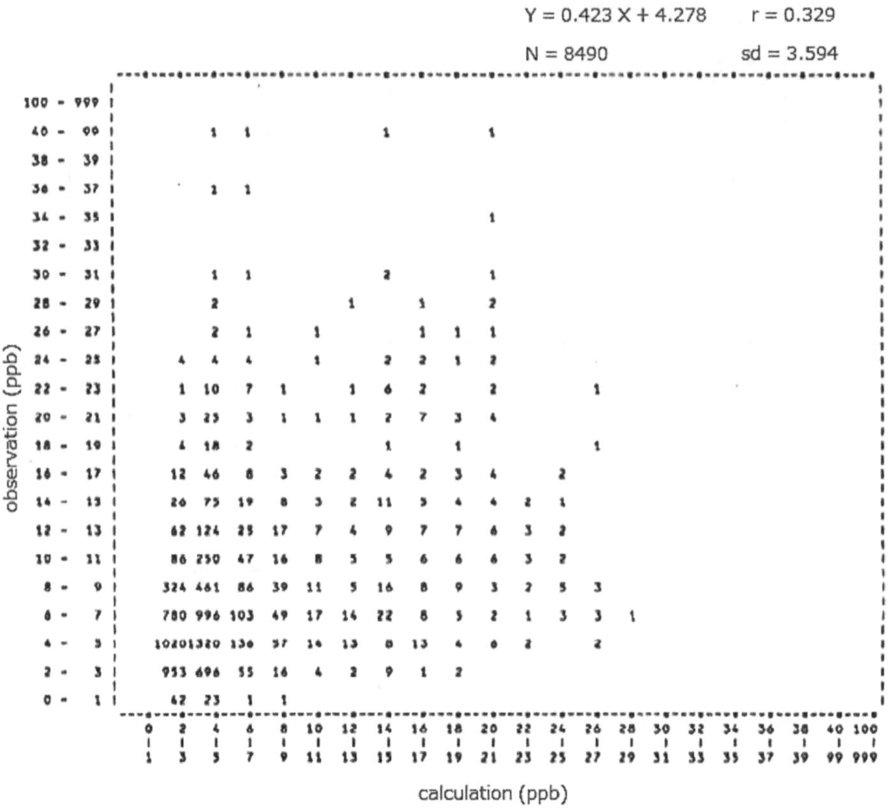

FIGURE 18.5 Scatter diagram of the hourly SO_2 concentrations at the No.108 monitoring station in the Kashima area. The numbers in the figure are frequencies of occurrence under each condition. The background concentration of 3.2 ppb was added for all of the calculated data.

Source: IPCAJ (1985).

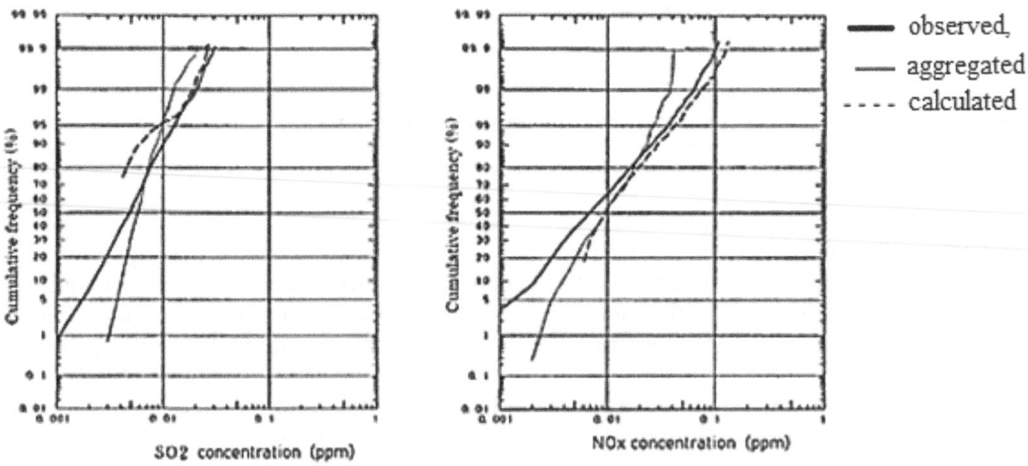

FIGURE 18.6 Observed, aggregated and calculated concentrations in the percentile-and-concentration graph.

Source: IPCAJ (1985).

curve used to determine whether the concentration distribution can be approximated by a log-normal distribution.

Based on the calculated concentration values for each condition category, one can draw a cumulative concentration distribution curve. Here, it is assumed that variation between condition categories is "variation between groups" and variation within the same category is "variation within the group". With this in mind, it can be said that the air quality model can only reproduce between-group variability, and the model cannot reproduce within-group variability. Next, let us replace every hourly concentration data during one year with the average value of the observed data corresponding to each condition category shown in Table 18.4. In other words, the replaced concentration is obtained by removing the within-group fluctuations from the observed data. If the cumulative concentration distribution curve (percentile-and-concentration graph) is drawn with this data, it can be seen that the maximum value becomes smaller, and the whole approaches the centre of the distribution. Let us call this result the aggregated value. In a sense, this is the limit that can be reproduced by the air quality model. This case of cumulative appearance frequency is shown in Figure 18.6. From these analytical results, it is possible to see the cause of the difference between the calculated and observed frequency distribution of concentrations shown in Figure 18.2.

18.4 Air Quality Model for Tall Stacks

In an earlier study on atmospheric dispersion, most were on near-surface emission sources. Pasquill (1961) proposed a method for quantitatively calculating concentrations by analysing the behaviour of smoke emitted from ground-level sources; this method was modified by Gifford (1961). Further, Turner (1967) improved on this scheme, which subsequently became widely used. It was known that this method had various problems to apply to dispersion from tall stacks, but until the advent of AERMOD, there was no suitable alternative. The biggest issue here is whether the plume spreads generated from near-surface experimental data can be used to predict smoke dispersion from tall stacks.

18.4.1 Early Studies in the United States

This section will explore the challenges and breakthroughs of an air quality model applicable to emissions from tall stacks. The Tennessee Valley Authority (TVA) was the first to undertake an extensive study on the plume from tall stacks, and the Large Power Plant Effluent Study (LAPPES) conducted by the US National Air Pollution Control Administration (NAPCA, later shifted to the US EPA) had continued since then (Schiermeier and Niemeyer, 1970). The data of the LAPPES project was obtained by measuring the SO_2 concentration while flying a helicopter in the plume from a power plant stack. The Keystone power plant is a coal-fired power plant with a 244-m stack. The SO_2 concentration data were measured to determine the cross-sectional shape of the plume at downwind distances of 4, 10 and 16 km. The concentration in the plume was measured by several round-trip flights at the same altitude, and the average value was calculated.

Using this data, the relationship between the plume spread (or, say, dispersion coefficient) of high-rise sources and the Pasquill stability was analysed. Here, the vertical plume spread, namely standard deviation σ_z was calculated as the square root of the second moment around the centre of gravity using the horizontally integrated concentration CIC (crosswind integrated concentration) at the same altitude. Figure 18.7 shows the result of comparing this plume spread with the value estimated from the Pasquill–Gifford (PG) diagram. In this figure, the plume spread obtained from the PG diagram is indicated by a dotted line. According to the results, at the downwind distance of 4 km, the estimated value and the measured value agreed relatively well for the neutral stability case (Pasquill stability D class), but the estimated

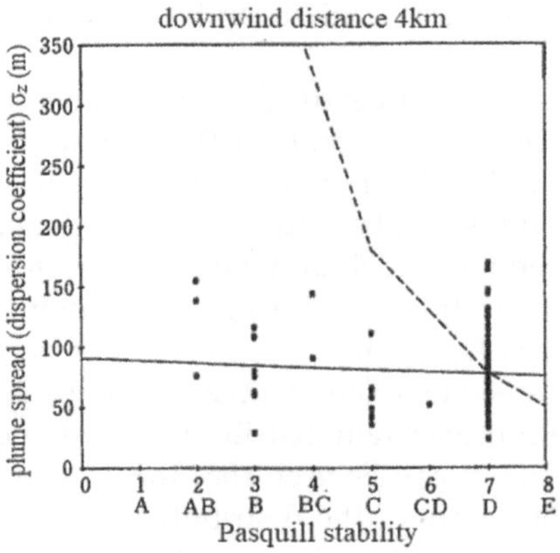

FIGURE 18.7 Relationship between the Pasquill stability and vertical plume spread (dispersion coefficient) σ_z obtained by LAPPES data. (The regression line, when regression analysis is performed in which, for convenience, the Pasquill stability is quantified, is shown by a solid line. In addition, the estimated value from the Pasquill–Gifford diagram is shown by the dotted line.)

Source: Adapted and modified from Okamoto and Shiozawa (1990).

spread by the PG algorithms showed a tendency of overestimation for the unstable case (A to C classes). This discrepancy was even more pronounced at a downwind distance of 10 km.

Since then, several research cases dealing with the dispersion of smoke from tall stacks have been reported, but no clear cases have been found that would lead to the meta-analysis and development of new prediction methods. However, such reported cases suggest that:

- Even if the stability class on the ground level is A to C, if the atmosphere of the upper layer is not convective, the plume spread σ_Z hardly increases within a few kilometers downwind distance from the stack.
- The plume spread σ_Z in the vicinity of a tall stack is almost determined by the fluctuation of the altitude of the plume rise.

This tendency can be confirmed, for example, by the observations of the STATE78 project by McElroy et al. (1981). At the beginning of the early 1970s, the LAPPES observations and the TVA observations strongly suggested these problems. Nevertheless, it seemed that, as a result of only using the Pasquill-Gifford algorithm to determine the consistency of the maximum concentration on the ground, the plume spread σ_Z in the stability A class could not be discarded. This is because the discussion focused only on the consistency of the model calculation values at the maximum (or second-highest) value in the concentration distribution rather than on the phenomenological consideration of what kind of dispersion conditions emerge under what kind of meteorological conditions. This may be thought of as an environmental policy issue of what concentration to estimate and what decision to make rather than an atmospheric environmental science issue.

18.4.2 Analysis of Air Tracer Dispersion Experimental Data in Japan

Similar atmospheric studies have been conducted in Japan since the late 1960s, including air tracer dispersion experiments for high-rise stacks. Some of these studies and the related data quality were described in several reports, such as Sakuraba (1970). Using these datasets, a study was conducted to evaluate the Gaussian-type air quality model using the Pasquill–Gifford plume spread estimation algorithm (Okamoto and Shiozawa, 1990). The dataset used

TABLE 18.5

Summary of Dispersion Experiments Used in the Analysis

Site	Release method	Source height (m)	Releasing duration (min)	Number of runs*
1	Helicopter	150	30	6
2	Tower	313	30	5
3	Stack stage	159	30	8
4	Stack injection	145–280**	30	8
5	Helicopter	300–500	60	5
6	Stack injection	190–298**	60	8
7	Helicopter	200–350	60	6
8	Tower	140	60	8
9	Stack injection	200–300**	60	1

* This is the number of valid cases, excluding cases where tracers could not be successfully collected due to sudden changes in meteorological conditions and so on.
** Effective stack height,

here was drawn from nine experimental programs conducted in seven districts, as shown in Table 18.5.

In the study, 55 cases (runs) of data were used except for seven cases that were inappropriate for the analysis. In the air tracer dispersion experiment using a tower and a helicopter, the tracer releasing height was used as the plume centreline height. In addition, in the experiment of tracer injection into a stack, the data of the plume centreline height measured by photographing or laser radar soundings were used. The Pasquill stability classes were determined using the surface wind and solar radiation observations, and the plume spread corresponding to the stability class was obtained from the Pasquill–Gifford diagram. Concentration calculations were done for all sampling points. In some cases, the main purpose of the experiment was to grasp the horizontal plume spread; hence the distance between the sampling points in the downwind direction was set large. However, in most of the cases, the sampling points were densely arranged in an arc shape having the same downwind distance. Here, for each of the 55 experimental cases, the tracer concentrations at all sampling points (which vary from experiment to experiment, but generally tens to 100) were calculated. In each case, the highest measured value among all the measured points and the highest concentration calculated value among all the calculated points (not necessarily where the highest concentration was observed) were extracted and compared. That is, the comparison was made for concentrations measured and calculated at the same time, but there was no correspondence between measured and calculated locations. This is the result of the comparison regardless of location within the domain. For each case, the concentration normalised by the tracer releasing intensity was used and classified by the atmospheric stability at the time of the experiment. The ratios of calculated to observed values were tabulated for each atmospheric stability class, as shown in Table 18.6.

The results indicated that the Gaussian plume model with the Pasquill–Gifford plume spread estimate algorithm used in this study had a tendency of overestimation that became more pronounced as the atmospheric stability became more unstable.

18.4.3 Comparison of Japanese Data and US Analysis Results

A comparison between the results of a Japanese and US studies on the evaluation of air quality models for emissions from tall stacks was conducted. In this comparison, instead of the study presented Section 18.4.1, results of the newer Plume Model Validation and Development (PMV&D) project were used. The PMV&D project is one of the largest-scale atmospheric

TABLE 18.6

Ratio of Calculated and Observed Maximum Ground Level Normalised Concentrations (C/Q) for Different Atmospheric Stability Classes

Atmospheric stability	Number of data	$C_{cal.}/C_{obs.}$ (mean ± SD)
A, AB	10	11.01 ± 9.59
B, BC	16	2.99 ± 4.77
C, CD	16	1.61 ± 1.45
D	13	1.41 ± 1.07

Note: These data were unpaired in location but paired in time. The calculation did not take into account the time correction of the plume spread σ_y due to the meandering.

dispersion experiment in this field. There were 1500 sampling points in the area of 50 kilometers around the thermal power plants, and about 200 of them were used to collect air samples according to the meteorological conditions. There were more than 300 hours of tracer (SF_6) data from Kincaid and Bull Run power plants, respectively (Moore et al., 1988). In conclusion, Bowne et al. (1983) stated that the model performance results demonstrated important shortcomings in all the models tested, including a general inability to estimate hour-by-hour ground-level concentrations.

Data in Table 18.6 showed that, for unstable conditions (that is, when Pasquill stability was A or AB observed), the calculated concentration values were about ten times higher than the observed. On the other hand, the results of the PMV&D project suggested that the Gaussian plume air quality model cannot predict the hourly concentrations at a given location but may be not necessarily too bad for the maximum concentration that could appear sometime somewhere. These results suggested that it may not be possible to make predictions under highly unstable conditions at designated times and locations by modelling physical properties such as atmospheric turbulence. However, it turns out that there is some usefulness in ignoring its physical properties and using Pasquill stability A to predict only the maximum concentration value for convenience. To predict the highest concentration at the designated location and time by the air quality model, many difficulties arise. For example, if the AERMOD is selected, the following questions would need to be answered; if the plume bifurcates (updraft plume and down-draft plume) in a convective mixing layer, how high is the surface level concentration compared to without it? Also, is there a case where bifurcation does not occur even in the convection mixed layer, or how likely is it to occur? For predictions that do not specify a location and time, such discussion may be quite unnecessary. However, this means that there is no difference from the method that used the conventional algorithms with Pasquill stability A. This is not an attempt to accurately model the dispersion process in the atmosphere but simply to match only the concentration values, even using elaborate model equations.

18.4.4 Needs for Further Considerations

There is a problem with how to express the plume spreads. Even today, the Pasquill method, proposed about 60 years ago, is widely used. Finally, the development of AERMOD can be regarded as a revolutionary alternative to atmospheric modelling. The plume spread can be expressed to some extent as a theoretical function of the turbulence of the wind (standard deviation of the wind speed component). However, it is difficult to estimate the hourly three-dimensional turbulence component up to the effective stack height from the limited measured values of available meteorological elements. Furthermore, if a complex atmospheric boundary layer model is adopted, the number of parameters to be estimated increases, and there is a concern that the accuracy of estimating the final plume spread may decrease. From a statistical point of view, the accuracy of the estimation of the final objective variable may be reduced because several steps of auxiliary variables to be estimated are inserted before the final objective variable is obtained. Consequently, what does it mean to use the turbulence component metric data instead of the Pasquill stability categorical data? If it is not possible to sufficiently explain at what meteorological condition the designated highest concentration appears, it seems to be thought that simple categorical data may be easier to understand, even if it does not fully explain atmospheric turbulence in the upper layer.

One of the weakest points of the recent model may be that the turbulence intensity is not clearly defined. The turbulence of the wind is changing both temporally and spatially. Even if a huge amount of data is accumulated in the world, it is not possible to make a unified comparison or meta-analysis for many of them because, for some of them, the averaging time and temporal resolution are unknown. In particular, when developing the latest air quality model, it has barely been discussed whether the averaging time of the concentration, which is the base of the dispersion coefficient (plume spread), and the averaging time of the wind velocity component, which is the base of the turbulence intensity, are consistent or not. There is a significant risk in incorporating such data as model input parameters.

The AERMOD evaluation process was the confirmation of consistency with the top several concentration data on the higher concentration side, similar to the validation study of the former regulatory models. The AERMOD incorporates a plume width prediction submodule based on the atmospheric boundary layer theory, which is different from the conventional stability index. However, it is not clear to what extent it explicitly contributed to the improvement of prediction accuracy. This is because it is an evaluation with a high concentration value that occurred somewhere someday. In other words, they do not understand the meteorological conditions when high concentrations actually occur and ensure that the dispersion phenomenon under those conditions is correctly modelled. In particular, it appears that no special consideration has been given to the parameter settings in the meteorological submodule to handle high concentration part (e.g., the top 25), nor is there a clear analysis result showing that the configured meteorological submodule can handle these conditions.

18.5 Summary

This chapter discussed the characteristics of the air quality concentration data produced by air quality models. Air quality models for regulatory purposes must meet the requirements for the position on the probability distribution (percentile values) of concentrations as well as the spatiotemporal resolution required by the administration.

In this chapter, the characteristics of air pollutant concentrations calculated by using an air quality model for the Kashima region, which had high-quality emissions inventories and meteorological data, were analysed. The chapter content focused on the characteristics of the concentration values and also the problems of the air quality model. The result strongly suggested that it is necessary to consider the settings of parameters in the air quality model separately when determining long-term average concentrations, such as annual average values, and when determining extreme values, such as hourly maximum concentrations, or percentile values on the high concentration side.

Several issues related to air quality models for predicting the plume dispersion from tall stacks were considered, suggesting that it is still not fully understood when, where and under what atmospheric conditions high concentrations occur, making it difficult to model atmospheric dispersion for location- and time-specific predictions, and further discussion on this point may be necessary. It may also be necessary to consider whether it is physically feasible to predict high concentrations that occur with extremely low frequency (the highest one-hour average concentration in a year) and whether there is a need to do so.

REFERENCES

Bowne, N.E., Londergan, R.J., Murray, D.R. and Borenstein, H.S., 1983, *Overview, results, and conclusions for the EPRI plume-model validation and development project: Plains site. Final report, EPRI EA-3074,* Electric Power Research Institute, retrieved on 5 October 2023 from https://www.osti.gov/biblio/6092793#biblio-full

Busse, A.D. and Zimmerman, J.R., 1973, *User's guide for the climatological dispersion model, environmental monitoring series EPA-R4-73-024 (NTIS PB-227346AS),* U.S. Environmental Protection Agency.

Gifford, F.A., 1961, Use of routine meteorological observations for estimating atmospheric dispersion, *Nuclear Safety*, 2(4), 47–57.

Industrial Pollution Control Association of Japan (IPCAJ), 1985, *The study on the behaviour of particulate matter, vol. 8: Report on dispersion simulation* (p. 350) (In Japanese), Japan Environmental Management Association for Industry, Tokyo, Japan.

Industrial Pollution Control Association of Japan (IPCAJ), 1986, *The study on the behaviour of particulate matter, vol. 9: Final report* (p. 163) (In Japanese), Japan Environmental Management Association for Industry, Tokyo, Japan.

Japan Environmental Agency, 1982, *A manual for nitrogen oxides total pollution load control* (In Japanese), Pollution Research and Control Centre, Tokyo, ISBN 978-4874880142.

Japan Ministry of the Environment, 2023, *Environmental standards for air pollution* (In Japanese), retrieved on 5 October 2023 from https://www.env.go.jp/kijun/taiki.html

Kim Oanh, N.T. and. Permadi, D.A., 2009, *Photochemical smog modeling for ozone air quality management, chapter 12 in the book modelling of pollutants in complex environmental systems,* vol. 1, ILM Publications, Hertfordshire, UK, ISBN 9781906799007

Liu, M.K. and Moore, G.E., 1984, On the evaluation of predictions from a Gaussian plume model, *Journal of Air Pollution Control Association,* 34, 1044–1050.

McElroy, J.L., Eckert, J.A. and Hager, C.J., 1981, Airborne downlooking lidar measurements during STATE78, *Atmospheric Environment*, 15, 2223–2230.

Moore, G.E., Milich, L.B. and Liu, M.K., 1988, Plume behaviors observed using Lidar and SF_6 tracer at a flat and hilly site, *Atmospheric Environment*, 22(8) 1673–1688.

Okamoto, S., 1995, A consideration on air quality models for environmental assessment, *Bulletin of Tokyo University of Information Sciences Journal of Management Information Science*, 8(2), 99–109, retrieved on 5 October 2023 from http://weather.ou.edu/~spark/AMON/v3_n3/FORUM/Okamoto/

Okamoto, S., 2004 December 11, *Lecture note on prediction methods of air quality, 2nd lecture for newly planned road, Shin-Minato Yokoto-cho line,* Supported by KCS Corp, Chiba City (In Japanese).

Okamoto, S. and Shiozawa, K., 1987, A trajectory plume model for simulating air pollution transients, *Atmospheric Environment*, 20, 2145–2152.

Okamoto, S. and Shiozawa, K., 1990, Verification of a Gaussian plume model with Pasquill-Gifford dispersion coefficient algorithm for an elevated point source, *Journal of Japan Society of Air Pollution*, 25(4), 256–262 (In Japanese).

Pasquill, F., 1961, The estimation of the dispersion of windborne material, *Meteorological Magazine*, 90(1063), 33–49.

Sakuraba, S., 1970/1969 October 27–30, Sensitivities of air quality prediction to input errors and uncertainties, *Proceedings of Symposium on Multiple-Source Urban Diffusion Models*, 200–224, Research Triangle Park, NC, APCO Publication no. AP-86.

Schiermeier, F.A. and Niemeyer, L.E., 1970, *Large power plant effluent study (LAPPES), vol. 1: Instrumentation, procedures, and data tabulation (1968),* U.S. Department of Health, Education, and Welfare, Public Health Service, Environmental Health Service, National Air Pollution Control Administration, NAPCA Publication no. APTD 70-2.

Tada, K., Hazama, H., Kobayashi, K. and Okamoto, S., 1989, Development and evaluation of a diffusion model for particulate matter – application to Kashima area, *Journal of Japan Society of Air Pollution*, 24(1), 64–73 (In Japanese), https://doi.org/10.11298/taiki1978.24.64

Turner, D.B., 1967, *Workbook of atmospheric dispersion estimates,* National Air Pollution Control Administration, NAPCA, PHS Publication no. 999-AP-26.

U.S. EPA, 1977, *User's manual for single source (CRSTER) model, EPA-450/2-77-013 (NTIS PB-271360),* U.S. Environmental Protection Agency.

U.S. EPA, 2004, *AERMOD: Description of model formulation, EPA-454/R-03-004, office of air quality planning and standards,* US. Environmental Protection Agency.

U.S. EPA, 2021, *Systems architects, air quality display model (AQDM),* U.S. Environmental Protection Agency, viewed on 2021, retrieved on 5 October 2023 from EPA Environmental Modeling Catalogue – Systems Architects – Google ブックス

U.S. EPA, 2023, *NAAQS table,* US. Environmental Protection Agency, retrieved on 5 October 2023 from https://www.epa.gov/criteria-air-pollutants/naaqs-table

Zannetti, P., 1990, *Air pollution modeling: Theories, computational methods, and available software,* Springer Science & Business Media, New York, United States, ISBN 978475744675

19

Evaluation of Air Quality Model Performance

The model evaluation is a must to ensure that the model performance is satisfactory for its intended application. This chapter addresses the issues related to the application of statistical analysis to assess the model performance. Ideally, air quality models utilised in environmental administration should be capable of accurately predicting the concentrations specified in the relevant regulations. This could be a percentile value of a certain probability distribution based on the average concentration over a regulatory mean time. However, many circumstances may dictate that a more convenient method be used instead. Even in these cases, a model appropriate for the conditions must be selected. Furthermore, such issues must be taken into account when evaluating the performance of the model.

19.1 Challenges in Model Performance Evaluation

If the prediction result is a numerical value, the prediction error can be evaluated quantitatively. The observed values are first organised in a systematic dataset that will be used to compare with the calculated values obtained by the air quality model. Certain statistics such as the residual variance, correlation coefficient, and so on can be calculated. If the prediction result is a categorical variable, then a discriminant analysis can be used to predict whether there will be a high concentration above an alarm level. In such cases, a contingency table can be used, and the accuracy rate can be calculated.

Additionally, it is also important to consider whether the results of the evaluation are consistent with our intuitive sense, while its evaluation sense may not always be logical. For example, consider a weather forecast for the next ten days. The results of the forecast by meteorological analysis companies A and B are shown in Table 19.1, and the question is which company produced the better prediction.

TABLE 19.1

Comparing the Ten-Day Weather Forecast with the Actual Situation

Day	Actual	Company A's forecast	Company B's forecast
1	Sunny	Sunny	Sunny
2	Sunny	Sunny	Sunny
3	Sunny	Sunny	Sunny
4	Sunny	Sunny	Sunny
5	Sunny	Sunny	Sunny
6	Sunny	Sunny	Sunny
7	Rainy	Sunny	Sunny
8	Sunny	Sunny	Rainy
9	Sunny	Sunny	Sunny
10	Sunny	Sunny	Sunny

DOI: 10.1201/9781003597216-21

Here, the probability of sunny weather in this season is about 90%, so if one simply predicts that it will be sunny without doing anything, then the expected accuracy rate would be about 90%. Thinking that way, Company A forecasted "all sunny". Meanwhile, Company B predicts that a cyclone will affect the area eight days later and that rain will occur on day 8 by a weather forecast model simulation. As a result, Company B's forecast is wrong for two days (days 7 and 8). Therefore, the correct answer rate for Company A is 90%, and that for Company B is 80%. Which do you prefer? Perhaps meteorologists would rate "Company B could predict a low-pressure passage through with one day's error". But do many people support it? Here is the complexity of the evaluation that requires suitable criteria to be set and the evaluation to be done against those criteria. Let us now consider the evaluation of the simulation model with these problems in mind. In particular, this chapter considers an attempt to evaluate an air quality model against monitoring data.

19.2 Fundamentals of Air Quality Model Performance Evaluation

Performance evaluation of the regulatory air quality models was one of the central issues of the US EPA, and several reports were released in the 1970s (Hayes, 1979; Fox, 1981a, 1981b; Willmott, 1982). The challenges remain today: how to measure model performance and what the appropriate evaluation criteria are for that purpose.

A real-world phenomenon such as the behaviour of an air pollutant in the atmosphere is so overwhelmingly complex that a mathematical model cannot practically simulate all the processes involved with high accuracy. Assumptions, omissions and simplifications are made accordingly, which result in the uncertainties of the simulation results. For example, errors may be related to the mathematical representation of the important processes and parameters. Errors and simplifications associated with the input data of emissions and meteorological fields are known to contribute significantly to the uncertainties of the model outputs. Therefore, once a model has been developed, it should undergo validation studies involving comparison of the model output with the field measurements or use the results of tracer experiments designed for the purposes as detailed in Chapters 17 and 18. Further, before a validated model is applied in a specific case study for a selected urban area with specific emission sources, topography and meteorology, it needs to be operationally evaluated. A statistical model evaluation relies on the analysis of the model output against the field monitoring data to ensure that the model satisfactorily performs for a particular application. A flowchart describing important steps of model evaluation and application is presented in Figure 19.1. It is important to note that the model can be used for air quality management only after its performance is satisfactorily evaluated for the intended applications.

The model evaluation process for air quality management is shown in Figure 19.1, and there are various ideas on how to use it in environmental administration. Model evaluation studies in the United States are intended to select recommended models to be approved by the EPA. Therefore, the user is not allowed to arbitrarily modify the contents of the approved model. Japan's total pollution load control requires users to develop a model, evaluate it by themselves and demonstrate that it meets the criteria in Table 18.2 (Chapter 18). Then, if the criteria are not met, as shown in the flow of Figure 19.1 after "NO", the process returns to reviewing the input data etc.

A primary focus of the air quality model evaluation studies in the United States has been on Gaussian-type regulatory models, exemplified by AERMOD, the latest US EPA regulatory model. As with the past evaluation studies for the other regulatory air quality models, the evaluation here does not seek a phenomenological correspondence between observed and predicted concentrations but rather focuses on concentration values of the higher side that may appear at some place and in some time. Here, two characteristic statistics are quoted from U.S. EPA (2021) and shown next.

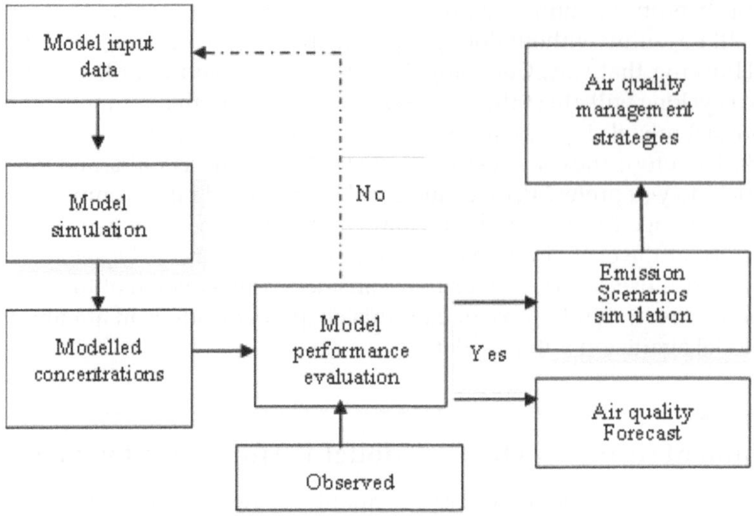

FIGURE 19.1 Framework of the model evaluation and application for air quality management.

19.2.1 Robust Highest Concentrations

Robust highest concentration (*RHC*) is calculated for each averaging period of each database. The *RHC* statistic is calculated as:

$$RHC = X(N) + [\bar{x} - X(N)] \times \ln[(3N - 1)/2]$$

where $X(N)$ is the N-th largest value, $\bar{x}$ is the average of $N - 1$ values, and N is the number of values exceeding the threshold value, usually 26.

19.2.2 Fractional Bias

The AERMOD evaluation involves screening to ascertain a better-performing model. The first step is to perform a screening test based on fractional bias:

$$FB = 2\{(O - P)/(O + P)\}$$

where FB is the fractional bias, O is the average of the highest 25 observed concentrations and P is the average of the highest 25 predicted averages. The fractional bias is also calculated for the standard deviation where O and P refer to the standard deviation of the highest 25 observed and predicted concentrations, respectively. This is done across all monitors and modelled receptors, unpaired in time and space for the 3-hour and 24-hour averaging periods.

The evaluation was conducted using 14 databases that cover a wide range of different dispersion conditions, including flat and complex terrain, with and without downwash (smoke entrainment by building wake), and urban and rural areas (U.S. EPA, 2021).

19.3 Performance Evaluation for 3D Photochemical Smog Models

The grid-type chemistry transport model (CTM) requires special consideration for the spatial resolution of the concentration values produced by the model. The use of this type of air quality model is essential for analysing photochemical air pollution.

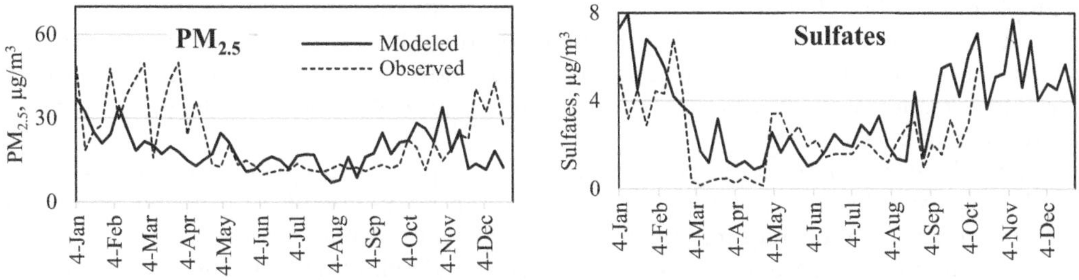

FIGURE 19.2 Time-series for simulated weekly $PM_{2.5}$ mass and sulphate component, averaged for a 2 × 2 km^2 grid, and the measurements of the suburban site located within the grid.

Source: Adapted from Ha Chi and Kim Oanh (2020).

TABLE 19.2

Definition of the US EPA Recommended Statistical Measures for Modelled Ozone Concentrations

Statistical measure	Equation	Suggested criteria
Mean normalised bias error (MNBE)	$\frac{1}{n}\sum_{i=1}^{n}\frac{C_p(x_i,t)-C_o(x_i,t)}{C_o(x_i,t)}$	±15%
Mean normalised gross error (MNGE)	$\frac{1}{n}\sum_{i=1}^{n}\frac{\lvert C_p(x_i,t)-C_o(x_i,t)\rvert}{C_o(x_i,t)}$	35%
Unpaired peak prediction accuracy (UPA)	$\frac{C_p(x,t)_{max}-C_o(x,t)_{max}}{C_o(x,t)_{max}}$	±20%

where n = number of monitoring stations;
$C_o\,(x_i,t)$ = observed values in monitoring station i for hour t;
$C_p\,(x_i,t)$ = predicted values at monitoring station i for hour t;
$C_o\,(x_i,t)_{max}$ = maximum hourly observed concentration over all hours and all monitoring stations;
$C_p\,(x,t)_{max}$ = maximum hourly predicted concentration over all hours and all surface grid squares.
Source: U.S. EPA (1991).

Statistical analyses used to evaluate the agreement between modelling and monitoring results include direct comparisons such as scatter plots, correlation coefficients, and time-series analysis. Figure 19.2 shows an example of the time-series of the simulated, 2 × 2 km^2 grid-averaged, $PM_{2.5}$ and sulphate concentrations in comparison with the observed values in a suburban site located within the grid. The CTM of WRF-CAMx model system (WRF: Weather Research and Forecasting and CAMx: Comprehensive Air Quality Model with Extension) was used in this case study. The time-series analysis helps to see how well the model can reproduce the observed trend of increasing or decreasing concentrations. Therefore, it is always useful to conduct the time-series analysis when evaluating the simulation results. Note that in a scatter plot between the calculated and observed concentrations, small trends of increase or decrease can be masked in larger fluctuations, making them difficult to see.

A wide range of statistical estimates are normally used together with the criteria that each estimate should be satisfied. Examples of the statistical measures used in the evaluation of the photochemical smog models by the US EPA are seen in Table 19.2. Other systematic statistical tools such as pattern recognition or cluster analysis, which focus on the statistical nature of model outputs rather than the individual values, are also used for the model evaluation (Kim Oanh and Permadi, 2009). With the development of monitoring technology of low-cost sensors and satellite remote sensing, more air pollution monitoring data become available for multiple uses, although they have certain shortcomings as mentioned in Chapter 8. The satellite data, for example, can be used to qualitatively track the air pollution plume movement, which is useful in large-scale air quality modelling studies. Densely distributed low-cost sensor networks, normally in urban areas, can be used for evaluation of model performance in the respective urban domains.

The model evaluation should be done for both temporal and spatial distributions of the considered variables, e.g., a concentration field of an air pollutant. In principle, the model performance evaluation requires intensive field measurements to collect appropriate data for the evaluation purpose. It is an intensive resource-demanding task; hence such monitoring data are not normally available, and this presents a huge challenge in conducting model performance evaluation. Besides, it is also important to note that not only do the modelling results have intrinsic uncertainties (data input errors and model formulation errors), but the monitoring data are not error-free (see details in Chapter 8). Thus discrepancies between the monitoring and modelling results are always expected. Selecting the right model, improvement in the input data, and proper model calibration and operation can help to reduce these discrepancies. In case of performance evaluation for grid-type models, it is important to keep in mind that the simulated results are averaged over an entire grid (a few to hundreds of square kilometers in area), whereas the monitoring data only characterise the concentrations at a specified site (or a few sites) within the area of a given grid, as seen in Figure 19.2. Thus the monitoring data are only comparable with the simulated gridded average if the site(s) are properly selected. For example, the monitoring data of CO at a busy roadside are normally much higher than the average CO concentration over the corresponding grid, say of 1×1 km^2, for which the model result is produced.

The evaluation of grid-type models requires special consideration of sub-grid scale phenomena as previously mentioned. This issue applies not only to measured values to be used for model evaluation but also to model-calculated values and is a matter related to the regional representativeness of concentration values, as explained in Chapter 15. For example, when evaluating the photochemical smog model performance, it is important to understand the ozone spatial distribution pattern. Higher ozone concentrations are principally observed at downwind locations of urban areas, while low ozone concentrations are observed in the plume released from large sources of NO such as a power plant with low-rise stacks that are normally located outside a city. Inside an urban area, ozone measured at the busy roadside is also lower than ozone measured in ambient sites due to the titration effect of NO released from on-road vehicles. The analysis shown in Section 21.4.5 of Chapter 21 also provides more insights into this phenomenon. Accordingly, the results shown in Table 21.8 can be used to find measuring stations having ozone levels affected by different factors, and this is useful for explaining why the model performance is worse for some stations than others.

In addition, comprehensive air quality models also encompass preprocessors for estimating meteorological parameters such as wind direction and speed, and submodules for calculating emissions discharged from pollution sources. Since the quantities produced by these submodules are input to the air quality model, they naturally affect the accuracy of the final concentration estimates. Therefore, evaluating the output of such submodules is also an important issue. Methods have been proposed to evaluate the input parameters to the air quality model, such as those for meteorology fields presented in Emery et al. (2001).

19.4 Features of the Statistics Used for Evaluating the Model Performance

This section considers the nature of the various statistics used for evaluating the model performance and also their consistency with the intended use of the air quality model. In doing so, it is better to provide concrete examples of applications rather than abstract descriptions. For this reason, the model performance evaluation using concentration values obtained from air quality simulations in the Kashima area was conducted. An overview of the model used here has been provided in Chapter 18, Section 18.3.1.

19.4.1 Criteria for Performance Evaluation

In the United States, when selecting recommended models, candidate models were selected based on statistical evaluation scores. In the performance evaluation of the air quality model, a predicted value by the model was compared with observed data. This method can be roughly classified into the following three cases:

1. Predicted and observed concentrations at specified locations and times,
2. Predicted and observed maximum concentrations under various conditions, in which there are no restrictions concerning location or time, and
3. Predicted and observed frequency distribution of the concentrations at a designated site (in percentile-concentration curve or Q-Q plots).

Here, the method of comparing predicted and observed values required in the first case (1) is the strictest evaluation. Many statistics have been proposed for this purpose. For example, the average and standard deviation of the residual (predicted value minus observed value), correlation coefficient, coefficient of variation, mean square error, and the like. Here, assuming that the observed value is Oi, the predicted value is Pi, and their average values are $\bar{O}$ and $\bar{P}$, the fractional bias (FB) and the normalised mean square error ($NMSE$) can be calculated by the following equations:

$$FB = (1/n) \quad \Sigma\,(2(Pi - Oi)/(Pi + Oi)) \tag{19–1}$$

$$NMSE = \left\{(1/n)\,\Sigma\,(Pi - Oi)^2\right\} / \left\{(1/n)\,\Sigma Pi \cdot \,1/n \;\, \Sigma Oi\right\} \tag{19–2}$$

where i is the data point number, and n is the total amount of data points. The fractional bias, FB is the statistic also used in the preceding AERMOD evaluation. Here, however, the calculation is performed using values from individual data points, as shown in Equation (19–1) rather than values calculated from a limited set of data on the higher concentration side. For comparison at the highest concentration, there is a method to specify the location and time. However, there is a method in which only the location is fixed, and the maximum value of the predicted concentration is compared with the maximum value of the observed concentration without setting a restriction on time. On the contrary, there is also a method of determining only the time and comparing the highest values regardless of the place.

19.4.2 Inter-Comparison of Evaluation Criteria

Although many criteria for evaluating air quality models had been proposed, their relationships were not considered so deeply. Kainuma et al. (1987) analysed the interrelationships of several statistics used to evaluate the performance of the air quality models. They used the calculated concentration values by the IPCAJ model and the observed values at the corresponding period and site, and the results are highlighted in this section.

The data used in this analysis were the observed and predicted (calculated) air quality concentrations in the Kashima area, which was taken from IPCAJ (1985). The outline of the air quality model used has been described in Section 18.3.1 of Chapter 18. Using this model, the hourly concentrations during one year from April 1980 to March 1981 were calculated. From these datasets, the analysis of evaluation scores was carried out using hourly data, particularly in the summer (July to August) and winter (December to January) periods. For this analysis, data were selected from 20 stations, including SO_2 data from 11 stations and NO_x data from nine stations. Here, monitoring stations with few missing data were selected, but, in some cases, both were measured by the same station, and each was counted as one station. Here, 15 different kinds of statistics were selected as indicators for measuring the performance of the air quality model. Almost all statistics adopted here are used for ordinary statistical analysis such as correlation coefficient and regression coefficient, and their meanings can be easily understood. Special indicators are explained here. For example, the index of agreement proposed by Willmott (1982).

Here, the index of agreement *D* and the mean fractional error *MFE* is calculated by the following equations:

$$D = 1 - \frac{\Sigma(O_i - P_i)^2}{\Sigma\left(\left|C'_{prei}\right| + \left|C'_{obi}\right|\right)^2} \quad (19\text{–}3)$$

$$MFE = (1/n)\Sigma\frac{(P_i - O_i)}{((P_i + O_i)/2)^{1/2}} \quad (19\text{–}4)$$

where P_i is the predicted concentration, and O_i is the observed concentration, and C'_{pre} and C'_{ob} are deviations from the observed mean.

In way of example, the 15 evaluation metrics have been calculated using the data of a typical monitoring station in the Kashima area, as shown in Table 19.3. The values of 15 evaluation scores were calculated for the summer and winter datasets at 11 SO_2 and 9 NO_X stations. Namely, a total of 40 values of the statistical evaluation scores were obtained. The correlation coefficients between 15 variables (evaluation scores) were calculated using these 40 data points. Principal component analysis was conducted to understand the similarities and differences

TABLE 19.3

Example of the Performance Evaluation Scores for the Summer and Winter Datasets at Monitoring Station K in the Kashima Area

Seq.	Symbol	Explanation of symbols	Summer	Winter
x1	r	Correlation coefficient	0.39	0.53
x2	ar	Regression coefficient (slope)	0.21	0.52
x3	br	Regression coefficient (intercept)	7.81	2.84
x4	cv	Coefficient of variance	0.45	0.75
x5	(P/O)	Mean of ratio (predicted/observed)	1.646	1.454
x6	s.d.(P/O)	Standard deviation of ratio (predicted/observed)	1.007	1.312
x7	d	Mean of differences	1.433	0.326
x8	Vd	Variance of differences	52.41	16.655
x9	Md	Mean of absolute differences	5.03	2.69
x10	D	Index of agreement	0.56	0.72
x11	MFE	Mean fractional error	−0.683	−0.197
x12	Root of RMS	Root of residual mean square	7.383	4.097
x13	RMSu/RMS	Systematic residual ratio	0.286	0.747
x14	Ratio (k1/N)	Percentage of data points where the difference between predicted and observed values falls within 50 ppb	0.999	1
x15	factor2 (k2/N)	Percentage of data points for which the ratio of calculated to observed values falls within the range of 1/2 to 2.	0.635	0.756
	N	Number of data	1475	1463

Source: Adapted and modified from Kainuma et al. (1987).

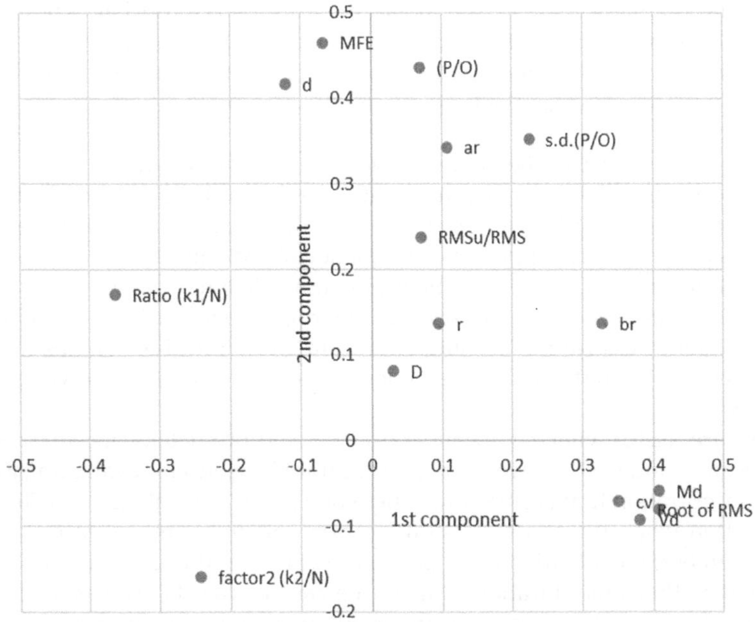

FIGURE 19.3 Result of the principal component analysis for the performance evaluation scores of the air quality model: scatter plots of the eigenvector of first and second components.

among evaluation metrics. The results of principal component analysis showed that the first and second eigenvalues were 5.7 and 4.2, respectively, with these two accounting for about two-thirds of the total variation. The distribution of the first and second eigenvectors is then shown in Figure 19.3.

The first principal component has positive and large values such as the square root of the mean squared residuals (root of RMS), mean of absolute differences (Md), and variance of the residuals (Vd), and is considered to represent the magnitude of the deviation between the predicted and observed values. In the second principal component, the values of *MFE*, mean of (P/O) ratio, mean of diffidences (*d*) etc. are large and are considered to represent the bias, namely whether the predicted concentration is overestimated or underestimated.

The result shown in Figure 19.3 also shows that *factor 2* and *Ratio* are quite different from the other scores. Such an analysis will provide some idea of the relationships among the evaluation scales. In addition, it is also important to consider what purpose the model will be used for and whether the evaluation scales are consistent with the purpose for which they will be used.

19.5 Reproducibility of Hourly Concentrations

It would be ideal if the output of the air quality model covered all ranges of the frequency distribution of concentration occurrence. However, as discussed in the previous chapter, some simplification is unavoidable when describing complex atmospheric phenomena with mathematical equations. There is also a limit to the input conditions that can be available. For these reasons, it is important to select criteria that can be used to evaluate the suitability of the model outputs for the intended use. And even if the long-term average concentration (in most cases, annual average) is the prediction objective, if the prediction performance of the smallest unit of the underlying prediction (in most cases, the one-hour value) is inadequate, it seems that the regulatory use of the model may not be appropriate. These concerns are raised due to the significant possibility that the calculated concentrations used to determine the contribution rates of individual pollution sources may not be appropriate or that adverse meteorological conditions, large fluctuations in emission intensity and so on may not have been sufficiently considered.

Let us consider how effective the model applied in the Kashima case study is for the prediction of the highest value of the hourly concentration and percentile value of the higher concentration side. Note that the model was developed with the target to calculate the annual average concentration. The cumulative appearance frequency of concentrations can be expressed by a percentile-concentration (PC) curve, as shown in Figure 18.6 in Chapter 18. Instead of pursuing a one-to-one correspondence, let us try to compare each percentile value on the cumulative appearance frequency distribution. Using data from all monitoring stations in the Kashima area, it is possible to obtain the percentile values of the predicted and observed concentrations on the PC curve.

Scatter plots are useful when comparing observed and predicted concentrations about the designated location and time. However, a Q-Q diagram is useful for evaluating a model using the highest concentration that occurs somewhere or a given percentage value. In other words, it is an evaluation when the constraints on location and time have been removed. In addition, it is an even better evaluation if the percentile values of the concentrations are obtained in many places at the same duration. This is the case of the IPCAJ Kashima study, which had the observed and predicted concentrations obtained at many monitoring stations during one year. These data were used to calculate the correlation coefficient between the observed and predicted first percentile value. Next, the correlation coefficient for the second percentile value was calculated and continued until correlation coefficients were calculated for the 98%, 99% and 100% values (maximum). Figure 19.4 shows a comparison of the observed and predicted values at all monitoring stations in the Kashima area for the 92, 95, and 98th percentile concentrations.

The summary of the overall results from the first to 99th percentile values is seen in Figure 19.5, which shows low correlation coefficients for low concentration values. Although the correlation coefficient represents the tendency between two variables and is not an absolute comparison, higher coefficients would indicate a better agreement. In the simulation in the Kashima area, a value of 3.2 ppb of the SO_2 background concentration due to the inflow contamination from the surrounding area was added to the simulation results, which may be the reason for the poor reproducibility in the low concentration part in Figure 19.5.

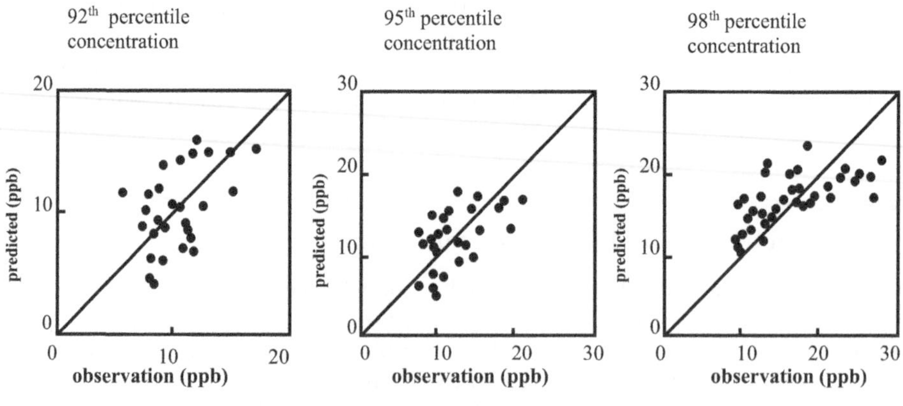

FIGURE 19.4 Scatter diagrams of the observed and predicted SO_2 concentrations at all 28 monitoring stations for the 92nd, 95th and 98th percentile values. (April 1980–March 1981, Kashima area).

Source: Adapted and modified from Okamoto (1992).

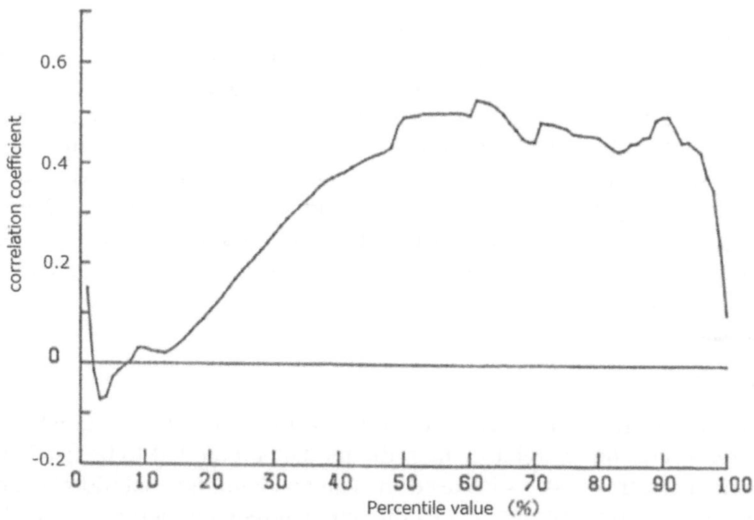

FIGURE 19.5 Correlation coefficient between the predicted value and observed value of each percentile concentration for the Kashima case study with 28 monitoring stations.

Source: Adapted and modified from Okamoto (1992).

Above the 95th percentile concentrations, the correlation between observations and predictions dropped rapidly. For the second highest concentrations, the correlation coefficient was about 0.2, as shown in Figure 18.4 in Chapter 18. After all, it can be seen that the reproducibility for the high-concentration part, which is the tail of the frequency distribution, is also poor in this model. On the other hand, in the range of about 50–95%, the correlation coefficient was high, indicating that this air quality model reproduced the atmospheric phenomenon properly. This in turn indicated that the method of dividing the meteorological conditions and the modelling of each condition were appropriate. In other words, it could be considered that the variation between groups was properly described and the variation within the group was suppressed to a certain extent. Nevertheless, it could be seen that the reproducibility of a few percent near the highest concentration was rapidly decreasing. The findings here can be explained by the use of air quality models with meteorological parameters corresponding to average conditions, rather than those aimed at reproducing anomalously high concentrations. That is, the highest concentration due to the variation of within-group variability for each subdivided meteorological condition could not be reproduced.

When trying to reproduce hourly concentration data, the most influential factor may be the input data of the wind direction. The plume spread in the crosswind direction is at most about 10°, so if the measured wind direction in 16 directions deviates by one direction, the calculated concentration will change by several digits or more. For this reason, Hanna et al. (2001) and Chang and Hanna (2004) have developed a method to evaluate the air quality model scaled as an arc-max concentration that can remove the uncertainty of wind direction and evaluate the concentration fluctuation due to other factors, especially for model performance at the maximum concentration. In this arc-max concentration method, the observed and predicted values of the maximum concentrations occurring at a location in the sampling points arranged in an arc are compared.

In addition, to perform an overall evaluation of the model performance instead of considering individual effects of different parameters, it is necessary to determine whether or not the frequency distribution of the concentration within a given certain period (or sites within a certain geographic range) can be reproduced by the model. In other words, this is an evaluation where the designation of the place and time is removed, and the comparison is done using both observed and predicted frequency distributions. An example is to compare the observed and predicted values in a cumulative frequency distribution, as shown in Figure 18.6, Chapter 18. Another example is a comparison in quantiles-quantiles (Q-Q) diagrams (U.S. EPA, 2003). In the Q-Q plot, the same sets of data as in the scatter plot are used, but the pairing is removed; instead, ranking of the observed and predicted data from the lowest to highest, respectively, was used. Accordingly, the lowest predicted concentration is first plotted against the lowest observed concentration. Then plot second, third and so on. The plot helps examine whether the model can generate a concentration distribution similar to that of the observed. The reproducibility of higher concentrations in the distribution can also be seen in the plot.

19.6 Summary

A variety of statistics are used to evaluate model performance. These statistics must be able to accurately represent the degree to which the model is intended to be used. The interrelationships among the various statistics used in such an evaluation should also be considered. When conducting statistical analyses to evaluate the agreement between model outputs and monitoring data, it is important to note that both datasets contain errors and uncertainties.

This chapter discussed the performance evaluation of air quality models using the observed data in two case studies. The first case study highlights the statistical criteria used for the performance evaluation of WRF-CAMx. The second case study presents the model performance evaluation for the Kashima area using the appropriate monitoring dataset. The second case study indicated that if a model had been developed targeting the annual average concentration, then its performance for the hourly highest concentrations may not be sufficient.

It is therefore important that the statistics used to evaluate the model performance are calculated by the concentrations with appropriate temporal and spatial resolution consistent with the intended use of the model. Further, it is necessary to ensure that the averaging time of the original data used for the evaluation matches the statistics required by the regulations.

REFERENCES

Chang, J.C. and Hanna, S.R., 2004, Air quality model performance evaluation, *Meteorology and Atmospheric Physics*, 87, 167–196, https://doi.org/10.1007/ s00703-003-0070-7

Emery, C., Tai, E. and Yarwood, G., 2001, *Enhanced meteorological modeling and performance evaluation for two Texas ozone episodes, prepared for the Texas natural resource conservation commission*, ENVIRON International Corp., Novato.

Fox, D.G., 1981a, Judging air quality model performance, *Bulletin of the American Meteorological Society*, 62(5), 599–609, https://doi.org/10.1175/1520-0477(1981)062<0599:JAQMP>2.0.CO;2

Fox, D.G., 1981b March 9–13, Judging air quality model performance-review of the Woods Hole workshop, *Extended Abstracts 5th symposium turbulence, diffusion, and air pollution*, 14–16, American Meteorological Society, Atlanta, GA.

Ha Chi, N.T. and Kim Oanh, N.T., 2020, Photochemical smog modeling of $PM_{2.5}$ for assessment of associated health impacts in crowded urban area of Southeast Asia, *Environmental Technology & Innovation*, 21, 101241, https://doi.org/10.1016/j.eti.2020.101241

Hanna, S.R., Egan, B.A., Purdum, J. and Wagler, J., 2001, Evaluation of the ADMS, AERMOD, and ISC3 dispersion models with the Kincaid, Indianapolis, Lovett, Sweeny, and Duke forest field data sets, *International Journal of Environment and Pollution*, 16(1–6), 301–314, https://doi.org/10.1504/IJEP.2001.000626

Hayes, S.R., 1979, *Performance measures and standards for air quality simulation models, EPA-450/4-79-032*, U.S. Environmental Protection Agency, retrieved on 5 October 2023 from https://nepis.epa.gov/Exe/ZyPDF.cgi/2000X0FL.PDF?Dockey=2000X0FL.PDF

Industrial Pollution Control Association of Japan (IPCAJ), 1985, *The study on the behaviour of particulate matter, vol. 8: Report on dispersion simulation* (p. 350) (In Japanese), Japan Environmental Management Association for Industry, Tokyo, Japan.

Kainuma, Y., Takahashi, K., Okamoto, S. and Shiozawa, K., 1987, On evaluation measures of air quality model performance, *Journal of Japan Society of Air Pollution*, 22(6), 416–424 (In Japanese).

Kim Oanh, N.T. and. Permadi, D.A., 2009, *Photochemical smog modeling for ozone air quality management, chapter 12 in the book modelling of pollutants in complex environmental systems*, vol. 1, ILM Publications, Hertfordshire, UK, ISBN 9781906799007

Okamoto, S., 1992, Regulatory air quality model and its evaluation, *Journal of Japan Society of Air Pollution*, 27(3), A44–A52 (In Japanese).

U.S. EPA, 1991, *Guideline for the regulatory application of the urban airshed model, U.S. EPA EPA-450/4-19-013*, U.S. Environmental Protection Agency, retrieved on 5 October 2023 from https://nepis.epa.gov/Exe/ZyPDF.cgi/10003WHQ.PDF?Dockey=10003WHQ.PDF

U.S. EPA, 2003, *AERMOD: Latest features and evaluation results, EPA-454/R-03-003*, U.S. Environmental Protection Agency, retrieved on 5 October 2023 from https://gaftp.epa.gov/Air/aqmg/SCRAM/models/preferred/aermod/aermod_mep.pdf

U.S. EPA, 2021, *AERMOD model formulation and evaluation, EPA-454/B-21-003*, U.S. Environmental Protection Agency, retrieved on 5 October 2023 from https://gaftp.epa.gov/Air/aqmg/SCRAM/models/preferred/aermod/aermod_mfed.pdf

Willmott, C.J., 1982, Some comments on the evaluation of model performance, *Bulletin of the American Meteorological Society*, 63(11), 1309–1313.

Part III

Applications for Data Analysis

Part III focuses on the statistical analysis of environmental monitoring data, emphasising multivariate analysis techniques. In addition to those theoretical foundations for several important methods that have been provided in Part I, Part III introduces additional techniques, including canonical correlation analysis and clustering. This section aims to navigate readers through the different methods suitable for analysing specific dataset types illustrated with examples and real-life case studies to aid understanding. The content does not delve into the algorithms behind the analysis methods but directs readers to specialised, advanced statistical texts when needed. The case studies are derived from the authors' published research, especially in air quality analysis, and extend to other areas like biogeography and marketing research.

Chapter 20 outlines the various types of data, with a particular focus on categorical data, which are crucial in many surveys such as epidemiological studies examining the health effects of environmental pollution and in the economic valuation of environmental goods. Chapter 24 details the methods for analysing such data. Chapter 21 provides examples of principal component analysis and factor analysis using various forms of data, and explains how these techniques are applied to different types of data in environmental monitoring. Chapter 22 introduces case studies that utilise multivariate analyses to predict ambient air pollutant concentrations. Chapter 23 presents a basic concept of receptor modelling with the case studies of source apportionment for particulate matter (PM) air pollution. Most of the data presented in the tables and figures in Part III are sourced from actual research reports. Some fictitious data has also been prepared to illustrate the analysis and calculation methods; in such instances, this is indicated accordingly.

DOI: 10.1201/9781003597216-22

20

Types of Statistical Data and Analysis Methods Suitable for Their Analysis

A sample extracted from a population is subjected to measurement to generate data, which are analysed to get information on the population. Measuring a sample will provide useful quantitative data in most cases. However, the human sense may be more appropriate in certain cases than quantitative measurements. A typical example of this case may be the evaluation of offensive odours. The sensory evaluation by inhabitants is often more appropriate for assessing the quality of the environment than the chemical analysis value of the malodorous component assumed by experts. This chapter provides details on the types of data obtained from environmental monitoring and the appropriate methods to be used for the analysis of different types of data.

20.1 Types of Data

The data may consist of numbers (numerical data) or group names/labels (categorical data). Numerical data can be divided into continuous and discrete data. The former can be obtained by measuring with some measuring instrument, such as length, weight, temperature etc. Discrete data can be obtained by counting, such as the number of trees in the garden. Categorical data consists of group names or labels and are further divided into unordered, which are mutually exclusive, and not ordered in any way, e.g., blood group types (A, B, O), and ordered, such as, e.g., socio-economic class, staging of a process.

A variety of ingenuity is used to collect data to know about the objects of interest. For example, if an emission from a chimney is thought to cause air pollution, first of all, it may be judged as to whether or not it is emitting black smoke, which can simply be done by visuality. To get more quantitative information, one can use the Ringelmann Smoke Chart, which was developed by French Professor Ringelmann in 1888 and has been used for the many years thereafter. It is also known as the Ringelmann scale, which is used for grading the opacity, appearance, density or shade of an emission. The chart had been accepted as the standard measure of smoke density in the standard code for power-plant testing that was formulated by the American Society of Mechanical Engineers (Grand Rapids Historical Commission, 2020). To measure smoke density, standard coloured papers are available with six levels of smoke density ranging from all white to all black. The observer decides the class of the smoke density by comparing the density of the coloured paper and the smoke.

To characterise the smoke emitted from the chimney, the following type of data may be considered.

1. Measurements/observations that show whether black smoke is emitted from the chimney (Yes or No)
2. Smoke density measurements on the Ringelmann smoke chart
3. Particulate matter concentrations in the flue gas using stack sampling and relevant analytical methods

DOI: 10.1201/9781003597216-23

TABLE 20.1

Types of the Data

Data type	Possible values	Examples	Distribution
Binary	0, 1	Binary outcome ("yes/no", "true/false", etc.)	Bernoulli
Categorical	1, 2,. . ., K	Classification of places of residence, educational background	
Ordinal	Integer	Order of arrival, order of size, order of occurrence frequency	
Count	Whole number (integers)	Number of products, number of damages, number of births	Binomial
Measured	Real number (floating-point arithmetic)	Length, weight, current of electricity	For example, normal (Gaussian)

The first data (1) is of the binary type. The second (2) is categorical data. The third data (3) is a numerical value. The first data is sometimes called linguistic data, qualitative data, or non-quantitative data. The name of "linguistic data" is based on the content of the data being written in everyday language (Yamada et al., 2002). The third data is sometimes called quantitative or numerical data. The second data occupies the middle position between the other two concepts.

Different data types require different mathematical procedures when using them for estimating the parameters of the population due to differences in probability distributions among others. Table 20.1 shows the types of characteristic data illustrated with examples. First, it can be divided into quantitative data and non-quantitative (linguistic) data. An example of statistical analysis of linguistic data is to analyse the number of people who answered the question of whether they like or dislike a particular product or the number of people who feel unwell or not. "Quantitative data" refers to data that records numerical values such as length, weight and speed. On the other hand, nominal scale data such as gender, occupation and "likes or dislikes of a thing" and rating scale data such as the five-level evaluation of learning performance are often referred to as qualitative data or categorical data.

Furthermore, consider inquiring about the preferences for three commodities (No. 1, No. 2, and No. 3) and organising the data accordingly. At this stage, the commodity numbers 1, 2, and 3 are merely labels, and any arithmetic operations such as addition, subtraction, multiplication and division are not permitted. Therefore, it is crucial to discern whether a numeral has a mathematical significance, even if it is represented by a numerical value.

20.2 Variables and Data

Variables that have no meaning other than mere labels, even if they are displayed as numbers, are sometimes called categorical variables, which can also be analysed using suitable statistical methods. In the data classification shown in Table 20.1 of the previous section, binary data in the form of "yes or no" can also be considered categorical data in a broad sense.

Generally, the data obtained by counting corresponds to the discrete type, and the data obtained by measurement corresponds to the continuous type. Typical examples of discrete random variables are binomial distribution or Poisson distribution. On the other hand, a typical example of continuous random variables is the normal distribution. The binary data just described follows the Bernoulli distribution, which can be considered a special case of the binomial distribution.

The following important definitions are extracted from Milton et al. (1997), and they are useful in following the content of this chapter.

- A random variable is a variable whose value is determined by chance.
- A discrete random variable is a random variable that can assume at most a finite or a countably infinite number of possible values.
- A continuous random variable can assume any value in some interval or intervals of real number, and the probability that it assumes any specific value is zero.

Let X be a continuous random variable. A continuous probability function for the random variable X is function f such that:

- $f(x) \geq 0$, for each real number x,
- The total area boundary by the graph of f and the horizontal axis is 1, and
- The probability of observing X with a value between any two points a and b is equal to the area of the region bounded by the graph f, $x=a$, $x=b$, and the horizontal axis.

In the case of a discrete random variable, if a sufficient number of data are obtained, and if certain conditions are satisfied, it may be possible to approximate a normal distribution. However, in the case of categorical data, special considerations may be required, which will be explained later.

Numerical data are data represented by numbers. It is also important that the number has a certain mathematical meaning, and that the number is not just only a label. Such data are further divided into discrete and continuous.

- ***Discrete data*** can generally be considered as data obtained by counting and are represented by an integer. Further, the ratio calculated by dividing discrete data by discrete data is also discrete data. For example, disease morbidity and defective product rates are discrete data, even if they are not integers.
- ***Continuous data*** may be generally considered as data obtained by measurement and are represented by a real number. It is also important to note that the number of significant digits in the measured value depends on the accuracy of the measuring instrument.

20.3 Analytical Methods for Various Types of Data

Quantitative data analysis methods were discussed in Part I. For categorical data, such as data as labels that have no mathematical meaning, different analysis methods are required. For example, suppose one wants to analyse the difference between the two population means using the t test; population 1 and population 2 are categories. In the one-way ANOVA, Level 1, Level 2 and Level 3 are also categories. Therefore, if a researcher wants to analyse the level names themselves as data, it is necessary to choose a suitable statistical analysis method.

Next, consider a regression model that analyses cause-and-effect relationships. If the explanatory (independent) variable contains a categorical variable, and it has only two groups (binary data), one can enter the values of 0 or 1 for this variable and calculate as usual. The result is the same as simply changing the constant terms in Group 1 and Group 2. An important issue discussed in this section is how to analyse cases where the objective (dependent) variable is a categorical variable. Note that, in regression analysis, terms such as "objective variable", "response variable", "dependent variable" and "explained variable" are used interchangeably. In this book, the term "objective variable" is mainly used.

As for the regression analysis, if the objective variable is categorical, it can be considered to be equivalent to discriminant analysis or cluster analysis, in which one will determine which group an individual data point belongs to. As described in Part I, this case can also be analysed by maximising the ratio of between-group and within-group variabilities or by using the Mahalanobis distance.

Such methods can also predict which group an individual will fall into. But before that, the problem of predicting the value of probability, such as the prevalence of a disease, is considered. An epidemiological case is used for illustration. In five towns, a questionnaire survey was conducted on 100 people to determine whether they had coughing or not and to determine the prevalence of respiratory diseases; the survey results are shown in Table 20.2. Suppose that one wants to develop a prediction model to estimate the morbidity by using the survey results in these five towns and then predict the morbidity in the sixth town with a pollutant concentration of 0.02 ppm. As illustrated in Figure 20.1, using an ordinary linear regression equation to predict morbidity in a town with a pollution concentration of 0.02 ppm results in a value of −0.0157. However, morbidity cannot be a negative value.

As can be seen from Figure 20.1, when the objective variable is probability, the regression line goes into the region with a morbidity of less than 0 and the region with a morbidity of 1 or more (i.e., >100%), which is impossible in principle. Therefore, it is preferable not to regress the relationship between the explanatory variables and the morbidity with a straight line but to regress with a sigmoid curve that changes in an S-shape between the morbidity of 0 to 1, as shown in Figure 20.2.

Next, let us examine the incidence rate of each person as shown in Table 20.3, not the morbidity within the town as seen earlier in Table 20.2. The predicted value in such a prediction model is the probability of the onset of the disease for each person. A person with a probability of 1 is a sick person, and a person with a probability of 0 is a healthy person. Furthermore, if this probability is above a certain threshold, it can be judged as a sick person. The logistic regression model can be usually thought to refer to such an analysis method.

Assume a study has been conducted to find a formula to predict whether asthma develops from past smoking history or air pollution conditions in the residence area. The medical

TABLE 20.2

Example of the Study on Respiratory Morbidity and Pollutant Concentration

Town	Respiratory diseases		Morbidity (%)	Pollutant concentration (ppm)
	Yes	No		
1	5	95	0.05	0.06
2	11	89	0.11	0.09
3	2	98	0.02	0.03
4	17	83	0.17	0.12
5	4	96	0.04	0.07

Note: This is a fictional case study example that uses a simple number to make it easier to understand the concept.

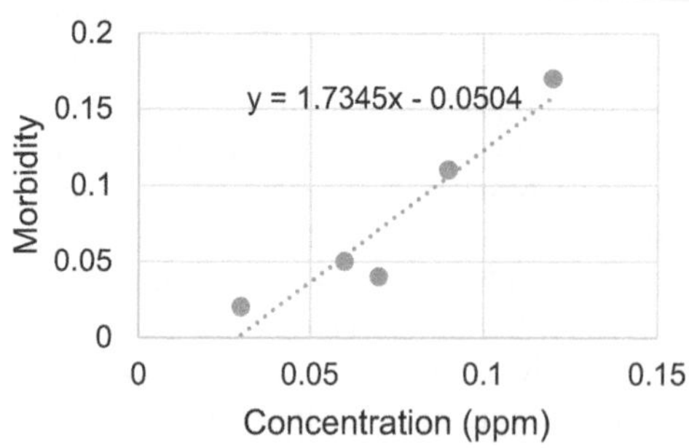

FIGURE 20.1 Scatter plots between morbidity and air pollutant concentration.

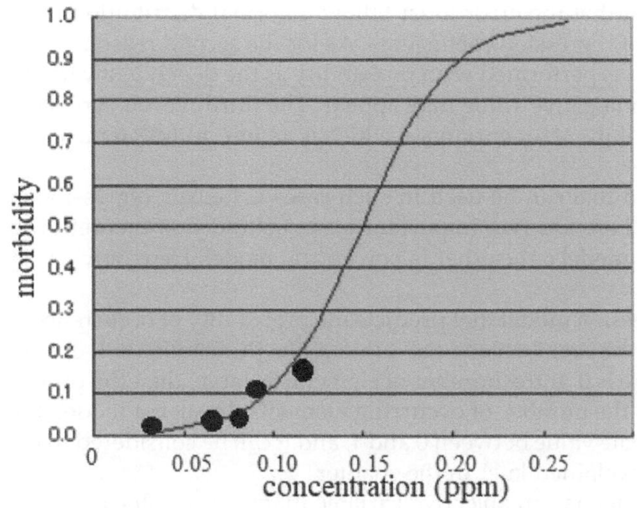

FIGURE 20.2 Estimating the morbidity by a sigmoid-type experimental model.

TABLE 20.3

Case on Medical History and Living Environment Data for 100 People

Personal number	Asthma status	Age	Pre-existing symptoms	Smoking period	Number of smoking	Air pollution concentration in the residential area	Heating types of the living room
No.	*y*	*x1*	*x2*	*x3*	*x4*	*x5*	*x6*
1	0	60	0	15	10	0.05	1
2	1	68	0	32	20	0.02	3
3	0	43	0	0	0	0.01	2
4	0	57	1	0	0	0.03	1
.							
.							
.							
100	0	38	0	0	0	0.04	1

Note: This is a fictional case study example, which is only for the illustration of the concept.

history and living environment data for 100 people are shown in Table 20.3. Now consider a regression equation as shown in Equation (20–1).

$$y = b_0 + b_1x_1 + b_2x_2 + \cdots \tag{20–1}$$

In this formula, the objective variable y is a categorical variable that takes a value of "1" if asthma has developed and "0" if it has not. However, if one finds an ordinary regression equation, the predicted value can be negative or larger than 1. The use of logistic regression can avoid the appearance of numerical values below 0 or above 1.

The reasons why ordinary linear regression cannot be used for categorical dependent variables are as follows:

- Error is not following the normal distribution,
- Predicted values sometimes become strange, and
- Error is not evenly distributed.

First, the condition that the error must follow a normal distribution is a prerequisite necessary for the t test of regression coefficients. As for the second reason, for example, if ordinary regression analysis is performed with probability as the dependent variable, a predicted value greater than 1 or a negative value may appear. The third reason is related to the assumption of the uniformity of the error variances, which is an important premise for obtaining accurate estimates.

A useful method that can be used in such cases is logistic regression analysis, also called logit regression. There are two important ideas behind this theoretical background: one is generalised linear models; the other is perceptron model. Here only a brief introduction will be presented.

Consider a regression model that predicts the probability of occurrence of a certain event. In this model, when the event occurs, the value of the probability is 1, and when the event does not occur, the value is 0. If the number of elements is large, then the probability of occurrence is calculated from the number of occurring elements. In such a model, the objective variable takes an intermediate value between 0 and 1, and it can be considered to belong to generalised linear models, as explained later in the chapter.

The simplest model has an objective variable that takes a value of "1" when a certain event occurs and a value of "0" when it does not occur. A regression model should predict which category each individual should belong to. In this case, the value taken by the objective variable is either 0 or 1, and no other value can be taken. This is an extension of the perceptron model discussed later.

20.3.1 Logistic Regression

Logistic regression is a statistical model that uses a logistic function to model a categorical dependent variable. The logistic regression estimates the parameters of a logistic model. Mathematically, a binary logistic model has a dependent variable with two possible values, for example, such as "sick or healthy" which is represented by an indicator variable, where the two values are labelled 0 and 1. In the logistic model, the log odds (the logarithm of the odds) are expressed by a linear combination of the independent variables (predictors).

Logistic regression is used in various fields such as medical fields and social sciences. For example, logistic regression may be used to predict the risk of developing a given disease, based on observed characteristics of the patient (age, sex, results of various tests etc.). The logistic curve is used as the sigmoid curve because it is often useful in this case. In the sigmoid curve in Figure 20.2, the appearance rate (probability p) is converted to log odds, i.e., logit, as follows:

$$\text{Logit (log odds): } l = \ln\left(p/(1-p)\right) = b_0 + b_1 x \tag{20–2}$$

Here, consider the format of the formula without considering the error distribution etc.; this is equivalent to a linear regression formula in which logit is the objective variable with the clues as the explanatory variables. Furthermore, if the number of explanatory variables is multiple, the right-hand side becomes a model of the same format as the multiple regression equation. This is called a linear logistic model, and regression analysis based on this model is called logistic regression analysis.

In regression analysis with probability as the objective variable, there is a method to analyse the appearance ratio of individuals with a certain attribute in a certain population, as shown in Table 20.2. On the other hand, as shown in Table 20.3, it is also possible to consider a model that estimates the probability of having a certain attribute (for example, morbidity) for each person using a certain risk factor as an explanatory variable. Here, if the relationship between risk factors and the presence or absence of disease development is displayed graphically, it will be shown in Figure 20.3.

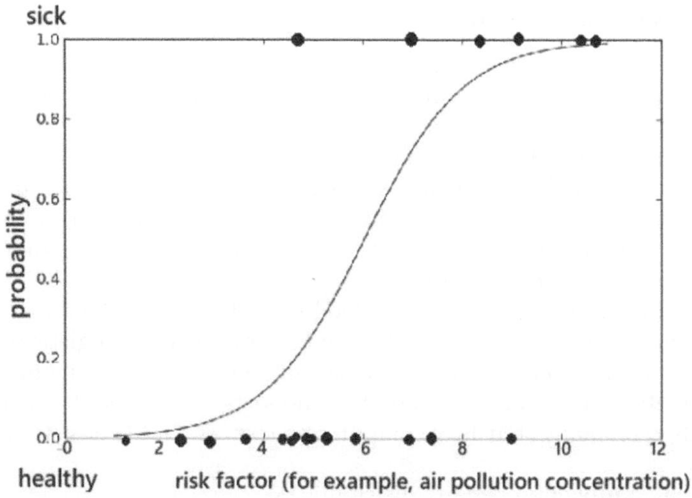

FIGURE 20.3 Relationship between the health status and risk factor.

In Figure 20.3, the horizontal axis is the explanatory variable, and the vertical axis is the objective variable. There are healthy people ($y = 0$) and sick people ($y = 1$) hence the measured values are only 0 and 1, but the estimated values are distributed between 0 and 1. A scatter plot can be used that considers person with an estimated probability of getting sick higher than a certain threshold value is a sick person, and a person with a probability of less than that value is a healthy person. This means that the target person is classified into two groups using the selected explanatory variables. In this way, one can see that logistic regression can be used for discrimination problems.

20.3.1.1 Logistic Regression as a Prospective Study

Logistic regression analysis can be used as a method for analysing risk factors that influence the development of a disease and predicting whether a subject will develop the disease before it actually develops. Therefore, in principle, this method is applied to the data obtained in the prospective study. In other words, in logistic regression, the explanatory variable (independent variable) is a fixed value that does not contain an error, and the objective variable (dependent variable) is a variable that contains an error term.

20.3.1.2 Discriminant Analysis as a Retrospective Study

Discriminant analysis, on the other hand, is a multivariate extension of diagnostic analysis. It is a technique for analysing why a subject had the disease after the onset of the disease. Therefore, this method is considered to be a retrospective study in principle, and it is assumed that only the explanatory variables contain the error term.

Overall, discriminant analysis assumes a multivariate normal distribution whereas logistic regression analysis, on the other hand, requires the assumption that there is a linear relationship between the logit or log odds ratio and the explanatory variables. Strictly speaking, this may be as difficult to implement as the assumption that the explanatory variables are multivariate normal distributions (Sugimoto, 2014). A good example of how to understand the characteristics of these analytical methods can be found in Antonogeorgos et al. (2009), which discusses and reports the differences and similarities between logistic regression and linear discriminant analysis when assessing factors associated with asthma prevalence in children.

20.3.2 Generalised Linear Model

Many useful statistical analysis methods, including ordinary regression models and logistic regressions presented here, can be explained within the same theory of generalised linear models, which is briefly presented in this section.

A generalised linear model (GLM) is a flexible generalisation of ordinary linear regression that allows response variables (dependent variable) with error distribution models other than the normal distribution. GLM generalises linear regression by allowing a linear model to be associated with a response variable through a link function and making the variance of each measurement a function of its predictor.

The generalised linear model was formulated by Nelder and Wedderburn as a way of unifying various other statistical models, including linear regression, logistic regression and so on (Ram, 2022). It was based on an iteratively reweighted least squares method for estimating the model parameters. However, maximum-likelihood estimation has become popular and is the default method on many statistical computing packages. GLM consists of three elements:

- An exponential family of probability distributions,
- a linear predictor, and
- a link function.

In generalised linear models, each outcome Y of the dependent variables is assumed to be generated from a particular distribution in an exponential family, a large class of probability distributions that includes the normal and binomial distributions. The mean μ of the distribution depends on the independent variables X through:

$$E(Y/X) = \mu = g^{-1}(X\beta) \tag{20–3}$$

where $E(Y|X)$ is the expected value of Y conditional on X; $X\beta$ is the linear predictor, a linear combination of unknown parameters β; g is the link function. The unknown parameter β is usually estimated by maximum likelihood or other methods. The typical forms of this model can be shown in Table 20.4.

For a more detailed explanation, see the appropriate literature, such as Pennsylvania State University (2022).

20.3.3 Perceptron

In machine learning, the perceptron is an algorithm for supervised learning of binary classifiers. A binary classifier is a function that can determine whether an input represented by a vector of numbers belongs to a particular class. This is a type of linear classifier. In other words, it is a classification algorithm that makes predictions based on a linear prediction function that combines a set of weights and a feature vector.

The perceptron algorithm was invented by Rosenblatt in 1958; it was intended to be a machine, not a program, but it was initially implemented in software (Blackmon, 2018). The perceptron is a simplified model of biological neurons. A complete understanding of neural

TABLE 20.4

Common Distributions with Typical Uses and Link Functions

Regression models	Data type	Distributions	Link function
Ordinary regression	Real	Normal	$X\beta = \mu$
Poisson regression	Integer	Poisson	$X\beta = \ln(\mu)$
Logistic regression	Binary	Bernoulli	$X\beta = \ln(\mu/(1-\mu))$

behaviour often requires the complexity of biological neuron models, but a perceptron-like linear model can also produce the behaviour found in real neurons.

In the modern sense, the perceptron is an algorithm for learning a binary classifier called a threshold function. It is a function that maps its input x (a real-valued vector) to an output value $f(x)$ (a single binary value).

$$f(x)=\begin{cases} 1, & \text{if } w\,x+b>0, \\ 0, & \text{otherwise.} \end{cases} \tag{20–4}$$

where w is a vector of real-valued weights, wx is the dot product Σwx, where the number of elements of the vectors w and x is m, m is also the number of inputs to the perceptron, and b is the bias. The bias shifts the decision boundary away from the origin and does not depend on any input value. The value of $f(x)$ (0 or 1) is used to classify x as either a positive or a negative instance, in the case of a binary classification problem.

20.4 Analysing an Interrelation among Categorical Variables

Here, it is supposed that a researcher investigated the number of patients with four important types of diseases in three towns in association with the effect of environmental pollution. The results are shown in Table 20.5. Further, if it is desired to discuss who is in a serious condition or who has no health problems, one needs to look at each person's score. However, if one just wants to know which towns have similar tendencies, alternatively, or to find out which diseases have similar tendencies of occurrence, the aggregated data in Table 20.6 are sufficient. The test for independence is a statistical method to find out whether the ratio can be said to be the same for any category, in this case, any town and any disease.

In this case one can study one variable at r levels (in this example, 3 towns), and the other c levels (in this example, 4 diseases), and make an $r \times c$ contingency table, as shown in Table 20.6.

TABLE 20.5

Health Survey Results for 900 People in Three Towns.

Personal number	Town 1	Town 2	Town 3	Disease 1	Disease 2	Disease 3	Disease 4
1	1			1			
2							
3							
4		1				1	
5							
6			1	1			
7			1		1		
8							
9							
10							
.							
900							
total	18	11	25	24	13	7	10
Number of respondents	300	300	300				

Note: This is fictional data. In this table, each subject is assigned a 1 if he/she has the relevant disease and left blank if healthy. Additionally, a 1 is placed in the column for the town where the sick individuals reside.

TABLE 20.6

Contingency Table That Aggregates the Data in Table 20.5

Parameter	Disease 1	Desease 2	Desease 3	Desease 4	Total
Town1	7	3	2	6	18
Town2	5	2	1	3	11
Town3	12	8	4	1	25
Total	24	13	7	10	54

The general null hypothesis to be tested via a contingency table is that there is "no association" between the two classification variables. The alternative is that there is an association between them. In this kind of study, the null hypothesis of "no association" is equivalent to a null hypothesis of independence (Milton and Arnold, 1995).

The contingency table with the occurrence frequency n_{ij} entered in each cell can be rewritten as a contingency table with the probability p_{ij} by dividing the frequency by the total sample size n. Accordingly, the expected cell frequency E_{ij} is expressed by Equation (20–5).

$$E_{ij} = n\, p_{ij} \tag{20–5}$$

The fact that the row-side variable and the column-side variable of the contingency table are independent means that Equation (20–6) holds.

$$p_{ij} = p_{i.}\, p_{.j} \tag{20–6}$$

One can use this property to think of a test of independence. Now consider the null hypothesis of independence, which is stated as:

$$\mathrm{H_o}:\ p_{ij} = p_{i.}\, p_{.j} \quad i = 1, 2, \ldots, r, \quad j = 1, 2, \ldots, c.$$

The alternative hypothesis is that $\mathrm{H_1}$: $p_{ij} \neq p_{i.}\, p_{.j}$ for at least one i and j.

Now, if one has sufficient data points, the statistic expressed by Equation (20–7) follows a chi-square distribution. The degrees of freedom are $(r - 1)(c - 1)$. This property can be used to test the stated hypothesis.

$$\sum_{i=1}^{r}\sum_{j=1}^{c} \frac{\left(n_{ij} - \hat{E}_{ij}\right)^2}{\hat{E}_{ij}} \tag{20–7}$$

where the expected value of cells i and j $\hat{E}_{ij}$ can be calculated as follows:

$\hat{E}_{ij}$ = (marginal row total) (marginal column total)/total sample size.

If the null hypothesis is rejected, the population behind this data obtained is not independent. That is, there is an association. In the preceding example, the two variables, disease and town, are not independent. In other words, there is an association between the two. Then, as a next step, one would like to know which diseases and which towns show similar tendencies. One will also want to know which towns (among the three) have similar patterns of occurrence of sick people. In such a case, a correspondence analysis is a useful statistical method.

TABLE 20.7

Data of Example 4–7 of Part I in a Contingency Table

(a) Results of health survey				(b) Expected values for each cell			
	Healthy	Illness	Total		Healthy	Illness	Total
A town	35	15	50	A town	40	10	50
B town	45	5	50	B town	40	10	50
Total	80	20	100	Total	80	20	100

The concept of goodness-of-fit test has been explained in Chapter 14. The purpose of the test is slightly different, but the idea behind the test procedure used is identical to the study mentioned in this example. The test statistic compares observed category frequencies with those expected with the assumption that the stated null hypothesis is true.

Section 4.5 of Part I discussed testing whether the proportions of a certain characteristic for the two groups are the same or different, based on samples taken from two populations. Such challenges can also be analysed by the test of independence described here. Because the discussion is about the presence or absence of certain characteristics for two populations, the raw data can be organised into a contingency table of 2 by 2. The data of Example 4–7 of Part I can be summarised in a contingency table shown in Table 20.7. Here, when the expected value of each cell is calculated from the marginal probabilities, the result is as shown in Table 20.7b.

The null hypothesis H_0 and the alternative hypothesis H_1 are:

$$H_0: p_A = p_B$$
$$H_1: p_A > p_B$$

Regarding the setting up of the alternative hypothesis, remember that this problem is a one-tailed test, as explained in Example 4–7 in Chapter 4. The chi-square statistic calculated by Equation (20–7) is:

$$\chi^2 = \frac{(35-40)^2}{40} + \frac{(15-10)^2}{10} + \frac{(45-40)^2}{40} + \frac{(5-10)^2}{10} = 6.25$$

Also, since the number of rows and the number of columns are both 2, the degrees of freedom are $(2–1) \times (2–1) = 1$.

Based on this result, let us make a judgment with a significance level of 5%. In this test, the alternative hypothesis is $p_A > p_B$, so it is a one-tailed test. Therefore, at the significance level of 5%, it is necessary to read the χ-square value with a cumulative probability of 90% from the left side with the statistical table. The 90% χ-square statistic with 1 degree of freedom is 2.71. The calculated value of the χ-square statistic is 6.25. Therefore, this null hypothesis is rejected. Namely, it can be said that there is a significant difference in the prevalence of the diseases between towns A and B and that the prevalence of town A is higher than that of town B.

Compare this result with the analysis procedure of Example 4–7 in Part I. One will find that the conclusions of the test are the same. In addition, make sure that the value of the test statistics and the rejection limit value also correspond. Now recall the explanation of the relationship between the normal distribution and the chi-square distribution in Chapter 3 of Part I. Specifically, it is the relationship shown in Equation (3–13) of Part I. The numbers are as follows:

$$1.645^2 = 2.71$$
$$2.5^2 = 6.25$$

One can understand these properties from the relationship of probability distribution explained in Part I, namely, in Figure 3.9.

Looking at Table 20.6, diseases 1, 2, and 3 have similar tendencies in the number of patients in each town. However, only disease 4 occurs frequently in town 1 and tends to be different from the others. Also, if one looks at the tendency in each town, it can be seen that town 2 and town 3 have similar proportions of patients with four diseases, but the proportions in town 1 are different. Correspondence analysis is a way to make such relationships visible in an easy-to-understand graph.

Correspondence analysis is a statistical analysis tool having the purpose of revealing similarities or differences between variables in the row and column directions of a contingency table, respectively. It focuses on whether the ratios shown in each cell of the contingency table are similar in the row and column directions. The results of this analysis are equivalent to the PCA using raw data before aggregation in Table 20.5. However, since the data are aggregated for each cell, it is not possible to obtain information that corresponds to the principal component score.

It is noted that correspondence analysis is a multivariate statistical method for categorical data. It was proposed by Hartley and later developed by Benzécri (Phillips, 1995). Conceptually similar to PCA, it provides a way to view or summarise a set of data in a two-dimensional graph format. The data used for this analysis is organised in the form of a two-dimensional spreadsheet. The most typical example is data on the frequency of occurrence of events corresponding to each cell, namely, the contingency table, and this method treats rows and columns equally. Many researchers in different research fields have proposed almost the same or similar analysis methods under different names. In addition, it has been clarified that the method already reported as the third class of Hayashi's quantification theory is an analysis method equivalent to correspondence analysis (Asano, 2021).

20.5 Measuring Scale for Subjective Sensuous Factors

Consider measuring human mind movements, health conditions, and so on. For example, in an epidemiological survey, to get information on the degree of affliction (for example, headache or nausea by illness), it is necessary to ask respondents a question, to get answers and assess them. The accuracy of the answers varies greatly depending on how the question is asked. In general, it is expected that the person in charge is familiar with the analysis method of the questionnaire survey. Especially, when measuring likes and dislikes of a product, it is also an important issue to measure the degree of likeness and non-likeness by the appropriate evaluation scale. For example, it may be necessary to consider how to get respondents to answer for multiple products or new product candidates that will be available in the future. Surveys that rely on human sensations require special attention not only regarding measurement scales but also sampling methods. For instance, in a study examining preferences for beer brands, the sequence in which the beers are sampled is significant: the initial brand tasted may be perceived as the best. This issue pertains to the sampling method. The evaluation scale and its measurement method used in such cases are discussed in this section.

20.5.1 Level of Measurement

Level of measurement or scale of measure is a classification that describes the nature of information within the values assigned to variables. Psychologist Stevens developed the best-known classification with four levels of measurement (Taylor, 2020), that is nominal, ordinal, interval and ratio. This framework of distinguishing levels of measurement originated in psychology and is widely used by scholars in other disciplines.

- ***Nominal scale:*** The data of variables represented by the nominal scale are classified into categorical data. Even if is the data are expressed numerically, they have no mathematical meaning and cannot be added, subtracted, divided or multiplied. For example, the lot number of the residence display etc.

- ***Ordinal scale:*** The ordinal scale is expressed in order. For example, one can rank these data first, second, third and so on.
- ***Interval scale:*** The interval scale orders the numbers in meaningful divisions. Temperature can also be said to be an interval scale. A temperature difference of 10 degrees between 15 and 25 degrees means the same as 10 degrees between 80 and 90 degrees.
- ***Ratio scale:*** The ratio scale is similar to the interval scale, with one major difference. The origin, the value of zero, has important implications. For example, 30 degrees is not twice as hot as 15 degrees. This is because 0 degrees is not the origin of the absolute scale. In other words, the Celsius scale is only the freezing point of water. Therefore, temperature is not a ratio scale.

20.5.2 How to Measure the Scale

The following methods can be used when collecting subject evaluation data for the surveyed candidates.

- ***Ranking method:*** Ask the respondents to place the candidates in the order they like (or, for example, to the extent that they feel the urge to buy).
- ***Scoring method:*** Ask the respondent to give a score (for example, 100 points for the most favourite item, 0 points for the worst case) for each candidate.
- ***Rating scale method:*** Respondents select the appropriate grade from the prepared grades for each candidate. For example, one can use a five-point rating, such as "like", "somewhat like", " can't say either", "somewhat dislike" or "dislike" or other rating scales.
- ***Paired comparison method:*** Present each two candidates at a time and ask the respondent which one he/she prefers or agrees with.

In the rating scale method, it is possible to know the degree of preference by scores, but it may not be possible to know the favourite order of candidate products. Meanwhile, even if the respondent is asked about the preferred order of the candidate products and the answer is returned, one cannot know whether the respondent likes each candidate product. The respondent was asked about the order, so he/she only answered the order. Therefore, it may be possible that the respondent hates all the candidates. In the case of asking for answers using the evaluation scale, there is an increasing tendency for the intermediate answer of "I can't say either" to be received. Even in such a case, one can know the subtle differences by listening to the order of the candidates.

Consider asking the respondents whether they like three choices: A, B and C (e.g., products). When asked about them in order, for both person of Case 1 and Case 2 (Figure 20.4), the answers were A, B and C. Next, ask them whether they liked or didn't like the products on the rating scale. Figure 20.4 shows the result of these rating scores on a straight number line. For a person in Case 2, the preference for both B and C is not so much different, but if asking the

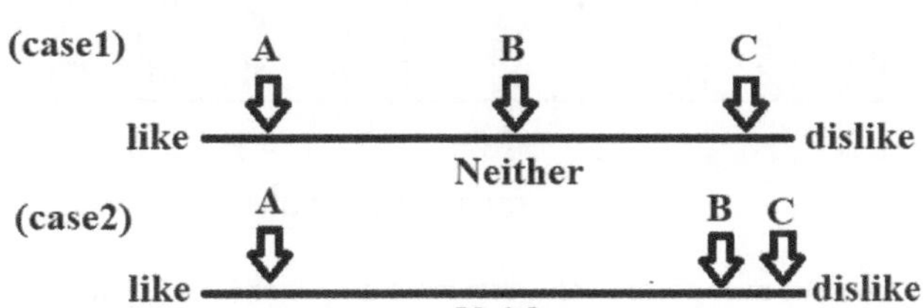

FIGURE 20.4 Respondent's answer to question, data of the rating and ranking scores on the line of scaling.

questions in the order of precedence, one can identify such a slight difference. Now, simply treat the ranking numbers obtained as metric quantities, there is a risk of amplifying the difference between B and C. Such cases are further introduced in Chapter 24.

The example introduced here is the result of a questionnaire survey conducted by Horie et al. (2017), which helps to explain some analysis examples of the characteristics of the response data. They presented the cards with 20 candidate products and asked for an evaluation. All cards were evaluated on a seven-point rating scale as follows: 7: want to buy very much, 6: want to buy, 5: want to buy a little, 4: can't say either, 3: don't want to buy a little, 2: don't want to buy, 1: never want to buy. Next, respondents were asked to rank the products from 1st to 20th in the order they wanted to buy. This survey was conducted for university students, and 123 respondents (out of 134) who did not make any mistakes or omissions were valid responses.

First, based on the answer data obtained by the ordinal scale for each respondent, the median (the product located at the centre of the whole when the candidate products are arranged in ascending or descending order) is calculated. Then the rating scale value of the corresponding product was searched. With this method, it is possible to compare the difference between the evaluation values by the ordering method and those by the rating scale method. The median is the average of the scores of the 10th and 11th cards on the rating scale method because they use cards for 20 candidate products. In addition, the intermediate answer (neither can be said) in the rating scale method is 4 points. Figure 20.5 shows the comparison between the ranking and rating scale data for the aggregated respondents.

This histogram shows the occurrence frequency of the rating scale value that corresponds to the centre of the ranking (ordering) scale data. As for a respondent with this rating scale value of 4, the answer was "neither" at the centre of the 20 cards. On the other hand, respondents with a median value of 1 mean that most of the 20 products are disliked. Even if the respondent ranks all of the 20 cards, there may not be any cards for the products that they want to buy. Note that because the number of choices is even, the median is the average of the two central data. So, in this case, the median is the average of the 10th and 11th values; hence the decimal digits appear.

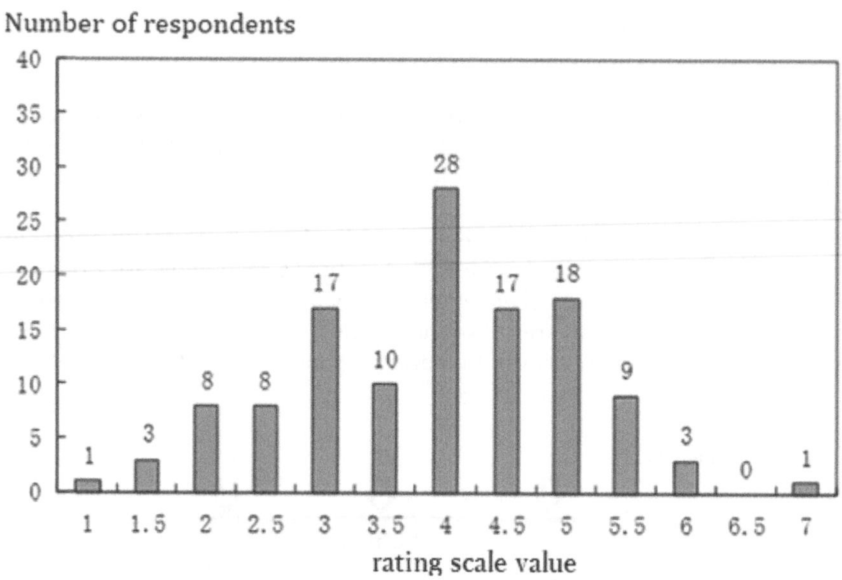

FIGURE 20.5 Histogram of the rating score value corresponding to the median in ranking data.

Source: Adapted and modified from Horie et al. (2017).

Although evaluation by the rating scale method is an absolute evaluation, there is a problem that it is difficult to discriminate subtle differences in whether or not the customer wants to purchase because the intervals between the scales are coarse. Therefore, each of the ranking methods and the rating scale method has advantages and disadvantages and can be used properly according to the purpose. This problem is also an important issue when measuring the preference of a commodity and considering various aspects of the commodity, which will be discussed further in Chapter 24.

20.6 Summary

Many difficult problems are associated with measuring attitudes toward commodities in marketing research and measuring symptoms in epidemiological research. Consequently, many related studies require various types of statistical data other than the quantitative values normally used in statistical analysis. Appropriate analytical methods are available for those types of data. In doing so, it is important to think about what and how to measure. Furthermore, in the questionnaire survey, it is important to properly design the questionnaire sheet because it greatly influences the accuracy of the final analysis result.

Responses in the questionnaire surveys are measured using a variety of scales, each has advantages and disadvantages. Ordinal data is good for measuring subtle differences, but the overall absolute scale remains unknown because the origin of the ruler cannot be determined. Note that, for the surveys, reducing the burden on the respondents is the most important concern. Excessive stress on respondents increases the number of sloppy responses and increases the probability of causing errors in the final analysis results.

REFERENCES

Antonogeorgos, G., Panagiotakos, D.B., Priftis, K.N. and Tzonou, A., 2009, Logistic regression and linear discriminant analyses in evaluating factors associated with Asthma prevalence among 10- to 12-years-old children: Divergence and similarity of the two statistical methods, *International J Pediatric, Open Access*, 2009, Article ID 952042, https://doi.org/10.1155/2009/952042

Asano, H., 2021, *Correspondence analysis, 10th lecture in SAS essay "marketing and data analysis"* (In Japanese), viewed on 2021, retrieved on 15 January 2021 from https://www.sas.com/ja_jp/industry/higher-education/fieldvoice/colum-asano/colum-asano–10.html

Blackmon, J.C., 2018, *Rosenblatt's perceptron*, published by Jcblackmon, https://jcblackmon.com/2018/01/08/rosenblatts-perceptron/Black

Grand Rapids Historical Commission, 2020, *Ringelmann smoke charts*, viewed on 2020, retrieved on 6 June 2020 from http://www.historygrandrapids.org/article/2211/ringelmann-smoke-charts

Horie, N., Shimamura, Y. and Okamoto, S., 2017, Consideration on the willingness-to-pay of the environmental attributes of a product by conjoint analysis, *J Japan Association for Management Systems*, 34(2), 225–233 (In Japanese).

Milton, J.S. and Arnold, J.C., 1995, *Introduction to probability and statistics principles and applications for engineering and computing sciences*, 3rd edition, McGrow-Hill, Inc., New York, NY.

Milton, J.S., McTeer, P.M. and Corbet, J.J., 1997, *Introduction to statistics*, WCB McGraw-Hill.

Pennsylvania State University, 2022, *Eberly college of science, STAT 504, analysis of discrete data, 6.1-introduction to GLMs*, viewed on 5 February 2022, https://online.stat.psu.edu/stat504/lesson/6/6.1

Phillips, D., 1995, *Correspondence analysis, issue 7, social research update*, University of Surrey, https://sru.soc.surrey.ac.uk/SRU7.html

Ram, P., 2022 updated, *Generalised linear model, what does it mean?*, Great Learning, https://www.mygreatlearning.com/blog/generalised-linear-models/

Sugimoto, N., 2014 last correction, *Logistic regression, chapter 10 of "introduction of statistics"*, Absurd Scrap Museum (In Japanese), retrieved on 6 January 2021 from http://www.snap-tck.com/room04/c01/stat/stat.html

Taylor, M., 2020, *Level of measurement, a classification that relates the values that are assigned to variables with each other*, Corporate Finance Institute, https://corporatefinanceinstitute.com/resources/data-science/level-of-measurement/

Yamada, K., Okamoto, S. and Ayano, K., 2002, *Quality control for non-engineering students*, revised edition, JUSE Publishing Co., Tokyo, Japan (In Japanese).

21

Applications of Principal Component Analysis and Exploratory Factor Analysis

This chapter presents the practical applications and clarifies the differences between principal component analysis (PCA) and exploratory factor analysis (EFA), which were not fully explained in Chapter 7. It features case studies that clarify the situations and ways these analytical methods are used. This chapter also explains essential technical terms to the extent necessary to enable readers to better understand the characteristics and various aspects of the case studies. The emphasis is on demonstrating examples of application using various datasets rather than on the detailed methodologies of each analysis. The quality of the data used is crucial; datasets containing numerous undetected (below detection limits) data points or questionable data will not yield reliable and interpretable results with any of these analytical methods.

21.1 Important Issues and Analytical Procedures for PCA and EFA

Important technical terms and analytical procedures needed for the case studies presented in this chapter are included in this section. However, there are slightly different views on these terms; hence the content here relies more on the general interpretation that is widely accepted among the researchers in the field. Having different objectives, PCA and EFA methods are similar because they both summarise the original variable set into a smaller, meaningful set of synthetic variables that maximises the retained information and variability derived from the original variable data (see more details in Chapter 7).

21.1.1 Terminology

The proportion of variance that can be explained by a set of factors common to many observed variables is called **communality**. The degree of communality provides information for determining whether a particular factor should be retained. There is also variance unique to that variable that is not explained by the factors, which is called **uniqueness**.

- ***Communality:*** This is calculated as the sum of squares of factor loadings explaining the observed variables and represents the extent to which the extracted factors as a whole explain the individual observed variables. The sum of the common variances of all the factors is the communality of the observed variables and takes values between 0 and 1. Therefore, subtracting the communality value of each observed variable from 1 gives the residual, unexplained variation. This value is called ***uniqueness***.
- ***Simple structure:*** Simple structure is a state in which the observed variables show higher factor loadings for factors with strong relationships and lower factor loadings for factors with weaker relationships, and each factor is more clearly structured.

DOI: 10.1201/9781003597216-24

21.1.2 Factor Extraction Process

One of the important procedures in EFA is factor extraction. There are several methods to perform the procedure, and they can be seen in the manual of SPSS (Statistical Package for the Social Sciences), a typical analysis software. The IBM SPSS (2022) includes several methods for factor extraction. Available methods are principal components, unweighted least squares, generalised least squares, maximum likelihood, principal axis factoring, alpha factoring and image factoring.

Most of the applications in the case studies presented in this chapter used the principal components method for factor extraction, which is referred to as factor analysis in principal component mode. A few case studies also used principal axis factoring.

- ***Principal components:*** Principal component analysis is used to obtain a solution for the initial factor.
- ***Principal axis factoring:*** This is a method of extracting factors from the original correlation matrix using the squares of the multiple correlation coefficients placed on the diagonal as initial estimates of communality. In some cases, the newly obtained communality is replaced with diagonal elements of the correlation matrix, and this is repeated until the difference from the previous value is sufficiently small.

21.1.3 Determination of the Number of Factors (Components) to Retain

For both EFA and PCA, it is necessary to decide how many factors to employ. In PCA and EFA, there are many possible components (factors) to be used for analysis, and, in principle, the maximum number of components can be an equal number of variables. Therefore, it is important to select meaningful factors (components) and retain them for the analysis.

Traditionally, it appeared that the principal components method was used for factor extraction, and the factors were often selected based on the Kaiser criterion, although there are various criticisms about this procedure. For further details, see specialised books in the field. In the context of this chapter, many, but not all use this traditional method.

- ***Kaiser criterion:*** Components with eigenvalues greater than 1 are adopted as having more information than one original variable.
- ***Scree test:*** The variance (eigenvalue) of the factors (or components) are listed in descending order, and the factor is censored at the point where its value is rapidly decreasing and only the earlier factors are adopted.
- ***Interpretability:*** Based on the loadings of a factor (or component), the meaning of the factor is interpreted and only those factors that can be interpreted are adopted.

It is important to decide how many factors to adopt. According to Jolliffe and Cadima (2016), in PCA, it is common to use a predefined percentage of the total variance listed to determine the number of principal components (PCs) to retain, e.g., 70% of the total variance is common at the subjective cut-off point. However, for graphic representation, often only the first two or three PCs are used for visual illustration. In most cases, PCA focuses on the first few PCs, but there are situations where the last few PCs are important, such as in outlier detection. This rule of thumb may also be applied to FA, in which the principal component method is applied to extract factors.

21.1.4 Rotation

In EFA, a new coordinate system is set up by the rotation so that the obtained results have a structure that is easy to interpret. There are two types of rotation: orthogonal rotation, in which the coordinate axes are orthogonal after rotation, and oblique rotation, in which the coordinate axes are not orthogonal after rotation. The orthogonal rotation method is a method of rotating with the factor axes orthogonal to each other. A typical method among them is varimax rotation. It has been used very often in various fields because it often gives easy-to-interpret results, which means that the structure of the factor is simple. In other words, the rotation of the factor axes can be thought of as a search for ***simple structures***. A simple structure is a pattern of results in which each variable has a high loading on only one factor.

21.2 A New Way of Thinking about PCA

PCA is a common multivariate analysis method that has been used for a long time in various fields. The method has found new applications in the recent development of information technology (IT) such as artificial intelligence (AI) and image analysis as reviewed by Jolliffe and Cadima (2016). They also indicated that the main uses of PCA are descriptive rather than inferential. Therefore, PCA as a descriptive tool needs no distributional assumptions and, as such, is very much an adaptive exploratory method that can be used on numerical data of various types.

The recent growth of Big Data science dealing with very large datasets in areas such as image analysis or machine learning has brought about important methodological advances in data analysis. In these new application areas, one can see attempts to explain PCA from a different perspective than in the past. The higher the number of dimensions in a dataset, the higher the computational cost, not only in data analysis but especially in machine learning, and PCA can be used as a solution. If the number of dimensions (PCs) can be reduced to three or less, it can also be used for data visualisation. PCA automatically finds features from trends in given data and performs dimensional compression to lower dimensional data that better represents the features; this is a type of machine learning that is classified as "unsupervised learning", especially in that it automatically finds features (Aidemy, 2023). In areas such as machine learning, PCA can be described as follows:

- PCA is an unsupervised, non-parametric learning algorithm that is used for dimensionality reduction in machine learning.
- PCA is a statistical process that converts the observations of correlated features into a set of linearly uncorrelated features with the help of orthogonal transformation.

Goonewardana (2019) also mentions the effectiveness of PCA in artificial intelligence. When the feature space is reduced by PCA, dimensionality reduction can be expected to improve model performance and efficiency without significantly impacting accuracy measures.

21.3 Relationship between PCA and EFA

PCA and EFA are both statistical analysis methods for reducing the dimensionality of the dataset, and there are many similarities in the analysis procedures. However, on the other hand, they are based on completely different conceptual models for the relationship between data and factors (components).

As for the difference between PCA and EFA, Brown (2009) explained that PCA analyses variance and EFA analyses covariance (common variance). Thus, when researchers want to analyse only the variance that is accounted for in an analysis (as in situations where they have a theory drawn from previous research about the relationships among the variables), they should probably use EFA to exclude unique and error variances, to see what is going on in the covariance, or common variance. When researchers are just exploring without a theory to see what patterns emerge in their data, it makes more sense to perform PCA (and thereby include unique and error variances), just to see what patterns emerge in all of the variance.

Alavi et al. (2020) mentioned that there are reasons that encourage researchers to use PCA rather than EFA. First, there are circumstances in which PCA could be considered a good approximation of EFA leading to yielding similar output statistics. Second, another reason for the increased use of PCA is that it is usually the default option in some statistical software packages, increasing its use despite other approaches. An awareness of the differences between PCA and EFA allows for alignment between the statistical approach and research objectives and ensures appropriate interpretation of results. In this regard, Suhr (2005) mentioned that PCA and EFA are both variable reduction techniques, and if communalities are large, close to 1.00, results could be similar and further discusses the differences between PCA and EFA.

In EFA the factors influence the data, whereas PCA considers the data as influencing the components. This is the view of researchers who see PCA and EFA as conflicting concepts. However, this may not be relevant to researchers who use PCA only to compress the amount of information. The relationship between PCA and EFA is actively discussed in fields such as psychology, behavioural sciences, and epidemiology. However, in many fields of science and engineering, if not all, only a few researchers have shown interest in EFA. All the causal relationships assumed in EFA are linear models. In such fields, if it is known that there is a clear linear relationship between the variables (factors) and measured data in a causal system, one might think that focusing on factor extraction is appropriate. However, if the relationship is known, a researcher might think that it would be more suitable to analyse the phenomenon physically or chemically rather than statistically.

21.4 Case Studies on the Application of PCA and EFA

Here are some examples of statistical analysis using PCA and EFA. These examples provide insights into the usefulness and challenges of these analysis methods. As mentioned previously, the recent focus of PCA use is for dimensionality reduction of the dataset. Traditionally, however, PCA is a typical multivariate analysis method, and its main purpose is to analyse similarities and heterogeneities among many variables, and most of the examples presented here using PCA analysis are based on this traditional viewpoint. Case 6 is presented as an example of analysis with EFA, assuming that latent factors behind the data can be assumed to some extent at the beginning of the analysis. Finally, Case 7 is presented as a contrast between PCA and EFA and emphasises the importance of the data quality subjected to the analysis. A summary of these Cases 1–7 is shown here.

1. Analysis of air quality data in Tokyo in the 1960s
2. Classification of meteorology patterns affecting air quality in Northern Thailand
3. Application of PCA for analysis of roadside air quality
4. Analysis of spatial homogeneity in wind field by cosine similarity
5. Analysis of the spatial distribution of photochemical air pollution (including detection of outliers)
6. Application of a binary dataset to analyse the biogeographic distribution of butterflies
7. Application of PCA and EFA to the analysis of PM elemental composition for identifying contributing sources

Cases 1, 2, 3, 5, and 7 are examples of analyses for common metric data. These analysis methods can be applied to various types of data. Case 4 describes the analysis of vector quantity of winds with cosine similarity. Case 6 presents a case study of binary data. In addition, Case 7 details an unsuccessful analysis and identifies its causes. By looking at these various analysis cases, the readers can gain insights into the advantages and disadvantages of these methods. Note that this chapter will not address the analysis of certain categorical data, such as that obtained from questionnaire surveys or sensory tests. These topics will be explored in greater detail in Chapter 24.

21.4.1 Case 1: Analysis of Air Quality Data in Tokyo in the 1960s

The first application of PCA discussed here is for the data collected in Tokyo about 50 years ago when many cities in Japan, Europe and the United States were still suffering from air pollution. Shiozawa et al. (1972, 1975a, 1975b) studied a statistical prediction system for air pollutant concentrations to prevent extreme air pollution. PCA and other methods were used to analyse the relationship between air pollutant concentrations and meteorological conditions. Air pollutant concentration data were available on an hourly basis for five air pollutants (listed in Table 21.1) measured in February 1968 at a site located in front of the former Tokyo Metropolitan Government Office (presently, near Yurakucho railway station). To exclude the effects of the diurnal variations of the concentrations, the daily average values (24-hour average) were calculated from the hourly data to examine day-by-day changes. First, these daily averages were analysed to reveal the relationship with the synoptic scale meteorological change.

It could be assumed that the daily activities that produce pollutants in the community were not very different day by day, the variations of the pollutant concentrations were mainly dominated by atmospheric factors such as meteorological conditions. Table 21.1 presents the correlation coefficients between the pollutant concentrations, which shows all positive values.

The positive correlations therefore indicate that the pollutants had similar variation trends, i.e., together they increase or reduce. Considering the emission sources, NO_x was mainly emitted by automobiles and various stationary sources, SO_2 was mainly emitted by stationary sources and CO was mainly from automobiles. Since the correlation coefficient between CO and SO_2 was positive, it could be considered that some common factors were related to both pollutant concentrations that may be linked to the meteorological condition.

PCA is generally a useful procedure whenever a study aims to determine the minimum number of independent new variables needed to be condensed from the original dataset of too many variables. The value of the weighting factor for finding the principal component score (PC score) is proportional to the correlation coefficient between the PC score and the original variable. This correlation coefficient is also called principal component loading (PC loading). The PC loading also plays an important role in understanding the meaning of the principal components.

The PCA for the daily average concentration of five air pollutants retained two PCs, which explained 80% of the total variances of the pollutant concentrations as seen in Table 21.2 (Shiozawa et al., 1971). The loadings for all pollutants on PC1 (the first principal component)

TABLE 21.1

Correlation Coefficients between the Daily Average Concentrations of Five Air Pollutants

	SPM/PM_{10}	SO_2	CO	NO	NO_2
SPM/PM_{10}	1.000	0.570	0.596	0.152	0.413
	SO_2	1.000	0.618	0.424	0.657
		CO	1.000	0.340	0.599
			NO	1.000	0.547
				NO_2	1.000

Notes: Sample size: $n = 28$; Threshold value in uncorrelated test with sample size 25 and 5% significance level, $r\ (25, 0.05) = 0.381$.

SPM (suspended particulate matter) is measured with a light scattering instrument and is considered to be equivalent to PM_{10} because it is calibrated with the measurement data of a low-volume air sampler with the size selection device for PM_{10}.

TABLE 21.2

Result of the PCA, PC Loading, and Its Variance

Component		1	2
PC loading	SPM	0.71	−0.559
	SO_2	0.83	−0.062
	CO	0.83	−0.223
	NO	0.605	0.706
	NO_2	0.841	0.246
Variance (% explained)		3.007 (60.1%)	0.926 (18.5%)

are positive, which indicates the similarity between various pollutants (also indicated by the positive correlation coefficients in Table 21.1). Note that since all the loadings are positive and have the same sign on PC1, it is considered to be an indicator of overall air pollution. In the PC2 (second PC), some loadings are positive, and others are negative, so the second component represents the difference between the pollutants. In the PC2, NO and NO_2 have positive loadings, which are different from other pollutants. These may be associated with the vehicle emissions that are discharged from near the surface. Note that these data were measured back in February 1968 and reflected the situation just before the onset of strict vehicle emission regulations.

To understand the meaning of each PC, the relationship between synopsis-scale meteorological conditions and PC scores was analysed. Accordingly, the weather map pattern was superimposed on the PC score distribution to reveal their relationship. Figure 21.1 is a scatter plot of the PC scores. The ordinate (*Y*-axis) indicates the first PC, and the abscissa (*X*-axis) indicates the second PC. The marks in the scatter plot represent the PC scores, and each of the attached numbers in Figure 21.1 indicates the date (in February 1968). By comparing these plots to daily weather charts, one can observe that the changes in PC scores correspond to synoptic-scale pressure patterns.

As shown in Figure 21.1, the typical pressure distribution in the study area in winter consists of high pressures that govern the Eurasia continent and low pressures over the Sea of Okhotsk. This is the basic pattern of atmospheric pressure during winter, in general, which changes periodically. In addition, after high pressure moves east, a trough of atmospheric pressure approaches, and the low pressure (cyclone) advances. As this cyclone travels along the southern coast of the Japanese archipelago, rain and snow fall in the Tokyo area. On the other hand, when the low pressure goes to the Sea of Japan, southerly winds become stronger and the weather worsens, but it does not rain as much. Regardless of which course it follows, when it moves to the Sea of Okhotsk or the Pacific Ocean, it will develop further, inducing a strong northwest monsoon. The interesting finding that the concentration of each air pollutant is changing along with this meteorological cycle provides a useful clue when constructing statistical prediction models. In the Tokyo metropolitan area, strong southerly winds cause an increase in sulphur oxide concentrations because of the presence of industrial areas on the south side facing Tokyo Bay. In addition, nitrogen oxide concentrations, a relatively high percentage of which is emitted from near the ground surface, tend to be lower due to higher wind speeds. Specifically, this is the case on February 19–20 and 28–29.

21.4.2 Case 2: Classification of Meteorology Patterns Affecting Air Quality in a Mountainous Area

A city located in a mountainous area in Southeast Asia experiences air pollution haze episodes around March each year. One of the most severe haze episodes was observed in March 2007 when the whole city was blanketed with haze for a few weeks and the number of respiratory

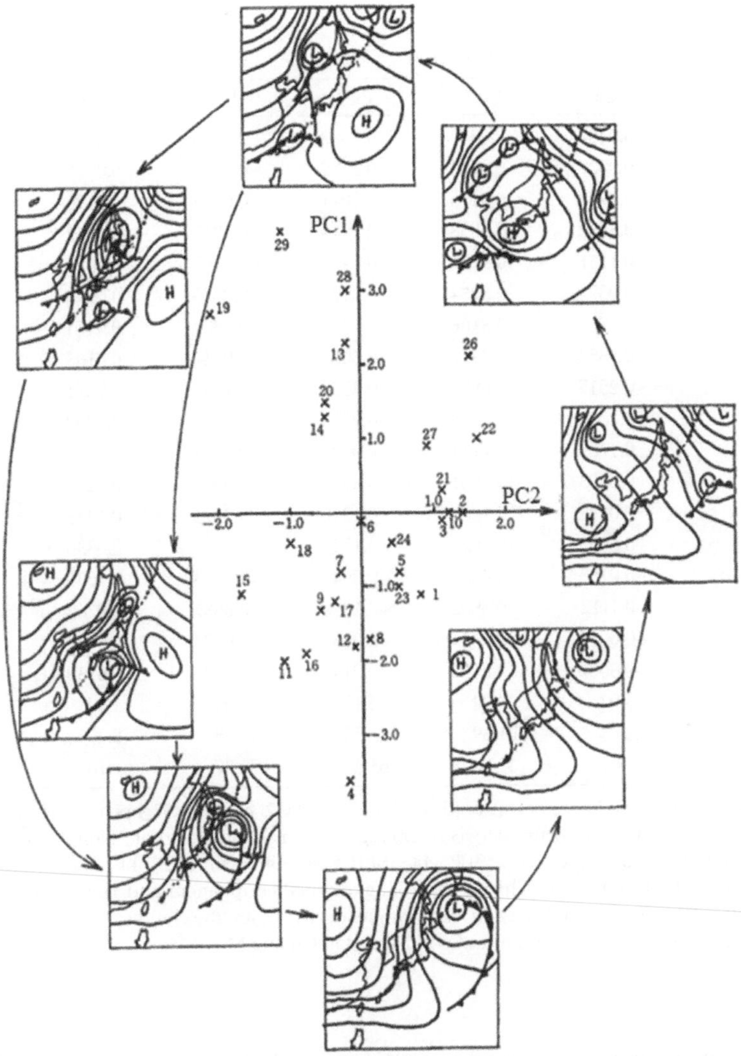

FIGURE 21.1 Scatter plots of the scores of the first and second principal components and the corresponding synoptic weather patterns. The numbers on the scatter plots represent the dates.

Source: Adapted and modified from Shiozawa et al. (1971).

patients also increased sharply in this period. Data from the ambient air monitoring stations of the government monitoring network showed that the 24 h average PM_{10} level in the city reached the peak of 396 µg/m^3. The climate of the area is influenced by a monsoon regime. The period from mid-February until the end of May is the transition period from Northeast monsoon (most prevalent in December–January) to Southwest monsoon (most prevalent in July–August). The hottest weather observed during March–April coincided with the presence of intensive thermal lows governing the area.

The final goal of this case study was to develop a statistical prediction model for PM_{10} in the area, which is detailed in Section 22.7, Chapter 22. This section presents the application of PCA to develop an automatic scheme to classify the meteorological conditions governing the region during the haze-prevalent months (February–April) into homogeneous patterns using PCA. The meteorological conditions play a key role in haze formation, not only

TABLE 21.3

Loadings on Five PCs with Varimax Rotation

Variable	PC1	PC2	PC3	PC4	PC5	Communalities
PG5_ybcm	**0.9402**	−0.0049	0.1692	−0.0585	0.0212	0.9165
PG4_whcm	**0.9337**	0.0286	0.1582	−0.1464	−0.0226	0.9197
SLP_ybC	**0.8700**	0.3476	0.2021	−0.0554	0.0196	0.9220
SLP_whC	**0.8552**	0.3838	0.1918	−0.1313	−0.0187	0.9330
SLP_cm	0.1161	**0.9723**	0.1466	−0.0107	0.0025	0.9804
SLP_bkk	0.0375	**0.8754**	0.3124	−0.0873	−0.0078	0.8730
SLP_np	0.1965	**0.8106**	**0.5374**	−0.0004	0.0174	0.9848
SLP_hkC	0.3853	**0.7411**	**0.5067**	−0.0444	0.0163	0.9567
PG1_bkkcm	−0.2515	**−0.6873**	0.3523	−0.1887	−0.0271	0.6960
PG3_npcm	0.2338	0.1320	**0.9273**	0.0171	0.0339	0.9333
PG2_nkcm	0.1767	0.2181	**0.9024**	0.0784	0.0712	0.9044
PG6_hkcm	**0.5193**	0.3001	**0.6885**	−0.0623	0.0242	0.8382
Cl_cm	0.0232	0.1863	0.1349	**0.7575**	0.1273	0.6435
Td_cm	−0.3093	−0.2823	0.1725	**0.6861**	0.0836	0.6829
MH_cm, morning	−0.0185	0.0566	−0.0048	**0.6653**	0.0066	0.4463
Vis_cm	−0.1112	−0.0802	−0.1840	**0.6087**	0.0262	0.4239
WDI_cm	−0.0015	−0.0324	−0.0653	0.0633	**−0.8934**	0.8074
Wsp_cm	−0.0007	−0.0195	0.0147	0.2930	**0.8130**	0.7475
Eigenvalue	3.97	3.94	3.15	2.05	1.49	
% Variance	22.08	21.89	17.53	11.38	8.29	
% Cum Var.	22.08	43.96	61.49	72.87	81.16	

Note: The high loadings (>0.5) are set in bold. PG1_bkkcm, PG2_nkcm, PG3_npcm, PG4_whcm, PG5_ybcm, PG6_hkcm are the pressure gradients between Chiang Mai (cm) and Bangkok, Nongkhai, Nakhonphanom, Wuhan, Yibin and Haikou, respectively. SLP_ybC, SLP_whC, SLP_cm, SLP_bkk, SLP_np, SLP_hkC are the sea level pressures at Yibin, Wuhan, Chiang Mai, Bangkok, Nakhonphanom and Haikou, respectively. Cl_cm, MH_cm, morning Td_cm, Vis_cm, WDI and Wsp_cm, respectively, are the cloud cover, morning mixing height, dewpoint, visibility, wind direction index and wind speed in Chiang Mai.

due to the dispersion and transport of pollutants but also in the occurrence and intensity of forest fires, which are a main cause of the haze in the area. It is therefore important to classify the meteorological conditions into patterns and analyse the PM_{10} levels associated with each pattern for identifying the haze-prone meteorological patterns. Subsequently, early warnings and, further, a prediction model for PM_{10} can be developed for the identified patterns.

The surface meteorological data at 7:00 LST (00 UTC) during February–April of 8 years (2001–2008) were collected from a regional weather station network of 7 stations: 4 stations in Thailand (Chiang Mai, Bangkok, Nongkhai and Nakhonpanom) and 3 stations in China (Yibin, Wuhan and Haikou), as detailed in Kim Oanh and Leelasakultum (2011). These included all weather variables at the stations plus the calculated difference between the sea level pressure (SLP) at each station and Chiang Mai, respectively, to represent the horizontal pressure gradient. The data processing procedure was similar to a previous study (Kim Oanh et al., 2005). First, data screening was done using PCA and K-mean clustering to select a set of the most suitable variables for the final PCA analysis. The final dataset selected for the analysis included 18 meteorological variables, shown in Table 21.3, characterising the thermodynamic properties of the air masses over the study area, which could be classified into three groups. The first group primarily included the meteorological variables of Chiang Mai. The second group included variables related to SLP in the selected stations, and the third comprised variables related to the pressure horizontal gradients between Chiang Mai and other stations.

The PCA, conducted on the final set of 18 meteorological variables, produced the first five PCs that collectively explain 81% of the total variance in the raw dataset. Table 21.3 presents the loadings of meteorological variables on each PC, where the values >0.5 were considered significant and bolded. PC1, explaining 22.1% of the data variance, represented the regional pressure field influencing the area as seen from the high positive loadings of SLP at two weather stations in central China (Yibin and Wuhan) and the horizontal pressure gradients between those stations and Chiang Mai, respectively. PC2, which explains around 21.9% of the data variance, had high loadings of SLP at four stations and a negative loading of the pressure gradient between Bangkok and Chiang Mai; hence PC2 appears to represent the pressure field at lower latitudes of around Northern Thailand. PC3 explained 18% of the variance and was linked mainly to the horizontal pressure gradient field. PC4 explained around 11% of the data variance and had high positive loadings of cloudiness, dewpoint temperature, morning mixing height, and visibility measured in Chiang Mai. Wind direction index and wind speed at Chiang Mai had high loadings on PC5, which explained around 8% of the data variance and suggested that this PC would be linked to dynamic characteristics of the local air mass.

In the next step, the K-means clustering technique was applied to the component score matrix (714 days, 5 PCs) which successfully classified the meteorological conditions of the 714 days into four patterns that appeared to be consistent with those observed on the synoptic charts. The average meteorological variables of the days belonging to each pattern were calculated to better explain the thermodynamic characteristics of the patterns (Kim Oanh and Leelasakultum, 2011). Figure 21.2 shows examples of the synoptic weather maps on the

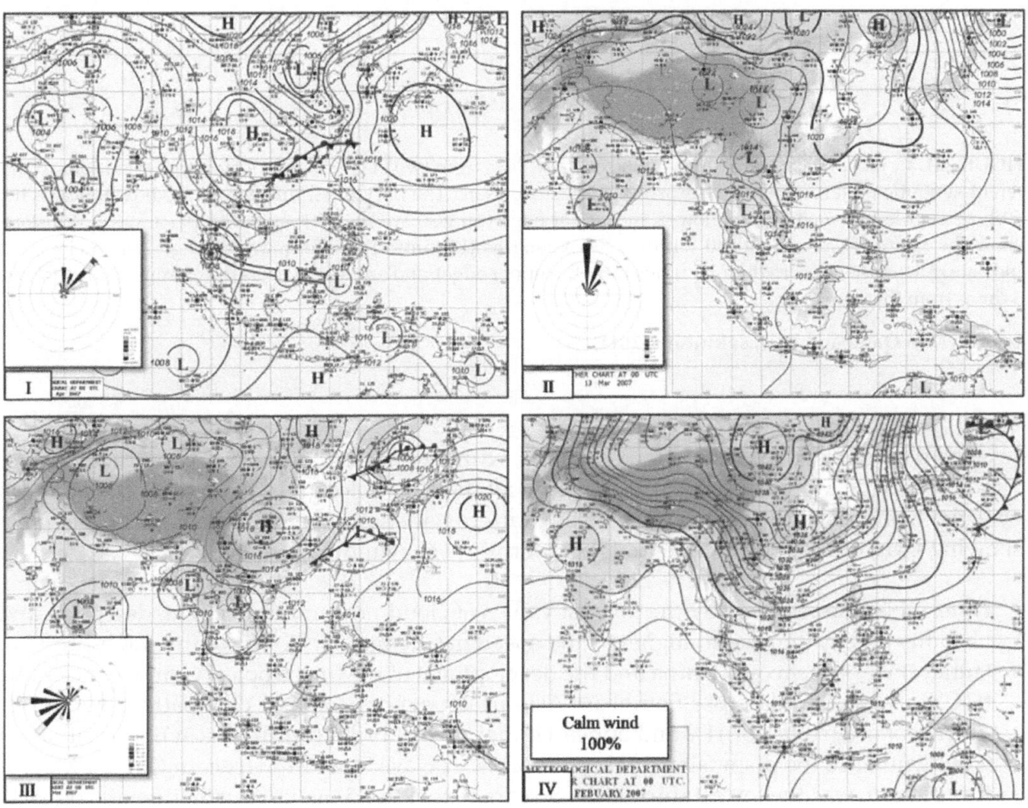

FIGURE 21.2 Examples of regional surface synoptic charts at 07.00 LST (00 UTC) with inserted wind roses: (I) Pattern I (April 30, 2007), (II) Pattern II (March 13, 2007), (III) Pattern III (March 27, 2007) and (IV) Pattern IV (February 1, 2007).

Source: Extracted from Kim Oanh and Leelasakultum (2011).

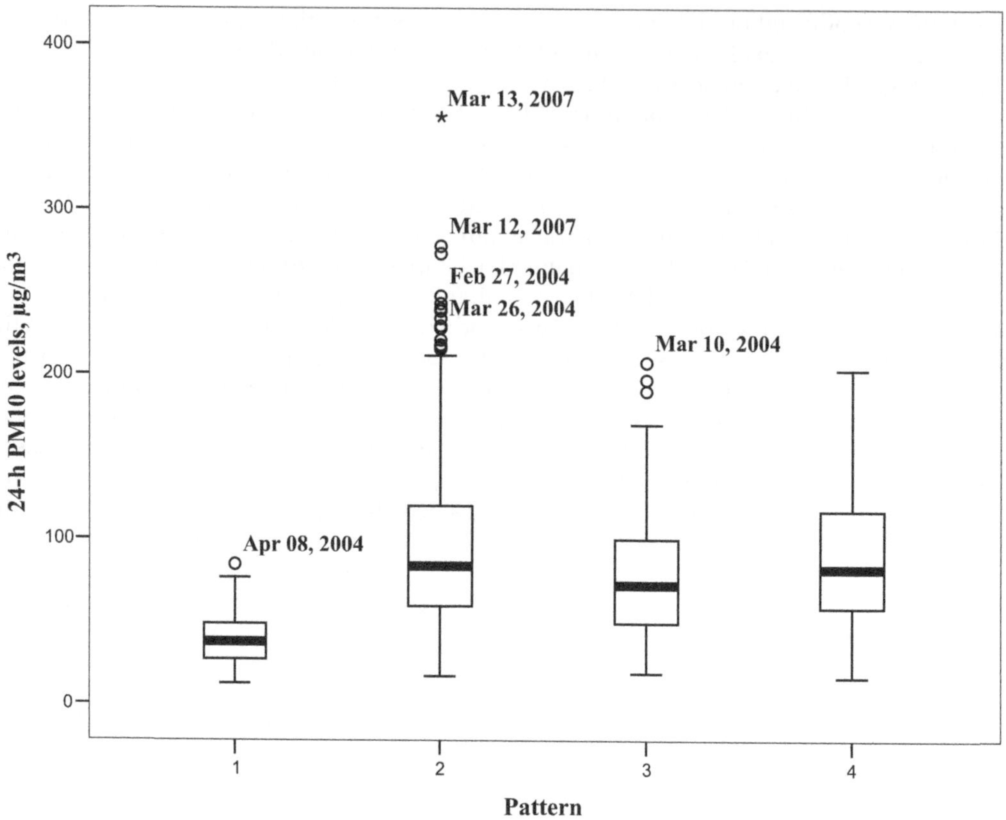

FIGURE 21.3 Box plots of the average value of daily PM_{10} from two of Chiangmai's monitoring stations in different synoptic patterns, February–April of 2001–2008. (*Note:* The horizontal line across the box indicates the 50th percentile (the median), the lower and upper edge (hinges) are the 25th and 75th percentiles. Whiskers show the range of values that fall within 1.5 box lengths (1.5 times the interquartile range). Open circles (°) indicate mild outliers (more than 1.5 times from the upper edge), and an asterisk (*) is the extreme outlier (above 3 times from the upper edge).

Source: Kim Oanh and Leelasakultum (2011).

selected days representing these four patterns, with the inserted wind rose constructed using winds at Chiang Mai measured at 7:00 LST on all the days belonging to the respective pattern over the 8 years. The characteristics of the patterns are summarised below while the plots showing PM_{10} levels associated with the patterns are shown in Figure 21.3.

- Pattern I is characterised by the highest values of cloudiness, mixing height, dewpoint and visibility and by the second highest wind speed in Chiang Mai. A weakening high-pressure (anticyclone) system over China and the development of the Intertropical Convergence Zone (ITCZ) are observed across Southeast Asia in the synoptic chart. This pattern had a frequency of occurrence of 11% and appeared mainly in April. Overall, this synoptic situation would enhance the dispersion of air pollutants. The average PM_{10} levels in Chiang Mai for this pattern, was the lowest during the 8 years with 40 ± 15 μg/m^3. Further details on PM_{10} associated with the four patterns are provided in Chapter 22 (Section 22.7) in a case study of developing a forecasting model for PM_{10} in Chiang Mai.
- Pattern II features a high-pressure ridge extending from Northeastern China toward Southeast Asia and the Pacific Ocean. The dry weather is observed in Chiang Mai with clear

sky, lowest morning mixing height and low wind speed (0.2 m/s). This pattern indicates stable conditions influenced by high-pressure systems that enhance the build-up of high air pollution levels. It has the highest occurrence frequency during the study period (49%), mainly observed in March and April. PM_{10} levels were the highest among the four patterns with an average of 97 ± 54 µg/m^3.

- Pattern III is shown with the highest wind speed, and the wind directions had the highest frequency of westerly to south-westerly winds. In the synoptic charts, a vast thermal low with several cells is observed over East and Southeast Asia. This pattern is not frequently observed (14%) and is fairly well distributed over the 3 months. The average PM_{10} was 79 ± 41 µg/m^3, which is the second lowest but still almost double that of Pattern I.
- Pattern IV is characterised by the highest SLP at all the stations. Synoptic charts typically show a strong stationary ridge dominating over Southeast Asia. Calm wind (100%), clear sky and dry air observed in Chiang Mai indicate the influence of stable anticyclone systems that limit pollutant dispersion. This pattern occurs quite often (26%) and is mainly observed in February when the study area is under the influence of the Northeast monsoon. The average PM_{10} was 89 ± 42 µg/m^3, which is the second highest to Pattern II.

21.4.3 Case 3: Application of PCA for Analysis of Roadside Air Quality

PCA transforms a large number of variables (observed data) into a meaningful small number of PCs, which are composite variables aggregating some amount of information. By carefully analysing the weight coefficients (loadings) and PC scores obtained, it is possible to reveal relationships among a large number of observed variables. This section presents two examples of the application of PCA from this perspective.

The dispersion pattern of road traffic emissions can be studied using the distance attenuation of pollutant concentrations from the road as explained in Chapter 17. Typical pollutants from automobile exhausts include carbon monoxide (CO), nitrogen oxides (NO_x), and volatile organic compounds (VOCs), as well as specific compositions in particulate matter. In particular, a sub-group of VOCs called BTEX (benzene, toluene, ethylbenzene and xylenes) contains the markers suitable for analysing the degree of influence of vehicle types such as diesel and gasoline vehicles.

21.4.3.1 Roadside BTEX Measurements and Traffic Flows in an Urban Area

Hourly concentrations of benzene, toluene, ethylbenzene, m&p-xylenes and o-xylene, plus CO, NO_x, and SO_2 were monitored at the roadside simultaneously with the hourly traffic volume during the dry season of 2004, in a city in Southeast Asia (Quynh Truc and Kim Oanh, 2007). The selected three roads included the TC street with a high traffic volume, DBP street with a low traffic volume and the NT street with a high traffic volume and running through an industrial area. Hourly NO_x, CO and SO_2 were recorded by automatic monitoring instruments while the hourly BTEX samples were conducted during the morning peak hours (6:30–9:30 h), the mid-day off-peak hours (11:00–13:00 h), and evening peak hour (16:00–19:00 h). BTEX samples were collected by SKC-coconut shell charcoal tubes and analysed by GC–FID with necessary QA/QC. BTEX levels were measured simultaneously at both sides of each street, whereas other gases were measured at the downwind side.

The pollution levels were significantly higher in the most heavily travelled street of NT with an hourly traffic volume of 15,000–24,000 veh/h during daytime (7:00 AM – 7:00 PM), followed by TC with 10,000–15,000 veh/h, and the least at DBP with 1500–2000 veh/h. Specifically, the BTEX measured in NT was the highest with the geometric mean of benzene of 120 µg/m^3, followed by TC with 65 µg/m^3, and the least with DBP with 30 µg/m^3. Motorcycles were the most dominant in the traffic fleet, sharing 94–96%. The VOC levels were found to decrease with increasing downwind distance from the road, as seen in Figure 21.4, which shows the monitoring data at 3, 10, 30 and 50 m from the TC road shoulder. Note that the measurements

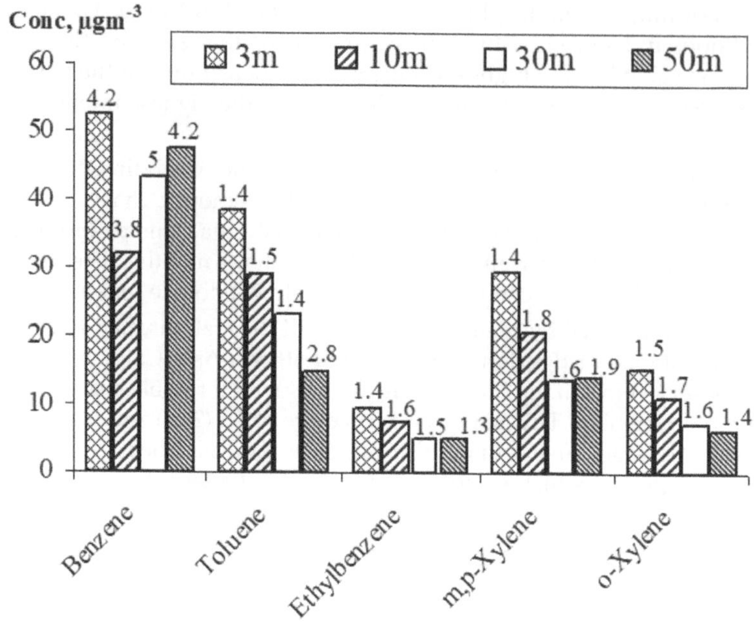

FIGURE 21.4 Geometric mean concentration and its geometric standard deviation of BTEX (shown values) at distances of 3, 10, 30 and 50 m downwind from the street (each data point is the mean of 14 measurements).

were done during the days with stable northeast wind; hence the wind direction was almost perpendicular to the street orientation. Concentrations of the BTEX decreased considerably with the distance except for benzene, which may be contributed by other sources such as a parking lot near the monitoring points.

PCA was performed on the obtained dataset using the software package SPSS to reveal the relationship between the hourly measured pollutants and the hourly traffic flows (number of vehicles per hour) of six vehicle categories (Table 21.4). To exclude the effects of wind direction on the roadside on the BTEX concentrations (windward vs. leeward), the average of the levels simultaneously measured at both sides of each street was used in the PCA analysis. In total, 1366 data points were included in the dataset after the exclusion of an abnormally high level of benzene in the NT roadside (>10,000 µg/m^3) which was most likely due to the intermittent influence of the industrial activities.

Three PCs were retained that collectively explain 77% of the total variance of the dataset (Table 21.4). PC1 has high loadings of xylenes, ethylbenzene, toluene, CO (all above 0.9), benzene (0.59) and motorcycle number (0.4), suggesting that these pollutants were likely to be associated with motorcycle emissions. Bus, van, taxi, car, motorcycle, and truck were possibly linked to NO, which all have significant loadings in PC2. The truck group and SO_2 have high loadings in PC3, which confirms the effect of high sulphur in diesel used by trucks at the time of the measurements for trucks, and therefore can contribute to SO_2 levels measured at the roadside. As expected, NO_2, a secondary air pollutant that is mainly formed in the atmosphere from the precursor of NO, was not linked explicitly to any vehicle groups.

PCA results are thus highly consistent with other findings of the study to indicate that the roadside levels of air pollution in the city were strongly influenced by traffic activities. It further indicates the need for appropriate traffic management, including improving vehicle technologies and controlling the traffic flow, to reduce the air pollution in the city.

TABLE 21.4

Loadings of Different Variables on Principal Components with Varimax Rotation

Variable	Component PC1	PC2	PC3
Benzene	**0.594**	0.030	**−0.411**
Toluene	**0.937**	0.254	−0.088
Ethylbenzene	**0.940**	0.291	−0.055
m,p-xylenes	**0.961**	0.212	−0.062
o-xylene	**0.959**	0.208	−0.039
Motorcycle	**0.400**	**0.639**	**−0.531**
Van	0.238	**0.806**	−0.066
Car	0.363	**0.721**	**−0.449**
Taxi	0.176	**0.774**	0.260
Bus	0.199	**0.849**	0.035
Truck	−0.217	**0.586**	**0.521**
CO	**0.911**	0.164	−0.029
NO	0.276	**0.727**	0.093
NO_2	−0.074	**−0.635**	0.261
SO_2	0.001	−0.007	**0.814**
%Variance	49	18	10

Note: Values ≥ 0.4 are significant and bolded.

21.4.3.2 Analysis of Particulate Matter Composition Measured at a Roadside in Tokyo

Particulate matter (PM) with the particle size range ≤11 μm samples were collected around Tokyo's Loop Road No. 8, which sees a traffic volume of approximately 100,000 vehicles per day. The composition of these samples, detailed in Table 21.5, was analysed to assess the impact of emission sources. This PM sampling was conducted over two continuous weeks during winter (December 10–24, 1998) and spring (April 7–21, 1999) in a Tokyo residential area (Okamoto et al., 2001). Sixteen identical air samplers (Andersen impactors) were positioned along the alley at the position of 0, 25, 50 and 100 m from the road's edge on both sides. At each location, two samplers were installed: one with a cellulose acetate membrane filter and the other with a quartz filter, ensuring ample filter material for all analyses. The findings indicated that PM concentrations diminished with increasing distance from the road, showing a decrease of approximately 20% at 100 m. The total PM composition in winter primarily consisted of carbon (a combination of EC and OC) at 40%, water-soluble ions at 30% and metals at 10%.

PCA was applied to the dataset comprising the chemical composition concentrations in all the PM samples, combining the concentrations in both fine and coarse fractions; hence the levels can be regarded as the compositions of PM with the particle size range ≤11 μm. The dataset contains 16 samples (8 sampling points and 2 seasons) and 25 components. Therefore, PCA was performed using a dataset of 25 variables and 16 data points. The result showed that three principal components were retained. The PC loading and cumulative contribution rate are shown in Table 21.5. For every PC score, the attenuation tendency according to the distance from the road was examined, and the results are shown in Figure 21.5. The first principal component (PC1) has large differences between the winter and spring seasons. PC2 has a large difference between the north and south sides of the road. PC3 has the most pronounced attenuation trend from the road.

TABLE 21.5

PC Loadings after the Varimax Rotation

Component	PC1	PC2	PC3	Average concentration ($\mu g/m^3$)
Al	−0.106	**0.849**	−0.037	0.66
Br	**0.928**	0.030	−0.336	0.020
Ca	**0.913**	0.323	−0.023	0.58
Ce	**0.377**	0.643	0.259	0.00084
Cl	**0.875**	0.134	−0.412	0.79
Cr	−0.508	−0.090	0.051	0.0029
Fe	**0.953**	0.125	−0.108	0.73
K	**0.935**	0.246	−0.108	0.50
La	**0.884**	0.344	−0.160	0.00044
Mg	**0.721**	0.137	−0.049	0.23
Mn	**0.934**	0.318	−0.131	0.020
Na	−0.929	0.050	0.178	0.67
Sb	**0.771**	−0.034	−0.310	0.015
Sc	**0.651**	**0.702**	0.018	0.00011
Si	0.247	**0.853**	−0.016	2.3
Sm	**0.858**	0.404	−0.227	0.00005
Ti	**0.771**	0.569	−0.041	0.046
V	−0.669	0.595	0.261	0.0042
Zn	**0.927**	0.000	−0.327	0.17
SO_4^{2-}	−0.286	0.085	**0.946**	4.9
NO_3^-	**0.950**	0.137	−0.065	7.4
Cl^-	**0.967**	0.117	−0.184	2.8
NH_4^+	**0.977**	0.023	−0.035	4.6
EC	**0.942**	0.145	0.206	17.2
OC	**0.951**	0.207	−0.102	6.9
Variance	16.1	3.8	1.8	
% cumulative variance	64.3	79.3	86.4	

Note: EC: elemental carbon; OC: organic carbon.

The factor loadings of PC1 are mostly large positive values (Table 21.5); hence it is thought to be a factor that represents the overall trend of increase or decrease in concentration. The factor scores in winter show a high value (Figure 21.5); therefore, PC1 indicates the seasonal variation. The high loadings of Br, K, Zn and OC suggest a large contribution of waste incineration or open burning, even at this busy roadside. PC2 has a high factor loading for such soil-related components as Ce, Sc, Si and Ti, which are larger at sampling points close to the roadside but also larger at locations farther away from the road on the south side of the road in winter (Figure 21.5). On the south side of this road, there was a park that included tennis courts and a baseball field; hence there could be a possibility that the soil dust is dispersed from these bare soil areas during the dry winter months to the sampling site. PC3 has a very high loading of SO_4^{2-} and tends to have fairly high V, Na and EC. At the time of this study, the sulphur content in diesel fuel was not yet strictly regulated and its main source was thought to be diesel vehicle exhaust. With the pronounced tendency of PC3 to attenuate from the road, PC3 thus can be attributed to diesel vehicles.

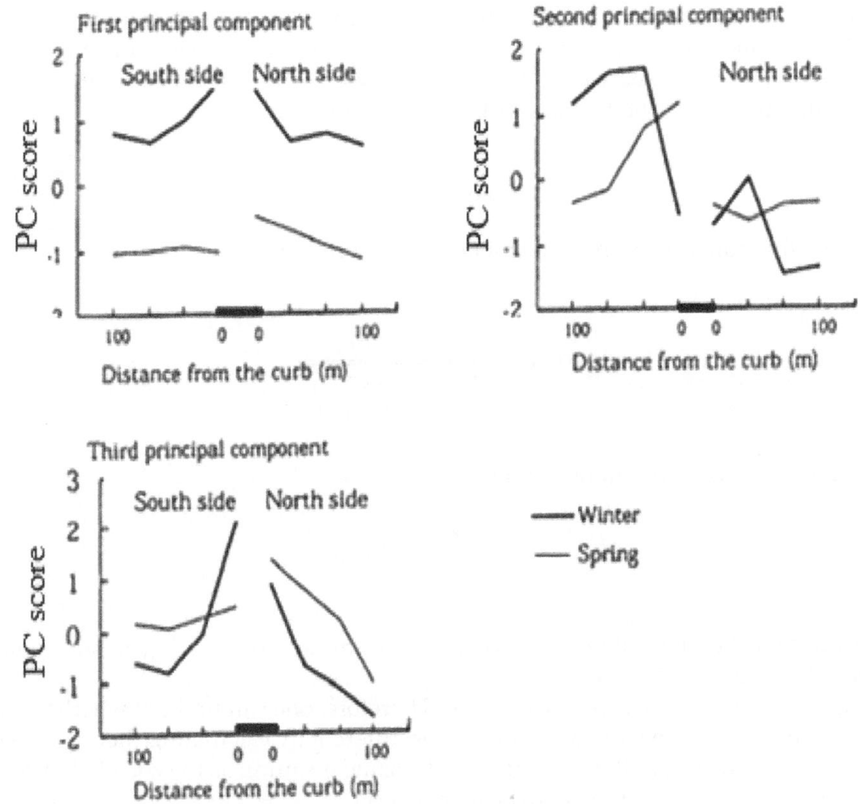

FIGURE 21.5 PC scores for PM data collected at the sampling points in different seasons.

21.4.4 Case 4: Analysis of Spatial Homogeneity in Wind Field by Cosine Similarity

This case study examines whether it is necessary to divide the region into separate areas for air quality simulation when wind data are available at numerous monitoring sites within the region of interest. Furthermore, if it is necessary to divide the region, which way would be most appropriate for the division? Because cluster analysis is useful for such a division, this case study also discusses how the results of PCA differ from those of cluster analysis. When calculating air pollution concentrations using an air quality model, the most influential input is the wind data. In the first step of the IPCAJ simulations for the Kashima area, a preliminary analysis was conducted to select the most suitable wind measurement data as there were 28 monitoring stations equipped with anemometers in the study area (IPCAJ, 1985). A useful statistic in such cases may be the cosine similarity, which was used in this case study.

21.4.4.1 Cosine Similarity

Cosine similarity and Pearson's correlation coefficient can be seen as a variation of the inner product of vectors. Suppose that we have two vectors, $\boldsymbol{a}$ and $\boldsymbol{b}$, and need to measure the similarity between them. The basic similarity function is the inner product:

$$\text{Inner}\,(\boldsymbol{a},\boldsymbol{b}) = \sum a_i\, b_i$$

If a tends to be high when b is high and a is low when b is low, we can say that the inner product is large, and the two vectors are more similar. The size of the inner product is infinite. The way to limit this value between –1 and 1 is to divide by the norm of the vector and calculate the cosine similarity. The cosine similarity of vectors a and b is:

$$\cos\theta = \frac{a \cdot b}{|a||b|} \tag{21–1}$$

If we rewrite this equation with the elements of the vector, we get:

$$\cos\theta = \frac{\sum_{i=1}^{n} a_i b_i}{\sqrt{\sum_{i=1}^{n} a_i^2}\sqrt{\sum_{i=1}^{n} b_i^2}} \tag{21–2}$$

Here, the correlation coefficient of x and y is as follows.

$$r = \frac{Cov(x,y)}{\sqrt{Var(x)}\sqrt{Var(y)}} \tag{21–3}$$

Considering that the elements of vectors **a** and **b** are data of x and y, it can be seen that Equations (21–2) and (21–3) are equivalent.

The wind data are a typical vector variable. Therefore, one can find out whether the winds observed at the two observatories are similar or different by calculating their cosine similarity. And since the cosine similarity value has the same meaning as the correlation coefficient, it can be applied to various other statistical analyses.

21.4.4.2 Wind Data of Kashima Area

In the Kashima coastal area, several air quality monitoring stations were installed in line with large-scale industrial development. Many monitoring stations were also equipped with anemometers to record wind direction and wind speed. Since there are 28 stations in the area, the air quality simulation had to determine which stations would be best used for wind data, or whether it would be better to divide the region into several areas and use different wind data for each. The available wind measurement data were for one year from April 1980 to March 1981 (IPCAJ, 1985). The wind distribution chart (wind rose) based on the hourly wind data is shown in Figure 21.6.

Regarding the wind vectors at points A and B, the cosine similarity was calculated by the following formula, where θ is the difference between the angles, as shown in Figure 21.7.

$$r(V_A, V_B) = \frac{\Sigma \overline{OA_i} \cdot \overline{OB_i} \cos\theta}{\Sigma \overline{OA_i} \cdot \overline{OB_i}} \tag{21–4}$$

21.4.4.3 Application of PCA

Cosine similarity has the same meaning as the correlation coefficient, and the relationship between all points can be displayed as a correlation matrix. A part of this matrix is shown in Table 21.6.

The area is flat and is not far from the coastal line, which may explain the small variations in the wind data between the sites and which is shown by large correlation coefficients between the stations, i.e., 0.68–0.96 with most values ≥0.9.

FIGURE 21.6 Wind roses in the Kashima area from April 1980 to March 1981.

Source: IPCAJ (1985).

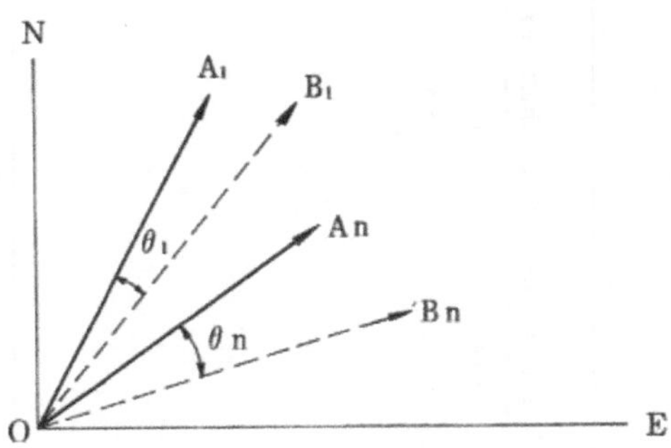

FIGURE 21.7 Coordinate system in vector representation of wind direction and speed data.

The PCA was carried out using the wind vector correlation matrix, namely cosine similarity, between measurement stations. The eigenvalues of the PCs are shown in Table 21.7. It can be seen that the first PC represents nearly 90% of all variation in the dataset. The eigenvalues of the third PC and below are less than 1.0 and can be interpreted that it carries only insignificant information from the original dataset. The scatter plot of the loading of the first and second PCs (Figure 21.8) shows that the loading of the first PC is ≥0.9. It is therefore indicated

TABLE 21.6

Part of the Correlation Matrix

Meteorological station	Kyuchu	Takamagahara	Shimohanawa	Izumikawa	4th youchien	Takamatu	Taya	Hirai	Toyozu	Kashima	Namino
Kyuchu	1	0.89	0.94	0.93	0.87	0.94	0.93	0.91	0.96	0.93	0.92
Takamagahara		1	0.93	0.92	0.68	0.88	0.94	0.95	0.87	0.91	0.96
Shimohanawa			1	0.95	0.77	0.92	0.93	0.92	0.93	0.93	0.94
Izumikawa				1	0.79	0.94	0.93	0.93	0.93	0.94	0.94
4th youchien					1	0.87	0.78	0.73	0.89	0.8	0.74
Takamatu						1	0.89	0.9	0.93	0.95	0.92
Taya							1	0.94	0.92	0.91	0.95
Hirai								1	0.9	0.93	0.96
Toyozu									1	0.91	0.9
Kashima										1	0.95
Namino											1

Source: Adapted and modified from IPCAJ (1985).

TABLE 21.7
Result of the PCA

Principal component	1	2	3	4	5
Eigenvalue	25.26	1.11	0.32	0.2	0.13

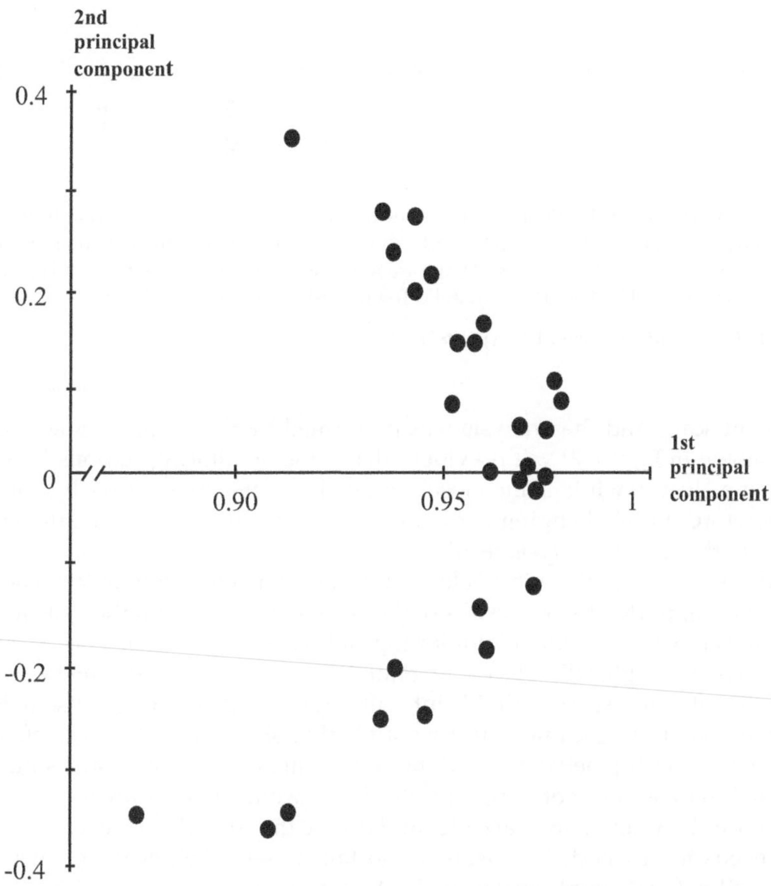

FIGURE 21.8 Results of plotting the PC loading of the first and second principal components.

that the use of meteorological data from one station is sufficient for air quality simulation in the target area.

21.4.4.4 Comparison with the Results of Cluster Analysis

Cluster analysis is one of the most common methods for classifying multiple individuals with certain characteristics into several groups (clusters). The hierarchical method, which will be described in Chapter 22 (Section 22.3), combines the two closest individuals or clusters to form new higher-level clusters one after another. As a measure of this proximity, there are Euclidean distance, correlation coefficient etc., and many methods such as the shortest distance method, the longest distance method and the group average method have been proposed as measurement methods.

In this case study, the monitoring stations were classified using the wind observation data at 28 sites in the Kashima area. The correlation coefficient (cosine similarity) was used as

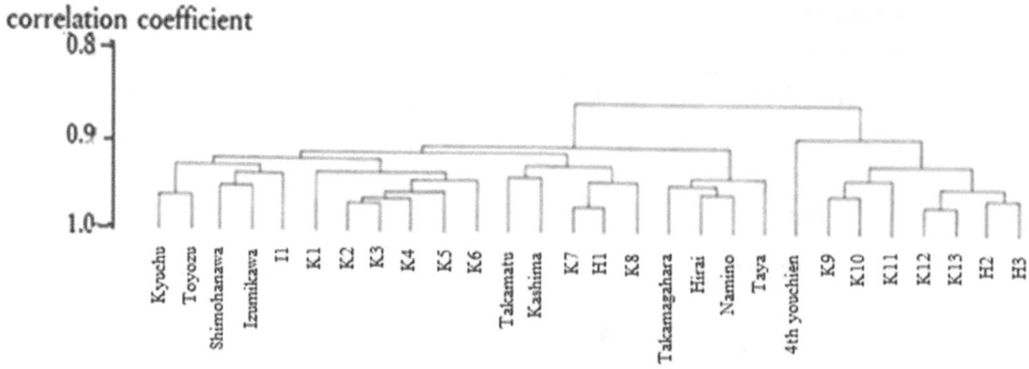

FIGURE 21.9 Dendrogram of the cluster analysis result for the wind field distribution in the Kashima area using cosine similarity data. (*Note:* 11 stations in Kashima Town (located in the north) are indicated by their names. The stations in Kamisu Town (central part) are indicated by K1 to K13. Stations in Hazaki Town (south side) are indicated by H1 to H3. The station in Itako Town (west side) is indicated by I1.

Source: Adapted and modified from IPCAJ (1985).

a measurement scale, and the analysis was performed by the group average method. From the results shown in Figure 21.9, it is evident that some monitoring stations located far apart belong to same cluster, while some closely located monitoring stations belong to different clusters. Therefore, it is challenging to divide the area for air quality simulation based on the dendrogram of the cluster analysis results.

This result suggests that the wind field is almost uniform throughout the whole area. This is not positively supported by the results of the cluster analysis but rather due to the negative reason that the crowds (clusters) of monitoring stations shown in the dendrogram cannot be separated when geographically placed. Overall, PCA may be more suitable if it is desired to find out whether it is necessary to divide into sub-regions rather than the discussion in case of how to divide from the beginning. This case study thus shows that PCA is an effective tool for analysing the relationship between wind measurement points using cosine similarity.

Cluster analysis may be more appropriate if one would like to proactively evaluate the slight differences between many variables and divide the subject. However, when discussing whether it needs to be divided, it seems important to see whether the element (or component) has a well-defined simple structure. Furthermore, if, as in this case, the contribution of the first PC (ratio of variance) is greater than 90%, there would be no need to consider every single variable, e.g., wind observation site but data from one station is enough to represent the wind field in the study area. PCA would be more suitable than cluster analysis for this type of discussion.

The analytical case presented here was part of a study to determine which wind measurement data from the available dataset should be used to calculate concentrations in the air quality simulation (see an overview of this air quality model in Chapter 18). There were 28 air quality monitoring stations that were also equipped with anemometers in the target area of this simulation. Therefore, regional classification was necessary to determine which meteorological data should be used for particular emission sources before the simulation. Based on the PCA results, it may be better to be considered as a uniform meteorological field for the entire area without dividing the area into sub-regions.

21.4.5 Case 5: Analysis of the Spatial Distribution of Photochemical Air Pollution

In this example, PCA was applied to reveal the spatial homogeneity of a study area and the need to divide it into smaller areas before developing a prediction model with discriminant analysis (see more details in Section 22.5, Chapter 22). In principle, both PCA and FA can be used to analyse the relationship between many variables, but PCA appeared to be more appropriate as it did not involve any conceptual model of the data in advance.

The environmental standard for photochemical air pollution in Japan is not set for ozone but for photochemical oxidants (Ministry of the Environment, 2023). Photochemical oxidants (O_x) are the total of oxidisable substances produced by photochemical reactions including ozone, peroxyacetyl nitrate (PAN) and so on. It is defined as the value obtained by measuring the amount of iodine released from a neutral potassium iodide aqueous solution; the nitrogen dioxide content is excluded, and it may be considered that more than 90% of the measured amount is ozone.

In developing the prediction model for photochemical air pollution, it was examined whether one prediction model could be applied to the whole area of Tokyo, or it should be divided into several areas and considered a different model for each area. First, the regional characteristics of O_x concentrations were analysed by PCA (Okamoto et al., 1985). The dataset used in this analysis was the daily maximum values from April to September 1982, i.e., a focus was on the summer period. During the day, the O_x concentration is the highest around noon to early afternoon; hence the maximum hourly values during a day is considered along with the daily average. In analysing such air quality data produced with a large monitoring network, it is sometimes necessary to consider whether it is better to exclude observation sites that contain missing values or to exclude the date when the missing values occurred. Here, we set up the following three conditions and considered a total of six cases, one for the analysis of daily average values and the other for daily maximum values.

1. Daily average for 153 days at 29 sites, excluding sites with many missing data,
2. Daily maximum for 153 days at 29 sites, excluding sites with many missing data,
3. Daily average for 125 days at 29 sites, excluding rainy days,
4. Daily maximum for 125 days at 29 sites, excluding rainy days,
5. Daily average for 103 days at 37 sites, excluding days with many missing data, and
6. Daily maximum for 103 days at 37 sites, excluding days with many missing data.

The contribution rate of the first principal component, PC1 is 84% in the case of the daily maximum oxidant concentration for 103 days at 37 measurement stations (Case 6). For Cases 1 to 5, it was 80%, 82%, 78%, 78% and 82%, respectively. Furthermore, the contribution rate of PC2 is about 3–5%, and the contribution rate of the remaining PCs is relatively small. Therefore, here, it was considered that one prediction model can be used for the entire Tokyo metropolitan area. It was also found that there was no clear difference between Cases 1 to 6. Therefore, the following discussion describes only Case 6, for which the results are shown in Table 21.8. Note that among 37 stations, there were 21 stations located in the Eastern Ward area (E) and 16 stations in the western Tama (W) area.

PC1 has high loadings at all stations, most of which are above 0.8, and has high loadings at stations with high average maximum O_x concentrations. PC2 has high positive loadings at stations located in the eastern part of Tokyo, i.e., the Ward area, and negative loadings in the western part of Tokyo, the Tama area. Therefore, PC2 can be used to distinguish between the eastern and western parts of Tokyo based on the O_x concentration distribution.

Overall, PC1 explained about 84% of the variance of the dataset of the daily maximum values. The fact that the variance (eigenvalue) of PC1 is overwhelmingly large as compared to the other PCs indicates that this PC can be used to determine the overall variability and that there is no need to divide the study area into sub-areas. A single model thus should be sufficient to predict O_x concentrations for the entire Tokyo area.

PC3 showed relatively high positive loadings in Arakawa, Shibuya, and Hachioji. Therefore, it can be assumed that this PC reflects the peculiarities of the location of the monitoring station. In particular, Arakawa and Shibuya had long been problematic due to the high position

TABLE 21.8

Average Values of the Daily Maximum O_x Concentrations at Each Monitoring Station and the PCA Result, PC Loadings

No.	Eastern or western area	Station	Average of daily maximum O_x (ppm)	PC1	PC2	PC3
1	E	Tocho-mae	0.022	0.90	0.20	−0.05
2	E	Harumi	0.023	0.88	0.26	−0.07
3	E	Minato	0.029	0.88	0.10	−0.11
4	E	Kokusetu-tokyo	0.045	0.96	0.06	−0.07
5	E	Bunkyo	0.034	0.88	0.24	0.09
6	E	Joto	0.034	0.91	0.30	−0.08
7	E	Shinagawa	0.045	0.95	0.06	−0.15
8	E	Meguro	0.044	0.93	0.20	0.01
9	E	Kojiya	0.037	0.89	0.14	−0.06
10	E	Setagaya	0.043	0.96	0.00	0.08
11	E	Shibuya	0.027	0.89	0.08	**0.19**
12	E	Nakano	0.045	0.95	0.05	0.08
13	E	Kugayama	0.039	0.95	0.01	−0.01
14	E	Arakawa	0.031	0.81	0.26	**0.42**
15	E	Itabashi	0.030	0.85	0.19	−0.10
16	W	Shakujii	0.042	0.93	−0.11	−0.15
17	E	Nerima-kita	0.049	0.92	−0.07	0.09
18	E	Adachi	0.040	0.88	0.29	0.09
19	E	Katsushika	0.042	0.91	0.28	−0.03
20	E	Edogawa	0.045	0.93	0.20	−0.04
21	W	Hachioji	0.026	0.79	−0.15	**0.52**
22	W	Tachikawa	0.040	0.95	−0.21	−0.08
23	W	Musashino	0.058	0.98	−0.08	0.05
24	W	Ohme	0.055	0.92	−0.19	0.09
25	W	Fuchu	0.046	0.94	−0.22	−0.06
26	W	Chofu	0.049	0.96	−0.12	−0.15
27	W	Machida	0.049	0.92	−0.22	0.02
28	W	Koganei	0.044	0.96	−0.17	−0.10
29	W	Kodaira	0.059	0.96	−0.18	−0.04
30	W	Tanashi	0.051	0.95	−0.20	−0.08
31	W	Fussa	0.056	0.92	−0.27	0.09
32	W	Komae	0.044	0.97	−0.04	−0.08
33	W	Higashiyamato	0.060	0.96	−0.23	0.00
34	W	Kiyose	0.054	0.95	−0.15	0.02
35	W	Tama	0.054	0.93	−0.28	0.03
36	E	Suginami	0.034	0.94	−0.07	−0.10
37	E	Kogaiken	0.016	0.80	0.20	−0.16
Eigenvalue				31.22	1.25	0.71

Note: E: eastern Tokyo, Ward area; W: western Tokyo, Tama area.

of the instrumental inlet. For this reason, nitric oxide, which destroys ozone through chemical reactions, may not be reachable. In addition, the daily maximum values of these stations were relatively low compared to other stations in the vicinity. PC3 had a small eigenvalue (0.71) and explained only 2% of the variance of the dataset, but this is useful in the search for the outliers; i.e., the monitoring station reflects the peculiarities.

21.4.6 Case 6: Application of a Binary Dataset to Analyse the Biogeographic Distribution of Butterflies

PCA is a descriptive statistical analysis technique that is unaffected by the distribution system of the data. Similarly, factor analysis can be used without too much concern about the distribution of the data (and its background population), unless the maximum likelihood method is used for extracting factors, or the analysis is subject to a statistical test. Taking advantage of this property, it is thought that FA can analyse various types of data. This section illustrates the use of FA for the analysis of a binary dataset obtained from a survey of butterfly distribution in Thailand.

Generally, when surveying the distribution of plants or animals over large areas, sampling methods do not follow statistically predetermined procedures, such as those described in Chapter 9. Often, to conduct such a study, understanding its habitat and habits is much more important than understanding statistical sampling theory. Moreover, the purpose of the survey is often not a high-priority activity. In many cases, sufficient data cannot be obtained due to cost constraints, and we may have no choice but to rely on amateur activities. This case introduces an example of analysis using limited data in such cases using the dataset of butterfly distribution.

21.4.6.1 Zoogeographic Zones of Thailand

Thailand, located in a tropical area, is covered by a zoogeographical Asian region. The country may be divided into several zoogeographic zones according to several researchers as seen in Figure 21.10 and detailed here.

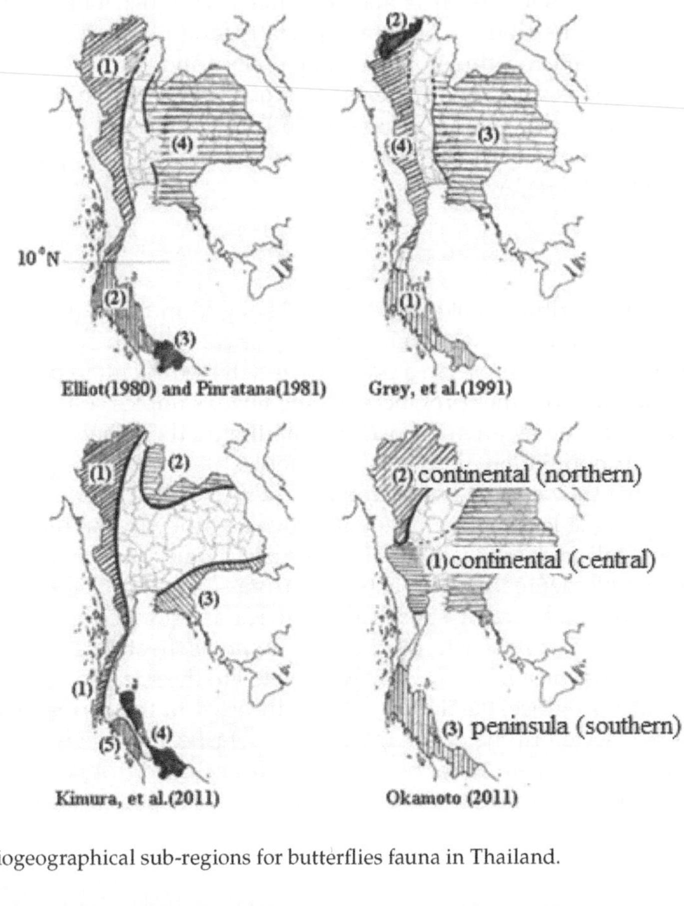

FIGURE 21.10 Biogeographical sub-regions for butterflies fauna in Thailand.

Elliot (1980) mentioned that Thailand could be divided into four zoogeographic zones in which the more sedentary butterflies, characteristic of a primary forest, tend to become a distinct subspecies. These four zones were the Indo-Burmese zone covering northwest and west upper 10N latitude, the Kedawian zone covering the peninsula below 10N latitude, the Malayan zone covering the extreme south-east confined to the Yala district, and the Indo-Chinese zone that covers the centre and eastern area and extending to Vietnam.

Grey et al. (1991) also considered the distribution of insect fauna in Thailand into four zones as follows: Sundaic zone covering the southern and peninsular area, Palearctic species dwelling zone covering the northern mountains including Doi Inthanon, Indochinese eastern zone, and Indian (or Indo-Burmese) western zone.

Kimura et al. (2011), however, indicated that this area could be divided into five sub-regions as follows: the western zone along the Myanmar border from Chiang Mai to the peninsular districts, the north-eastern zone including Nan district along the Laos border, the eastern zone along the Cambodian border, the extreme south of the peninsula (east coast), and Tarutao Island and west coast of peninsula adjacent to Kedah of Malaysia. Central areas including Phu Khieo and Dong Phayayen-Khao Yai Forest complex are considered as the mixture of the previous three zones.

According to the PCA results by Okamoto (2011), the difference between the northern (continental) and southern (peninsular) regions is noticeable. This trend is thought to be formed by the information of many common species that are found everywhere. However, many species that influence this decision are often of little interest to experts in the field. Expert judgment may not be quantitative but qualitative, and it is thought that subjective information on the species of interest is particularly important.

Butterflies are often used as a marker of biodiversity, because Lepidoptera, is an order of insects comprising butterflies and moths, due to their greater stability or less mobility compared with other animals such as birds and mammals. The distribution of Lepidoptera is highly dependent on the distribution of flora, especially food plants. Therefore, by analysing these data, one can get some insight into the zoning of flora and fauna. Recently, several excellent databases became accessible and can be used for quantitative analysis of lepidopterans in Southeast Asia, especially in Thailand. This case study introduces an example of the analysis done by Doolgindachbaporn et al. (2019) using the latest database and explains how to analyse categorical measurement data by FA.

21.4.6.2 Dataset

Let us investigate the distribution of butterflies in Thailand in terms of their inhabitation at the provincial level. The sample size (number of samples collected) here is not very important. Even with a large number of samples in a province, it will be difficult to prove that the habitat density is higher than that of other provinces having fewer samples collected. This is because this case study did not employ an appropriate sampling method that can correctly estimate the conditions of their population. If none of the samples were taken in a particular province, it may infer either that they are not living in that province or that they are not being investigated. However, getting at least one sample will prove that they live in the province.

The data of the Papilionidae (Swallowtail butterflies) family in Thailand were used. The survey data were supplemented by data extracted from several websites. The data is binary for the presence or not of a butterfly species in a particular province of the country. The raw data is thus composed of 0 or 1 in each cell of the Excel spreadsheet. The column of the spreadsheet corresponded to the province. The row was related to the species or subspecies. If a species was recorded in a province, the datum "1" was inserted in the corresponding cell. If not, the datum "0" was inserted in the cell. A part of the datasheet is shown in Table 21.9. There are 77 provinces in Thailand, but about two-thirds were excluded from the analysis due to the lack of suitable data. Finally, the dataset of 91 species (and subspecies) and 30 provinces were prepared for the analysis.

TABLE 21.9

Part of raw data, species and subspecies used in the analysis for Papilionidae.

	Mea Hong Son	ChiangRai	ChiangMai	Nan	Lampang	Phrae	Tak	Phetchabun	Mukdahan	Ubon Ratchathani	KhonKaen	Chaiyaphum	Nakhon Ratchasima	Kanchanaburi	Ratchaburi	Petchaburi	PrachuapKiriKhan	Saraburi	NakhonNayok	Sa Kaeo	Chonburi	Chanthaburi	Trat	Ranong	Phuket	Trang	Satun	Surat Thani	Nakhon Sri Thammarat	Yala
Bhutanitis lidderdalii ocellatomaculata	0	0	1	0	0	0	0	0	0	0	0	0	0	0	0	0	0	0	0	0	0	0	0	0	0	0	0	0	0	0
Troides helena cerberus	1	1	1	1	1	1	1	1	0	0	0	1	0	0	0	0	0	0	1	0	1	1	1	1	1	1	1	1	1	0
Troides aeacus aeacus	1	1	1	1	1	0	0	0	1	0	1	1	1	1	1	1	0	1	1	1	1	1	0	0	0	0	0	0	0	0
Troides aeacus malaiianus	0	0	0	0	0	0	0	0	0	0	0	0	0	0	0	0	0	0	0	0	0	0	0	0	0	1	0	0	1	0
Troides amphrysus ruficollis	0	0	0	0	0	0	0	0	0	0	0	0	0	0	0	0	0	0	0	0	0	0	0	1	0	1	1	0	1	1
Troides cuneifera paeninsulae	0	0	0	0	0	0	0	0	0	0	0	0	0	0	0	0	0	0	0	0	0	0	0	0	0	0	0	0	1	0
Atrophaneura sycorax egertoni	0	0	0	0	0	0	0	0	0	0	0	0	0	0	0	1	0	0	0	0	0	0	0	1	0	0	0	1	1	0
Atrophaneura zaleucus	1	1	1	1	1	1	1	1	0	0	0	1	1	1	1	1	0	0	1	0	1	1	0	0	0	0	0	0	0	0
Atrophaneura varuna varuna	0	0	0	0	0	0	0	0	0	0	0	0	0	0	0	0	0	0	0	0	0	0	0	1	1	1	1	1	1	1
Atrophaneura varuna astorion	0	1	1	0	0	0	0	0	0	0	0	0	0	0	1	0	0	0	0	0	0	0	0	0	0	0	0	0	0	0
Atrophaneura nox erebus	0	0	0	0	0	0	0	0	0	0	0	0	0	0	0	1	0	0	0	0	0	0	0	0	0	0	1	0	0	0
Atrophaneura aidoneus	0	1	1	0	0	0	0	0	0	0	0	0	0	0	0	0	0	0	0	0	0	0	0	0	0	0	0	0	0	0
Byasa dasarada barata	0	1	1	1	1	0	1	0	0	0	0	1	0	0	0	0	0	0	0	0	0	0	0	0	0	0	0	0	0	0
Byasa polyeuctes polyeuctes	0	1	1	1	0	1	1	0	0	0	0	0	0	0	0	0	0	0	0	0	0	0	0	0	0	0	0	0	0	0
Byasa adamsoni adamsoni	0	0	1	1	0	1	1	0	1	0	0	0	0	0	0	1	0	0	0	0	1	1	0	0	0	0	0	0	0	0
Byasa adamsoni takakoae	0	0	0	0	0	0	0	0	0	0	0	0	0	0	0	0	0	0	0	0	0	0	0	0	0	0	0	1	0	0

Source: Data for Khon Kaen and SaKaeo were taken from Doolgindachbaporn et al. (2019).

TABLE 21.10
Part of the Correlation Matrix

	Mea Hong Son	Chiang Rai	Chiang Mai	Nan	Lampang	Phrae	Tak	Phetchabun	Mukdahan
Mea Hong Son	1	.537	.590	.620	.550	.432	.537	.550	.201
Chiang Rai	.537	1	.592	.636	.434	.373	.471	.434	.036
Chiang Mai	.590	.592	1	.682	.451	.389	.543	.506	.211
Nan	.620	.636	.682	1	.564	.569	.537	.508	.260
Lampang	.550	.434	.451	.564	1	.365	.612	.533	.286
Phrae	.432	.373	.389	.569	.365	1	.482	.488	.327
Tak	.537	.471	.543	.537	.612	.482	1	.493	.218
Phetchabun	.550	.434	.506	.508	.533	.488	.493	1	.286
Mukdahan	.201	.036	.211	.260	.286	.327	.218	.286	1

In this analysis, a province was considered to be a variable; to analyse the spatial differentiation or similarity between the provinces in terms of the butterfly's distribution. A correlation coefficient between two provinces was calculated, which is shown in part in Table 21.10.

Several regional characteristics have been identified from previous studies of butterflies in the region. Therefore, if it is possible to assume some factors are related to regional characteristics before starting the analysis, FA is more appropriate than PCA for that analysis. Besides, it would be sufficient if the tool could identify the characteristics related to the distribution of the species in the areas. The FA of the principal component mode was employed for extracting initial factors, adopted components with eigenvalues greater than or equal to 1, and conducted varimax rotation to obtain the simple structure.

21.4.6.3 Results and Discussion

Factor analysis was performed with the IBM-SPSS package PASW Statistics version 19. The eigenvectors were calculated and only five factors having eigenvalues ≥ 1.0 were retained. Varimax rotation was conducted, and the rotated factor loadings are shown in Table 21.11. From the loadings, it is easily interpreted that factor 1 represents the eastern region; factor 2 is the northern region; factor 3 is the peninsular region; factor 4 is the western region.

The provinces that have higher factor loadings are plotted in Figure 21.11. As shown in this figure, sampling sites (provinces) that have similar positions within this scatter diagram show an adjacent position in a real geographical situation. Therefore, one can easily understand the meaning of each factor. The scatter plot between factors 1 and 2 indicates that most of the provinces, except for the peninsula region, are continuously distributed on this graph, mirroring their actual geographic locations.

The FA results may support Kimura's hypothesis that central areas such as Phu Khieo, Phu Kradueng and Khao Yai can be explained as a mixture of Indo-Myanmar and Indo-Chinese sub-regions. In the scatter diagram of factor loadings (Figure 21.11), the positions of the southern provinces were strongly isolated from that of the northern (continental) portion. This result suggested that zoo-geographical sub-regions are consistent with the distribution for a corresponding pair of the two subspecies, one is continental, and another is peninsular.

The provinces with large factor loadings for each factor are mapped in Figure 21.12, in which the distribution could be seen explicitly. Factor loadings of factor 2 for the northern provinces such as Chiang Mai, Chiang Rai, Nan, Mae Hong Son and Lampang are high, indicating this factor is related to the northern species. Likewise, factor 3, with large loadings for Trang, Satun, Yala, Surat Thani and Ranong, is related to the southern area.

In summary, by conducting FA, five factors were obtained, and the meanings of the former four factors could be explained. The obtained results of the geographical distribution of

TABLE 21.11

Factor Loadings after Varimax Rotation, Variables (Provinces) Are Rearranged according to the Values of Factor Loadings, Factor Loadings Higher Than 0.50 Are Shown in Bold Character

Province/factor	1	2	3	4	5	Communality
Mukdahan	**0.792**	0.036	−0.026	−0.091	0.321	0.74
Chanthaburi	**0.747**	0.32	0.039	0.29	−0.01	0.75
Chonburi	**0.746**	0.281	0.097	0.034	0.109	0.66
Sa Kaeo	**0.741**	0.171	0.127	0.334	−0.131	0.72
NakhonNayok	**0.738**	0.37	0.032	0.424	0.067	0.87
Saraburi	**0.726**	0.118	−0.048	0.296	0.076	0.64
Trat	**0.709**	0.311	0.098	−0.055	−0.317	0.71
Nakhon Ratchasima	**0.684**	0.241	0.008	0.461	0.236	0.79
Ubon Ratchathani	**0.678**	−0.013	0.05	0.131	0.519	0.75
KhonKaen	**0.633**	0.396	0.123	0.339	0.226	0.74
Chaiyaphum	**0.61**	**0.52**	0.013	0.286	−0.127	0.74
Nan	0.211	**0.791**	−0.241	0.1	0.054	0.74
ChiangRai	0.057	**0.771**	−0.127	0.019	−0.113	0.63
Tak	0.147	**0.754**	0.108	0.14	0.024	0.62
ChiangMai	0.178	**0.744**	−0.327	0.042	0.067	0.70
Mea Hong Son	0.251	**0.735**	0.036	0.231	0.107	0.67
Lampang	0.153	**0.692**	0.131	0.154	0.374	0.68
Phetchabun	0.397	**0.651**	0.121	−0.05	0.114	0.61
Phrae	0.426	**0.569**	0.003	0.036	−0.312	0.60
Trang	0.065	0.001	**0.854**	−0.142	0.099	0.76
Ranong	−0.031	−0.016	**0.795**	0.028	−0.086	0.64
Yala	−0.057	−0.118	**0.786**	0.254	−0.031	0.70
Surat Thani	0.07	−0.085	**0.777**	0.069	0.208	0.66
Satun	−0.003	−0.168	**0.745**	−0.025	−0.145	0.60
Nakhon Sri Thammarat	0.114	0.071	**0.694**	−0.003	0.134	0.52
Phuket	0.157	0.193	**0.653**	−0.332	−0.139	0.62
PrachuapKiriKhan	0.221	0.054	−0.044	**0.845**	−0.128	0.78
Petchaburi	0.3	0.156	−0.043	**0.744**	0.118	0.68
Kanchanaburi	0.366	0.339	0.059	**0.711**	0.256	0.82
Ratchaburi	0.388	0.421	0.055	0.13	**0.596**	0.70
(Variance)	6.57	5.47	4.34	3.01	1.49	20.87

Papilionidae in Thailand appear to be in line with the classification of zoo-geographical sub-regions proposed by other researchers. This case study also shows that, even with a binary dataset (whether or not a province is the habitat of a certain species), the analysis could give meaningful results, i.e., similar to the case of numerical datasets.

21.4.7 Case 7: Application of PCA and EFA to the Analysis of PM Elemental Composition for Identifying Contributing Sources

Analysis of constituent data in particulate matter (PM) measured at a given site, and the use of receptor models to quantify the contributions from emission sources have received much attention in the field of atmospheric science. The PCA application on PM composition data has been presented in Case 3 in Section 21.4.3.2, and the EFA application will be further detailed in Chapter 23. This case study focuses only on the aspects of using PCA and EFA.

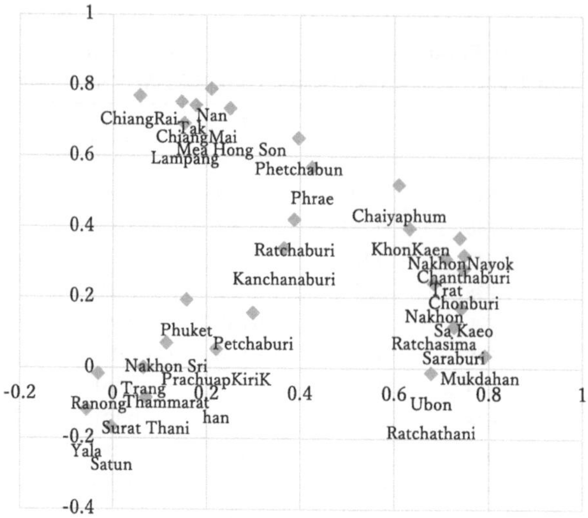

FIGURE 21.11 Scatter plot of varimax rotated factor loadings (factors 1 and 2).

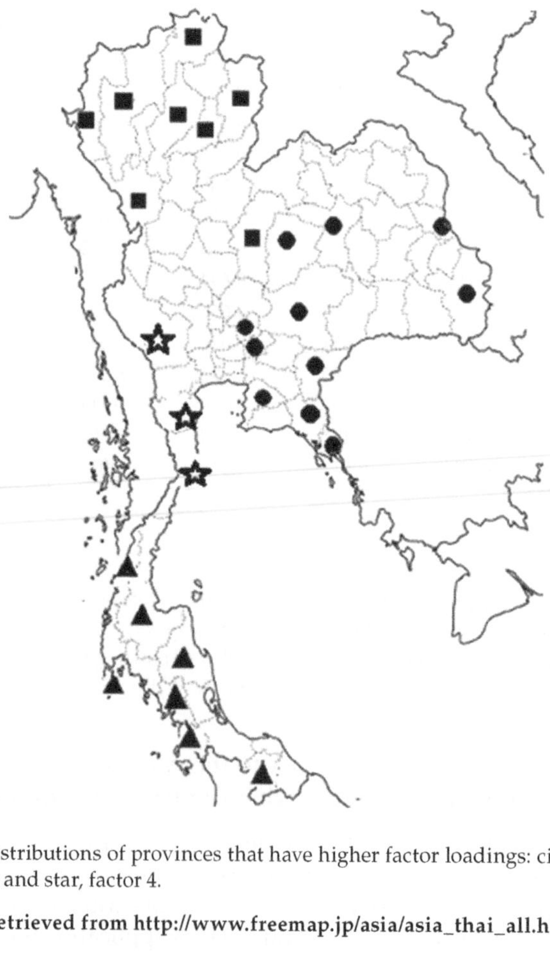

FIGURE 21.12 Spatial distributions of provinces that have higher factor loadings: circle, factor 1; rectangular, factor 2; triangle, factor 3; and star, factor 4.

Source: Map data were retrieved from http://www.freemap.jp/asia/asia_thai_all.html

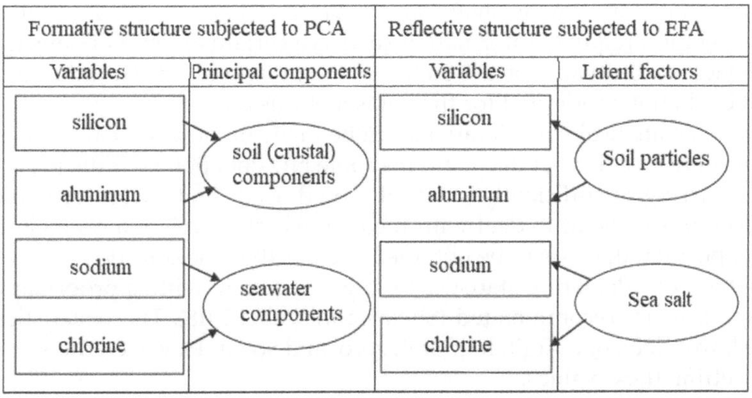

FIGURE 21.13 Conceptual diagram of the PM composition analysis created concerning the diagram showing the relationship between components/factors and variables in the PCA and EFA approaches.

Source: The frame of this conceptual diagram is taken from Alavi et al. (2020).

Alavi et al. (2020) explained the relationship between EFA and PCA clearly and straightforwardly, as using an easy-understanding schematic diagram. Let us use this schematic diagram to illustrate the relationship between EFA and PCA in PM composition analysis. Accordingly, EFA identifies two underlying/latent factors that explain the variation in the variables of PM composition, whereas PCA reduces a large amount of measurement data to two principal components.

The receptor model for analysing elemental data in PM is, in principle, equivalent to the EFA model, and if it can be successfully analysed, it can be a very powerful analytical tool as further detailed in Chapter 23. In the model shown in Figure 21.13, the factor loadings correspond to the pollution contribution rates of the emission sources corresponding to the latent factors. Then the factor loadings cannot be negative. Technically, one would rotate the factor axes in various ways to look for non-negative loadings, but this often does not work in practice.

When analysing PM elemental data and using a receptor model to estimate pollution contributions, it is recommended that the analysis first be performed by EFA. If a simple structure can be obtained by rotating the factor axes, and if the majority of the factor loadings are non-negative, then the analysis has a high probability of success. If not, the choice of variables to employ (in this case, elements) should be reconsidered. In many cases, it is better to exclude elements with a large amount of data below the detection limit. Also, there should be more than several elements, at least, more than two (which is called a marker element) corresponding to the latent factor (in this case, the emission source) subject to the analysis.

In the case of the elemental composition data of PM, the appearance of several below-the-detection-limit data points (non-detectable) is unavoidable due to several restrictions (in many cases, analysis cost restrictions). This section discusses some of the precautions that should be taken when using such datasets. Details of the dataset used in this case study were from the Kashima SPM study (IPCAJ, 1986; Okamoto and Kobayashi, 1991), which have been presented in Chapter 13. The dataset contained several non-detected data points hence the first step was to consider how to handle these data points.

21.4.7.1 Study of the Impact of Non-Detected Data Points in PCA

First, the method of processing data below the detection limit, non-detected (ND) data was examined. Recently, the handling method of such ND data has been discussed mainly in fields such as safety evaluation of toxic chemical substances (Shoari and Dube, 2018; ITRC, 2022). However, this method requires information such as a certain amount of detected data points (above the detection limit) and/or the status of the probability distribution of the assumed data; hence it could not be adopted for the dataset of this case study.

To some extent, statistical considerations can be made on how to deal with estimating the population mean from a dataset that contains some ND data points. If the proportion of non-detect data occurring is sufficiently small, it would seem that various techniques may be made. However, it is difficult to deal with the case where correlation coefficients with other elemental composition data are to be obtained such as the case considered here. In these references, a method of calculating statistics by replacing them with appropriate values is also used, although it is not recommended (Shoari and Dube, 2018). These statistics include, for example, the lower detection limit, half of it, zero, and so on. However, there is no particular rationale for setting these values.

In this analysis, PCA was performed on the PM composition data before EFA was performed. Five treatment methods for the ND data points are presented here, which accordingly formed five different datasets for the analysis.

1. Data points below the detection limit are treated as missing data.
2. Data points below the detection limit are assumed to be 1/2 of the detection limit.
3. Data points below the detection limit are assumed to be 1/10 of the detection limit.
4. Data points below the detection limit are assumed to be the same as the lower detection limit.
5. Same as method 3 but taking the logarithm of the value.

The dataset subjected to this analysis contains 37 variables (mass, chemical components of elements or ions) and 60 measurement data (species) of the coarse and fine fractions of the PM sampled by Andersen impactor (IPCAJ, 1986). The 60 measurement data points covered four seasons of observations at 15 sites (excluding one site that changed location during the experiment).

The PCA results for the five different datasets show that there is not much difference in the magnitude of eigenvalues for the first PC and in the number of eigenvalues greater than 1 for the Kaiser criterion. In the scatter diagram for the loadings of the first and second PCs, some differences were observed in the plots of elements that have many ND data. Only a few PCs were able to interpret the meaning based on their loadings, but all were not satisfactory.

As previously indicated, since there was not much difference among the five cases, 1/10 of the limit of detection was adopted in subsequent analyses, bearing in mind that none of the ND treatment methods was considered satisfactory (IPCAJ, 1986). In particular, chemical compositions with large amounts of ND data points should be excluded from subsequent analyses, especially for receptor modelling.

21.4.7.2 FA as a Preliminary Step to Identify PM Source Factors

FA was performed in principal component analysis mode for the same prior PM composition dataset. First, the correlation matrix is orthogonally decomposed. Next, factors with eigenvalues greater than 1 were retained based on the Kaiser criterion. Then the factor axes were varimax rotated to obtain more interpretable factors, and the results are shown in Table 21.12. The descending order of the eigenvalues before the rotation exhibited a slowly decreasing trend, hence it was not easy to select several factors to be retained based only on the eigenvalues. Although there were about 10 factors with eigenvalues of 1 or more, the interpretable factors for coarse and fine particles were 4 and 6, respectively (IPCAJ, 1986). The ratio of the total variance explained by these factors was about half, but none of the factors was satisfactorily interpreted.

TABLE 21.12

Results of the EFA for Kashima PM Andersen Impactor Dataset (varimax rotation)

	Number of factors with eigenvalues of 1 or greater	Interpretation of the factors						Cumulative contribution of interpretable factors
		1	2	3	4	5	6	
PM coarse fraction	9	Soil	Sea salt	Steel	Secondary particle	—	—	56%
PM fine fraction	11	Soil	Secondary particle	Anthropogenic	Anthropogenic	Anthropogenic	Steel	54%

For example, Table 21A-1 in the chapter appendix shows an example of the varimax rotated analysis results: fine fractions retained 11 factors with the eigenvalues ranging from 5.1 to 1.4. Furthermore, looking at the factor loadings, in one factor, interpretable sources may correspond to the positive side and the negative side, respectively. It is important to note that factor loadings are quantities that are proportional to the pollution contribution rate of the emission source corresponding to that latent factor. The presence of a mixture of both positive and negative factor loading within a single factor means that the appropriate factor model has failed to be constructed in the first place. Accordingly, for this case study, it was difficult and almost impossible to estimate the source contribution based on the results of FA performed on this dataset. The most important reason may be the large amount of data below the detection limit. This result suggests that it is important to select appropriate elements for preparing input data for FA.

Additionally, in principle, for FA, all data points must be extracted from the same population assumed in the causal model. In the dataset used here, several components have mixed data with different trends. Sulphate and nitrate ions in PM are produced by chemical reactions in the atmosphere and thus exhibit complex behaviour and correlations that vary with the season. Datasets containing such components can adversely affect the results of factor analysis. As for the fine fraction, secondary particles such as sulphate and nitrate are the most important components. In the study area, SO_4^{2-} in spring and summer was high, but in winter the NO_3^- was much higher than in other seasons. Correlation between the SO_4^{2-} and NO_3^- was quite different in the fine and coarse fractions. (See Figure 1 of Okamoto et al. [2012].) For the three seasons except for summer, the correlations between sulphate and nitrate were positive and seemed to be similar. Therefore, as for the fine fraction of the Kashima dataset, it might be extracted from two populations, namely summer and another three seasons. In such a situation, both positive and negative factor loadings can be seen in one factor, as shown in Table 21A.1.

If the factor analysis results in a negative value for the factor loading, the contribution concentration cannot be estimated using a regression model with factor scores as the explanatory variables. Accordingly, if the factors cannot be interpreted, subsequent quantitative assessments may not be possible and may require basic review, including resampling and other analysis ways.

This result suggests that analysis with aggregated datasets including many ND data points or questionable data that may be extracted from different populations will not yield good results. Chapter 23, which describes the receptor model, will present examples of analysis results using other datasets with carefully selected components. The positive matrix factorisation (PMF), one type of receptor model, that can handle such a difficult dataset is also detailed in Chapter 23.

21.5 Summary

This chapter discussed the applications of PCA and EFA through several case studies, considering their characteristics and challenges. Cases 1–5 are the applications of PCA; for them, the choice of the PCA method was appropriate because there were no assumptions or hypotheses behind the measured variables. The main objective of these PCA applications is to analyse the interrelationships among the variables. There is also variation in the aim of PCA applications; for instance, the PCA used in Case 2 was close to its original purpose of dimensionality reduction, as the PC with aggregated information was predetermined to be used for cluster analysis. Cases 4 and 5 (Sections 21.4.4 and 21.4.5) both aim to address whether PCA can be applied to treat objects (in these cases, monitoring points), represented by many variables as a single group, or whether they need to be separated. Such problems can be dealt with by other multivariate analysis methods, but in PCA, it is attractive to be able to evaluate the validity of the judgment by the size of the variance (eigenvalue) of the principal components.

Case 6 is an application of EFA. In this case (Section 21.4.6), it was found that there are several theories about the regionality of the species from the existing literature, which may not necessarily be a strict concept, but FA was chosen for the analysis. However, since the principal components method was used to extract the factors, the analysis procedure is not much different from PCA. Case 7 (Section 21.4.7), highlighted the importance of the quality of the dataset used for PCA and EFA illustrated with the PM compositional data. The quality of the analysis results was significantly degraded when a large number of non-detected data or questionable data for any reason were present in the dataset subjected to the analysis. In factor analysis, it appears that considering whether the data are of sufficient quality to withstand the expected analysis would be as important as or more important than the choice of how to extract the factors or rotate the factor axes.

REFERENCES

Aidemy Co. Ltd., 2023, Principal component analysis and eigenvalue problem, *Aidemy Magazine* (In Japanese), viewed on 25 September 2023, https://aidemy.net/magazine/672/

Alavi, M., Visentin, D.C., Thapa, D.K., Hunt, G.E., Watson, R. and Cleary, M., 2020, Exploratory factor analysis and principal component analysis in clinical studies: Which one should you use?, *J Advanced Nursing*, 2020(76), 1886–1889, https://onlinelibrary.wiley.com/doi/pdf/10.1111/jan.14377

Brown, J.D., 2009 January, Principal components analysis and exploratory factor analysis – definitions, differences, and choices, in statistics corner, questions and answers about language testing statistics, *Shiken: JALT Testing & Evaluation SIG Newsletter*, 13(1), 26–30, https://hosted.jalt.org/test/PDF/Brown29.pdf

Doolgindachbaporn, T., Choosai, C., Aoki, T., Yamaguchi, S. and Okamoto, S., 2019 June 13–15, A study on the distribution of Papilionidae in Thailand based on statistical analysis, *The 5th environmental Asia international conference on "transboundary environmental nexus: From the local to regional perspectives"*, Chiang Mai, Thailand, also available *Environment Asia* (2019), 12(Special issue), 10–17, https://doi.org/10.14456/ea.2019.58

Elliot, J.N., 1980, *Preface of the book "butterflies in Thailand vol. 4: Lycaenidae"* (ed. by Pinratana, A., pp. vi–vii), The Viratham Press, Bangkok.

Goonewardana, H., 2019 March 1, PCA: Application in machine learning, *Apprentice Journal*, https://medium.com/apprentice-journal/pca-application-in-machine-learning-4827c07a61db

Gray, D., Piprell, C. and Graham, M., 1991, *National parks of Thailand, industrial finance corporation of Thailand*, Communications Resources (Thailand) Ltd., Bangkok.

Industrial Pollution Control Association of Japan (IPCAJ), 1985, *The study on the behaviour of particulate matter, vol. 8: Report on dispersion simulation* (p. 350) (In Japanese), Japan Environmental Management Association for Industry, Tokyo, Japan.

Industrial Pollution Control Association of Japan (IPCAJ), 1986, *The study on the behaviour of particulate matter, vol. 9: Final report* (p. 163) (In Japanese), Japan Environmental Management Association for Industry, Tokyo, Japan.

International Business Machines Corporation, 2022, *IBM SPSS statistics base 28*, viewed on 28 April 2022, https://www.ibm.com/docs/en/SSLVMB_28.0.0/pdf/IBM_SPSS_Statistics_Base.pdf

Interstate Technology and Regulatory Council (ITRC), 2022, *Chapter 5.7 managing nondetects in statistical analyses, in "groundwater statistics for monitoring and compliance: Statistical tools for the project life cycle"*, viewed on 9 May 2022, https://projects.itrcweb.org/gsmc-1/Default.htm#Welcome.htm%3FTocPath%3D_____1

Jolliffe, I.T. and Cadima, J., 2016 April 13, Principal component analysis: A review and recent developments, *Philosophical Transactions of the Royal Society A, Mathematical, Physical and Engineering Sciences*, https://doi.org/10.1098/rsta.2015.0202

Kim Oanh, N.T. and Leelasakultum, K., 2011, Analysis of meteorology and emission in haze episode prevalence over mountain-bounded region for early warning, *Science of the Total Environment*, 409, 2261–2271.

Kim Oanh, N.T., Piromyaporn, C., Ekbordin, W. and Wangwongwatana, S., 2005, Meteorological pattern classification and application for forecasting air pollution episode potential in a mountain-valley area, *Atmospheric Environment*, 39(7), 1211–1225.

Kimura, Y., Aoki, T., Yamaguchi, S., Uemura, Y. and Saito, S., 2011, *The butterflies of Thailand based on Yunosuke Kimura collection*, vol. 1, Mokuyosha Publishing Co., Tokyo.

Ministry of the Environment, 2023, *Environmental quality standards in Japan – air quality*, viewed on 12 August 2023, https://www.env.go.jp/en/air/aq/aq.html

Okamoto, S., 2011, A study on the distribution of Lepidoptera in Thailand based on statistical analysis, *Evolutionary Sciences*, 16, 93–116, https://www.nodai.ac.jp/rieb/list/research_paper/16.html

Okamoto, S., Ito, M., Hayashi, M. and Shiozawa, K., 1985, Development and evaluation of statistical prediction model for photochemical smog, *J Japan Society of Air Pollution*, 20(1), 33–39 (In Japanese), https://doi.org/10.11298/taiki1978.20.33

Okamoto, S. and Kobayashi, K., 1991, Characterization of aerosols in the Kashima area – a summary of IPCAJ Kashima SPM study, *Atmospheric Environment*, 25B(1), 133–142.

Okamoto, S., Takahashi, K. and Minakawa, N., 2001/2000 June 5–8, Study on suspended particulate matter in vicinity of roadway, *Paper presented at 9th international scientific symposium transport and air pollution*, Avignon, France, also available *International J Vehicle Design*, 27(1–4), 173–182.

Okamoto, S., Wangkiat, A., Pongkiatkul, P., Nakkhwan, C. and Kim Oanh, T.N., 2012, Comparative study on the CMB-8 and PMF models for a coastal industrial area, *International J Environmental Protection*, 2(9), 1–6.

Quynh Truc, V.T. and Kim Oanh, N.T., 2007, Roadside BTEX and other gaseous air pollutants in relation to emission sources, *Atmospheric Environment*, 41, 7685–7697.

Shiozawa, K., Ohtaki, A. and Okamoto, S., 1971, A study on the relation between air pollutants concentration and air pressure distributions, *Bulletin of Science & Engineering Research Laboratory, Waseda University*, 53, 131–138 (In Japanese).

Shiozawa, K., Ohtaki, A. and Okamoto, S., 1972, Study on the prediction of sulfur dioxide concentration based on meteorological factors, *Memoirs of the School of Science & Engineering, Waseda University*, 36, 57–75.

Shiozawa, K., Ohtaki, A. and Okamoto, S., 1975a June 15–20, Statistical prediction modeling for regional air pollution prevention, *68th annual meeting of the air pollution control association, paper #75-53.4*, Boston, USA.

Shiozawa, K., Okamoto, S. and Ohtaki, A., 1975b, On the air pollution prediction systems, *J Sangyo Kogai (J Industrial Pollution Control)*, 10(12), 2371–2386 (In Japanese).

Shoari, N. and Dube, J.S., 2018, Toward improved analysis of concentration data: Embracing nondetects, *Environmental Toxicology and Chemistry*, 37(3), 643–656, https://setac.onlinelibrary.wiley.com/doi/pdf/10.1002/etc.4046

Suhr, D.D., 2005 April 10–13, Principal component analysis vs. exploratory factor analysis, *SAS conference proceedings SUGI30, statistics and data analysis, paper 203-30*, Philadelphia, Pennsylvania, https://support.sas.com/resources/papers/proceedings/proceedings/sugi30/203-30.pdf

Appendix

TABLE 21A.1

Varimax Rotated Factor Loadings for the Fine Fraction of Kashima Data.

	Factor 1	Factor 2	Factor 3	Factor 4	Factor 5	Factor 6	Factor 7	Factor 8	Factor 9	Factor 10	Factor 11
AL	0.889	0.0	0.0	0.0	0.0	0.0	0.0	0.0	0.0	0.0	0.0
MASS	0.797	0.0	0.0	0.0	0.0	0.0	0.292	0.0	0.0	0.0	0.0
SC	0.717	0.0	0.0	0.0	0.0	0.0	0.0	0.0	0.0	0.0	0.0
AG	0.708	0.0	0.0	0.0	0.0	0.0	0.0	0.0	0.0	0.0	0.0
S1	0.673	0.0	0.275	–0.355	0.0	0.0	0.0	0.0	0.0	0.278	0.0
S04$^-$	0.0	–0.902	0.0	0.0	0.0	0.0	0.0	0.0	0.0	0.0	0.0
V	0.0	–0.806	0.316	0.0	0.0	0.0	0.0	0.0	0.0	0.0	0.0
CL	0.491	0.781	0.0	0.0	0.0	0.0	0.0	0.0	0.0	0.0	0.0
CL$^-$	0.588	0.729	0.0	0.0	0.0	0.0	0.0	0.0	0.0	0.0	0.0
NO_3	0.391	0.630	0.365	–0.301	0.0	0.0	0.0	0.0	0.0	0.0	0.0
BR	0.0	0.620	0.0	0.582	0.0	0.0	0.0	0.0	0.0	0.0	0.0
NA	0.0	0.0	0.805	0.0	0.0	0.0	0.0	0.0	0.0	0.0	0.0
S	0.0	0.0	0.773	0.0	0.0	0.0	0.0	0.411	0.0	0.0	0.0
CA	–0.384	0.367	0.654	0.0	0.0	0.0	0.0	0.272	0.0	0.0	–0.253
K	0.458	0.0	–0.513	0.0	0.0	0.0	0.0	0.0	0.0	0.489	0.0
CU	0.0	0.0	0.0	0.769	0.0	0.0	0.0	0.0	0.358	0.0	0.0
SB	0.0	0.0	–0.321	0.753	0.0	0.0	0.0	0.0	0.0	0.0	0.0
AS	0.0	0.0	–0.371	0.533	0.313	0.0	0.0	0.420	–0.267	0.0	0.0
CS	0.0	0.0	0.0	0.0	0.785	0.0	0.0	0.0	0.0	0.0	0.0
PB	0.0	0.0	0.0	0.0	0.689	0.0	0.0	0.0	0.0	0.0	0.0
FE	0.380	0.0	0.0	0.0	0.650	0.437	0.0	0.0	0.0	0.0	0.0
CR	0.0	0.0	0.0	0.0	0.0	0.928	0.0	0.0	0.0	0.0	0.0
MN	0.256	0.0	0.316	0.0	0.420	0.709	0.0	0.0	0.0	0.0	0.0
ZN	0.486	0.455	0.0	0.0	0.0	0.505	0.367	0.0	0.0	0.0	0.0
LA	0.0	0.0	0.0	0.0	0.0	0.0	0.849	0.0	0.0	0.0	0.0
CE	0.0	0.0	0.0	0.0	0.0	0.0	0.746	0.0	0.0	0.0	0.0
TH	0.0	0.0	0.0	0.0	0.0	0.0	0.0	0.819	0.293	0.0	0.0
HF	0.0	0.0	0.0	0.0	0.0	0.0	0.0	0.0	0.802	0.0	0.0
NI	0.0	0.0	0.0	0.0	0.0	0.0	0.0	0.0	0.722	0.0	0.0
CO	0.0	0.0	0.0	0.0	0.0	0.0	0.0	0.0	0.0	0.788	0.0
TI	0.350	0.0	0.0	o.o	0.0	0.0	0.0	0.0	0.0	0.0	0.708
LU	–0.255	0.0	0.0	0. 340	0.453	0.0	0.0	0.262	0.0	–0.379	0.308
SM	0.465	0.0	0.0	0.0	0.0	0.0	0.0	0.0	0.0	–0.354	0.0
CD	0.279	0.333	0.0	–0.488	0.0	0.0	0.436	0.0	0.0	0.0	0.0
SE	0.0	0.0	–0.381	0.0	0.0	0.0	0.0	0.0	–0.331	0.341	0.0
W	0.0	0.0	0.0	0.0	0.0	0.0	0.0	0.498	0.0	0.328	0.481
BA	0.0	–0.366	0.295	0.463	0.0	0.0	0.0	0.0	0.0	0.0	0.436
var.	5.130	4.372	3.171	2.806	2.272	2.165	2.093	1.907	1.881	1.663	1.404

Note: The factor loadings of 0.25 or less were replaced by 0, and those of 0.5 or more were sorted in descending order. Var. is the variance of the PC.

Source: IPCAJ (1986)

22

Statistical Analyses for Air Quality Predictions

Prediction of air pollutant concentrations is one of the most important issues of air quality management. Air quality models using dispersion equations require the emission and meteorology input data. In the early stages of air quality management when emission inventory data have not been sufficiently completed, it may be useful to investigate the relationship between air pollutant concentrations and the related factors such as meteorological conditions and to build a statistical prediction model for air pollutant concentrations. Even when an appropriate air quality model, i.e., dispersion model is available, the analysis of the relationship between air pollutant concentrations and meteorological conditions is necessary to gain insights into and improve the model performance under various conditions, especially for the prediction of air pollution episodes. Statistical methods for air quality prediction may be applied when the emissions input data are not readily available. This chapter introduces applications of multivariate analysis techniques to develop models for predicting air quality. Important algorithms that have not been detailed in Part I will be briefly reviewed here and these include canonical correlation analysis, cluster analysis and the artificial neural network. A few theoretical backgrounds will be explained through case studies of statistical air pollution analysis.

22.1 Challenges with Statistical Air Quality Predictions

In general, it was generally perceived that air quality models relying on statistical methods exhibit several shortcomings, particularly in their application for predicting future concentrations. The following issues have been of concern for those who have a pessimistic view of such statistical methods.

- Statistical models are not suitable for future projections. The reason is that pollutant concentrations are predicted based on existing normal emission conditions, and future emission intensities and conditions are not reflected in the historical data.
- The accuracy of parameter estimation in multiple regression models or other statistical models is not satisfactory because the data used to determine model parameters are not representative of future situations but have been obtained in the current or past situations.
- Due to insufficient information at the tail of the frequency distribution, multiple regression models or other similar models are not suitable to predict peak values such as those in episodes of high concentrations

For these reasons, physico-mathematical air quality models, i.e., dispersion models, are more popularly used in air quality management than their statistical counterparts. However, if the monitoring data, especially concentration data in the concerned areas, are sufficient for the analysis, then applications of statistical models may be less resource-consuming than the dispersion modelling because the latter requires detailed emission input data. Nevertheless, to improve the prediction using statistical modelling, such as regression analysis, it is also necessary to understand not only the effect side, namely, the pollutant concentrations to be predicted but also the causal side data, such as emissions and atmospheric conditions.

DOI: 10.1201/9781003597216-25

Generally speaking, the following two issues may be considered for predicting air quality.

1. Predicting air quality conditions several years in advance for the regulation and licensing of emission sources in the environmental managing plans of local governments and other organisations.
2. Predicting air quality conditions several hours to a few days in advance to activate emergency measures, such as for the public to avoid excessive exposure and for the authority to take actions to control emissions from specific sources. Thus, even if the control of source emissions cannot be done in time, the predicted concentration ranges can be used to alert residents to stay indoors and not do exercises outdoors when ambient air pollution is high, hence avoiding significant health damage.

As discussed in Chapter 18, the predictive performance of air quality models using dispersion equations is significantly reduced when the location and timing of the prediction are specified. In many such cases, although no concrete examples of comparative applications are quoted here, it seems that the statistical models may provide better concentration estimates at specified locations and times. Thus the two modelling approaches are complementary to improve the quality of the predictions; hence both are important. This chapter therefore examines various aspects of the statistical predicting and/or forecasting methods.

In Japan, air pollution became a serious problem in the late 1960s, and the development of monitoring systems in conjunction with information systems using computers, which were beginning to spread at that time, progressed rapidly. However, the performance of the computers of the time was not sufficient to handle the operation of complex air quality models. Therefore, research began on a statistical prediction system that could be supported using the computer resources available at the time (Ogino, 1967). Moriyama et al. (1977) reported the construction of an air quality monitoring network of Chiba Prefecture that included monitoring equipment in the emission source; they reasoned that accumulating emission data from stacks would be invaluable for the development of future air pollution prediction systems.

In some Japanese prefectures and cities, when ambient concentrations exceed alarm thresholds and are expected to continue for more than a few more hours, large factories that have signed agreements are asked to reduce the emissions. The response to such requests is also confirmed by the real-time monitoring network of the emission sources (Okamoto, 2019). Kawasaki City, once one of the most polluted cities in Japan, had established a comprehensive monitoring network system linking seven air quality monitoring stations in the city with 42 major factories and had begun full-scale operation in 1972. These plants contributed nearly 95% of the city's sulphur dioxide emissions. A total of 272 source monitoring devices, including oil meters, sulphur dioxide concentration meters of the flue gas, and gas flow-meters, were installed at these 42 plants to immediately monitor emissions (Murase et al., 1973). In addition, several prefectures and large cities, such as Kawasaki City, have signed agreements with their major plants to determine what to do in the event of an alarm or warning of high air pollution levels. Overall, sulphur dioxide emissions could be reduced by 10–20% by suspending operations of less important facilities and switching to higher-quality fuels. Initially, it seemed that it was envisioned that a concentration prediction system could be incorporated into such a system, but such attempts have never been implemented until today. This system, still in effect today in Japan, shows that during the last about four decades the emergence of high ambient SO_2 concentrations exceeding the alarm level has never occurred.

This chapter provides an overview and examples of statistical analysis techniques related to predicting air pollution concentrations. First, it provides an overview of canonical correlation analysis and cluster analysis, which do not directly predict air pollution concentrations but are also used for preliminary data processing related to statistical air quality prediction. Several examples of the application of neural networks and discriminant analysis to photochemical air pollution prediction are also introduced. Finally, two case studies are presented that used surrogate variables to represent the source's emissions and hence did not rely on the explicit emission input data, thus avoiding the need to prepare extensive emission inventory datasets. The first case study focused on the air pollution problem in the Tokyo area about 50 years ago, and the second is the case of the haze problem expressed by PM_{10} in a mountainous area in Southeast Asia area. The former case used dummy variables as the diurnal fluctuations and weekly fluctuations of the emission source intensity. In the latter case, hotspot fire satellite data, an indicator of biomass open burning, and air mass trajectory data were used as the surrogate emission data of particulate matter.

22.2 Canonical Correlation Analysis and Its Application

The behaviour of air pollutants is complex, which is the result of atmospheric processes such as advection, diffusion and deposition, along with physical transformation and chemical reactions among the pollutants. Air pollution conditions are expressed by using multiple pollutant concentrations. Weather conditions affecting air pollution are also associated with several measurable meteorological elements. If one considers multiple pollutant concentration data as variables in the resultant system and multiple meteorological element observation data as variables in the causal system, canonical correlation analysis is a method that allows the analysis of the relationship between these two groups of variables. If one considers the variables of the resultant system as a whole, this method can also be considered as falling in the category of statistical predicting methods in a broad sense. It is useful to understand the relationship between two groups, each of which can be described by multiple variables, and to reveal the relative importance of each variable (Shiba, 1967; Pennsylvania State University, 2023).

22.2.1 Algorithms

Let the meteorological observation vector be $(x_1, x_2, \ldots x_p)$ and the air pollutant observation vector be $(y_1, y_2, \ldots y_q)$. Then let the vectors u and v be the linear combination of these vector elements.

$$u = \alpha' x$$
$$v = \beta' y$$

Here, $\boldsymbol{x}$ and $\boldsymbol{y}$ are the previously defined observation vectors, and $\boldsymbol{\alpha}$ and $\boldsymbol{\beta}$ are weight coefficient vectors.

$$\alpha' = (\alpha_1, \alpha_2, \cdots\cdots, \alpha_p)$$
$$\beta' = (\beta_1, \beta_2, \cdots\cdots, \beta_q)$$

As described, the elements of the weight vector are determined so that the correlation coefficient between the linear combination u and v, which is the overall characteristic of each group, is maximised. It can be viewed as representing the relative magnitude of the influence between individual elements of multiple meteorological factors that affect multiple air pollutant concentrations. The correlation coefficient maximised under such conditions is called a canonical correlation coefficient and is represented by Rc, which is the square root of the eigenvalue λ, as seen in Equation (22–1).

$$\left.\begin{aligned} R_c &= \sqrt{\lambda_i} \\ \left(\boldsymbol{R}_{22}^{-1}\boldsymbol{R}_{21}\boldsymbol{R}_{11}^{-1}\boldsymbol{R}_{12} - \lambda_i \boldsymbol{I}\right)\beta' &= 0 \\ \alpha'_i &= \left(\boldsymbol{R}_{11}^{-1}\boldsymbol{R}_{12}\beta'_i\right) / \sqrt{\lambda_i} \end{aligned}\right\} \quad (22\text{–}1)$$

The Matrix I is a unit matrix, and if a correlation matrix between meteorological factors and pollutants is R, and the matrices, $\boldsymbol{R}_{11}$, $\boldsymbol{R}_{12}$, $\boldsymbol{R}_{21}$, and $\boldsymbol{R}_{22}$ are small matrices formed by the following division:

$$R = \left(\begin{array}{c|c} R_{11} & R_{12} \\ \hline R_{21} & R_{22} \end{array}\right) \tag{22–2}$$

where:
$\boldsymbol{R}_{11}$ is the correlation matrix of meteorological factors,
$\boldsymbol{R}_{12}$ is the correlation matrix between the meteorological factors and pollutant concentrations,
$\boldsymbol{R}_{21}$ is the transpose matrix of $\boldsymbol{R}_{12}$, and
$\boldsymbol{R}_{22}$ is the correlation matrix of pollutant concentrations.

22.2.2 Application

In this case study, canonical correlation analysis was performed to determine which meteorological factors had a significant effect on the concentration of which pollutant. The analysis was done using the same winter 1968 data used for the analysis in Section 21.4.1, i.e., monitoring data from the former Tokyo Metropolitan Government Office station in February 1968 (Shiozawa et al., 1972). The hourly meteorological data from the Tokyo Meteorological Observatory and the pollutants concentration were extracted four times a day at six-hour intervals with a total of 112 data points.

The results of the canonical correlation analysis are shown in Table 22.1, which do not show a clear trend compared with the results of the PCA shown in Section 21.4.1. This may be due to the confounding effects of the large synoptic-scale meteorological phenomena and diurnal fluctuation of the meteorological parameters.

The first canonical variate, in the meteorological group, had positive values of the temperature, dew point temperature, and wind direction (north–south component), suggesting that these variables may contribute to weather conditions with high air pollution potential. Further, during the daytime, the temperature is high, the humidity is low, and the wind direction is often blowing from the south, these weather conditions are associated with the high values of the first canonical variate.

The coefficients in the first canonical variate pollutant group show the negative value for SPM (equivalent with PM_{10}), which is different from other pollutants. In the PCA analysis using the daily average (see Table 21.1), all pollutants were positively correlated. The reason for the difference between the two cases lies in the datasets used for the analyses. The canonical correlation analysis here used every six-hour concentration data that reflected the diurnal variations, i.e., the within-the-day fluctuations. The diurnal variation of SPM was different from other pollutants. SPM levels were higher in the early morning when other concentrations were not as elevated or had not yet peaked. Furthermore, SPM was highly correlated with several metrological variables, such as relative humidity and visibility, and exhibited similar diurnal variation trends.

Thus it may have been more appropriate to use daily averages, which exclude diurnal variations (variability within a day), in the analysis to see an overall trend and its relationship to synoptic-scale meteorological conditions. Nevertheless, some relationships between many meteorological variables and air pollution concentrations have been revealed by this analysis, and this suggested that various meteorological observations can be used as explanatory variables when predicting air pollutant concentrations using statistical models.

TABLE 22.1

Canonical Correlation Coefficient and Weighting Coefficient Vector of Canonical Variate at the Former Tokyo Metropolitan Government Office Station in February 1968

	1st canonical variate	2nd canonical variate
Canonical correlation coefficient	**0.843***	**0.830***
Meteorological factors		
Air pressure	0.19	0.23
Air temperature	0.34	−0.24
Vapor pressure	−0.15	0.08
Relative humidity	−0.45	−0.60
Wind speed	0.17	0.18
Cloudiness	0.04	−0.06
Weather	−0.03	0.04
Precipitation	0.01	−0.03
Visibility	−0.15	0.12
Pattern of atmospheric pressure gradient	0.05	0.15
Atmospheric pressure gradient	0.05	0.02
Dew point temperature	0.27	0.65
Hours of daylight	0.01	−0.02
Wind direction (S–N)	0.50	−0.03
Wind direction (E–W)	−0.07	−0.24
Pollutants		
NO_2	0.32	−0.06
NO	0.23	−0.70
CO	0.42	0.64
SO_2	0.49	0.04
SPM (equivalent with PM_{10})	−0.66	0.31

Note: The significance level of 5% is indicated with an asterisk (*).

22.3 Cluster Analysis

Cluster analysis is a useful analytical technique for partitioning various data and variables, as will be discussed later in this section. Repeated partitioning of data with various explanatory variables to search for conditions under which air pollution concentrations would be high may be used to predict high pollution concentrations. It is also possible to apply cluster analysis to select the explanatory variables for input into other prediction methods such as regression analysis. The discussion in this section focuses more on cluster analysis as a prediction method in a broad sense.

Clustering is the division of a set of classification objects into subsets that achieve internal cohesion and external isolation. It is also called cluster analysis in the field of statistical analysis and multivariate analysis and is often used in data mining as a basic data analysis method. Each divided subset is called a cluster. There are several types of partitioning methods. When all classification targets are elements of exactly one cluster, it is called a hard or crisp cluster, which is a focus of this section. On the other hand, when one cluster has multiple elements that partially belong to other clusters at the same time, it is called a soft or fuzzy cluster.

Generally, four items need to be decided at the start of cluster analysis:

1. ***Target of grouping:*** Whether to classify samples or variables
2. ***Classification type (type, generation):*** Hierarchical or non-hierarchical method

3. ***Distance (similarity) between objects used for classification:*** Euclidean distance, Mahalanobis distance, cosine distance, etc.
4. ***Cluster merging method (measurement method of distance between clusters):*** Ward method, group average method, shortest distance method, longest distance method and so on

"Target of grouping" is automatically determined when the purpose is decided. In "Classification type", most Big Data analyses use non-hierarchical methods. It is not necessary to consider the "Cluster merger method" when using the non-hierarchical method. The most difficult matter may be to select "Distance between objects used for classification". It is said that cluster analysis is a method for collecting similar things, but it is difficult to define how to measure the similarity and to decide whether the objects are "similar" or "not similar". In particular, it is not an overstatement to say that it is still a grey area in cluster analysis and has not been developed at the pace to match the advancement of the data analysis methods.

There are many types of parameters to be specified in cluster analysis, and there is no answer for selecting the best way or the most appropriate cluster definition. For example, there is also no correct answer to the question of how to determine the optimal number of clusters. Another common problem with non-hierarchical methods is that the result depends on how the initial values are set in the analysing procedure. Interested readers can find detailed explanations of the calculation methods in various statistics books such as Wierzchon and Klopotek (2018).

22.3.1 Hierarchical Method

At first, the technique examines all objects (samples) and combines the most similar ones into a group. This process can be expressed as a hierarchy shown in a dendrogram in Figure 22.1.

First, given data consisting of n objects, an initial state with n clusters containing only one object is created. Starting from this state, find the distance between the targets. The distance between clusters C1 and C2 is expressed as d (C1, C2), and calculated by a predefined distance (dissimilarity). In addition, the two clusters with the shortest distance are successively merged. Then, this merge is repeated until all the targets are merged into one cluster. By this, the hierarchical structure between all objects is acquired.

A dendrogram (tree diagram) shown in Figure 22.1 is a binary tree in which each terminal node represents a target, and the merged cluster is represented by non-terminal nodes. The horizontal axis of non-terminal nodes represents the distance between clusters when they

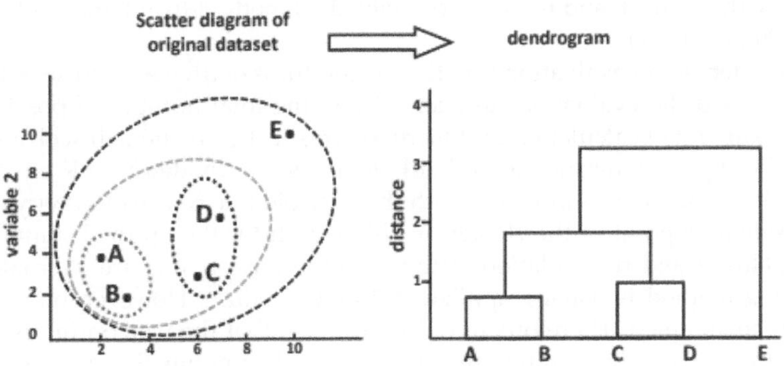

FIGURE 22.1 Hierarchical cluster image.

Source: Prepared with reference to Albert Co. (2020).

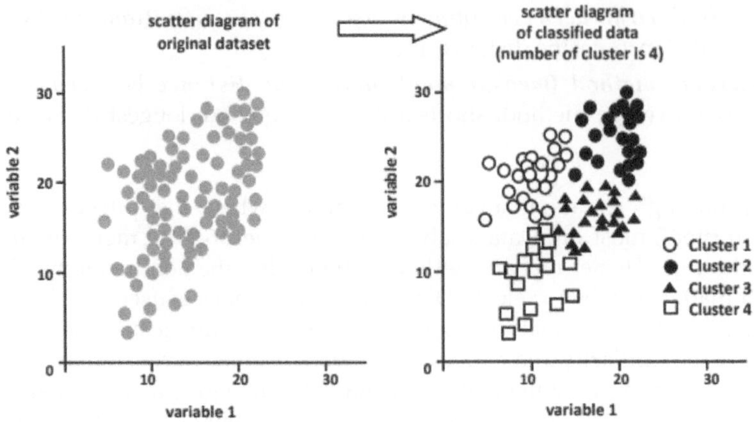

FIGURE 22.2 Non-hierarchical cluster image.

Source: Prepared with reference to Albert Co. (2020).

are merged. Various methods depend on the difference of the distance function *d* (C1, C2) between the clusters C1 and C2. The four methods used for the calculation of the distance between clusters are:

1. Nearest neighbour method,
2. Furthest neighbour method,
3. Group average method,
4. Ward's method.

22.3.2 Non-Hierarchical Method

Non-hierarchical cluster analysis is one of the methods to create clusters by gathering those with similar properties from a mixture of objects with different properties. This is a method in which the number of clusters to be divided is determined in advance and the sample is divided into a fixed number of groups (exclusive subset), which is schematically shown in Figure 22.2. Unlike hierarchical cluster analysis, this method is suitable for analysing Big Data with a large number of samples. However, the number of clusters is decided in advance subjectively by the analyst, and there is no established method to automatically calculate the optimal number of clusters.

This method defines an evaluation function of goodness of division and searches for a division that optimised the evaluation function. When the total number of possible partitions increases, the amount of calculation will increase drastically, so one will seek a semi-optimal solution. In the typical K-means method, all elements are divided into K clusters so that a given evaluation function is minimised with the centroid (the centre of gravity of the cluster) as the representative point of the cluster. In this search for the optimal solution, allocation to the target cluster and recalculation of the representative points are alternately repeated. A hill-climbing method is usually applied in this procedure. However, only local optimal solutions can be obtained. Therefore, to reduce the risk that the true optimum value cannot be searched, the initial value is randomly changed, or data points that are distant from each other are selected as the initial representative points. Using these procedures, the result that minimises the evaluation function may be selected.

In summary, this section briefly described cluster analysis. It is often used as a basic data analysis tool in fields such as data mining. Of the several types of division methods – hard or soft clustering – here the description was given only for the cases of hard clusters where all classification targets were exactly the element of one cluster. Examples of cluster analysis have already been provided in Chapter 21 in conjunction with principal component analysis: non-hierarchical cluster analysis in Section 21.4.2 and hierarchical cluster analysis in Section 21.4.4.

22.4 Artificial Neural Network as Compared to Linear Regression Model: Case Study of Prediction of Maximum Hourly Ozone in an Urban Area

Artificial neural networks (ANNs) are computer programs designed to emulate biological neural networks (such as the human brain) in terms of learning and pattern recognition. ANNs have been under development for many years for use in a variety of disciplines to extract meaning from complicated data and to make predictions. The most popular ANN is a feed-forward back-propagation, multilayer perceptron (MLP) neural network. The major building block for any ANN architecture is the processing element or neuron. An ANN generally consists of three or more layers: an input layer, one or more hidden layers, and an output layer. A more detailed explanation of the algorithms of the artificial neural network can be found in a specialised book in the field.

The case study presents the development of a four-layer feedforward back-propagation neural network architecture to predict ozone concentrations in a city. The architecture of this ANN model is illustrated in Figure 22.3 (Nghiem and Kim Oanh, 2009).

The input neurons receive data from external sources to the system, the hidden neurons receive signals from all neurons in the preceding layer, and the output neurons send information out of the system. These neurons are connected by lines of communication called connections. Associated with each connected pair of neurons is an adjustable value or weight. This case study selected the feed-forward back-propagation MLP to develop the ANN model. Eight variables, detailed in the description of the "Characteristics of Input Data", were used as inputs, and the output from the MLP model was the daily maximum 1 h O_3 level over the city (one value per day in the study area). All input variables were normalised to provide values between 0.05 and 0.95 using Equation (22–3).

$$O'_i = \frac{0.9(O_i - O_{i,min})}{O_{i,max} - O_{i,min}} + 0.05, \tag{22–3}$$

Where O'_i is a transformed observed value, O_i is the actual observed value, $O_{i,\,min,}$ and $O_{i,\,max}$ is the minimum and maximum values among all observed values over the Bangkok domain.

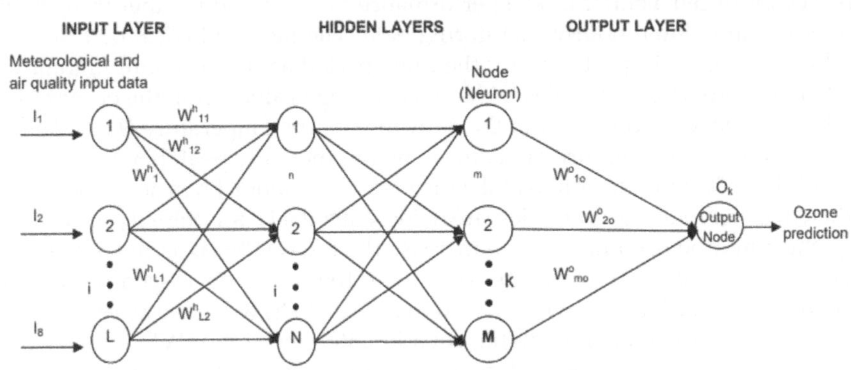

FIGURE 22.3 Architecture of two hidden layer feed-forward neural networks.

Normalisation of input data was performed for two reasons: to provide a commensurate data range so that the models were not dominated by any variable that happened to be expressed in large numbers, and to avoid the asymptotes of the sigmoid function. Once the best network is found, all the transformed data were converted back into their original values using the following equation,

$$O_i = \frac{(O_{i,max} - O_{i,min})(O'_i - 0.05)}{0.9} + O_{i,min} \qquad (22\text{–}4)$$

In the training process, the number of hidden layers and hidden nodes and connection weights between neurons of the MLP network were determined by an iterative process in the training stage with the training subset (361 patterns in this study) until the training error, measured by a set of performance indicators, is lower than the acceptable threshold. The initial values of the weights were randomly selected, and they could be both negative and positive values. The activation function used in the hidden and output layers was determined by the required degree of accuracy of the problem under study. In this case study, the learning algorithm used was Levenberg–Marquardt's back-propagation of the MATLAB Neural Network Toolbox. The activation function selected for the layers was logistic sigmoid for hidden layers and linear for the output layer. The number of hidden layers and hidden neurons (nodes) was increased systematically, checking each time if the prepared neural network produced a stable performance error in the performance plot until the best MLP was found. The trained MLP network model was then validated with the second data subset of 120 patterns. The resulting predictions were compared with the observed data, and performance indicators were determined.

22.4.1 Characteristics of Input Data

There were 13 automatic ambient air quality stations in the city, operated by the Pollution Control Department (PCD) at the time of the study. The monitoring data for the city were collected on an hourly basis, and O_3 monitoring data were available at 11 stations. The O_3 data used for the analysis were the highest values of daily maximum 1-h concentrations observed among all the 11 monitoring stations, i.e., the peak O_3 in the city (one value per day over the domain). This study focused only on the period of high O_3 pollution in the city which is from January to April (winter and local summer) (Zhang and Kim Oanh, 2002). For each day in the period from January 1 to April 30 during 2000–2003, the highest hourly peak O_3 observed in the city was used for the analysis. Concentrations of O_3 precursors including non-methane hydrocarbon (NMHC), NO, and NO_2 used in the analysis were the averages of observed data of the respective pollutant from all monitoring stations in the study area during the morning hours, from 6:00 to 9:00. Note that this period includes the morning rush hours with the high traffic emissions of the day. These average values were used in the final model as they produced a better model performance as compared to other trials with various combinations of individual hourly monitoring data. The meteorological data were obtained from the Meteorological Department for the same period as the air quality monitoring data. The selected meteorological variables include the average values from the observations in the morning hours (from 6:00 to 10:00) and the daily maximum temperature (T_{max}). The data used for the prediction are the measurements made before the daily peak hourly ozone occurs in the city at 13:00–15:00 with the aim to make the prediction about 3 h in advance. Nevertheless, the predicted meteorological variables are widely available for the city in a few days in advance, which may be used not only to increase the forecast horizon but also to reduce the dependence of the availability of the observed data. To remove the discontinuity of the wind direction (WD) angle at 360°, the wind direction index (WDI) was used to represent the wind direction, which was calculated using the equation, $WDI = 1 + \sin(WD + \pi/4)$, where WD is the wind direction (with 0° corresponding to the North). The total dataset used for the modelling included 481 lines (1 per day) and 9 variables:

1. NMHC, non-methane hydrocarbons,
2. NO, nitric oxide,
3. NO_2, nitrogen dioxide,
4. O_3, ozone concentration,
5. WS, wind speed,
6. WDI, wind direction index,
7. RH, relative humidity,
8. T_{max}, daily maximum temperature, and
9. GRAD, global radiation.

22.4.2 Performance of the Multilayer Perceptron (MLP) Model in Comparison with the Linear Regression (LR) Model

The iterative process in the training stage, using 75% of the dataset, found that the architecture of the best MLP network should be 8-10-14-1; i.e., it should have the input layer containing 8 neurons, the first hidden layer containing 10 neurons, the second hidden layer containing 14 neurons, and the output layer with 1 neuron. The scatter plots of predicted and observed O_3 concentrations for the training and testing datasets are illustrated in Figure 22.4a and b, respectively.

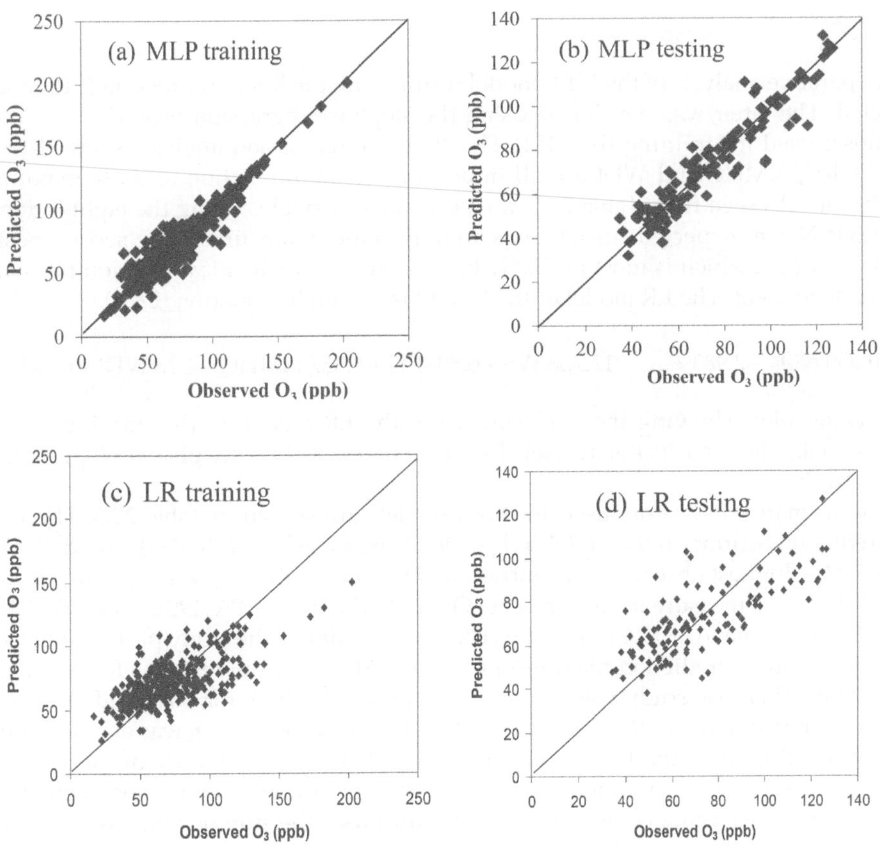

FIGURE 22.4 Scatter plots of observed vs. model predicted values: (a) training dataset for MLP model, (b) testing dataset for MLP model, (c) training dataset for LR model, and (d) testing dataset for LR model.

TABLE 22.2

Stepwise Regression Results

Steps	Set of variables	R^2
1	NO_2	0.200
2	NO_2, T_{max}	0.273
3	NO_2, T_{max}, WS	0.315
4	NO_2, T_{max}, WS, RH	0.351
5	NO_2, T_{max}, WS, RH, NMHC	0.371
6	NO_2, T_{max}, WS, RH, NMHC, WDI	0.396

TABLE 22.3

Performance Indicators for the Developed Models, Multilayer Perceptron (MLP) Model, and Linear Regression (LR) Model

Prediction method	MLP		LR	
Parameters \ Data subset	Training	Testing	Training	Testing
RMSE (ppb)	11.8	13.5	21.4	22.4
R^2	0.89	0.85	0.39	0.34
UPA (%)	−26 to 29	−24 to 23	−58 to 70	−63 to 56

A comparative analysis of the MLP model results with the linear regression (LR) model was conducted. The latter was developed using the stepwise regression procedure on the same data subset used for training the MLP. The stepwise regression analysis showed that NO_2, T_{max}, WS, RH, NMHC and WDI are all important for the prediction of daily maximum 1-h O_3 in the city. As seen from Table 22.2, the best single variable among the eight independent variables is NO_2 as expected since NO_x is an important O_3 precursor. The second-best single variable was T_{max}, which is most probably because the photochemical reaction rates are temperature dependent. The LR model of the best fit is shown in Equation (22–5).

$$O_3 = 0.650\ NO_2 + 2.983\ T_{max} - 12.738\ WS - 0.806 RH + 6.367\ NMHC - 4.26\ WDI + 6.121 \quad (22\text{–}5)$$

The scatter plots showing the performance of the LR model on the training and testing subsets of data (the same as the datasets for MLP) for prediction are given in Figure 22.4c and d, respectively.

The performance indicators used for both models are shown in Table 22.3. The unpaired peak prediction accuracy (*UPA*), $UPA = (P_i - Oi)/Oi$, was used to evaluate the model's ability to reproduce the highest observed concentration anywhere in the study area, which is a recommended statistical measure for urban-scale O_3 modelling (U.S. EPA, 1991). The US EPA guidelines (1991) recommend a *UPA* range of ± 20%. For each day of the study period, there was one pair of observation-prediction maximum O_3 levels. Since the interest was for the maximum hourly ozone, this case study used a cut-off value of 60 ppb O_3 for observed O_3 in the UPA calculation, meaning the UPA was calculated only when observed O_3 was above 60 ppb.

As compared to MLP, the LR model underestimated the peak hourly ozone values in the domain. The reason for the underestimation is that the fitting of regression coefficients was solved using the least squares method, which minimises the sum of squared errors. In principle, the LR does not capture the extreme values but aims at modelling the "average" behaviours for the predictand (output) variable.

In summary, the prediction of maximum ozone concentrations by the MLP model appears to be more accurate than the LR. It is also clear that the regression analysis results were not sufficient for predicting extreme values away from the centre of the distribution. The MLP model thus has a potential application in forecasting extreme hourly O_3 levels in the study area. An accurate prediction of such high ozone levels in advance is important from the public health protection perspective.

22.5 Application of Discriminant Analysis for Forecasting Photochemical Air Pollution in Tokyo

As explained in Section 21.4.5, in Japan the environmental standards for photochemical air pollution are based on photochemical oxidants O_x (O_3, PAN, and other oxidising substances, see Section 21.4.5), not ozone alone. Nowadays, high O_x concentrations still occur sometimes in Tokyo, but the problem was certainly much more serious during the 1980s when photochemical air pollution warnings were issued quite frequently. For this reason, research to predict O_x concentration from the concentration of primary pollutants such as NO_x and the meteorological variables had attracted much attention then. An important issue here was whether high O_x concentrations exceeding the warning issuance level would appear. Therefore, the application of discriminant analysis (detailed in Chapter 7) was considered suitable to answer the question.

The characteristics of the target area were analysed in advance using PCA (detailed in Section 21.4.5), which confirmed that the target area can be treated as a single domain; hence only one prediction model would be required.

22.5.1 Input Data and Its Preliminary Analysis

Before constructing a predictive model for the Tokyo metropolitan area, preliminary analyses were conducted to select useful variables (Ito et al., 1984). The relationship between meteorological variables and O_X concentrations was examined for high-concentration days, i.e., with O_x of 0.10 ppm or above, which is the threshold for issuing caution information to schools in advance. In this case, the highest hourly O_x concentration of the ambient air quality monitoring stations in Tokyo each day was used. The histogram of the occurrence frequency of the maximum temperature and the average daily wind speed during the data period of six months, from April to September 1982, is shown in Figure 22.5.

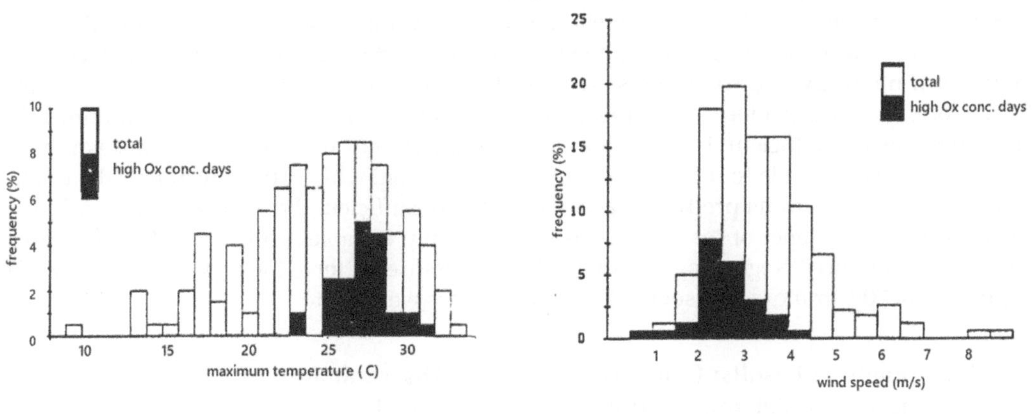

FIGURE 22.5 Occurrence frequency of daily maximum temperature and daily average wind speed, comparison of overall and oxidant high concentration days.

Source: Adapted and modified from Ito et al. (1984).

In these diagrams, the frequencies of days with high concentrations of O_x (≥0.10 ppm) are darkened. It can be seen that high O_x concentration events did not occur on days with a maximum temperature ≤ 23°C or a daily average wind speed ≥4.5 m/s.

Oxidant concentrations often peak around noon to afternoon, so it is necessary to predict for some hours in advance. Therefore, the preparation for the input data such as concentrations of the precursor pollutants (NO_2 and NMHC), meteorological measurements, or its forecast values, should be done using the data in the early morning of the same day. Therefore, the prediction model should operate using only the input data available by early morning on the forecast day. Note that the meteorological data can be generated by numerical weather forecast models that are normally available in advance with sufficient accuracy.

First, the days were divided into two groups according to the O_x concentrations. The analysis was carried out using two discrimination standards. One is 0.12 ppm of hourly O_x concentration as the standard threshold for issuing a photochemical smog warning. The other is 0.10 ppm as the standard for issuing warning information to schools (Ito et al., 1984). In way of example, this section focuses on the days with $O_x \geq 0.10$ppm (Okamoto et al., 1985). The dataset used to create the prediction model consisted of 144 days. which had no missing data out of the total of 183 days of the 6 months. The number of days when a warning was issued for the public was 16 days, and the number of days when the school information was issued was 28 days.

The analysis was done by both multiple regression analysis and discriminant analysis. For the multiple regression, the predicted objective variable was the maximum hourly O_x concentration that appeared in any of the 37 monitoring stations throughout Tokyo. The following seven explanatory variables were used for both methods.

1. Daily maximum temperature
2. Occurrence of sea breeze
3. Maximum mixing depth (MMD)
4. Oxidant (O_x) concentration of the previous day
5. NO_2 concentration of the same day morning time
6. Non-methane hydrocarbon (NMHC) concentration of the same day morning
7. Type of weather map (atmospheric pressure pattern)

The concentrations of NO_2 and NMHC in the early morning were calculated as the arithmetic mean of the hourly values at 7:00, 8:00 and 9:00 at three typical monitoring stations. When using this prediction model, the observed meteorological data of the day were not fully available; hence the numerical weather forecasted values were used. However, the correlation coefficient between the observed value and predicted value by weather forecast in the same day (in 1982) was good for maximum temperature (0.952), but for maximum wind speed and minimum humidity, the coefficients were relatively low, i.e., 0.658, and 0.780, respectively. Therefore, it was considered that the prediction accuracy other than the maximum temperature was not sufficient; hence other parameters were not used as explanatory variables. Besides, it was desirable to include also sunshine duration, which strongly affects the photochemical reaction, but its prediction accuracy was not sufficient; hence it was also excluded. The presence or absence of sea breeze was set as dummy variables with "1" and "0" for "presence" and "absence". Regarding the atmospheric pressure pattern, the type in which high O_x appeared at 20% or more was set to "1", and the others were set to "0".

22.5.2 Prediction Results: Comparison between Discriminant Analysis Model and Multiple Regression Model

The discriminant analysis was used to predict whether or not concentrations exceeded the warning level while the multiple regression analysis was used to estimate the hourly

TABLE 22.4

Contingency Table

		Prediction (estimation)	
		Low	**High**
Observation	Low	*a*	*b*
	High	*c*	*d*

maximum O_x concentration of each day. The objective (dependent) variable in the multiple regression analysis is the highest value that appeared at any of the more than 30 stations in Tokyo, and it is one data point per day. The same evaluation criteria were used for both the regression analysis and the discriminant analysis results. The prediction results are evaluated using a contingency table as shown in Table 22.4, where *a, b, c* and *d* are the appearance frequencies for each condition.

Here, the accuracy rate (or score) is calculated by the following formulae: Score 1 is the overall accuracy, and Score 2 is an index that focuses only on the occurrence of high concentrations to see how much of those occurrences could be predicted.

$$\text{Accuracy rate 1 (score 1)} = (a+d)/(a+b+c+d) \tag{22–6}$$

$$\text{Accuracy rate 2 (score 2)} = d/(c+d) \tag{22–7}$$

For the threshold limit of 0.10 ppm, two cases were analysed. Case 1 included all explanatory variables, while Case 2 excluded the maximum mixing depth (MMD). The reason is as follows: although MMD was said to be quite closely related to photochemical air pollution at the time, the coefficient of the discriminant equation and partial regression coefficient were too small in this analysis. Case 3 was also set with a discrimination standard of 0.12 ppm, which was considered in the discriminant analysis but not in multiple regression analysis. The results presented in Table 22.5 and Table 22.6 show that the MMD was not a useful variable in this case.

To see the overall trend, an accuracy rate of 1 (score 1) is suitable. However, accuracy rate 2 (score 2) seems to be appropriate from the viewpoint of reliably issuing an alarm on a high-concentration day. As for the evaluation of the result by multiple regression analysis (Table 22.6), score 1 shows a higher correct accuracy rate than score 2, which is the overall evaluation. However, discriminative models tend to have higher scores 2. In other words, this is the correct answer rate when the evaluation focuses only on days with high concentration.

In regression analysis, one can set aside some of the pre-obtained data and evaluate the fitness of the regression equation using this data that was not used to estimate the regression coefficients. In this case study, the evaluation was done on the data from the previous year (1981), which was not used to estimate the coefficients. Since the maximum daytime temperature and the presence of sea breezes were not yet observed at the time of the forecast for that day, predicted values from weather forecasts were also used as input for the model evaluation. The evaluation results are shown in Table 22.7 with three types of evaluations:

1. Evaluation result using the 1982 observed data; this is considered to be an outcome of interpolation.
2. Validation result using the 1981 observed data; actual meteorological values were used as input data for the day.
3. Validation result using the 1981 data, including the forecast data provided by the Meteorological Agency; for the meteorological input data, the predicted value from the weather forecast was used.

TABLE 22.5

Results of the Discriminant Analysis in the Summer Season 1982

Case	Case 1	Case 2	Case 3
Critical O_x concentration for discrimination	0.10 ppm	0.10 ppm	0.12 ppm
Mahalanobis distance	105.6	102.4	65.6
Ratio of variances, between group/total	0.65	0.65	0.56
Correlation coefficient between O_x concentration and discriminant function	0.70	0.72	0.62
Weights of discriminant function			
1. Daily maximum temperature	0.35	0.33	0.32
2. Occurrence of sea breeze	0.62	0.6	0.52
3. Maximum mixing depth	−0.15		
4. O_x concentration of the previous day	0.19	0.24	0.46
5. NO_2 concentration	0.24	0.19	0.47
6. NMHC concentration	0.21	0.25	−0.03
7. Type of weather map (synoptic pattern)	0.18	0.14	0.03
Evaluation	%	%	%
Score 1	83	85	85
Score 2	89	89	94

TABLE 22.6

Results of the Regression Analysis in the Summer Season 1982

Case	Case 1	Case 2
Multiple correlation coefficient	0.74	0.73
Root mean square error (ppm)	0.0241	0.0242
Partial regression coefficient		
1. Daily maximum temperature	0.24	0.26
2. Occurrence of sea breeze	0.41	0.44
3. Maximum mixing depth	0.12	
4. O_x concentration of the previous day	0.24	0.21
5. NO_2 concentration	0.18	0.22
6. NMHC concentration	0.06	0.03
7. Type of weather map (synoptic pattern)	0.12	0.15
Evaluation	%	%
Score 1	90	90
Score 2	54	61

TABLE 22.7

Comparison of the Predictive Performance for the Discriminant Model and Multiple Regression Model

Model types / Evaluation types	Evaluation score	Discriminant model	Multiple regression model
Estimation based on data	Score 1	0.85	0.90
of 1982	Score 2	0.89	0.61
Prediction by observed	Score 1	0.81	0.89
data of 1981	Score 2	0.82	0.50
Prediction by forecasted	Score 1	0.80	0.87
data of 1981	Score 2	0.91	0.45

Fairly good results were obtained by the discriminant model with a higher accuracy rate (score 2) for the high-concentration days. The accuracy rate (score 1), which is the overall accuracy with full concentration range, of the multiple regression model is higher than the discriminant model. This is because the discriminant model misclassified (predicted) some low-concentration days as high-concentration days. Furthermore, the reason may be that the discrimination points of the high-concentration day and the low-concentration day are much higher than the overall average. In the multiple regression model, many predicted values are gathered around the average value, and it seems that it was difficult to predict the high-concentration day in the tail of the distribution. Since the parameters, i.e., regression coefficients are estimated in regression analysis to reduce the residual sum of squares, regression analysis may likely not have been able to respond well to data classification at discriminant points far from the centre of the overall distribution.

In summary, it appears that the discriminant model is better than multiple regression for predicting high-concentration days that exceed the warning issuance level. This may be because the discriminant point in the two-group discrimination was much higher than the overall average; i.e., the point is far away from the centre of the distribution.

22.6 Case Study of Predicting SO_2 Concentration in Tokyo in the 1960s

The case introduced dates back to air pollution in Tokyo in the 1960s, which was also explained in Chapter 21, Section 21.4.1. As previously explained, the effects of meteorological conditions on air pollution concentrations are significant. Air pollution varies significantly from day to day by far more than the changes in the daily emissions. Therefore, if it can be assumed that the daily emission is nearly stable, then a statistical predicting model with meteorological conditions as the explanatory variable can be developed. This case study presents the development of a multiple regression model for prediction of the 24-hour-ahead concentration of SO_2, which was the most important issue at that time (Shiozawa et al., 1972).

22.6.1 Characteristics of Data Used and Setting of Explanatory Variables

In this regression model, the objective variable (dependent variable) was air pollution concentration, and this pollution dataset consisted of SO_2 concentrations measured at five air quality monitoring stations in Tokyo in the winter of 1969. Most of the explanatory variables (independent variables) were meteorological data, primarily obtained from observations at the Japan Meteorological Agency Tokyo Observatory.

It was found that air pollutant concentrations were mainly dominated by the synoptic-scale meteorological condition, and it changed periodically, as shown in Section 21.4.1. Therefore, the synoptic patterns of the very day and the day before were introduced into the model as the dummy variables. The diurnal variations and the weekly variations of the emission source activity were introduced into the model instead of the emission intensity. Table 22.8 shows the explanatory variables introduced into the model. The synoptic meteorological patterns (weather map) were classified in detail for the first time, and the patterns shown in Table 22.8 were consequently introduced into the model.

The results of the principal component analysis for the daily average concentrations, as detailed in Section 21.4.1 of Chapter 21, indicated that the air pollution concentrations in the area were closely related to the synoptic-scale meteorological conditions. Furthermore, as described in Section 22.2, diurnal variation (within-the-day fluctuation) in concentrations was also examined using canonical correlation analysis. For statistical models such as regression analysis when emission inventory is not available, a simple regression analysis method could be used that relies on surrogate emission data to model the diurnal variation. The surrogate variables for pollutant emissions are represented by variable numbers 20 through 25 in Table 22.8. That is, they are dummy variables with the values assigned to show that the

TABLE 22.8

Explanatory Variables of the Regression Model

Variable No.	Precedent hours	Explanatory variables
0		Constant
1	48	Atmospheric temperature
2		Relative humidity
3		Wind direction (S-N) 1 if W. D. containing S component 0 if others
4		Wind speed
5		Cloudiness
6		Weather
7		Precipitation
8		Hours of daylight
9	24	Atmospheric temperature
10		Relative humidity
11		Wind direction (S-N)
12		Wind speed
13		Cloudiness
14		Visibility
15		Hours of daylight
16	The day before	Winter type of pressure pattern 1 if winter type 0 if others
17		Front of migratory anticyclone
18		Back of migratory anticyclone
19		Trough of atmospheric pressure
20		Holiday 1 if holiday 0 if others
21		Hourly variation 1 if 8:00–19:00, 0 if others
22		Hourly variation (20:00–24:00)
23		Interaction between 20 and 21
24		Interaction between 20 and 22
25		Interaction between 20 and 19
26	The day	Winter type of pressure pattern
27		Front of migratory anticyclone
28		Back of migratory anticyclone
29		Trough of atmospheric pressure

Note: WD is wind direction.

emission levels are higher during the daytime when industrial and daily activities are more active and lower in the early morning and nighttime.

22.6.2 Modelling Results and Evaluation

To validate the developed regression model, hourly data on SO_2 concentrations at five monitoring stations in Tokyo in February 1969 were used. The multiple correlation coefficients were 0.71 to 0.77, and the results of the analysis of variance showed that the model was significant at a 0.01 significant level at all monitoring stations. The significant regression coefficients

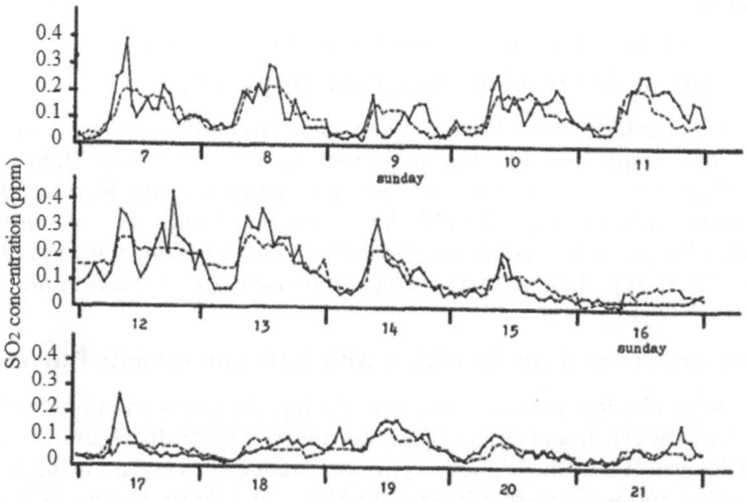

FIGURE 22.6 Comparison of observed (solid line) and predicted (dashed line) SO_2 concentrations from February 7 to 21, 1969, at the Kojiya monitoring station.

Source: Adapted and modified from Shiozawa et al. (1972).

were the atmospheric pressure pattern, the diurnal variations, and the weekly variations at most of the monitoring stations. The significant meteorological factors were, in general, the atmospheric temperature, the south-north components of wind direction, and the hours of daylight. The wind direction was important because it relates to the transportation of the pollutant from the emission sources. The hours of daylight corresponded to the diurnal variations of the concentration; it may play a role as a surrogate variable for the important within-the-day fluctuation of the concentrations. The regression coefficient of a dummy variable indicates the difference of the dependent variable between two set levels of the independent variable (explanatory variable). The diurnal variations of the concentration indicate higher values, ranging from 0.013 to 0.014 ppm in the daytime, compared with the nighttime. The comparison of the SO_2 concentrations between the observed and predicted by the regression equation is shown in Figure 22.6 for the Kojiya monitoring station near Haneda Airport in way of example.

Figure 22.6 shows that the predicted values have some agreement with the observed values. It also shows that there was not so much deviation between the predicted and observed values on February 8, 10, 11, and 14, when the low-pressure system was moving northward over the Japanese archipelago. A similar situation was observed at other stations as well.

For most monitoring stations, including Kojiya station, the partial regression coefficients for the atmospheric pressure patterns were relatively large, and the difference depending on the location was not so large. It was suggested that the synoptic-scale meteorological condition had a greater impact on air pollution in the wide area of the city. Therefore, to some extent, this model can be applied to predict concentrations 24 hours ahead. However, the accuracy of the maximum concentration during the peak time was not satisfactory. This was due to the limit of the ordinary statistical prediction method that the estimation accuracy is reduced at the tail of the distribution apart from the average value.

In this analysis, the problem of multicollinearity may not have been fully considered at the initial stage of the variable selection. For this reason, the choice of this explanatory variable may not have been appropriate and may need further consideration.

22.7 Haze Prediction Using Meteorology Clarification and Stepwise Regression Modelling Approach

This case presents the latter part of the study on the haze conditions in a mountainous urban area. The former part, presented previously in Section 21.4.2 of Chapter 21, has classified the meteorological patterns governing the area into four patterns using PCA and K-mean clustering techniques. In this section, the PM_{10} levels associated with these patterns are further analysed to develop an early haze warning system in the area, along with a prediction model developed for the most polluted pattern using a stepwise regression technique.

22.7.1 Characteristics of Haze Associated with Different Synoptic Patterns

Out of four meteorological patterns prevalent during the study months, the highest PM_{10} was observed in Pattern II as evident from the statistics given in Figure 21.3 and detailed in Table 22.9. Pattern II had the highest occurrence frequencies (49%) during the eight years of the data period and was mostly observed in March and April. Pattern IV had the second highest occurrence frequency (26%) and observed more in February. The other two patterns had lower occurrence frequencies (11, 14%) during the period (Kim Oanh and Leelasakultum, 2011). Pattern II had the highest PM_{10} levels observed at the two stations in the area, 97 ± 54 μg/m³, with 28% of the days belonging to this pattern having 24-h PM_{10} exceeding the NAAQS. Pattern I had the lowest PM_{10} levels, and none of the days belonging to this pattern had the pollution exceeding NAAQS (Table 22.9).

22.7.2 Construction and Evaluation of Statistical Prediction Model

The prediction model for 24-h PM_{10} levels in the area was developed for the days belonging to Pattern II. The stepwise linear regression procedure was applied using the data collected from February to April of 2007–2009 with a total of 100 days. The input data consist of all available meteorological variables at 7:00 LST of the forecast day, which are the same as those used for the classification scheme presented in Section 21.4.2; the previous 24-h PM_{10} averaged for two stations in the area; as well as the hotspot counts on the forecast day and the previous day around the area (Table 22.10). The stepwise procedure examines all entered variables and removes any variable with an F statistic below a significant level. In this case study, an F value of 0.15 was selected as the significant level for retention of a variable. The procedure retained only four independent variables as shown in Equation (22–8), which has the maximum coefficient of multiple determination (R^2).

TABLE 22.9

Occurrence Frequency of the Synoptic Patterns and Associated 24-h PM_{10} Levels in a mountainous urban area, 2001–2008

	Synoptic Pattern			
Parameters	**I**	**II**	**III**	**IV**
Total days of occurrence, day	75	351	96	185
Occurrence frequency, %	11	49	14	26
Average and 1 STD of 24-h PM_{10} levels, μg/m³	40 ± 15	97 ± 54	79 ± 41	89 ± 42
Frequency exceeding 24-h NAAQS (120 μg/m³) at any of the two stations in the urban area, %	0	28	16	24

Source: Adapted from Kim Oanh and Leelasakultum (2011).

TABLE 22.10

Independent Variables Used in the Stepwise Regression Analysis

Variable	Description of variable
Ta	Surface temperature at Chiang Mai Airport (meteorology station)
Td	Dew point temperature at Chiang Mai airport
Vis	Visibility at Chiang Mai airport
Ws	Wind speed at Chiang Mai airport
WDI	Wind direction index at Chiang Mai airport
Cl	Cloud cover at Chiang Mai Airport
Hum	Humidity at Chiang Mai Airport
SLP	Sea level pressure at Chiang Mai airport
PG1	The pressure gradient between Bangkok and Chiang Mai Airport
PG2	The pressure gradient between Nongkhai and Chiang Mai airport
PG3	The pressure gradient between Nakhonphanom and Chiang Mai station
PG4	The pressure gradient between Wuhan and Chiang Mai airport
PG5	The pressure gradient between Yibin and Chiang Mai airport
PG6	The pressure gradient between Haikou and Chiang Mai airport
P_PM_{10}	The average level of 24-h PM_{10} at the City Hall station and Yupparaj stations of the previous day
Hs80	Hotspot within 80 km from Chiang Mai airport
P_Hs80	Hotspot within 80 km from Chiang Mai airport of the previous day
Hs150	Hotspot within 150 km from Chiang Mai airport
P_Hs150	Hotspot within150 km from Chiang Mai airport of the previous day

Note that the emissions data are important to include in the regression analysis. However, due to the lack of the daily emissions data, the daily hotspot counts were used to represent the biomass open burning activity, which was one of the most significant emission sources during the haze episodes. The hotspot counts were found to be highly correlated with 24-h PM_{10}, especially the daily counts within a 150 km radius surrounding the Chiang Mai airport or Hs150 (Kim Oanh and Leelasakultum, 2011). However, the stepwise regression analysis excluded these variables, most probably due to a strong correlation between the hotspot counts and other explanatory variables that have been retained in the model, e.g., visibility and 24-h PM_{10} of the previous day in Equation (22–8). On the other hand, the contribution of the regional transport of air pollution to Chiang Mai is important and should be considered. This contribution was qualitatively analysed using the five-day HYSPLIT (Hybrid Single-Particle Lagrangian Integrated Trajectory model) backward trajectories, which were found to have a specific pathway for each identified pattern (Kim Oanh and Leelasakultum, 2011). Thus it indicates that the regional transport has already been adequately incorporated into the pattern identification which has also been reported previously (Pongkiatkul and Kim Oanh, 2007).

$$[PM_{10}] = -4044.53 + 0.568\,P\_PM_{10} - 0.008\,Vis - 1.948\,Hum + 4.258\,SLP \qquad (22\text{–}8)$$

where

$[PM_{10}]$ is the 24-h PM_{10} ($\mu g/m^3$) on the forecast day, averaged for two stations in the urban area;

P_PM_{10} is the 24-h PM_{10} ($\mu g/m^3$), averaged for two stations in the urban area, on the previous day;

Vis is the visibility (m);

Hum is the relative humidity (%);

SLP is the sea level pressure (millibar, mb)

(All observed meteorology data were at Chiang Mai airport, 07:00 LST on the forecast day.)

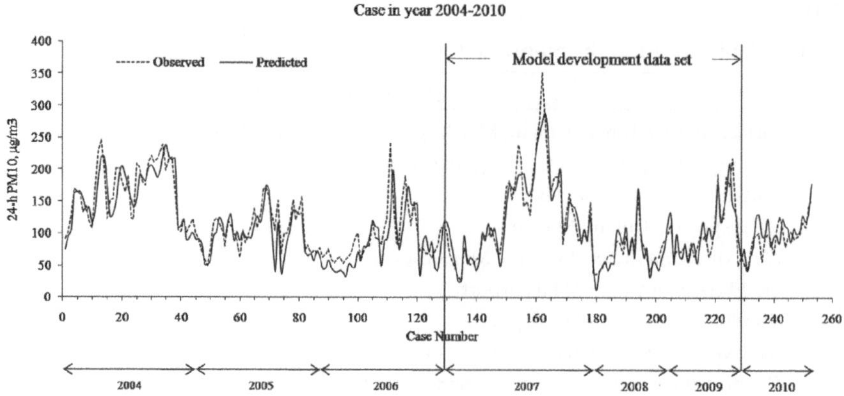

FIGURE 22.7 Observed vs. predicted 24-h PM_{10} in the urban area for the days in Pattern II during February–April of 2004–2010 using the tested dataset (left) and model development dataset (right).

The model performance was tested first with the dataset used for the model development (2007–2009), which showed satisfactory predictions, with $R^2 = 0.87$ and RMSE of 22 µg/m³. To evaluate the model robustness, the model performance was tested by using an independent dataset from the February–April months of 2004–2006 and 2010 (4 years), which also showed a relatively high R^2 of 0.81 and RMSE of 23.3 µg/m³. The model was also evaluated to assess its forecast skill in terms of correct and incorrect predictions of the PM_{10} levels above or below the 24-h NAAQS (120 µg/m³) following the category forecast verification statistics recommended by U.S. EPA (2003). The results showed that the model was able to achieve an accuracy of $A = 90.4\%$ (Kim Oanh and Leelasakultum, 2011).

The model performance for the seven years is presented in Figure 22.7, highlighting the comparison between the observed and estimated 24-h PM_{10} for the total of 250 days of Pattern II. The extremely high observed values were under-predicted, which is anticipated as the statistical regression models tend to predict means better than the tails of distribution, as previously discussed.

22.7.3 Early Warning of Haze

A scheme for early warning of haze in the study area has been proposed, which consists of three steps:

1. Identification of synoptic pattern of a given day using the automated meteorological classification scheme,
2. Issue a warning signal if it would be Pattern II, and
3. Predict the PM_{10} levels in the urban area using Equation (22–8) for the days of Pattern II.

This procedure can forecast the synoptic patterns and 24-h PM_{10} level in advance if the corresponding predicted meteorology is available for use. Thus, if the next day meteorology is used, then the forecast span would be 24 h. Accordingly, the identification of synoptic patterns can be made 24 h in advance, and a warning signal could be issued if the classification scheme using the forecasted meteorological data indicates that Pattern II is likely to occur next day.

22.8 Summary

In this chapter, the prediction and forecast of air pollution concentrations using various statistical methods have been presented and illustrated with real-life case studies. Statistical methods may be very useful for improving the accuracy of predictions especially when the processes of atmospheric dispersion in the study areas are not fully understood (a common case) and when the emission data are not readily available. It is also expected that such statistical analysis should advance our understanding and help improve the performance of the physico-mathematical dispersion models.

While statistical models require less computation time, they must deal with large datasets, so multivariate analysis methods are used in the model development. In addition, if the dimensions of the dataset can be reduced by making use of PCA, EFA etc., ordinary regression models can be developed. The statistical prediction and/or forecast models were the most popular in the past to meet the demand for estimating the concentrations several hours to a day ahead from the perspective of preventing serious health damage. Examples of forecasting systems using regression equations and artificial intelligence (neural networks) can be found in academic papers, and some of them have also been presented in this chapter. The use of the prediction results to reduce the emissions from a factory or stop open burning during the high concentrations periods would depend on the effectiveness of environmental administration. Nevertheless, accurate predictions serve as useful information to residents to take measures for avoiding excessive exposure such as refraining from outdoor activities or providing properly equipped cleaned rooms for school children and vulnerable individuals.

REFERENCES

Albert Co., 2020, *Basic knowledge on data analysis* (In Japanese), viewed on August 2020, https://www.albert2005.co.jp/knowledge/data_mining/cluster/cluster_summary

Ito, M., Udagawa, M., Sohfuku, T. and Okamoto, S., 1984, *Photochemical smog forecast by discriminant analysis, annual report of the Tokyo metropolitan research institute for environmental protection* (pp. 15–19) (In Japanese), Tokyo Metropolitan Government Office, Tokyo, Japan.

Kim Oanh, N.T. and Leelasakultum, K., 2011, Analysis of meteorology and emission in haze episode prevalence over mountain-bounded region for early warning, *Science of the Total Environment*, 409, 2261–2271.

Moriyama, S., Andou, J., Kuchikata, A., Sawachi, Y., Kitada, H. and Akiba, T., 1977 August 1–5, Stack gas monitoring system for warning and control, *IFAC Symposium on Environmental Systems Planning, Design and Control, Kyoto, Japan*, 10(7), 129–134, https://doi.org/10.1016/S1474-6670(17)66825-3

Murase, A., Oikawa, N., Itazaki, Y., Toba, K., Ogura, T., Okada, A. and Shimabukuro, K., 1973, Air pollution and source monitoring system in Kawasaki city, *Fujitsu*, 24(1), 20–42 (In Japanese).

Nghiem, L.H. and Kim Oanh, N.T., 2009, Comparative analysis of maximum daily ozone levels in urban areas predicted by different statistical models, *ScienceAsia*, 35, 276–283.

Ogino, S., 1967, Outline of the Osaka prefecture pollution monitoring centre establishment plan, *Production and Technology*, 19(7), 17–24 (In Japanese).

Okamoto, S., 2019 June 13–15, The transition of air pollution and its countermeasure in Japan, *The 5th environment Asia international conference on "transboundary environmental Nexus: From the local to regional perspectives"*, Chiang Mai, Thailand, also available of a revised version in Doolgindachbaporn, T. and Okamoto, S., 2020, Lessons learned from countermeasures against air pollution in Japan, *Environment Asia*, 13(2), 1–14, https://doi.org/10.14456/ea 2020.26, ISSN 1906-1714, ISSN Online 2586-8861

Okamoto, S., Ito, M., Hayashi, M. and Shiozawa, K., 1985, Development and evaluation of statistical prediction model for photochemical smog, *J Japan Society of Air Pollution*, 20(1), 33–39 (In Japanese).

Pennsylvania State University, 2023, *Eberly college of science, Canonical correlation analysis, lesson 13 in applied multivariate analysis*, viewed on 8 April 2023, https://online.stat.psu.edu/stat505/lesson/13

Pongkiatkul, P. and Kim Oanh, N.T., 2007, Assessment of potential long-range transport of particulate air pollution using trajectory modelling and monitoring data, *Atmospheric Research*, 85(1), 3–17.

Shiba, S., 1967, *Canonical correlation method, chapter 11 in "correlation analysis methods for behavioural science"*, The University of Tokyo Press, Komaba, Tokyo (In Japanese).

Shiozawa, K., Ohtaki, A. and Okamoto, S., 1972, Study on the prediction of sulphur dioxide concentration based on meteorological factors, *Memoirs of the School of Science & Engineering, Waseda University*, 36, 57–75.

U.S. Environmental Protection Agency (U.S. EPA), 1991, *Guideline for the regulatory application of the urban airshed model, U.S. EPA EPA-450/4-19-013*, Office of Air Quality Planning and Standards, Research Triangle Park, NC.

U.S. Environmental Protection Agency (U.S. EPA), 2003, *Guidelines for developing an air quality (ozone and PM2.5) forecasting program, EPA-465/R-03-002*, http://www.epa.gov/airnow/aq_forecasting_guidance-1016.pdf

Wierzchon, S.T. and Klopotek, M., 2018, *Modern algorithms of cluster analysis*, https://doi.org/10.1007/978-3-319-69308-8, https://link.springer.com/book/10.1007/978-3-319-69308-8

Zhang, B.N. and Kim Oanh, N.T., 2002, Photochemical smog pollution in the Bangkok metropolitan region of Thailand in relation to O_3 precursor concentrations and meteorological conditions, *Atmospheric Environment*, 36, 4211–4222.

23

Receptor Models

Receptor models are the analytical tools for estimating the quantitative contributions of the pollution source or source category using the monitoring data at the receptor site. The receptor models thus do not directly describe the physical processes in the atmosphere such as advection, diffusion and deposition of air pollutants that are the fundamental principles of dispersion models discussed in Chapter 18. Receptor models have emerged as one of the most attractive tools in the analysis of air pollution caused by suspended particulate matter. Receptor models can also be used for substances other than particulate matter such as volatile organic compounds (VOCs). However, this chapter will focus on the application of receptor models to analyse the particulate matter data. Knowledge of statistical analysis is essential for understanding the concept of receptor modelling. In particular, multivariate analysis, such as factor analysis, is closely related to the algorithms of many commonly used receptor models.

23.1 Basic Concept of Receptor Models

23.1.1 What Is a Receptor Model

The air quality models discussed in Chapter 18 are mathematical models that describe the spread of smoke and are commonly called diffusion or dispersion models. Apart from that, there is another group of models called receptor models. Both modelling approaches help to establish the source receptor relationship, which is required for air quality management. These two modelling approaches are commonly considered to supplement each other to thoroughly reveal such a relationship in the air pollution studies. The receptor models primarily utilise pollutant data collected at a receptor site to quantify the contributions from various emission source categories to the pollution levels measured at a receptor site. This is the reason why these models are named receptor models, which is used to contrast with the term "dispersion models". The latter calculate the concentrations at the receptors using the information of the source side (e.g., emission sources) and hence are also called the source model. The term "source model" is now rarely used; however, it is useful in understanding the emergence of the term "receptor model". Furthermore, researchers who focus on atmospheric dispersion do not typically use the term "source model". Instead, if they encounter it, they may view it as a submodule or preprocessor that estimates emission intensity within dispersion calculations. Fully grasping the relationship between these two modelling approaches remains challenging.

If many sources of pollutant emissions affect a target location, a dispersion model can be used to calculate the contributing concentrations for each source and sum them to obtain the resulting concentration at that location. Therefore, it is possible to know the contribution of each source to the pollution at the target location. On the other hand, regarding air pollution caused by particulate matter (PM), if sufficient information such as particle size, the most important chemical composition, and its shape can be obtained at the receptor site, it is possible to use receptor models to seek the relationship between pollution at the receptor site and the emission sources based on that information. Thus both approaches can be used to solve the problem, and their results often supplement each other to reveal the source–receptor relationship.

DOI: 10.1201/9781003597216-26

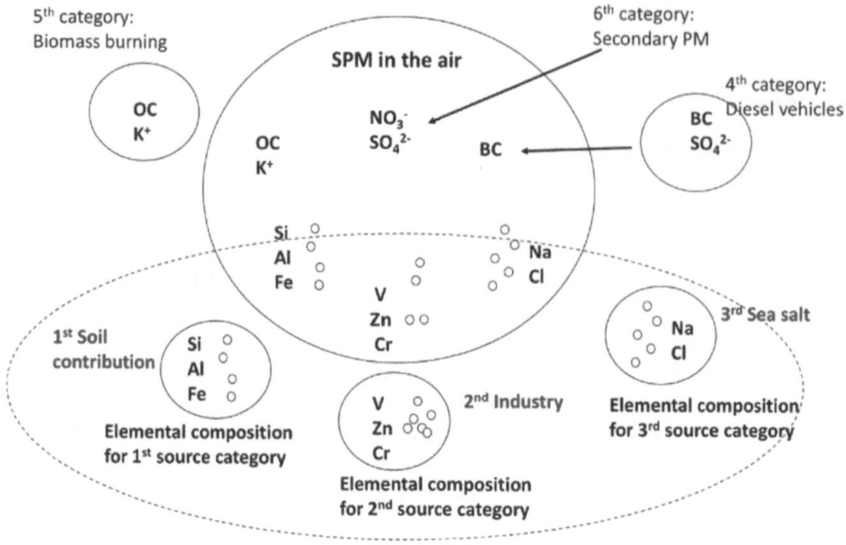

FIGURE 23.1 Characteristics of particulate matter in the atmosphere.

This chapter focuses on receptor models that rely on various methods using different computational algorithms that require certain statistical knowledge to comprehend. In air pollution research, various types of information can be obtained at the receptor site regarding PM pollution. The known chemical composition and/or morphological characteristics of PM allow the identification of the emission source types and their quantitative contributions to the measured PM mass at the site. The microscopy technique is used to gain the morphological characteristics of PM and is highly beneficial for analysing asbestos, mineral crystals or plant spores and pollen.

Particulate matter in the atmosphere is a collection of particles generated from various sources, and some parts are generated from soil, sea salt and so on. On the other hand, particles are generated by the combustion of fuels, and some are released into the air during transportation and handling of ore, coal, flour and so on. Here, the former are called particles of natural origin, and the latter are called particles of human-made or anthropogenic origin. It also contains secondary particles that are formed in the atmosphere from precursor gases such as SO_2 and NO_x (see Figure 23.1). Many receptor models assume a linear relationship between source-side composition and receptor-side contribution. Therefore, phenomena described by non-linear models such as chemical reactions in the atmosphere cannot be handled. For this reason, many receptor models treat it as an unknown contribution. However, some statistical models, such as factor analysis models, can accommodate these secondary particles conveniently. Details are given in the relevant sections in this chapter.

23.1.2 Basic Concept of a Chemical Mass Balance (CMB) Model

Several types of receptor models exist, as detailed later in this chapter. The models that calculate source contributions based on the concept of mass balance are called chemical mass balance models (CMB models). First, let us explain the basic concept of the calculation algorithms of the CMB model, which is the most typical receptor model using the following example. Suppose the mass and the elemental composition of airborne particulate matter collected at a certain place is as follows:

Aluminium, Al: 1.0 μg/m^3,
Zinc, Zn: 0.1 μg/m^3,
Sodium, Na: 0.6 μg/m^3,

TABLE 23.1

Chemical Composition of Particulate Matter of Emission Source Types

	Composition (%)		
Element	**Soil**	**Factory**	**Sea salt**
Al	**10.0**	1.0	0.0
Zn	0.5	**5.0**	0.0
Na	1.0	1.0	**30.0**
Ca	2.0	5.0	1.0
Others	86.5	88.0	69.0

Calcium, Ca: 0.4 μg/m^3,
Mass concentration: 20.0 μg/m^3.

In this instance, three source types are taken into account: soil, factory dust and sea salt particles. Table 23.1 presents the presumed proportion of each element in the particles originating from these sources, also known as the source profile data. The source profile refers to the chemical composition of particles discharged from a source of emission.

The simplest idea is to choose only one indicator element that represents each source category and that the effects of other sources on that element can be neglected. Let us call the receptor model based on this idea of the indicator element method. In the indicator element method, it is necessary to select only one indicator element representative of each source.

The indicator elements for each emission source are highlighted in Table 23.1. Their ratios are represented as relative values with the total equating to 1 and are presented as follows:

Soil Al (0.10)
Factory dust Zn (0.05)
Sea salt Na (0.30)

Therefore, the contributing concentration of soil, factory and sea salt can be easily calculated:

Soil: $1.0 \div 0.10 = 10.0$ μg/m^3
Factory: $0.1 \div 0.05 = 2.0$ μg/m^3
Sea salt: $0.6 \div 0.30 = 2.0$ μg/m^3

In reality, there may be other emission sources than these three indicated target sources that should also be included in the source apportionment. In any case, the resultant concentration of each indicator element is equal to the sum of the contributing concentrations from all sources. A model can be developed, which is a set of equations for the number of indicator elements, with the contributing concentration of each source as an unknown variable. These unknown variables can be solved by these equations accordingly. In addition, this indicator element may be called a marker element.

In this simultaneous equation model, let the contributing concentration of soil be x μg/m^3, the contributing concentration of factory dust be y μg/m^3, and the contributing concentration of sea salt be z μg/m^3. The mass balance for the indicator elements Al, Zn and Na is expressed by Equations (23–1), (23–2) and (23–3).

$$\text{Al: } 0.10x + 0.01y = 1.0 \tag{23–1}$$
$$\text{Zn: } 0.005x + 0.05y = 0.1 \tag{23–2}$$
$$\text{Na: } 0.01x + 0.01y + 0.30z = 0.6 \tag{23–3}$$

Solving the simultaneous equation yields x = 9.9, y = 1.01, and z = 1.64. Therefore, the contributing concentration of soil is 9.90 µg/m³, the contributing concentration of factory dust is 1.01 µg/m³, and the contribution of sea salt is 1.64 µg/m³.

If the number of indicator elements is larger than the number of sources, the least squares method can be used. Here, suppose that Ca had been analysed in addition to Al, Zn and Na, which are indicator elements for the three types of sources previously mentioned. Assuming that the contributions of soil, factory and sea salt are x, y, z µg/m³ as in the previous case, the concentration of each element is expressed by Equations (23–1) to (23–4).

$$\text{Al}: 0.10x + 0.01y = 1.0 \qquad (23\text{–}1)$$

$$\text{Zn}: 0.005x + 0.05y = 0.1 \qquad (23\text{–}2)$$

$$\text{Na}: 0.01x + 0.01y + 0.30z = 0.6 \qquad (23\text{–}3)$$

$$\text{Ca}: 0.02x + 0.05y + 0.01z = 0.4 \qquad (23\text{–}4)$$

In this set of simultaneous equations, the number of unknowns is 3, and the number of equations is 4, so in general, a solution that satisfies all of the Equations (23–1) to (23–4) cannot be found.

Here, assuming that the left sides of the Equations (23–1) to (23–4) are C_1, C_2, C_3 and C_4, the differences between the estimated value and the observed value for each element are C_1–1.0, C_2–0.1, C_3–0.6 and C_4–0.4. These differences can be positive or negative; therefore the square is taken and summed up to evaluate the overall absolute difference.

$$(C_1 - 1.0)^2 + (C_2 - 0.1)^2 + (C_3 - 0.6)^2 + (C_4 - 0.4)^2$$

Therefore, if the values of x, y and z are determined so that this sum of squares is minimised, that value, respectively, will be the contributing concentration from each source that is most suitable for explaining the measured value of each elemental concentration. In such calculations, there may be a large gap of several orders of magnitude or more between the elemental concentrations. For this reason, it may be better to give the same importance to the elements with the lower elemental concentration and the elements with the higher elemental concentration, by dividing the difference by the concentration of each element and calculating the sum of squares. Expressing this sum as S is as follows:

$$S = \left(\frac{C_1 - 1.0}{1.0}\right)^2 + \left(\frac{C_2 - 0.1}{0.1}\right)^2 + \left(\frac{C_3 - 0.6}{0.6}\right)^2 + \left(\frac{C_4 - 0.4}{0.4}\right)^2$$

Recall that the left sides of the Equations (23–1) to (23–4) are C_1, C_2, C_3 and C_4.

$$C_1 = 0.10x + 0.01y$$
$$C_2 = 0.005x + 0.05y$$
$$C_3 = 0.01x + 0.01y + 0.30z$$
$$C_4 = 0.02x + 0.05y + 0.01z$$

Here, the values of x, y and z must be determined so that S is minimised. Solving this problem is easily done by a computer. As a result, the obtained values of x, y and z are:

$$x = 10.94\ \mu\text{g/m}^3$$
$$y = 1.04\ \mu\text{g/m}^3$$
$$z = 1.63\ \mu\text{g/m}^3$$

TABLE 23.2
Comparison of the Calculated Results by Several Methods

	Indicator element method	Simultaneous equations	Weighted least squares
Soil	10.00	9.9	10.94
Factory	2.00	1.01	1.04
Sea salt	2.00	1.64	1.63

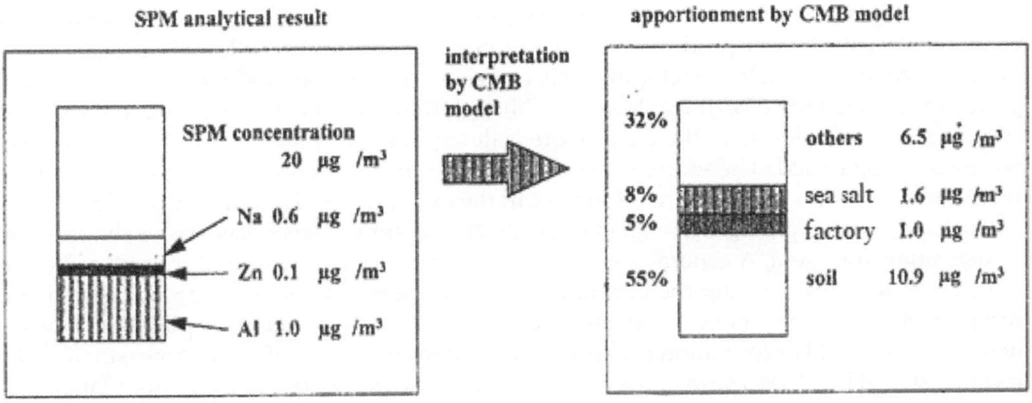

FIGURE 23.2 Source apportionment of the particulate matter by the chemical mass balance (CMB) model.

Table 23.2 presents the outcomes of these calculations. Minor variations in the results may occur based on the method of calculation used. The preceding discussion clarifies that, of the three methods, least squares estimation possesses the greatest robustness.

In summary, as shown in Figure 23.2, the CMB model can be considered a tool that translates chemical composition data into source contribution.

23.2 Feature of Receptor Models

Research on the receptor model has been active since the late 1970s, and many research papers can be seen. These relatively early findings can be found in the US Environmental Protection Agency-sponsored workshop Quail Roost II. The major papers reported at this conference have been featured in the Journal of *Atmospheric Environment*, with an overall overview reported by Stevens and Pace (1984). There are also many excellent review papers commenting on the study of the receptor model. For example, Watson (1984), Hopke (2016), Pongkiatkul and Kim Oanh (2012). In addition, related research and published software in the US EPA can be accessed from the EPA website (U.S. EPA, 2022).

As seen previously, the basic idea of the CMB model is to find the contribution rate of the emission source based on the information of the components measured at the receptor site. There are many ways to calculate this contribution rate, and some of the most representative ways are discussed in this section.

As can be seen in the CMB model described in the previous section (Section 23.1), the model formula for calculation is the same as the simultaneous equations. The CMB model can be utilised for source apportionment even with single dataset comprising approximately 10–20 elemental (or chemical) compositions at a receptor. Other models that utilise statistical methods of multivariate analysis such as a factor analysis are also used for source apportionment studies. However, these models require a dataset with enough data points to derive a correlation coefficient matrix between each component. For example, with about 20–30 chemical components analysed, the number of data points or samples required is about 40–50. This receptor modelling approach thus gives some overview of the plausible source types or their categories affecting pollution at the receptor site, even if the information on the source side is not necessarily sufficient.

The CMB model can take advantage of when the source profiles are available. However, difficulties may arise in the CMB applications because the changes in source emission rates, emissions compositions and meteorological conditions such as wind direction may cause large uncertainties in the source apportionment results. Also, there is the collinearity problem when some source compositions are very similar. In contrast, multivariate models can identify unknown sources without intensive prior information on source compositions but need an ambient dataset produced by a large number of samples collected at a site. Some multivariate models such as factor analysis can show several sources and associated source profiles related to factor loading. However, although the amount corresponding to the source contribution associated with the factor score is determined, its quantitative absolute value is not directly generated. Therefore, some mathematical manipulation is required to estimate the source contribution as illustrated further in the example provided in Section 23.5.

The fundamental hypothesis in all mathematical receptor models is based on a simple mass conservation argument. A general assumption in receptor modelling is that source profiles do not change with time during the period from the emission till pollutants are measured at a receptor site. If several sources exist, and there is no interaction among air pollutant sources, neither any removal or formation processes, the total mass of an air pollutant measured at the receptor would be a linear sum of the contributions of the individual source. This relationship is the same for both the CMB model and the multivariate analysis model and is expressed by a mathematical formula as shown in Figure 23.3.

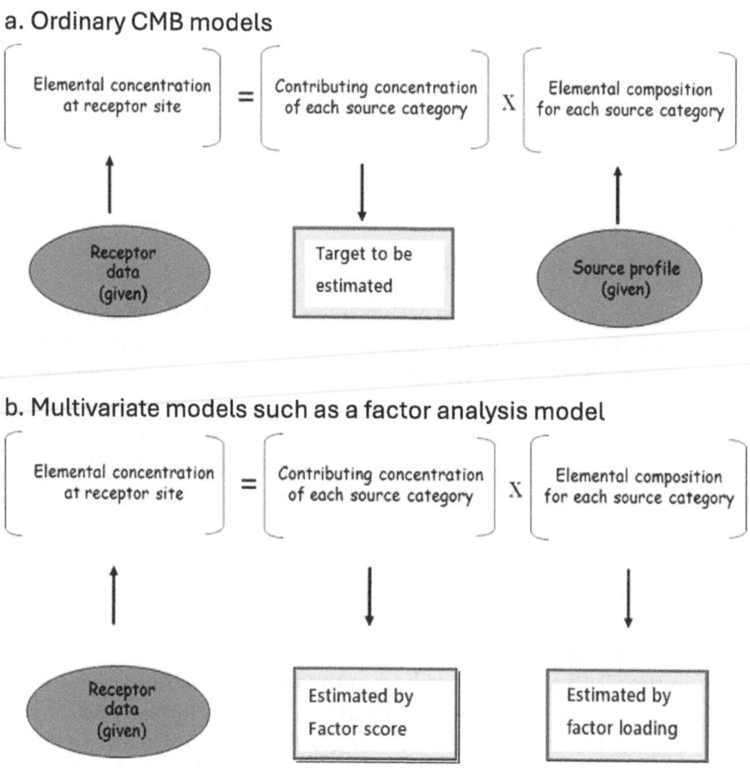

FIGURE 23.3 Comparison of the ordinary CMB models with multivariate models such as a factor analysis model.

TABLE 23.3

Strengths and Limitations of Some Common Receptor Models

Models	Strengths	Limitations
Chemical mass balance models		
For example, chemical mass balance version 8 (CMB8)	Mathematically easier to implement Does not require a large number of samples Produces quantitative source contributions directly Has built-in performance fitting measures User-friendly software	Requires comprehensive advanced information on source profiles Faces collinearity problem when the source composition profiles are almost similar Cannot incorporate the temporal and spatial variation in ambient concentrations and source emissions Cannot identify sources with unknown profiles
Multivariate models		
For example, Positive matrix factorisation (PMF)	Does not require comprehensive advanced information on source composition profiles Incorporates the time variation Uses non-negativity constraints, which is more physically realistic Rotation (determining the direction of coordinate axes in factor analysis) can be controlled by the user	Requires a large number of ambient samples Faces difficulty in identification of several sources Employs the trial-and-error method Needs rescaling to produce final quantitative source contributions Needs background knowledge on source composition profile for result interpretation Inherent rotational ambiguity of the factor analysis

Source: Adapted from Pongkiatkul and Kim Oanh (2012).

Now compare the characteristics of ordinary CMB models with multivariate models. Some statistical models such as a factor analysis model allow us to estimate both contribution rates at the receptor site and source profiles from a large amount of data for a large number of chemical compositions. Figure 23.3 shows a comparison between these two types of receptor models.

Advanced multivariate models such as positive matrix factorisation (PMF) can generate source profiles and provide information that leads to the quantification of source contributions. However, the number of factors (number of sources) must be estimated in advance. The concepts and characteristics of the PMF are described in Section 23.6. Table 23.3 summarises the advantages and limitations of some commonly used receptor models.

Now let us explain algorithms such as how to solve concrete equations for some typical receptor models. The factor analysis model (FA-MR model) and positive matrix factorisation (PMF) are covered in their respective sections; hence they will only be briefly mentioned in this section.

23.2.1 Least Squares Model

The CMB model is a mathematical model used to estimate the value of the source contribution S_j from the measured value of the component concentration C_i when the value of the component composition F_{ij} of the emission source is known. Here, the subscript i represents the i-th element, and the subscript j represents the j-th emission source.

With the concentrations C_i of the n chemical components measured, to estimate the contribution S_j from m sources, a system of equations (simultaneous linear equations) can be derived (Equation 23–5). Note that the m sources are not the number of m chimneys but the number of source groups (source types or source categories, e.g., factories, diesel vehicles or rice straw burning). The individual emission sources within each group, e.g., coal-fired industrial boilers, discharge PM having very similar chemical components.

$$\left.\begin{aligned} C_1 &= F_{11}\ S_1 + F_{12}\ S_2 + \ldots\ldots + F_{1m}\ S_m \\ C_1 &= F_{21}\ S_1 + F_{22}\ S_2 + \ldots\ldots + F_{2m}\ S_m \\ &\vdots \quad \vdots \\ &\vdots \quad \vdots \\ C_n &= F_{n1}\ S_1 + F_{n2}\ S_2 + \ldots\ldots + F_{nm}\ S_m \end{aligned}\right\} \tag{23–5}$$

Rewriting Equation (23–5) as a matrix gives the following equation:

$$\boldsymbol{C} = \boldsymbol{FS} \tag{23–6}$$

The numbers m and n must satisfy the following equation $\{n > m\}$. Now, applying the case described in the previous chapter, that is, the cases shown in Equations (23–1) to (23–3), the elemental values of the matrix are as follows:

$$\mathbb{C} = \begin{pmatrix} c_1 \\ c_2 \\ c_3 \end{pmatrix} = \begin{pmatrix} 1.0 \\ 0.1 \\ 0.6 \end{pmatrix}$$

$$\mathbb{F} = \begin{pmatrix} F_{11} & F_{12} & F_{13} \\ F_{21} & F_{22} & F_{23} \\ F_{31} & F_{32} & F_{33} \end{pmatrix} = \begin{pmatrix} 0.10 & 0.01 & 0 \\ 0.005 & 0.05 & 0 \\ 0.01 & 0.01 & 0.30 \end{pmatrix}$$

In the least squares method, a source contribution vector **S** to be estimated is expressed by Equation (23–7).

$$\hat{\boldsymbol{S}} = \left(\boldsymbol{F}^t \boldsymbol{F}\right)^{-1} \boldsymbol{F}^t\ \boldsymbol{C} \tag{23–7}$$

Here, the superscript t indicates that it is the transposed matrix of that matrix. Also, the superscript –1 indicates that it is an inverse matrix.

In the ordinary least squares method, all data are treated equally, so only the components having a large absolute value in the measured data contribute to the magnitude of the residual sum of squares. That is, in actual data, there is a concentration difference of more than a thousand times between Al, Fe, Na and the like having relatively high component concentrations and Ce, Sc, V and the like having low component concentrations. For this reason, in the model of Equation (23–7), the information of these elements with small component amounts is ignored as a result.

To cope with this problem, a method of equally evaluating information on all chemical components has been considered. That is, the residual is divided by the measurement value to calculate its sum of squares. Further, there is a difference in the difficulty of analysis depending on the chemical component, and therefore, the accuracy of the analysis data is also generally different. In such a case, a method of obtaining as accurate a solution as possible by applying a bigger weight to more accurate data is also conceivable. Here, dividing by the measured value means that the reciprocal of the measured value is used as the weight. A method is conceivable in which, instead of the reciprocal of the concentration (measured value), the reciprocal of the analysis error (variance of repeated measurement values) is used as the weighting coefficient. A matrix in which the weighting factors are arranged on diagonal elements is represented by Equation (23–8).

$$W = \{w_{ij}\}, \quad w_{ij} = \begin{cases} \dfrac{1}{\sigma_{ci}^2}(i = j) \\ 0(i \neq j) \end{cases} \tag{23–8}$$

The weighting coefficient is the reciprocal of the measurement error variance and is better from the viewpoint of reducing the weight of elements with poor measurement accuracy. Here, the measurement error variance is σ_C^2. Using the weighting factors shown in Equation (23–8), the estimated source contribution is expressed in Equation (23–9).

$$\hat{S} = \left(F^tWF\right)^{-1} F^tWC \tag{23–9}$$

23.2.2 Effective Variance Model

In the ordinary CMB model, the value of F_{ij} is given as a known amount, and the value of the contribution S_j of the j-th source is estimated from the measured value C_i of the chemical component i. However, the value of F_{ij} is also measured and is considered to include some errors. Therefore, when considering the measurement error as the weight in the weighted least squares method, it is necessary to consider the error not only for C_i but also for F_{ij}. A least squares method considering such weighting factors has been proposed by Watson et al. (1984), and is called an effective variance method. The variance, namely the square of standard error in the case of the effective variance model, is defined by Equation (23–10).

$$\sigma_{EVi}^2 = \sigma_{ci}^2 + \sum_{j=1}^{m}\left(S_j \cdot \sigma_{Fij}\right)^2 \tag{23–10}$$

23.2.3 Ridge Regression Model

In the calculation by the CMB model, different sources j and j' should be differentiated by F_{ij} and F_{ij}'. For the sources in which $F_{ij} = F_{ij}'$ for all chemical components, that is, sources having the same component composition, their contributions cannot be calculated by the CMB model. In addition, actual measurement data may be obtained with various errors. Therefore, even if the compositions of the components are not completely the same if both are very similar, it becomes very difficult to identify their source contributions. This phenomenon is statistically called multicollinearity, which has already been explained as one of the important issues when using multiple regression analysis in Chapter 6 of Part I.

When the problem of multicollinearity occurs, the calculation result will be very unstable. Ridge regression is one of the numerical methods that can be used for solving systems containing such multicollinearity data. By slightly changing the basic equation, the multicollinearity is forcibly avoided, and an approximate solution is obtained. More specifically, Equation (23–11) is solved by introducing a parameter k (> 0). Since this result is not an unbiased estimator, the setting value of this parameter k should be as small as possible, and it may be better to reject the solution if it is difficult or impossible to support a theoretical interpretation.

$$\hat{S} = \left(F^tWF + kI\right)^{-1} F^tWF \tag{23–11}$$

Here, $\boldsymbol{I}$ is a unit matrix. Details of this method are introduced in the relevant textbooks such as Chatterjee and Price (1977).

23.2.4 Linear Programming Model

The basic equation of the CMB model is expressed in Equation (23–5). The solution can be obtained by applying linear programming to this model. The problem requires a mathematical algorithm to determine a non-negative variable, known as the contribution factor, which maximises the objective function – the sum of contributions from known sources. The task must adhere to the constraint where the product of the source profile's component ratio and the source's contribution remains below the component concentration at the receptor, as governed by a linear inequality. In other words, maximising the sum of the contributions of the considered sources is also minimising the contribution of the unknown sources.

The advantage of this model is that, unlike least squares, the solution obtained is non-negative. However, even if the calculation results are derived from incorrect data, the solution looks plausible at first glance. If the analytical value of a particular component, which is critical to the constraints, has a significant error, the consequences can be quite severe, and there is a chance that the analyst may overlook this issue. The issue arises from the fact that even when utilising data with significant errors, one can still obtain non-negative results that appear plausible.

23.2.5 Factor Analysis Model

Among a group of multivariate analyses, a method called factor analysis (see Chapter 21) is widely used in the source apportionment study. The basic formula of the chemical balance model explained at the beginning is consistent with the basic mathematical model of factor analysis, as previously explained. In addition, a regression analysis model with factor scores as explanatory variables will be detailed in Section 23.5.

23.2.6 Positive Matrix Factorisation (PMF)

Positive matrix factorisation was developed by Paatero and Tapper (1994) and became a well-known receptor model, which is based on the multivariate approach. Non-negativity factorisation, both loadings and scores, which is built into the PMF, is the major difference from other multivariate receptor models. In addition, PMF does not rely on information from the correlation matrix but utilizes a point-by-point least squares minimisation scheme. Therefore, the profiles produced can be directly compared to the input matrix without any further transformation. These advantages can make PMF different from other multivariate methods (Pongkiatkul and Kim Oanh, 2012).

23.2.7 Structural Equation Modelling (SEM)

Structural equation modelling (SEM) can be regarded as a comprehensive statistical model that includes almost all multivariate models (Toyoda et al., 1992). The basic concept of the SEM will be explained in Chapter 24. Multiple regression analysis, which uses factor scores as explanatory variables, seems to be one of the most typical applications. Therefore, it can be considered that factor analysis–multiple regression model analysis is very similar to SEM. In SEM analysis, PM concentration is determined by the sum of the contributions of several PM sources, which correspond to the latent factors. This source profile, which corresponds to the factor loading, is defined by each chemical composition, and these relationships can be expressed by a path diagram. An example of a PM source apportionment study using the SEM can be found in Sakurai et al. (2003).

23.3 Application Procedure of the CMB Model Calculations

This section describes the general procedure for calculating source contributions to PM using the ordinary weighted least squares CMB method, which is the most commonly used receptor model. In the explanation, the dataset obtained in the Kashima SPM study described in Chapter 13 will be used.

23.3.1 Calculation Procedure

The general ordinary calculation procedure using the least squares CMB receptor model is presented here. In reality, due to various constraints, it may not always be possible to do everything according to this procedure However, understanding this procedure may be useful in many situations.

23.3.1.1 Step 1: Prepare the Source Profile Data

The data used here is usually the elemental composition of PM released from emission sources that are considered important in the study area where the receptor site is located. It is desirable to collect and analyse particulate matter emitted directly from the sources. If this is not possible, existing research material or literature can be used to prepare the source profile data. Table 23.4 shows the source data used in the examples shown here (Okamoto, 1996), which is commonly referred to as the source profile.

23.3.1.2 Step 2: Prepare Environmental Data (Chemical Composition Data of Particulate Matter)

Sufficient quantity and appropriate quality of PM element concentration data are essential for analysis by receptor models. In addition, it is also important to confirm that the appropriate marker elements are included in the dataset. A measured value below the detection limit does not mean the concentration is zero, but it may be due to a limitation in the measurement

TABLE 23.4

Example of a Source Profile

Element	Road dust	Soil	Sea salt	Steel	Oil burning	Incineration	Diesel	Biomass burning
Al	6.1	5.85	3.0E-05	0	0.18	0.42	0.061	0.12
Fe	3.7	3.3	2.9E-05	22.3	2.9	0.62	0.036	0
Mn	0.076	0.078	5.8E-06	0.41	0.061	0.033	0.0012	0.03
Na	1.5	1.35	30.4	0	2.8	12	0.015	0
V	0.0085	0.0097	5.8E-06	0.0092	0.33	0.0027	0.00075	0.0004
Zn	0.014	0.0168	0	0.96	0.18	2.6	0.00041	0.000057
Pb	0.004	0.0048	8.7E-06	1.09	0.033	1.7	0.049	0
EC	4	0	0	0	50.3	6.7	68	10.2
OC	0.7	0	0	0	6.5	1.7	10	44.7
Sc	0.0017	0.0014	1.0E-07	1.3E-04	9.0E-06	4.60E-05	0	1.0E-07
Br	0.01	0.015	0.19	0.014	0	0.083	0.017	0.035

Note: EC is elemental carbon, OC is organic carbon, and unit is % mass.
Source: Adapted from Okamoto (1996).

methods (see Chapter 8). However, due to resource restrictions, in some cases these are unavoidable. It is preferable to delete such data below the detection limit even if this procedure is available. However, when a large number of elements are deleted, the calculation of the least squares method may become difficult. Therefore, it is convenient to assign a very small weighting factor to the data below the detection limit to have an effect substantially equivalent to deleting the data. When such a value appears, it will be necessary to take the following actions.

- Replace these data with half, one-tenth, or other appropriate values of the data related to the detection limit.
- The contribution is calculated by setting a smaller weighting coefficient w in the mathematical expression described in the algorithm.

23.3.1.3 Step 3: Conduct Screening to Prepare the Ambient Data Input

Beside the non-detected values, suspicious data points are also possible, for example by a failure in the sampling method, sample handling method or analysis method. Proper elimination of such inappropriate data is essential for obtaining good-quality analysis results. The data screening shown here is quoted from Pongkiatkul and Kim Oanh (2012).

- ***Raw data check:*** Mainly, check the relationship with the lower limit of detection. It also includes examining the confirmation items on the technicians' record forms during sampling (logbook).
- ***Time-series analysis:*** If the sampling is done in time-series, create a time-series graph. By doing so, it is possible to grasp the appearance of abnormal values.
- ***Scatter plots:*** If there are relevant combinations between the analysed components, use those data to create a scatter plot. If there is something wrong with the placement, there may be something wrong with the analysis of either component or both.
- ***Signal to noise:*** The signal-to-noise ratio (S/N ratio), presented in Equation (23–12), is generally applied to screen out the species that have poor measurement results. The calculation should be made for each species. A small S/N ratio (less than 0.2) indicates that the species has a high noise level (Paatero and Hopke, 2003) and should be discarded from the dataset.

$$\text{S/N ratio} = \frac{\text{Sum of species concentration}}{\text{Average of detection limit} \times \text{Number of below detection limit}} \quad (23\text{–}12)$$

- ***Reconstructed mass:*** If all abundant components such as carbon, ionic components and crustal components are measured, the total of unanalysed elements can be estimated to some extent. In particular, crustal components often exist as oxides, so it is possible to estimate the amount of oxygen that is not normally measured. By doing so, it is possible to verify the validity of the analytical value as a whole by grasping the expected shape of the compound and comparing it with the mass concentration of the particulate matter.

23.3.1.4 Step 4: Prepare the Analysis Software of the Receptor Model

Useful software such as CMB8 and PMF can be downloaded from the US Environmental Protection Agency's website (U.S. EPA, 2022). If a statistical analysis software for regression analysis or a mathematical software that calculates the inverse matrix is available for use, one can use such software, but some related knowledge is required. The choice of receptor model to use also depends on the availability of prepared data. When using multivariate analysis models such as PMF, it is necessary to have a sufficient amount of data.

23.3.1.5 Step 5: Set the Necessary Coefficients and Other Quantities

It is necessary to organise the information that can be inferred from the observed data for the contributions of sources that are not included in the source profile, i.e., cannot be modelled in the ordinary CBM algorithms. Also, setting the weighting coefficients for weighted least squares and estimating the contribution of secondary particles. For instance, one could consider the method outlined here. In this method, the weighting coefficient is determined by multiplying the reciprocal of the coefficient of variation (which is the standard error divided by the mean value, $\sigma T/\bar{X}$) for each element, as obtained from the repeated measurement test detailed in Table 13.12 of Chapter 13, by the concentration of the element.

The contribution of secondary particles generated by chemical reactions in the atmosphere cannot be determined by an ordinary CMB model. If analytical data of ion components are available, we can use these data. In addition, when there are emission source data, there may be cases where sulphate ions and the like are included in this source profile data, and it is necessary to take measures to prevent double counting the influence. However, the effects are considered to be minor, and they may be often ignored. Here, the following method could be used. That is, it is calculated by ($NH_4^+ + SO_4^{-2} + NO_3^- - S$). The value of S is the sulphur component concentration other than the secondary particles and is determined using the numerical value of the sulphur component composition and the value of the estimated result of each source contribution concentration. For example, in Okamoto (1996), the value of S was calculated from each source contribution obtained by the CMB model as shown in Equation (23–13).

$$\begin{aligned} S = {} & \text{soil} \times 0.0001 + \text{sea salt} \times 0.026 + \text{oil burning} \times 0.096 + \text{diesel exhaust} \times 0.032 \\ & + (\text{steel} + \text{incineration} + \text{biomass burning}) \times 0 \end{aligned} \tag{23–13}$$

23.3.1.6 Step 6: Evaluate the Calculation Result

It is necessary to make sure that the calculation result is consistent with relevant experience and knowledge. If there is a sufficient number of pairs, namely observed and calculated data points, they can be evaluated statistically. When a plurality of sources with similar chemical compositions are selected, it is also possible to confirm the difficult situation as follows.

When the non-orthogonality between explanatory variables is extremely strong as just described, this state is referred to as a problem of collinear data or multicollinearity, and it is very difficult to overcome this situation. This is because multicollinearity is not due to improper modelling, but rather an imperfection of the data itself. The measuring score for the degree of the effect of multicollinearity described above is the VIF value. The variance inflation factor (VIF) in statistics quantifies the severity of multicollinearity in the least squares regression analysis. It provides a measure of how much the variance of the estimated regression coefficients has increased due to multicollinearity. It is said that when the VIF value is larger than 10, multicollinearity becomes a problem. In this study, the weighted least squares method was applied, so the VIF value was calculated using the source profile data with the weighting coefficients. Note that the value of the weighting coefficient w_i applied to the element i was the reciprocal of the concentration value of the element i at the receptor.

$$\text{VIF}(j) = 1/(1 - Rj^2) \tag{23–14}$$

where j is the source number, and Rj^2 is the contribution (square of multiple correlation coefficient) when the j-th explanatory variable will be estimated by the other explanatory variables. From Equation (23–14), it is considered that as Rj^2 is closer to 1, the larger the VIF value is, the component composition of a certain source can be represented by a linear combination of the component compositions of other sources. Also, Chatterjee and Price (1977) explain that the VIF value is a diagonal element of the inverse matrix of the correlation matrix.

23.3.1.7 Step 7: Handling of Abnormal Values and Their Treatment

When calculating the source contribution using actual measurement data, various calculation inconveniences such as the existence of abnormal values may occur, so it is necessary to predetermine a method for dealing with such cases.

In the least squares method, since the partial regression coefficient is determined to minimise the residual sum of squares (the sum of squares of the difference between the measured value and the estimated value), a negative value may occur in the regression coefficient. However, this coefficient has a meaning of the contributing concentration in the least squares CMB model, so some measures may need to be taken to overcome this inconvenience. Although the occurrence of the negative values seems unavoidable, the negative contributing concentration is practically impossible and is treated as an abnormal value. When a negative calculated value appears, we can consider that it may be a sign that a multicollinearity problem has occurred. Look at a concrete example of how this phenomenon happened as detailed in Section 23.3.2.

23.3.2 Case Studies of Result Evaluation and Handling of Abnormal Values

To deeply understand the procedures that have been described so far, especially steps 6 and 7, it is better to illustrate them with actual examples. This example used the data obtained from the Hivol dataset in the Kashima area, as described in Chapter 13. Let us use the following data of 11 major marker elements: Al (72.3), Br (6.95), Fe (193), Mn (12.0), Na (622), Sc (0.02), V (7.66), Zn (51.5), Pb (89.2), Elemental C (4000), Organic C (2300). Here, the number in parentheses is the concentration in ng/m^3 units. The source contributions were calculated using the source profile data shown in Table 23.4. The results are shown in Table 23.5. In this example, the VIF value is over 10 for road dust and soil. From this calculation result, it was found that there was multicollinearity between the road dust and the soil. In such a case, the most realistic countermeasure is to delete either of the emission sources from the source profile data, but this does not mean that the contribution of the deleted source is necessarily zero. Table 23.6 shows the calculation results when road dust is removed.

It is appropriate to interpret that the calculated contribution of the source remaining in the model includes that of the deleted source from the source profile. In this case, the source data of the road dust having a negative value was deleted, and the calculation was performed again using seven source profiles. Road dust and soil are very similar in component composition. Therefore, the calculated value of the soil contribution in Table 23.6 also includes the contribution of road dust; hence it is considered to be the sum of the contributions of both.

Instead of calculating VIF values, the suitability of source data can also be determined by calculating correlation coefficients between source profile data. In the seven-source data

TABLE 23.5

Calculated Contributing Concentrations and the VIF Values for the Eight-Source Data

Source	Contributing concentration (μg/m^3)	VIF value
Road dust	0.48	134.41
Soil	−0.45	133.72
Sea salt	1.13	5.51
Steel	0.52	1.46
Oil burning	2.18	2.02
Incineration	2.05	2.26
Diesel	3.44	1.52
Biomass burning	4.42	1.96

Source: Adapted from Okamoto (1996).

TABLE 23.6

Calculated Contributing Concentrations and the VIF Values for the Seven-Source Data Excluding the Source Profile of Road Dust

Source	Contributing concentration (μg/m³)	VIF value
Soil	0.99	2.37
Sea salt	1.12	1.51
Steel	0.53	1.46
Oil burning	2.17	2.00
Incineration	2.04	2.24
Diesel	3.47	1.51
Biomass burning	4.40	1.94
Secondary*	7.11	
Unknown	1.49	
Total	23.32	

* The contribution of secondary particles was calculated by ($NH_4^+ + SO_4^{-2} + NO_3^- - S$) in mass unit.
Source: Adapted from Okamoto (1996).

excluding road dust, there is no combination of sources with a large absolute value of the correlation coefficient; therefore, this source profile seems to be appropriate for the source apportionment calculation.

23.4 Evaluation of the CMB Calculations

This section will describe the evaluation methodology and how well the calculated results of the receptor model fit with the observed data. Here again, the dataset from the Kashima SPM study is used. The data presented here were retrieved from IPCAJ (1986) and were generated using the Andersen impactors sampling. Although this dataset had some deficiencies, such as the fact that carbon content was not analysed and the source profile excluding soil was obtained from the literature, the results of this analysis are worth reporting here. Further, a subsequent re-examination of the dataset revealed some important things. The effects of the cement factory may not be so large, and the composition of the components is similar to that of the soil. As a result of various considerations, IPCAJ adopted the source profile data shown in Table 23.7 for use in the source apportionment study (IPCAJ, 1986). However, later some discussions suggested that the contributions calculated as cement industries were likely to be fine debris on road pavements or road dust (Okamoto et al., 2007).

For the trial calculation, in the source profile setting, 16 elements were selected based on the seven indicator elements from the seven major sources plus the other important elements, and 16 elements were finally adopted for. However, in the final calculation, three elements marked with a star in Table 23.7 (Cl, As and Se) were excluded based on the trial calculation results. In this case study, a subset of the elements data that were not used for least squares fitting in CMB calculations were also arranged. These data points can be used to evaluate the consistency of the model expression. This is because, in multiple regression analysis, in addition to the data used to estimate the partial regression coefficients, additional data may be used to validate the results predicted by the regression equation (IPCAJ, 1986).

First, it was examined to what extent the replacement of source data affected the contribution calculations. Here, three kinds of source profiles shown in Table 23.8 were considered. These three source profiles will be referred to as 7×7, 7×16 and 8×13, respectively. For this CMB calculation, data from the Andersen impactor was used. This data corresponds to almost PM_{10}

TABLE 23.7

Source Profile of the IPCAJ Source Apportionment Study

Elements		Soil	Sea salt	Steel	Incineration	Oil burning	Auto	Other ind.♦	Cement
Elements for fitting	Al	0.0634	0	0.01	0.0042	0.0021	0.0015	0.026669	0.024
	Na	0.01236	0.3042	0.014	0.12	0.045	0.00027	0.042252	0.004
	Mn	0.000966	0	0.049	0.00033	0.00024	0.000036	0.007912	0
	Zn	0.000136	0	0.0096	0.026	0.0018	0.0015	0.040077	0
	V	0.000121	0	0.00013	0.000027	0.01	0.000002	0.00083	0
	Pb	0.000058	0	0.021	0.017	0.00051	0.037	0.038988	0
	Sb	0.000001	0	0.00009	0.00061	0.000007	0.000002	0.001799	0
	Ca	0.01428	0.012	0.045	0.011	0.0082	0	0.031706	0.46
	Cr	0.000037	0	0.00225	0.00085	0.00089	0.000014	0.005159	0
	Sc	0.000016	0	0.000001	0	0	0	0.000002	0
	Ti	0.00432	0	0.001	0.0009	0.00074	0	0.002399	0.00144
	Fe	0.0388	0	0.083	0.0062	0.028	0.0041	0.010166	0.0109
	Br	0.00003	0.0019	0.00014	0.00083	0.000009	0.0062	0.001079	0
Elements without fitting	S	0.000123	0.026	0.048	0.13	0.096	0	0.082528	0
	Cl♦	0.000034	0.5505	0.034	0.27	0.00092	0.0034	0.023267	0
	K	0.0134	0.011	0.013	0.2	0.00085	0.0012	0.023101	0.0053
	Co	0.00001	0	0.000044	0.000021	0.000031	0.000002	0.000335	0
	Ni	0.000099	0	0.0029	0	0.0077	0.000039	0.008824	0
	Cu	0.000168	0	0.0037	0.0036	0.0036	0.00019	0.005302	0
	As♦	0.000006	0	0.0001	0.00015	0	0	0.002459	0
	Se♦	0.000001	0	0.000051	0	0.000048	0	0.000683	0
	Ag	0.000001	0	0.000054	0.00015	0	0	0.000093	0
	Cd	0.000003	0	0.00025	0.0005	0.00024	0	0.010128	0
	Cs	0.000002	0	0	0.000012	0	0	0.000009	0
	Ba	0.000403	0.000001	0	0.00039	0.00092	0.00064	0.021618	0
	La	0.000012	0	0.00001	0.000008	0	0	0.000015	0
	Ce	0.000028	0	0.000069	0.00017	0	0	0.000117	0
	Sm	0.000003	0	0	0	0.000008	0	0.000002	0
	Lu	0	0	0	0	0	0	0.000011	0
	W	0.000005	0	0.000047	0	0	0	0.000621	0
	Th	0.000004	0	0	0	0	0	0.000007	0

Note: The cross-shaped star mark (♦) represents the element or source considered in the trial calculation.
Source: Adapted from IPCAJ (1986).

with particles less than 11 μm of 100% cut. The number of data is 64 in total, with 16 sites and 4 seasons. Before evaluating the calculation result using individual data, we investigated how much the calculation result changes depending on the source data (source profile) used.

This comparison was a preliminary step using only the overall average values, not individual data. Table 23.9 shows the results of calculations using different source profiles. Table 23.10 shows the calculation results for different calculation methods using 8-sources-and-13-elements source profile data. The three calculation methods used here are least squares (LS), weighted least squares (WLS), and effective variance method (EV).

Around 1980, when the IPCAJ study had been conducted, the proportion of antimony (Sb) in the source profile for incinerators was not reliable because the effects of flame retardants contained in plastics were not properly understood (Okamoto, 1994). In addition, antimony in the source profile for other industries was probably overestimated due to that of the glassworks which has higher Sb contents. These deficiencies make the calculated contributions unstable

TABLE 23.8

Three Kinds of Source Profiles Prepared for Trial Calculation

		1	2	3	4	5	6	7	8	9	10	11	12	13	14	15	16
	Source	**Al**	**Na**	**Mn**	**Zn**	**V**	**Pb**	**Ca**	**Sc**	**Cr**	**Sb**	**Ti**	**Fe**	**Br**	**Cl**	**As**	**Se**
1	Soil																
2	Sea salt																
3	Steel																
4	Incineration																
5	Oil burning																
6	Automobile																
7	Cement																
8	Other industry																

Note: Source profiles 7 × 7 are denoted by a shaded area, those 7 × 16 are distinguished by a thick black border, and profiles of size 8 × 13 are identified by a lighter shade of colour.

TABLE 23.9

Results of WLS Calculations Using Different Source Profiles (μg/m³)

Source	**7×7**	**7×16**	**8×13**
Soil	15.8	16.3	16
Sea salt	2.1	1.4	1.9
Steel	0.3	0.2	0.1
Incineration	1.1	1.3	0.8
Oil burning	0.7	0.4	0.6
Automobile	1	0.6	0.4
Cement	0.8	0.8	0.8
Other ind.			0.4
Total	21.8	21.1	21.1

TABLE 23.10

Results for Different Calculation Methods (μg/m³)

Source	**LS**	**WLS**	**EV**
Soil	16.6	16	16.3
Sea salt	0.7	1.9	1.8
Steel	0.8	0.1	0.1
Incineration	5.1	0.8	1
Oil burning	1.4	0.6	0.6
Automobile	1.7	0.4	0.4
Cement	0.9	0.8	0.8
Other ind.	−2.8	0.4	0.4
Total	24.5	21.1	21.4

for these sources. The reason for this defect can be inferred from the fact that the antimony results were extremely poor in the evaluation of individual elemental data shown later (see Table 23.11).

Once the source profile data have been finalised, it is possible to calculate the contribution of each source using the PM data measured at a receptor site. One set of about 10–20 elemental concentrations analysed in one PM sample collected at a receptor site is referred to as one piece of data. In the case of a least squares CMB model, the degree of fitness of the model can be evaluated based on the magnitude of the residual sum of squares for one piece of the data. The value of the residual sum of squares can be obtained for each data. If there is an extremely erroneous large value in some elements, it may result in a very serious situation. Therefore, it is also important to find the elemental components that may have failed the analysis, if necessary, before starting the source apportionment calculation.

The evaluation of the Kashima dataset (64 data), first for each data, and then for each element is presented next.

23.4.1 Consistency of Each Piece of Data

If there is a certain number of sets of receptor data, the calculation result for each element can be summed up, and the overall goodness of the receptor model can be evaluated for each element. Of course, it is also possible to evaluate the goodness of the fit for each piece of data individually. The consistency of each data can be evaluated by the following statistics.

- The number of data exceeding the detection limit for each component, n,
- Correlation coefficient between calculated and observed concentrations, r.

The square root of mean residual sum of squares calculated by Equation (23–15).

$$\varepsilon = \sqrt{\frac{1}{n}\sum_{y_i > ND} \frac{1}{\sigma_i^2}(x_i - y_i)^2} \tag{23–15}$$

where x and y are calculated and observed elemental concentrations, $1/\sigma^2$ is a weighting coefficient of the weighted least square model. The suffix i is the elemental number. The value of σ^2 is analytical uncertainty that was calculated from the repeated analysis as shown in Table 13.12.

The evaluation results quoted from IPCAJ (1986) are considered here. Since there is a difference of several orders of magnitude between high-concentration elements and low-concentration elements, the correlation coefficient is calculated here after taking their logarithm values. Even so, the correlation coefficient was 0.95 or higher in all 64 cases, and in nearly half of the cases, the correlation coefficient was about 0.99. It should be noted that the correlation coefficients are naturally higher when the marker elements include elements of the order of micrograms and ones of the order of nanograms. The square root of mean residual ε is a value of about 1–4 and tends to be slightly smaller when the correlation coefficient is high. The number of non-detected (ND) data in a piece data was small enough and was shown only in the spring data. In other seasons there were no cases of ND data. Note that even a ND data was included in one data piece, it did not seem to have much effect on the increase in ε value.

23.4.2 Consistency for Each Element

Based on the correlation coefficient or other appropriate statistics between the calculated and observed values for each element, it is possible to evaluate the validity of the selection of indicator element and/or the selection of the source profile data. In the evaluation for each element, the average of the ratio between the calculated and observed values *CAL/OBS* and the standard deviation of the residual *RMS* was calculated:

$$CAL/OBS = \frac{1}{n}\Sigma(x_i/y_i) \tag{23–16}$$

$$RMS = \sqrt{\frac{1}{n}\Sigma\left(d_i - \bar{d}\right)^2}\ :\ d_i = x_i - y_i \tag{23–17}$$

where, x_i is the i-th calculated value, and y_i is the i-th observed value. These evaluations were calculated using only data above the detection limit. Therefore, the number of data is 64 for elements that do not contain any ND. The results are shown in Table 23.11.

In this evaluation, we also included the additional eight elements (S, Cl, K, As, Se, Sm, Ni and Cu) that were not used in the least squares fitting. Here, the amount of data above the detection limit N and the standard error divided by the average of the observed values $RMS/\overline{OBS}$ are also shown in Table 23.11. Here, a scatter diagram of calculated and observed values for aluminium and sodium that are used for fitting in the least squares calculation is shown in Figure 23.4. Figure 23.5 shows the scatter plot for samarium and arsenic, which were not used in the fitting calculations. As previously explained, in regression analysis, the

TABLE 23.11

Evaluation of the Consistency for Each Element

Elements	Number of data	CAL/OBS	Correlation coefficient	RMS/ $\overline{OBS}$
Al	64	0.95	0.988	0.0743
Na	64	0.931	0.985	0.1012
Mn	64	0.837	0.907	0.4007
Zn	64	0.991	0.775	0.5216
V	64	0.973	0.995	0.0317
Pb	64	0.956	0.622	0.2658
Sb	64	0.789	0.384	0.4443
Ca	64	1.034	0.987	0.0917
Cr	61	1.305	0.624	0.9958
Sc	64	1.124	0.944	0.2051
Ti	61	1.045	0.879	0.2393
Fe	62	0.858	0.845	0.2946
Br	64	1.252	0.817	0.3987
S	64	0.319	0.138	0.3997
Cl	64	0.865	0.953	0.2361
K	56	2.68	0.422	0.8664
As	64	0.53	0.211	0.407
Se	52	0.277	0.099	0.4162
Sm	58	0.718	0.873	0.2394
Ni	26	1.529	−0.019	0.7463
Cu	50	1.07	0.047	1.0146

Source: Adapted from IPCAJ (1986).

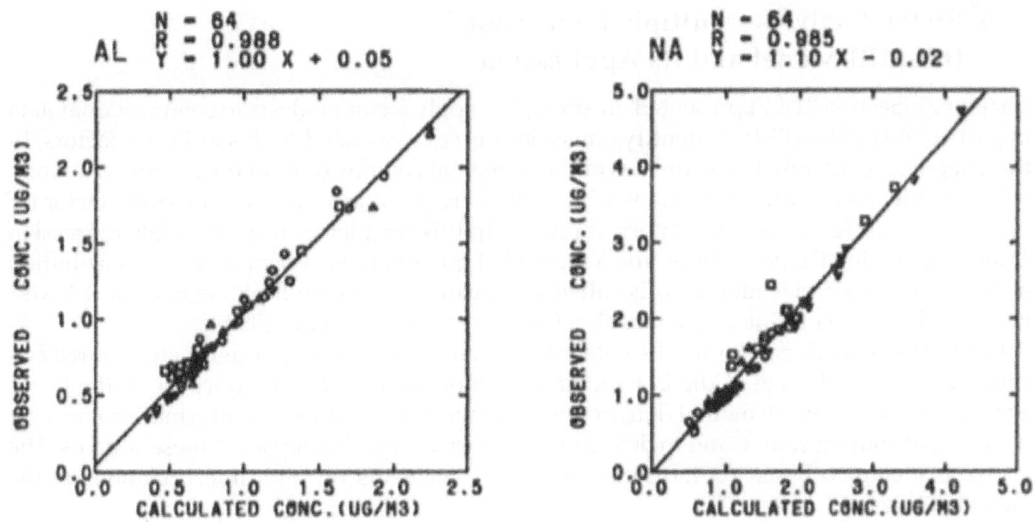

FIGURE 23.4 Scatter diagrams of calculated and observed values for aluminium and sodium used for fitting in the least squares calculation.

Source: Retrieved from IPCAJ (1986).

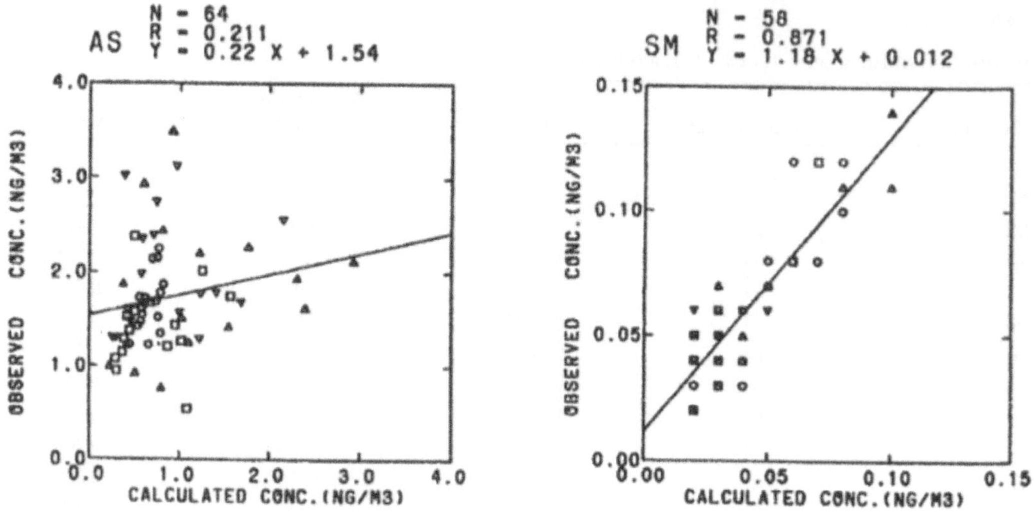

FIGURE 23.5 Scatter diagrams for samarium and arsenic that were not used for the fitting calculation.

Source: Retrieved from IPCAJ (1986).

obtained equation is sometimes evaluated by examining the extent to which the values of auxiliary data, which are not used to estimate the regression coefficients, are reflected in the regression equation. Samarium and arsenic can be interpreted as elements that play this role. In particular, since samarium is abundant in soil, it serves as supporting evidence that the soil contribution rate estimated from aluminium, scandium, iron etc. is valid.

Thus, regarding the CMB calculation, it is feasible to reserve some data for extra elements not employed in fitting and utilise them to validate the model results. These validation studies can make the model more convincing.

23.5 Factor Analysis–Multiple Regression (FA-MR) Model and Its Application

Chapter 21 (Section 21.4.7) presented an attempt to apply factor analysis to compositional data in particulate matter (PM) to identify emission sources associated with significant factors. In this chapter, the quantitative estimates of the elemental compositions of the identified sources will be made. As explained in Chapter 21, multiple regression analysis can use the factor (or component) scores as the explanatory variables. In this way, let us use a multiple regression analysis following factor analysis, and a method of quantitatively estimating the contribution of the source categories identified is called the factor analysis–multiple regression (FA-MR) model. Below is an outline of the model, referencing Okamoto et al. (1987a).

The factor analysis can be used to identify the emission source (more exactly, source category) at the initial stage of the PM source apportionment study. The purpose of the factor analysis of particulate elemental data in this field is to obtain necessary information concerning the contributing source and to determine the variation of the impact of these sources. The source is identified based on indicator (or marker) elements through interpretations of the factor loadings.

23.5.1 Formulation of the FA-MR Model

It is considered that the calculated factor score corresponds to the identified source contribution. However, the factor score is a normalised value for each factor, and the source contribution cannot be directly estimated by this factor score. Spengler and Thurston (1985) introduced the absolute score method to receptor modelling by applying the principal component scores to regression analysis. Let us now consider modelling that applies factor scores based on this method. A factor analysis (FA) and multiple regression (MR) model based on the absolute score was developed to estimate the quantitative source contribution in terms of the mass concentration:

- Identification of the contributing emission sources of particulate matter by FA, and
- Quantitative estimation of the identified source contribution by MR.

The basic model that is being applied to PM compositional data assumes that the concentration of a given element is a linear sum of the fractional contributing concentrations from each source (Equation 23–18).

$$C_{ij} = \sum_{k=1}^{p} a_{jk} F_{ik} \tag{23–18}$$

where C_{ij} is the measured concentration (in µg/m^3) of element j for sample i, a_{jk} is the average mass fraction (in %) of element j in the particles from source k, and F_{ik} is the particle mass contribution (in µg/m^3) of source k for sample i. p is the number of emission sources; more precisely, it is the number of source categories. m is the number of elements. The factor analysis model can be expressed in Equation (23–19).

$$C_{ij}^{s} = \sum_{k=1}^{p} \alpha_{jk} F_{ik}^{s} + e_{ij} \tag{23–19}$$

where, $C^{s}ij$ is the normalised value for the concentration C_{ij} for each element, α_{jk} is the factor loading, $F^{s}ik$ is the factor score, and e_{ij} is the residual.

Equation (23–18) is similar to the model of factor analysis shown in Equation (23–19). It was physically ascertained that the variation of the calculated factor score corresponds to the variation of the actual source contribution. Then the relation between the source contribution F_{ik} and the factor score $F^{s}ik$ is expressed by Equation (23–20).

$$F_{ik}^{s} = \frac{F_{ik} - \overline{F_k}}{\left(\overline{F_k^2} - \overline{F_k}^2\right)^{1/2}} \tag{23–20}$$

Equation (23–20) is applicable only if all the k factors are identified. This identification can be accomplished through an interpretation based on the factor loading of a particular element and a comparison between the variation of the factor score and the related variables such as the other pollutant concentrations or meteorological observations.

Since the factor score $F^{s}ik$ is the normalised value, the quantitative source contribution cannot be directly estimated by this score. Therefore, the method used for a quantitative determination of the contribution was proposed. The absolute factor score can be calculated through the next three steps (Equations 23–21 to 23–23).

1. Calculation of the normalised value $(C^{S}0)j$ for the case that all elemental concentrations are 0.

$$(C_0^{s})_j = \frac{0 - \overline{C_j}}{\sigma_j} \tag{23–21}$$

2. Calculation of absolute zero score P_{0k}:

$$P_{0k} = \sum_{j=1}^{m} W_{jk}(C_0^s)_j \tag{23–22}$$

where W_{jk} is the weighting coefficient for the calculation of the factor score.

3. Calculation of absolute factor score $(AFS)_{ik}$:

$$(AFS)_{ik} = F_{ik}^s - P_{0k} \tag{23–23}$$

Although the absolute factor score is proportional to the source contribution, the score is not in units of mass concentration. Thus a transformation to mass concentration units is necessary to obtain the contribution quantitatively. These proportionality constants can be determined by a multiple regression analysis for estimating the mass concentration by using the absolute factor score as the explanatory variables, as expressed in Equation (23–24).

$$Y_i = \sum_{k=1}^{p} z_k (AFS)_{ik} + z_0 \tag{23–24}$$

where Y_i is the mass concentration of sample *i*, and z_k is the partial regression coefficient. The product of z_k and $(AFS)_{ik}$ is the estimated contribution (in mass concentration unit) of source *k* for sample *i*. The value z_0 represents the regression constant, also known as the intercept, and signifies the contribution to the source that is not accounted for by the factors just mentioned, essentially representing the unexplained portion in the model.

23.5.2 Data Used in the FA-MR Modelling

The data used in this case study was retrieved from the report of IPCAJ (1986). Note that this is not the same dataset based on the PM samples taken by the Andersen impactor described in the previous section (detailed in Table 23.11), but a sub-dataset produced by Hivol samplers at a limited number of intensive monitoring sites. This dataset was intended to provide an understanding of daily PM concentration variability rather than regional distributions. Furthermore, it also included the carbon components and polycyclic aromatic hydrocarbons (PAHs). Two high-volume air samplers collected 24-h PM samples at the Itako and Kamisu monitoring stations, respectively, in the Kashima industrial area, from July 31 to August 15 (summer) and from October 28 to November 12, 1983 (autumn). The samples were collected by using the size-selected sampler with a preset cyclone; hence they are considered to be a fine fraction (less than about 2.5 μm in diameter) (IPCAJ, 1986), that have been previously described in Chapter 13. The number of samples used was 60. The number of chemical components (elements and ions) analysed here was 41, and the average elemental and PM mass concentrations during the period are shown in Table 23.12.

The FA-MR model was applied for the analysis of this dataset. The variables used in this analysis were selected according to the following selection criteria:

- The importance of the element as a marker of PM emission sources,
- The analytical precision for each element, and
- The number of valid samples with concentrations above the detection limit.

According to these considerations, 22 chemical components were selected for statistical analysis. These chemical components are labelled with an asterisk in Table 23.12.

TABLE 23.12

Average Component Concentration of the Hivol Dataset during IPCAJ's Summer and Autumn Field Monitoring Programs

Selection	Component	Concentration	Selection	Component	Concentration
	Ag	0.24	*	Sc	0.015
*	Al	64.1	*	Se	0.663
*	As	1.41		Sm	0.008
	Ba	4.53		Th	0.006
*	Br	7.35		Ti	8.08
	Ca	198	*	V	5.21
	Ce	0.5		W	0.235
*	Cl	969	*	Zn	49.4
	Co	0.08		Cd	4.17
*	Cr	1.41	*	Pb	59.9
	Cs	0.07		S	1590
	Cu	54.5	*	Si	581
*	Fe	168		Cl^-	331
	Hf	0.008	*	NO_3^-	757
	K	154	*	SO_4^{2-}	2920
	La	0.39	*	NH_4^+	1290
	Lu	0.004	*	EC	4420
*	Mn	10.1	*	OC	2940
*	Na	367	*	BaP	0.436
	Ni	3.23	*	BghiP	0.626
*	Sb	3.47		mass	21000

* The chemical composition used in the analysis.

Notes: EC: elemental carbon, OC: organic carbon, BaP: benzo (a) pyrene, BghiP: benzo (ghi) perylene. Unit: ng/m^3.

23.5.3 Source Apportionment Result

The analysis result shown here has been extracted from Okamoto et al. (1987a). Four factors could be identified with the varimax rotated factor loadings shown in Table 23.13. The first factor appears to represent particles from the iron-steel industry since the factor loading of the first factor was high for the elements of chromium (Cr), iron (Fe) and manganese (Mn). These elements are highly enriched in emissions from the iron-steel industry. The second factor depends upon NO_3^-, SO_4^{2-} and NH_4^+, which are the major chemical components of the secondary inorganic aerosol.

The secondary aerosol is formed in the atmosphere from gaseous primary pollutants, such as sulphur dioxide and nitrogen dioxide. In the second factor, the factor loading of vanadium (V), which is a marker element of oil combustion, was also high. It is suggested that the major origin of the secondary aerosol may be the sulphur dioxide produced by oil combustion. The third factor has been identified as automobile emissions and may contain the open burning of refuse and biomass, whereas the fourth factor is considered as soil particles as interpreted in Table 23.14.

The Kashima region faces the Pacific Ocean, and sea salt is the major component in suspended particulate matter, as confirmed in a previous study (Okamoto and Kobayashi, 1991). However, in this analysis for the fine fraction, sea salt could not be identified. This is an explainable result because sea salt particles are present mainly in the large size range,

TABLE 23.13

Varimax Rotated Factor Loading

Chemical component	Factor 1	Factor 2	Factor 3	Factor 4	Communality
Al	−0.072	0.073	0.33	0.768	0.709
As	0.297	0.594	0.19	0.23	0.531
Br	0.332	0.365	0.647	0.208	0.705
Cr	0.678	0.152	0.352	0.382	0.752
Fe	0.914	0.092	−0.043	0.083	0.852
Mn	0.942	0.196	0.03	−0.022	0.927
Na	−0.011	0.356	−0.232	−0.219	0.228
Sb	0.063	−0.022	0.606	0.045	0.373
Sc	0.504	0.13	0.013	0.614	0.648
Se	0.464	0.663	0.115	0.139	0.688
V	0.398	0.625	0.079	−0.028	0.556
Zn	0.815	0.23	0.327	0.14	0.843
Pb	−0.093	0.142	0.014	−0.002	0.029
Si	0.252	−0.006	0.127	0.712	0.587
Cl	0.389	−0.02	0.787	0.058	0.774
NO_3^-	0.152	0.557	0.632	0.092	0.741
SO_4^{2-}	−0.013	0.916	−0.05	0.055	0.845
NH_4^+	−0.017	0.953	0.204	0.092	0.958
EC	0.079	0.248	0.753	0.547	0.933
OC	0.058	0.039	0.765	0.519	0.86
BaP	0.728	−0.084	0.221	0.136	0.604
BghiP	0.676	−0.104	0.398	0.118	0.64
(variance)	4.923	3.748	3.675	2.437	14.783

TABLE 23.14

Interpretation of the Factor

Factor	Typical component	Interpretation
1	Cr, Fe, Mn and Zn	Iron steel industry
2	S, V, NO_3, SO_4 and NH_4	Secondary aerosol (oil burning)
3	Br, Cl^-, NO_3, EC and OC	Automobile exhaust and open-burning
4	Al, Sc and Si	Soil particle

whereas the data used here was the fine fraction. The absolute factor score was calculated for each sample, and each factor was previously identified. Then the contributing concentration for each sample was estimated by a multiple-regression analysis according to Equation (23–24) using the absolute factor scores as explanatory variables.

The results of a multiple regression analysis are shown in Tables 23.15 and 23.16. Although the regression coefficient of the iron-steel industry was not significant, the other three identified sources had significant regression coefficients. Since the particulate emission from a steel mill is very typical and can be easily distinguished, this contribution can be identified by the factor analysis. However, the contributions of iron steel industries were too small to obtain a significant regression coefficient in the multiple regression analysis. The average source contribution is shown in Table 23.17. It appears that about 70% of the mass concentration has been identified.

TABLE 23.15

Evaluation of Regression Equation by Analysis of Variance

Factor	Sum of square	Degrees of freedom	Mean square	Fo
Regression	6226	4	1556.4	28.6*
Residual	2998	55	54.5	
Total	9223	59		

** Significant at the 1% level.

TABLE 23.16

Partial Regression Coefficient

Factor	Regression coefficient	Standard error	Fo
Factor 1	0.05	0.97	0.002
Factor 2	9.34	0.97	92.2**
Factor 3	3.69	1.00	13.6**
Factor 4	2.52	1.03	6.0*
constant	6.29	2.06	

Note: * Significant at the 5% level.

** Significant at the 1% level, multiple correlation coefficient $R = 0.82$.

TABLE 23.17

Average Contributing Concentrations during the Monitoring Period

Source	Contributing concentration (μg/m^3)	Contribution (%)
Iron steal	0.03	0.14
Secondary aerosol	9.5	45.4
Automobile exhaust	2.1	10.0
Soil particles	3.5	16.5
Other sources (constant term)	6.3	30.0

For the fine fraction in this area, human-made particles predominate, compared with natural origin. In the Kashima area, secondary particles and automobile exhaust contributed 45.4 and 10.0%, respectively. The constant term (intercept) in the regression equation is the contribution of the unknown source that was not identified by the factor analysis. The result of this analysis was about 30%.

23.5.4 Evaluation and Discussion of FA-MR Model Results

This section examines whether the source categories identified by factor analysis were appropriate. The correlation with the other variables such as the gaseous pollutants concentration and the meteorological observations can give a hint to examine the relationship with emission sources assumed from factor loadings. The correlation coefficients between the calculated contribution and wind speed and nitric oxide (NO) concentration are shown in Table 23.18. The meteorological data and gaseous pollutant concentration data were averaged values, whose averaging time was consistent with the sampling period of the air sampler.

The temporal variation of the calculated source contributions at the Kamisu town hall monitoring station in the summer is shown in Figure 23.6. The arrows in this figure indicate the predominant wind direction during the sampling period of each day. The contributions of

TABLE 23.18

Correlation between the Calculated Contribution and Other Variables

	Iron	Secondary	Automobile	Soil
Wind speed	0.39**	−0.33**	−0.43**	−0.28*
NO concentration	0.03	−0.13	0.76**	0.33**

Notes: * Significant at the 5% level.
** Significant at the 1% level, number of data is 60.

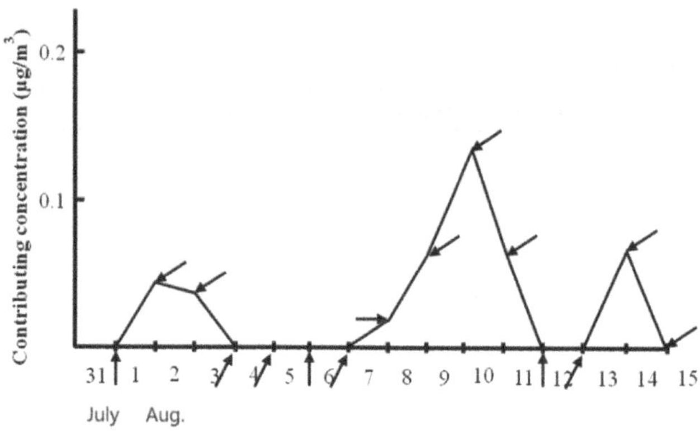

FIGURE 23.6 Time-series of calculated contributing concentration for the iron steel industry (arrows indicate the predominant wind direction).

Source: Adapted and modified from Okamoto et al. (1987a).

the iron-steel industry are large when the wind direction is north or northeast. This station is situated in the SSW direction of the Kashima industrial area. A positive correlation can be seen between the wind speed and the contribution of the iron-steel industry. The reason for this tendency is that the wind speed is high in the case of the NE direction (namely, this wind direction is from the sea) and the effective stack height (the height of the plume rise) becomes lower in a high-wind condition. In other words, this suggests that one of the reasons is a decrease in the effective stack height in strong wind conditions. The dataset used here is for fine fractions. The source of these fine particles is likely to be emissions from high chimneys rather than suspension from the surface. This information may be useful in interpreting the factors.

The contributions of automobile exhaust and soil particles are increased in the case of a lower wind speed. These sources are widely distributed, and the source heights are near ground level. The NO concentration is highly correlated with the contribution of automobile exhaust. A scatter diagram of this relationship is shown in Figure 23.7. Since the emission source of NO is mainly automobiles, this tendency seems to endorse the interpretation that the third factor shown in Table 23.14 corresponds to automobile exhaust. A correlation between NO concentration and soil particles is also high. This correlation suggests that the resuspended soil particles from road surfaces (road dust) play an important role in the total quantity of soil particles for a site near a roadway even in a fine fraction.

From the result of factor analysis, it was clear that there was a contribution from the steel industry, but the regression coefficient in multiple regression analysis was very small, and the contribution rate was extremely small. The calculation results presented here pertain to fine particles, which are likely not suspended from ore yards or other land but emissions from chimneys installed in various furnaces. Additionally, the high correlation with wind speed

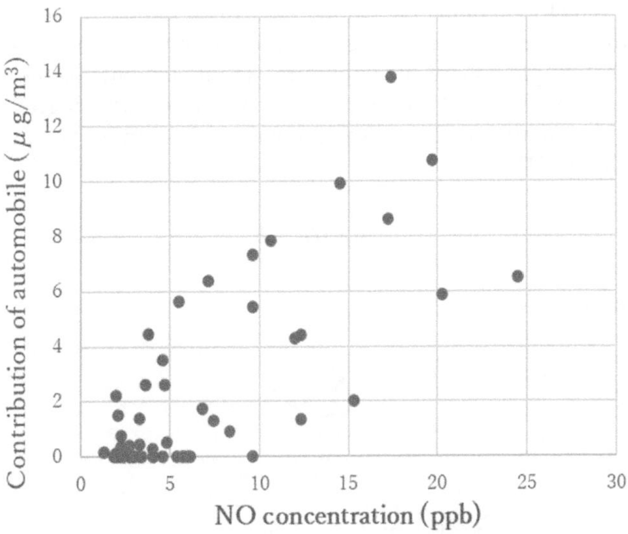

FIGURE 23.7 Scatter diagram between the NO concentration and contributing concentration for automobile exhaust.

Source: Adapted and modified from Okamoto et al. (1987a).

suggests that particles emitted from tall chimneys, such as sintering furnaces, may have experienced the suppression of plume rise under strong wind conditions.

In summary, the analytical results of the factor analysis and multiple regression FA-MR model suggested that it can be a useful tool for identifying the sources and estimating their contribution rates. This model was applied to data on the fine PM fractions obtained from the IPCAJ Kashima SPM study. The contribution of the identified sources could be determined by multiple regression analysis based on absolute factor scores. Other variables such as gaseous pollutant concentrations and meteorological observations are useful for the interpretation of the contributing concentrations and their quantitative estimation.

23.6 Positive Matrix Factorisation (PMF) and Its Application

Positive matrix factorisation is a receptor model based on multivariate analysis that has received much attention. A feature of PMF is that it is a non-negative factorisation of both loading and score. PMF does not rely on the correlation matrix and uses a point-by-point least squares minimisation scheme. Therefore, the generated profile can be compared directly with the input matrix without further transformation. In this section, the PMF calculation results will be compared with those of the CMB or other receptor models. Such an attempt should be useful not only for evaluating the model calculation results but also for examining the appropriateness of the source profile input to the CMB model calculation. However, not many such studies are found in the literature. This section introduces the analytical results of the AIRPET Bangkok dataset (see Chapter 12) and the Kashima dataset (see Chapter 13).

23.6.1 Basic Concept and Algorithm of the PMF

Many explanatory materials on PMF are also available from the US EPA website. A brief overview is provided in Pongkiatkul et al. (2010). In PMF, if the number of samples is n and the number of species is m, matrix X of data with dimensions of n rows and m columns can be formed. PMF aims to factorise matrix X into two matrices, namely G ($n \times p$) and F ($p \times m$), and the unexplained part of X (E), where p represents the number of factors extracted. Hence, each row of F represents a single source of pollutants, while elements of the row show the concentrations of the species. The first column in matrix G shows the degree of concentration of pollutants from the first source collected each day or each sample. The model can be written in Equations (23–25) and (23–26).

$$\boldsymbol{X} = \boldsymbol{GF} + \boldsymbol{E} \tag{23–25}$$

or in the component form of

$$x_{ij} = \sum_{k=1}^{p} g_{ik} f_{kj+e_{ij}} \tag{23–26}$$

when

$$i = 1,....,n;\ j = 1,....,m;\ k = 1,....,p$$

The product of G and F can explain the systematic variations in **X**, whereas **E** is the difference between the measurement **X** and the model **Y** as a function of factors **G** and **F**. This method is similar to that of factor analysis (FA). However, the common FA and principal component analysis (PCA) focus only on minimising the unweighted sum of squares of the residuals in matrix **E**. In contrast, the PMF model emphasises information from all samples. The model minimises the sum of the squares of the residuals weighted inversely with the standard deviations of the data points (Equation 23–27). Most of the species with higher concentrations have larger absolute standard deviations; thus their weight coefficients are smaller than those in unweighted models such as PCA.

The objective of the PMF analysis is to minimise the sum of the squares of the residuals weighted from each data point Q and constrains all elements of $\boldsymbol{G}$ and $\boldsymbol{F}$ factors to be non-negative values. The Q can be defined in Equation (23–27).

$$Q = \sum_{i=1}^{n}\sum_{j=1}^{m} \frac{e_{ij}^2}{s_{ij}^2} \quad \text{subject to } g_{ik} \geq 0 \text{ and } f_{kj} \geq 0,\ k = 1,...,p \tag{23–27}$$

where s_{ij} is the standard deviation of x_{ij}, while g_{ik} and f_{kj} are elements of **G** and **F**, respectively. The residuals e_{ij} are defined by Equation (23–28).

$$e_{ij} = x_{ij} - \sum_{k=1}^{p} g_{ik} f_{kj} \tag{23–28}$$

The most important feature of this PMF is the control of the rotational state of the solution by using a parameter called FPEAK. The central rotation is FPEAK = 0.0. A basic idea of this rotation is to subtract the values on one side (G factor or F factor) and simultaneously increase the values on another side without any change in the solution. Forcing FPEAK on the positive side causes subtractions of the **F** factors to **G** factors from each other and thereby yields a more physically realistic solution (Lee et al., 1999).

23.6.2 Application to a Multi-Site Dataset

In most prior studies using the PMF, the input datasets mainly consist of temporal variability, e.g., data collected at a site over a year. However, the research presented here attempted to

TABLE 23.19

Dominant Ingredient of Each Factor and Its Interpretation for Fine and Coarse Fractions of the Particulate Matter for the Kashima Area

Factor	Dominant ingredient	Interpretation
(a) Fine fraction		
1	Cr, Fe, Mn, and Zn	steel mill
2	Al, Fe, and Si	soil
3	Cl^-, NO_3^-, Br and Zn	Cl^- and NO_3^- rich secondary particles
4	Na and SO_4^{2-}	Aged sea salt
5	SO_4^{2-} and V	SO_4^{2-} rich secondary particle
6	K, NO_3, Sb and Zn	Refuse burning
(b) Coarse fraction		
1	Br, Cl^- and Na	Sea salt
2	Br, Fe, Mn, SO_4^{2-} and V	Industry (oil burning)
3	Al and Si	Soil
4	K and Zn	Refuse burning.
5	NO_3^-, SO_4^{2-} and Na	Secondary particles (including aged sea salt)
6	Cr, Mn and Ti	Steel mill

use a different type of dataset than any other studies, including both spatial and temporal variability, that is, a single dataset with multiple sites and different seasons. Here, the consideration also extends to whether the PMF can respond well to such issues.

A case shown here is an analytical result on the PM dataset with two particle sizes, coarse particles and fine particles, collected by Andersen impactor in the Kashima area. Each sample represents an average of around two weeks at a location, as shown in Chapter 13. The 16 sites are located in more or less one region (within a 30-km radius) which is likely affected by the same source types. The combined dataset consisting of 64 samples (16 sites, 4 seasons) was used.

A preliminary data screening excluded 12 elements by Instrumental Neutron Activation Analysis (INAA) because most data for these species were below detection limits or very small. PM mass and their uncertainty (10% of mass were assigned) were included in the input data to constrain the model to mass. The missing values in the data series of a considered species were replaced with its average measured concentration, while four times the average concentration was assigned for uncertainty of the missing value data point (Polissar et al., 1998). Uncertainties for a non-missing data point of elements were used as in the original dataset.

The results of the PMF analysis for this dataset show six meaningful factors for both fine and coarse fractions as shown in Table 23.19. As for the fine fraction, the PMF could identify the following emission sources: steel mill, soil, Cl^- and NO_3^- rich secondary particles, aged sea salt, SO_4^{2-} rich secondary particles, and refuse burning. Factor 6 is refuse burning that contains high concentrations of K, NO_3, Sb and Zn. Some parts of Br and Sb in fine fractions may come from fireproofing additives in waste products. In this area, a large incineration plant operated by a local government equipped with efficient dust collectors, and the combustion conditions were usually good, and therefore its contribution was not so large. Contrary to this situation, the contribution of small-size incinerators and open burning in backyards seemed to be large, and its composition was similar to that of biomass burning (Okamoto et al., 1990). As for the coarse fraction, the PMF could identify six source categories: sea salt, industry (oil burning), soil, refuse burning, secondary particles (including aged sea salt, $NaNO_3$), and a steel mill.

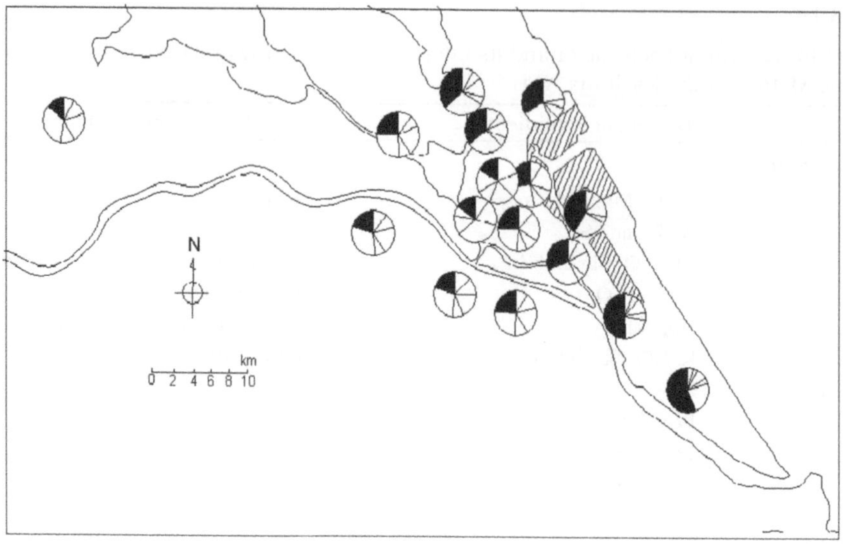

FIGURE 23.8 Regional distribution of the contribution of sea salt particles to the coarse fraction estimated by PMF (the black part in the figure is the contribution of sea salt particles).

Source: Adapted and modified from Pongkiatkul et al. (2010).

Since this dataset, unlike the analysis cases in many previous studies, consists of data from multiple locations, we will also look at the regional distribution of the analysis results. Here, the spatial distribution for the estimated contributions for each emission source category was examined. The sampling sites Nos. 5 and 8 (see Figure 13.1) faced trunk roads, and the contribution of soil was higher than those of other sites. This result suggests that the estimated source profile of soil contains road dust.

For coarse particles, the contribution of sea salt particles was found to be particularly large. Let us examine the regional trend of sea salt, especially that the closer to the coast, the larger its contribution. The regional distribution of its four-season averages is shown in Figure 23.8.

The most characteristic spatial distribution estimation results are related to the contribution of steel mills. The receptors near the steelworks have a higher contribution. Details of these calculation results are shown in Pongkiatkul et al. (2010). As for the fine fraction, the four-season average contribution of a steel mill at sampling site No. 3 is 27%. In the winter season alone, when this site was at the downwind side of the steel mill, the contribution rose above 30%.

Next, examine the characteristics of the source contribution rate by season. The analysis results of coarse fractions that can identify the source type more clearly are shown. For the six identified sources, the average of the contributions estimated at all 16 sites was arranged by season. The results are shown in Table 23.20. In spring and autumn when the NE wind direction was dominant, the contribution of sea salt was very high: 24% in spring and 30% in autumn. However, this northeast wind often involves rain and dampens the ground, thereby preventing the suspension of soil particles; hence the contribution of soil was decreasing. This may be a reason why elements related to sea salt particles and elements in the crust have a negative correlation coefficient.

23.6.3 Discussion of Source Profile Estimation Results

Multivariate receptor models such as factor analysis, which, of course, include PMFs and can simultaneously estimate not only source contribution rates but also source profiles (the composition of particles emitted by the source). This is theoretically possible, as already explained

TABLE 23.20

Comparison of Seasonal Contributions to Coarse Particles When Viewed as Average Values for 16 Sites (unit: %)

Estimated source	Winter	Spring	Summer	Autumn
Sea salt	11	24	7	30
Industry (including oil burning)	20	12	19	11
Soil	27	18	26	15
Refuse burning	17	4	7	7
Secondary Particles (including aged sea salt)	11	35	28	25
Steel mill	14	7	13	12

in Section 23.2 of this chapter (see Figure 23.3). However, there are not many reports that mention the estimation of the source profile in academic papers reported as application examples of the receptor model. Perhaps the main reason seems that the quality of the data used may not be sufficiently high to withstand advanced analysis, leading to results that did not meet expectations. Fortunately, in this PMF analysis, the estimation of the source profile was successful. Table 23.21 shows the results of estimating the source profile (composition) of coarse particles using PMF. Next, these results are compared with those obtained from the FA-MR model. Note that the FA-MR model failed to analyse fine particles of the Andersen sampler (which are not the Hivol data in Section 23.5), so only coarse particles are considered here.

See Table 23.22 for comparison; this is the analysis result using the FA-MR model explained in the previous section. However, this is the result of analysing the same PM dataset as the analysis result shown in Table 23.21. For details, see the footnote of Table 23.22. Looking at this result, the estimation of the source profile, in which the correlation coefficient between the major indicator elements showed a negative value, produced a negative component value. In this analysis, the factor corresponding to soil and the factor corresponding to sea salt particles were negatively correlated. For such factor structures, it seems unlikely that it is reasonable to apply a varimax rotation that maintains the orthogonal relationship of the factor axes. However, PMF seems to adequately address this issue, as shown in Table 23.21, which shows the estimated source profiles using PMF.

23.6.4 Comparison with the Source Apportionment Results by Different Receptor Models

The results obtained by analysing the PMF model are compared with the results calculated by CMB or other receptor models. By comparing the results, it is possible not only to see the degree of agreement between the calculated values of the contribution rates but also to evaluate the quality of the source profile data used in CMB. The software used here was CMB-8.2, the latest version of the US EPA's receptor model during study time (Watson et al., 1997). Hereafter in this section, this model is called the CMB model. Two case studies are included: one is a study on the Bangkok dataset, and the other is that for the Kashima dataset.

23.6.4.1 Case Study for the Bangkok Dataset

The AIRPET project, described in Chapter 12, conducted PM monitoring at several sites in the Bangkok Metropolitan Region (BMR) from 2002 to 2008. The obtained dataset was used for source apportionment study by Kim Oanh et al. (2014).

To reveal the differences in the source contributions produced by different models, CMB and PMF were applied for the PM datasets collected at two sites in BMR. In this case study, data produced by 24-h PM samples collected in the dry season of 2002–2004 at the Asian

TABLE 23.21

Estimated Source Profile for Course Fractions by the PMF (unit: %)

	Sea salt	Industry	Soil	Refuse burning	Secondary	Steel Mill
Al	0.01	0.05	**20.98**	0.03	0.01	0.02
Br	8.36	**10.52**	0.03	0.06	4.38	0.10
Cl^-	**51.37**	0.02	0.02	0.04	0.01	0.03
Cr	0.00	0.00	0.00	0.00	0.00	0.09
Fe	0.10	**20.64**	0.01	0.01	0.00	0.01
K	0.05	0.00	0.00	**2.82**	0.09	0.00
Mn	0.00	0.42	0.00	0.00	0.08	0.25
Na	**28.10**	0.16	2.25	0.05	7.10	0.07
NO_3^-	0.01	0.01	0.01	0.02	**20.92**	0.01
Sc	0.00	0.00	0.00	0.00	0.00	0.00
Si	4.94	0.05	**111.84**	0.13	0.18	0.08
SO_4^{2-}	3.90	**6.61**	0.02	0.08	**10.77**	0.12
Ti	0.00	0.00	0.05	0.00	0.00	**1.92**
V	0.00	0.07	0.00	0.01	0.01	0.00
Zn	0.00	0.00	0.00	**1.06**	0.00	0.00

Major markers are shown in bold in this table.

TABLE 23.22

Estimated Source Profile for Coarse Fraction by the FA-MR Model (unit: %)

Component	Soil	Sea salt	Industry	Secondary
Al	**9.52**	−3.61	2.18	−1.55
Br	0	0.12	0	0
Ca	1.66	7.18	11.66	14.02
Ce	0	0	0	0
Cr	0	0	0.04	0
Fe	**6.57**	−2.51	8.39	−2.43
Mn	0.22	0	0.84	0.22
Na	−4.25	**36.62**	1.13	14.01
Sc	0	0	0	0
Sm	0	0	0	0
Th	0	0	0	0
Ti	0.52	0	0.43	−0.29
V	0	0	0.04	0
Zn	0.13	−0.04	0.16	−0.36
Si	**45.6**	−12.28	17.78	−16.06
Cl^-	−8.09	**60.36**	2.87	**16.69**
NO_3^-	−2.04	3.69	1.6	**23.43**
SO_4^{2-}	−0.37	5.83	9.06	**9.25**

Note: The case used the PM dataset obtained in the Kashima SPM study explained in Chapter 13. Although the case described in Section 22.5 was the analysis result for the auxiliary dataset sampled with the cyclone-equipped Hivol, the case shown here in Tables 23.21 and 23.22 was that for the main dataset sampled with the Andersen impactor. In the factor analysis used here, the principal axis factoring was used for factor extraction and the varimax method was used for the factor axis rotation. Major markers are shown in bold in this table.

Source: Adapted and modified from Okamoto et al. (1987b).

Institute of Technology (AIT), a suburban area, and Bang Na, a mixed urban site, were used (Kim Oanh et al., 2009). For illustration, the results of source apportionment studies for the AIT site are presented in Table 23.23.

A part of this dataset was also used for calculations by another receptor model, COPREM. This is a hybrid physical receptor model developed by Wahlin (2003), which can fit chemical species to get better-fitting source profiles. Upadhyay (2002) presents the COPREM results at three BMR sites for PM collected between March and May 2002. In way of example, Table 23.24 shows the results for the Bang Na site.

The results of the source apportionment in this case study, although still far from perfect, show that all three receptor models of COPREM, CMB and PMF used for analysis of the PM compositional data collected at the selected sites in this study area revealed similar major contributors to the $PM_{2.5}$ pollution. These include secondary particles (ammonium sulphates and nitrates), diesel vehicles and biomass burning, which collectively made up 90% of the $PM_{2.5}$ mass. COPREM was useful in producing the source profiles that can be used in CMB modelling. The source categories identified by COPREM and CMB were similar because of the similar source profile data used in the calculation, but their relative contributions were slightly different. PMF, in principle, would be able to reveal the sources that could not be identified by the other two models because the model does not rely on the a priori information of the categories of sources and their associated profiles for the input data. In this case study, PMF can reveal some minor industry sources in the coarse particles as seen in Table 23.23.

23.6.4.2 Case Study for the Kashima Dataset

The PMF analysis result for the Kashima dataset was shown in Tables 23.19 through 23.21, which are compared with those of the analyses by CMB and FA-MR conducted for the same dataset. Note that for the fine fraction, the contributions could not be estimated by FA-MR because we failed to extract the simple structure of the factors. Therefore, as for FA-MR, only the result for coarse fraction was considered (Okamoto et al., 1987b).

A comparison of individual data points obtained by the PMF and CMB models was conducted (Okamoto et al., 2007). The CMB model, unable to directly estimate secondary particle contributions, considered only a limited number of source categories. The soil mass estimated by both the CMB and PMF models demonstrated a high correlation; however, the CMB's estimate was approximately four times greater than that of the PMF. The contributions from factories and steel mills, as determined by the PMF, are significantly larger than those identified

TABLE 23.23

Source Contributions (% of mass) Estimated by CMB and PMF Using the PM Data during Dry Seasons of 2002–2004 at the AIT Site

Fine fraction $PM_{2.5}$			Coarse fraction $PM_{10-2.5}$		
Sources	**CMB, %**	**PMF, %**	**Sources**	**CMB, %**	**PMF, %**
Diesel	30	29	Construction	15	8
Secondary*	33	27	Soil	42	47
Biomass	28	31	Biomass	21	27
Oil burning	3	(–)	Secondary**	18	5
Soil	4	(–)	Sea salt	1	9
NH_4^+ & Na^+ rich-source	(–)	9	Oil burning	1	(–)
Zn-rich source***	1	2	Zn-rich source	1	4
Pb-rich source****	(–)	2			

Note: (–) not revealed by PMF or no source profile input for CMB; *including $(NH_4)_2SO_4$ and NH_4NO_3 in $PM_{2.5}$; **mainly $NaNO_3$ in $PM_{10-2.5}$; ***Zn-rich profile (industries); ****Pb-rich profile (industries).

TABLE 23.24

Source Contributions (%) Estimated by CMB and COPREM Receptor Models for the PM Data of 2002 at the Bang Na Site

	Fine fraction $PM_{2.5}$		Coarse fraction $PM_{10-2.5}$	
Sources	**CMB, %**	**COPREM, %**	**CMB, %**	**COPREM, %**
Traffic (diesel)	36	36	0	<1
Biomass	34	33	0	0
$(NH_4)_2SO_4$	22	23	0	3
NH_4NO_3	2	3	0	0
$NaNO_3$	0	0	20	9
Sea salt	<1	1	15	6
Oil burning	2	2	3	1
Soil	1	1	26	11
Industry*	1	<1	<1	2
Construction	3	<1	34	68

Note: *Industry with a Zn-rich profile for $PM_{2.5}$ and industries with Zn-, Pb-, and Tin-rich profiles for coarse particle, $PM_{10-2.5}$.

by the CMB. Identifying soil-related components that show similar fluctuations to steelwork-related elements may be appropriate. These could be considered contributions from steelworks and industrial activities, distinct from those inherent to the soil. Scatter diagrams for steel mills indicate a consistent trend for both fine and coarse particles, with a significant correlation as illustrated in Figure 23.9. Nevertheless, the mass calculated by the CMB is roughly one-fourth of the estimate provided by the PMF.

The source profile used in the CMB calculation was derived from some limited source data in particulate facilities such as a coke oven, a smelter, and a blast furnace, therefore, its data may not represent the entire steel mill operations. The profile used is likely to not adequately count the dust, such as ore, steel slag and other soil-like components from bare land in the steel mill. Such particles were accounted for as soil particles rather than as a steel mill origin. In other words, as for the source profile of the entire steel mill, it is possible that the contribution of the facilities that were assumed to be the largest contribution was not so large.

Next, examining the results for the coarse fraction of sea salt particles, which were derived from three different models. When comparing the average contribution of sea salt particles across all sites during the four seasons, the findings are as follows: the contribution rates from the CMB, PMF and FA-MR models are 14%, 18% and 16%, respectively. Additionally, the scatter plot displaying individual data points indicates a strong correlation. As a result, it can be said that satisfactory analysis results were obtained for the coarse fraction of sea salt.

It seems that many previous PMF studies may not provide much discussion about the estimated source profiles, but it may be an important clue when evaluating the calculation result, and it seems to be an indispensable task. Regarding the coarse fraction analysed by PMF and CMB, the estimated contributions for sea salt were nearly equivalent (correlation coefficient of 0.84), suggesting that both results could be considered appropriate. This can be explained using the following reasons: the composition of sodium (Na) and chlorine (Cl) in the source profile of the coarse fraction of sea salt particles estimated by PMF was 28% and 51%, respectively. The NA and Cl compositions estimated by the FA-MR model were 37% and 61%, respectively, as shown in Table 23.22. Looking at the component composition estimates of the factors estimated to be sea salt particles by FA-MR, the composition of soil-related elements was negative, which may have offset the higher estimated values of Na and Cl.

The composition of aluminium in the estimated source profile of coarse components of soil was about 20%, and that of silicon was slightly above 100%. A typical PMF study imposes

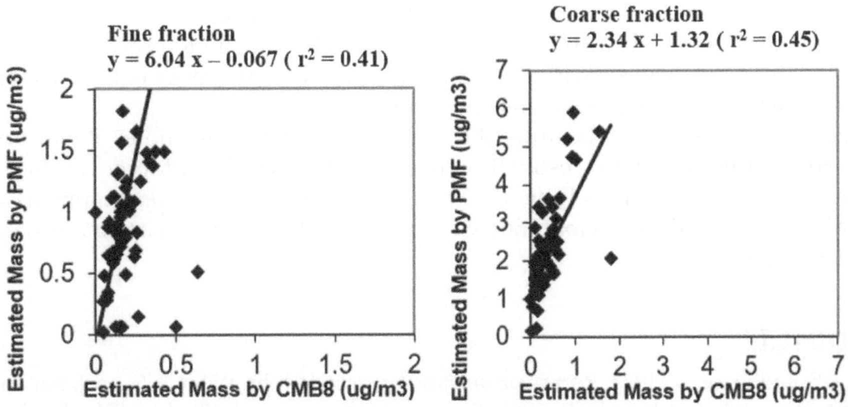

FIGURE 23.9 Comparison of the PMF and CMB calculations concerning the contribution of the fine and coarse fractions for steel mills.

Source: Adapted and modified from Okamoto et al. (2007).

non-negative constraints on the estimated source profile but does not set a condition that it should not be greater than or equal to 1 (100% or greater).

23.6.4.3 Highlights of Comparison of the Results Produced by Different Models

For the Bangkok and Kashima datasets, the source apportionment calculation was conducted by PMF, and the results were compared with that of the CMB model. Overall, it can be said that the results of both were in agreement. In particular, unlike the other cases presented by many researchers, the study using the Kashima dataset attempted to examine the performance of the PMF on multiple-site datasets. This dataset consists of 64 fine and coarse fractional data, namely at 16 sampling sites, for four seasons. The PMF result for coarse fraction was similar to that of the FA-MR model. As for the fine fraction, the FA-MR model could not work well due to a negative correlation between some componential concentrations. However, the PMF could overcome this difficulty and successfully estimated the emission source profile and the contribution of different sources.

23.7 Summary

This chapter described receptor models that use chemical composition data in ambient particulate matter to estimate source contributions. Typical methods are the chemical mass balance (CMB) model and multivariate analysis models which were explained in this chapter and illustrated with actual examples.

Since multivariate models such as factor analysis and PMF could estimate the source contributions and their source profiles at the same time, the validity of the estimated source profile could be evaluated by comparing it with the known elemental composition of a particular emission source. By comparing such an estimated source profile with that prepared for the source apportionment calculation with the CMB method, it is possible to evaluate the appropriateness of the source profile data and refine the profiles for use in the CMB calculation. However, the quality of the PM composition data used for the analyses must be sufficiently high.

As already explained in Chapter 21, statistical analysis on datasets that contain many non-detect data or data with unsuccessful measurements will not yield good results. Such datasets are, of course, also unsuitable for source apportionment analysis; if a factor analysis of PM component data fails to yield a simple structure, it will be difficult to proceed to quantitative source apportionment analysis with that data. In many cases, the cause of failure is the use of poor-quality data for the analysis by receptor modelling.

Although factor analysis can be utilised in source apportionment studies, it may not always lead to a successful quantitative estimation of source contributions. Moreover, for the effective application of receptor modelling to obtain quantitative source apportionment outcomes, having information on the principal source categories contributing to particulate matter at the study site is crucial.

REFERENCES

Chatterjee, S. and Price, B., 1977, *Regression analysis by example*, John Wiley & Sons, Japanese translation by Sawa, T. and Kanoh, S., 1981, *Practice of regression analysis*, Shinyosha Publishing Co., Tokyo, Japan.

Hopke, P.K., 2016, Review of receptor modeling methods for source apportionment, *J Air & Waste Management Association*, 66(3), 237–259, https://doi.org/10.1080/10962247.2016.1140693

Industrial Pollution Control Association of Japan (IPCAJ), 1986, *The study on the behaviour of particulate matter, vol. 9: Final report* (p. 163) (In Japanese), Japan Environmental Management Association for Industry, Tokyo, Japan.

Kim Oanh, N.T., Pongkiakul, P., Upadhyay, N. and Hang, N.T., 2014, *Monitoring particulate matter levels and composition for source apportionment study in Bangkok metropolitan region, chapter 9 of "improving air quality in Asian developing countries, compilation of research findings"* (ed. by Kim Oanh, N.T.), NARENCA, Vietnam Publishing House of Natural Resources, Environment and Cartography, Vietnam.

Kim Oanh, N.T., Pongkiatkul, P., Upadhyay, N. and Hopke, P., 2009, Design a monitoring program of ambient particulate matter for source apportionment study by receptor modeling, *Atmospheric Environment*, 43, 3334–3344.

Lee, E., Chan, C.K. and Paatero, P., 1999, Application of positive matrix factorisation in source apportionment of particulate pollutants in Hong Kong, *Atmospheric Environment*, 33, 3201–3212.

Okamoto, S., 1994, *Report on studies on the prediction of air pollutant concentrations, particulate matter source apportionment study (II)* (p. 70), Tokyo University of Information Sciences (In Japanese).

Okamoto, S., 1996, *Report on studies on the prediction of air pollutant concentrations, particulate matter source apportionment study (III)* (p. 65), Tokyo University of Information Sciences (In Japanese).

Okamoto, S., Hayashi, M., Nakajima, M., Kainuma, Y. and Shiozawa, K., 1987a October 20–23, A factor analysis-multiple regression model for source apportionment of suspended particulate matter, *Proceedings international conference on quality control 1987 – Tokyo ICQC'87*, 345–350, Tokyo, Japan.

Okamoto, S., Hayashi, M., Nakajima, M., Kainuma, Y. and Shiozawa, K., 1990, A factor analysis multiple regression model for source apportionment of suspended particulate matter, *Atmospheric Environment*, 24A, 2089–2097.

Okamoto, S. and Kobayashi, K., 1991, Characterization of aerosols in the Kashima area – a summary of IPCAJ Kashima SPM study, *Atmospheric Environment*, 25B(1), 133–142.

Okamoto, S., Nakajima, M., Kainuma, Y. and Shiozawa, K., 1987b. A study on the estimation of source contribution for the suspended particulate matter – application of the factor analysis – multiple regression analysis model, *J Japanese Society for Quality Control*, 17, 294–300 (In Japanese).

Okamoto, S., Wangkiat, A., Pongkiatkul, P., Nakkhwan, C. and Kim Oanh, T.N., 2007 September 9–14, Source apportionment study by the CMB-8 and PMF models for a coastal industrial area, *CD-ROM proceedings 14th IUAPPA world congress 2007*, Brisbane, Australia.

Paatero, P. and Hopke, P.K., 2003, Discarding or downweighting high-noise variables in factor analytic models, *Analytica Chimica Acta*, 490, 277–289.

Paatero, P. and Tapper, U., 1994/1992 August 17–21. Positive matrix factorisation: A non-negative factor model with optimal utilization of error estimates of data values. *Paper presented in 4th international conference on statistical methods for environmental sciences, Environmetrics*, Espoo, Finland.

Polissar, A.V., Hopke, P.K., Paatero, P., Malm, W.C. and Sisler, J.F., 1998, Atmospheric aerosol over Alaska 2. Elemental composition and sources, *J Geophysical Research*, 103(D15), 19045–19057.

Pongkiatkul, P. and Kim Oanh, N.T., 2012, *Receptor modeling for air pollution source apportionment study, chapter 2 of "integrated air quality management Asian case studies"* (ed. by Kim Oanh, N.T.), CRC Press, Boca Raton, FL.

Pongkiatkul, P., Kim Oanh, T.N., Wangkiat, A. and Okamoto, S., 2010, Further analysis and PMF application to the chemical composition data of IPCAJ's Kashima SPM study, *International J Environment and Pollution*, 40(4), 337–350.

Sakurai, N., Uchida, O., Hayashi, M., Garivait, H. and Okamoto, S., 2003 February 12–14, A statistical analysis for identifying emission sources of suspended particulate matter, *Proceedings of the 2nd regional conference on energy technology towards a clean environment*, vol. 2, 626–632, Phuket, Thailand.

Spengler, J.D. and Thurston, G.D., 1985, A quantitative assessment of source contributions to inhalable particulate matter pollution in metropolitan Boston, *Atmospheric Environment*, 19, 9–25, https://doi.org/10.1016/0004-6981(85)90132-5

Stevens, R.K. and Pace, T.G., 1984, Overview of the mathematical and empirical receptor models workshop (Quail Roost II), *Atmospheric Environment*, 18(8), 1499–1505, https://doi.org/10.1016/0004-6981(84)90374-3

Toyoda, H., Maeda, T. and Yanai, H., 1992, *Statistics for searching the cause, an introduction to covariance structural analysis, blue backs series*, Kodansha Publishing Co., Tokyo, Japan (In Japanese), ISBN 4-06-132926-X

Upadhyay, N., 2002, *Source apportionment of fine and coarse fractions of particulate matters in Bangkok metropolitan region by receptor modeling*, Master's thesis of Asian Institute of Technology.

U.S. Environmental Protection Agency, 2022, *Support center for regulatory atmospheric modeling (SCRAM), air pollutant receptor modeling*, viewed on 3 March 2022, https://www.epa.gov/scram/air-pollutant-receptor-modeling

Wahlin, P., 2003, A multivariate receptor model with a physical approach, *Atmospheric Environment*, 37, 4861–4867.

Watson, J.D., Cooper, J.A. and Huntzicker, J.J., 1984, The effective variance weighting for least squares calculations applied to the mass balance receptor model, *Atmospheric Environment*, 18, 1347–1355.

Watson, J.G., 1984, Overview of receptor model principles, *J Air Pollution Control Association*, 34(6), 619–623, https://doi.org/10.1080/00022470.1984.10465780

Watson, J.G., Robinson, N.F., Lewis, C. and Coulter, T., 1997, *Chemical mass balance receptor model version 8 (CMB8) user's manual, document no. 1808.1D1*, Desert Research Institute.

24

Statistical Analysis Methods for Questionnaire Survey Data: Case Studies on Environmental Valuation

Most of the environmental parameters can be measured quantitatively to directly assess the environmental quality, such as the concentration of pollutants in air and water. However, even if it is possible to physically or chemically measure such environmental issues as odours or noise, that measurement often does not match our perception of environmental goodness. A questionnaire survey is the best way to measure the quality of such a sensory environment. Questionnaires are also used to determine whether consensus has been reached on various policies including those related to environmental issues. Recently, questionnaires have also been used to assess the economic value of the environment. For example, suppose a survey is done to determine how much difference is appropriate to pay for two rooms with the same conditions, except that the view from the window is different. The purpose of the survey is to estimate the economic value of the environmental factors involved in the window view. Thus the concept of willingness to pay (WTP) has been proposed as a method of assessing the economic value of environmental elements that cannot be purchased directly in the market. This chapter introduces the evaluation of environmental value, including the concept of WTP. The data collected from such questionnaire surveys are often categorical data, which can be analysed using different methods as described in Chapter 20. In the case of multivariate data, correspondence analysis and discriminant analysis can also be applied. More recently, conjoint analysis and structural equation modelling (SEM) have attracted increasing attention; hence they are the focus of this chapter.

24.1 Conjoint Analysis

Many widely used software packages for statistical analysis now include conjoint analysis. Conjoint analysis may be better captured as an analysis method or a group thereof, depending on the purpose of the analysis rather than a method divided from other methods by well-defined statistically explainable classification criteria. In other words, it may be considered a processing method or algorithm that suits the purpose of use rather than classification by a mathematical statistical analysis method. The explanation given next provides the generally understood terminology of the conjoint analysis.

Luce and Tukey (1964) proposed a method that decomposes the effects of individual independent variables when the dependent variable, which is determined by the combined effect of the independent variables, is measured by the ordinal scale. It was called conjoint measurement. After that, many researchers involved in marketing research became interested in this method, proposed variations for parameter estimation, and called all these analysis methods or algorithms conjoint analysis. In earlier papers, only the method close to Luce and Tukey's basic model in which individual preference evaluation was given by ordinal scale was often named conjoint analysis. However, the recent semi-synthetic utility model and stochastic utility model (having a general idea as a random utility model) are also considered as a part of the conjoint analysis in a broad sense (Okamoto, 1999).

DOI: 10.1201/9781003597216-27

Louviere (1994) lists three important research areas as the origin of conjoint analysis:

1. Consumer theory by Lancaster (1966),
2. Decision theory in psychology and behavioural science, and
3. The theory of random utility model.

Lancaster (1966) presented a new way of thinking about the economic approach to consumer behaviour. That is, a product is represented by a combination of its attributes, the attributes can be measured, and the utility of the consumer can be described as a function of the attributes. And consumers act to maximise utility under budget constraints.

Next, in the field of decision theory in psychology, research related to information processing in consumer behaviour analysis is the object of attention. Multi-attribute utility theory (MAUT) by Keeny and Raiffa and Saaty's hierarchical analysis (AHP) analytically treat the judgment of a small number of high-level policy makers. In such an analysis, it is necessary to measure the preferences of the subject in detail. On the other hand, in the marketing field of many market types of research, the analysis of the judgment that the consumer makes in daily purchasing behaviour is the object of interest. The method used in this situation can be seen as conjoint analysis. Here, it is not possible to spend too much time collecting information from individual subjects, but the number of samples is relatively large, maybe by several tens or more. In this area, rather than the details of a particular individual decision process, the emphasis is placed on figuring out how a particular commodity (or virtual product) is evaluated in the entire market consisting of many consumers (or specific segments). Consumers choose or buy a product with the highest subjective utility, but their subjective utility always fluctuates. The model that selects the frequency of purchasing of the option or candidate commodity as a random variable can be regarded as being based on random utility theory. Although some researchers say that this idea is different from the conventional theory of conjoint analysis, it is also attractive because it provides results that are more directly related to purchasing decisions, such as market share analysis. Models based on this theory discuss whether to choose an option or not (buy or do not buy). Since it is unthinkable to buy a half-passenger car or a one-third pencil, it is sometimes called a discrete choice model. Generally, it is formulated as a multinomial logit model or a probit model. In many cases, a sufficient amount of data is needed to obtain stable parameter estimates of the models.

In the field of economics, Beggs et al. (1981) developed this theory of stochastic utility. Here, the utility of the target (product) is measured on an ordinal scale, and the partial utility or part worth for each attribute of the target is discomposedly estimated from the measured value. Furthermore, in the field of marketing research, Louviere and Woodworth (1986) and several others proposed a new conjoint analysis based on discrete selection data. Here, the experimental design method is used to assign the level of each attribute when analysing the partial utility (or part worth) for each attribute of interest. Such a method is called an experimental choice analysis or a choice-based conjoint analysis.

As one can see from this explanation, it seems difficult to definitively answer the question, "What is conjoint analysis?" But considering the application in the most widely used marketing research, the conjoint analysis may be expressed in three steps as follows.

1. Collect basic data by inquiring about preferences or attitudes toward goods (i.e., products or services) (e.g., degree of willingness to purchase).
2. The data are obtained by various methods – for example, questionnaire surveys, question answers on PC displays, mock store experiments etc.
3. Then estimate the value of the factors that make up the goods, and infer why the goods are preferred or why they should be purchased.

The output of the analysis is the degree of influence of each factor and its optimum level. Here, this factor is called an **attribute**, and the specific set value of the attribute is called a **level**.

Various methods have also been proposed regarding the method of collecting basic data. There are also many suggestions on how to generate options to present to respondents and how to estimate parameters (such as partial utility values). Many conjoint analysis software packages, such as SPSS, are devised to make good use of the knowledge of experimental design so that each attribute (factor) is orthogonal, even if the number of attributes and levels to be examined is too large for a factorial design ANOVA to handle. As a result, the number of conjoint cards (alternatives) can be significantly saved. This not only makes analysis easier but also has the advantage of reducing the burden on the respondents and collecting reliable data. In many cases, an algorithm that does not consider the interaction between attributes is used. Therefore, if it is expected that there will be an interaction, it will be necessary to devise some responses accordingly. For more information, refer to a technical book that describes in detail how to assign factors in an orthogonal experimental design.

When the collected data is an ordinal scale, monotonic regression analysis was once frequently used as an analytical method that emphasises the ranking. Recently, rating scale values and evaluation scores have become widely used as measurement scales for data, and least squares models suitable for such data have also become widely used. Furthermore, in the case of binary data such as "likes or dislikes" or "buying or not buying", logit analysis is used. If the ordinal scale data is regarded as continuous selection data, logit analysis can be performed even with the ranking data. Further details of these methods can be found in respective technical books. This chapter does not discuss the details of the algorithm but aims to illustrate, through examples of analyses using SPSS Conjoint analysis, the effectiveness of conjoint analysis for the analysis of which type of survey data.

24.2 Structural Equation Modelling (SEM)

Structural equation modelling (SEM) is widely used in various fields such as social sciences and psychology. An important origin of the basic idea of this modelling is believed to be the path analysis developed by genetics researcher Wright (Hox and Bechger, 1999; Matsueda, 2011). The most important contributor to this development is Jöreskog of Uppsala University, who applied the iterative maximum likelihood method to the path analysis. The model was called LISREL (Linear Structural RELations) (Jöreskog and Sorbom, 1988), which has development progressing with the evolution of computers. The SEM introduced here has similar terms, such as the covariance structure analysis model and causal analysis model. Let us explain the relationship between them. SEM and covariance analysis models can be considered roughly equivalent. For more information on causal models, see the following explanation.

Causal modelling: A causal model is an abstract quantitative representation of real-world dynamics. Hence a causal model attempts to describe the causal and other relationships, among a set of variables. The best-known form of causal modelling is path analysis, which was originally developed in genetics but was adopted to sociology as a technique in the 1960s. Most causal modelling is associated with survey research (Author of Oxford Reference, viewed in 2025).

According to a review article on causal models by Greenland and Brumback (2002), several methods derived from path analysis have been proposed, and there are several methods of interest in the field of epidemiology. Among them, SEM is highlighted as the preferred method in the field of social sciences.

24.2.1 SEM Concept

SEM includes a variety of mathematical models, computer algorithms, and statistical methods to analyse the data comprehensively. SEM expresses a causal relationship between the variables using two types of equations, the measurement equation, and the structural equation. This is considered to be a generalised statistical algorithm that includes almost all of the multivariate analysis methods. For example, a representative example of a model containing only measurement equations is principal component analysis, and a representative example of a model containing only structural equations is regression analysis.

In many cases, one can apply a measurement model that defines the latent variables using one or more observed variables and a structural model that infers the relationships between the latent variables. Links between the structural equation model components can be estimated by a regression analysis or more complex approaches. SEM is sometimes called covariance structure analysis. The reason is that by analysing the "structure of covariance between observed variables", one introduces latent variables that cannot be directly observed and analyse the structure of their causal relationships.

SEM uses the following three kinds of variables: observed variables, latent variables and error variables. Observed variables are those that can be measured directly. The fluctuations that cannot be directly observed and variations that cannot be explained by the causal relation being considered are error variables. When considering the causal relationship between many variables, it is effective to introduce the model of structure concepts. If two or more variables have contents that are similar to each other, that is, if the correlation coefficient between the two or more variables is high, there should be a common cause affecting the observed variables. This is the principle of the concept of structure. A variable called a latent variable is used when such a structure concept is handled.

When the hypothesis model settings shown in the path diagram and related data are input to SEM software and analysed, and a path coefficient that represents the value of each path is obtained. For the path diagram and path coefficient, see the following explanation. In addition, a model goodness-of-fit, that shows how consistent the model is with the data, is also displayed. From these outputs, it is possible to examine the suitability of the preset model, which may be represented in the path diagram, and the strength of the relationships between the variables. One of the features of SEM is a comprehensive model that can combine and analyse many statistical models such as factor analysis, multiple regression analysis and so on. Repeating factor analysis and regression analysis several times would risk accumulating calculation errors. On the other hand, it is said that SEM can analyse the entire model at once and can evaluate the fitness of the entire model. However, achieving this advantage depends on the proper data acquisition and sufficient knowledge that can be appropriately represented in the path diagram.

As explained previously, while SEM is very convenient, there are concerns that it can be very dangerous if misused. Tarka (2018) mentioned that even if many related books carefully present the limitations of SEM, practitioners have a prevailing tendency not to notice or even ignore these limitations. Hence he said that the use of SEM in the social sciences is full of overuse, incorrect interpretation and overinterpretation. The risks are not posed by the SEM itself but by the researchers' lack of awareness of the SEM's complexity and limitations.

24.2.2 Path Analysis

Path analysis is an analysis that uses path diagrams to clarify relationships between variables. Although this is the concept that became the prototype of SEM, it is now sometimes considered a part of SEM.

A path diagram is useful for organising relations between observed variables, latent variables and error variables and communicating the contents to others. This diagram shows the causal relationship by describing the relation between latent variables and observed variables, with the causal relationship indicated by an arrow. The following are the basic rules for creating a path diagram.

- Observed variables are enclosed in squares.
- Enclose a latent variable with a circle or ellipse.

- Error variables are not enclosed.
- Write an arrow from the influencing variable to the affected variable.
- For variables that have no causal relation but have correlation, write the arrows in both directions.

A specific example is used here to explain the path diagram. Assume that a factory emits SO_2, NO_x and PM, and the residents around that factory have frequent illnesses of types A and B. Now consider the following hypothesis.

- **The first model:** If the ambient concentrations of SO_2, NO_x and PM were measured as the severity of smoke from the factory, and the number of patients with A and B diseases caused by the smoke was measured, the causal model shown in Figure 24.1a can be considered.
- **The second model:** Assume concentrations of SO_2, NO_x and PM were measured in the vicinity of the factory to determine the extent of air pollution in the surrounding area. It can be considered that the severity of air pollution represents the degree of air quality, which determines the incidence of disease A and disease B. The causal model in this case is shown in Figure 24.1b.

The SO_2, NO_x and PM concentrations and the number of patients with disease A and disease B are observed variables, and air pollution and air quality are latent variables. In the first model shown in Figure 24.1a, the three mathematical formulas that show the relationship between the concentrations of SO_2, NO_x and PM and smoke from the factory are considered to be simple regression equations, and the path coefficient λ is equivalent to a simple regression coefficient. On the other hand, in the second model shown in Figure 24.1b, the mathematical formula for estimating the severity of air pollution from the concentrations of SO_2, NO_x and PM is the same as the multiple regression equation, and the path coefficient γ is equivalent to the partial regression coefficient.

The value of each variable is standardised so that the degree of relationship between the variables can be easily compared from the magnitude of the path coefficient. This allows us to process the data without worrying about what scale each variable was measured.

As explained previously, path analysis is a statistical analysis technique that analyses the dependencies among many variables. In path analysis, the relationships between variables

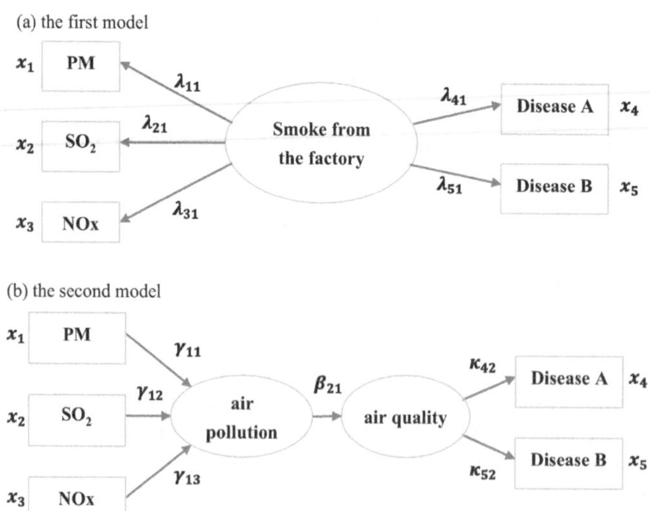

FIGURE 24.1 (a) upper diagram: Path diagram of the first model. (b) lower diagram: Path diagram of the second model.

are represented by path diagrams as described. Path analysis can be viewed as a special case of SEM and does not have measurement equations, only structural equations. Understanding the concept of path analysis is essential when using SEM to analyse data.

24.3 Measuring Consumer Perceptions of Environmentally Friendly Products

This section introduces methods for analysing consumer perceptions of environmentally friendly products. In the case study presented here, the analysis methods used are PCA, conjoint analysis, and SEM, using input data generated from a questionnaire survey.

As many consumers tend to select a product on its consciousness of the environment, the environmental aspect of the product has become increasingly important in marketing strategy today. These products are sometimes called green products. However, if people are not aware of its environmental effectiveness, the product will not be able to achieve the desired competitive position in the market. The most determinative factor of this system may be the consumer's recognition of the environmental aspect of the product. As for the marketing strategy, sometimes the perceptual environmental quality of the product is more important than its actual environmental performance. Therefore, a study on the consumer's perceptual evaluation of an environment-conscious product has been carried out. Because automobiles are one of the most popular industrial products and have a relatively large environmental impact throughout their lifecycle, this case study focuses on automobiles.

24.3.1 Consumer's Recognition of Green Products – Perception Mapping by PCA

A product may become an available indicator of the environmental consciousness of the company that made it. Many consumers recently choose products or services based on their environmental efficiency. Although Eco-labels can give a good indication of the environmental efficiency of the product, it is sometimes questionable whether the products with such labels can get an advantageous position in the market. Perhaps the most important reason may be the difference between the manufacturer's technical evaluation of the product and the consumer's perceptual evaluation and may also involve ignorance, prejudice or too high a price. Sometimes, the sensory feature is more important than its physical efficiency for environmental consciousness.

In this case study, a sensory (perceptual) positioning analysis was conducted with the primary objective of understanding the relationships between various environmental elements as vaguely perceived by the respondents.

Perception map: This is an important tool for investigating a consumer's brand image. First, the concept in the case of a perception map is explained as follows. In marketing research for product development, one may examine how consumers feel about existing products already on the market. In such cases, it is necessary to clarify the arrangement of existing products and/or the target products in the space of the consumer's image, which is commonly referred to as the positioning of the product. In this case, it is also necessary to clarify what factors are important to evaluate the target products by the consumer. In general, it is possible to ask how much consumers can agree on multiple competing products concerning the factors that characterise them. Specifically, the respondents are asked whether or not all target products (brands) are favourable concerning individual factors. For example, in the case of environmentally friendly home appliances, a question to ask would be, "Do you think that Brand X is positive for power saving?" Therefore, a dataset on L brands, M evaluation factors and N respondents can be obtained. Subsequently, the PCA or factor analysis for M variables can be performed with L by N samples (data points). The principal component score (average value of N people) of the L brand can be shown in a diagram arranged on a two-dimensional space consisting of two principal component axes. This picture is called a perception map.

TABLE 24.1

PC Loadings after Varimax Rotation

Environmental factor	Component 1	Component 2
Prevention of global warming	0.77	0.15
Saving the energy and resources	0.29	0.48
Prevention of oceanic pollution	0.67	0.40
Protection of natural forest	**0.80**	0.35
Prevention of acid rain	**0.85**	0.26
Ozone layer protection	**0.84**	0.22
Maintaining biodiversity	0.71	0.45
Prevention of soil pollution	0.66	0.54
Reduction of waste	0.14	**0.83**
Prevention of environmental pollution	0.67	0.43
Conservation of the natural environment	0.72	0.48
Prevention of public nuisance	0.68	0.46

Note: PC loadings of 0.8 or higher are displayed in bold.

The perception map analysis also provides good insights into the structure of consumer preferences for those product (or brand) groups. Okamoto and Suzuki (1998a, 1998b) analysed the consumer's perception structure of environmental consciousness for automobiles and related products. Here, 12 environmental factors, listed in Table 24.1, related to the target product group are presented. The questionnaire asked whether each of the six products was positive to the selected 12 factors or not. The respondents answered with a 7-point rating scale and the answer "I think very much" was quantified as 7 points, and "I don't think at all" as 1 point. The respondents to this survey were 166 university students.

The six products featured in this case study are:

1. Ultra-low sulphur gasoline
2. Vehicles with low-emission engine
3. Electric car
4. Hybrid car
5. High recycling rate for automobiles
6. Recycled resin automobile parts

These products were widely advertised in Japan in the late 1990s, such as in television commercials, and were considered familiar to the respondents. Here, it is interesting to show the relationship among environmental factors that affect the evaluation of these products, not the evaluation results of these products themselves. Therefore, the detailed overall results are not included. First, the correlation coefficients between 12 perceptual evaluation items were calculated, and then PCA was performed on this correlation matrix. The varimax rotated PC loading is shown in Table 24.1 (Okamoto and Suzuki, 1998b).

This result suggests that environmentally friendly products have two important characteristics. One is the pollution control factor represented by the first component, and the other is the energy and resource saving factor represented by the second component.

24.3.2 Estimating the Consumer's Attitude by Conjoint Analysis

Two important concepts, pollution prevention, and resource conservation, have been extracted from the perception map of automobiles and automobile-related products. Next, let us use

TABLE 24.2

Questionnaire for Conjoint Analysis, and Estimation Result of Choice Probability for Each Candidate Product

	Candidate product			Choice probability		
	Ratio of reusable parts	Reduction of emissions	Price increase	The number of respondents who chose first-order	Observed probability	Estimated probability by logit model
	(x_1)	(x_2)*	(x_3)			
No.	%	%	Yen			
1	0	50	10,000	2	0.013	0.016
2	20	0	20,000	0	0	0.022
3	5	20	0	5	0.032	0.022
4	20	20	10,000	5	0.032	0.067
5	10	20	20,000	1	0.006	0.012
6	0	20	50,000	2	0.013	0.001
7	0	100	20,000	6	0.038	0.036
8	10	0	10,000	2	0.013	0.013
9	20	100	0	100	0.641	0.556
10	10	100	50,000	5	0.032	0.021
11	10	50	0	8	0.051	0.086
12	5	0	50,000	2	0.013	0.001
13	20	50	50,000	3	0.019	0.016
14	5	100	10,000	6	0.038	0.108
15	5	50	20,000	0	0	0.016
16	0	0	0	9	0.058	0.007

* A vehicle with a 100% reduction in emissions is classified as a zero-emission vehicle.

conjoint analysis to quantitatively analyse how consumers evaluate these two environmental attributes when determining the value of automobiles. Note that this survey was conducted more than 25 years ago when electric vehicles were not yet widely used.

To estimate utilities of the two environmental aspects (resource conservation x_1 and pollution prevention x_2) and price increase x_3, conjoint cards, i.e., a set of virtual products, were produced by the SPSS using the orthogonal experimental design. Thus some concepts for low-pollution or zero-emission vehicles were presented to the respondents. Here, four levels are set for each of the three attributes. A total of 16 conjoint cards were generated. It is not shown directly here, but it will be introduced at the same time when explaining the results of the share analysis in Table 24.2. Now, one wants to estimate the value of the three variables (attributes), but one cannot buy the corresponding products only for each variable on the market. By inserting specific levels of values into the three variables, it is possible to construct an image of the final product, so that everyone can recognise the value of the product. In this case, the value of the individual variable is calculated, and the value of this individual variable is called the part worth as explained in Section 24.1.

Here the variable of the ratio of reusable parts is representative of the resource conservation, and the variable of reduction of emissions is representative of the pollution prevention. The 166 respondents were asked to determine their preference for each virtual product in 7-point rating and ordinal scales. Part worth functions derived from aggregated all-person data were estimated using ordinary least squares (OLS) with SPSS CONJOINT software. The contribution for each attribute was estimated from the ratio of the range (SPSS, 1991) and shown in Figure 24.2. This result suggests that respondents prefer pollution prevention over resource-saving and have also much interest in a price increase. The estimates of the preference for each level of price increase are shown in Figure 24.3.

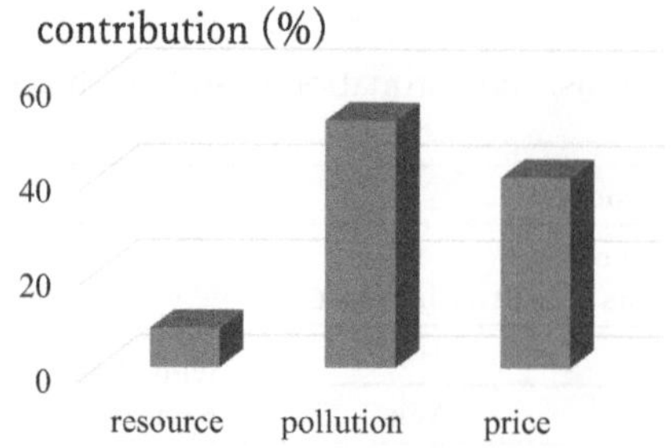

FIGURE 24.2 Contribution of each attribute: resource conservation (reusable parts), pollution prevention (toxic substance reduction) and price increase.

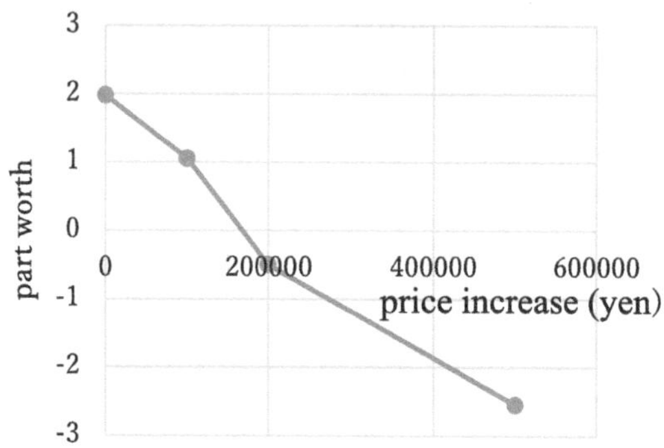

FIGURE 24.3 Part worth for each level of price increase.

It was found that, although the 200,000 yen decrease in the part worth of the price increase was not so large, the estimate of part worth of 500,000 yen was very low. In Table 24.2, a 100% reduction in emission means zero-emission vehicles, i.e., electric vehicles. Since many respondents prefer these products, the zero-emission concept may not be ignored in the market of green products.

Market share simulation: Conjoint analysis allows us to estimate market share of each profile (either products and/or virtual products listed on the conjoint card) based on the estimated preferences. For example, three simulation methods are prepared in the SPSS CONJOINT (SPSS, 1997): (1)the maximum utility model, (2) the BTL model and (3) the logit model. The maximum utility model is simply the probability of choosing a profile as the most preferred. The BTL (Bradley–Terry–Luce) model computes the probability of choosing a profile as the most preferred by dividing the profile's utility by the sum of all the simulation total utilities. The logit model is similar to BTL but uses the natural log of the utilities instead of the utility values. The maximum utility model, also known as the first-choice rule model, is applicable to high-priced products such as furniture and electrical appliances. Conversely, the BTL model is said to apply to everyday shopping commodities that are purchased frequently.

Conjoint analysis can also be applied to predict market share. In this scenario, the first-choice rule applies, indicating that consumers will select only the product they most prefer, leaving no opportunity for other options to be chosen. By using this rule, a simple logit regression model can be applied to estimate the choice probability for each brand or concept. The derived logit regression equation is shown in Equation (24–1):

$$\ln\{p/(1-p)\} = 0.894x_1 + 1.104x_2 - 1.123x_3 - 3.887 \quad (24\text{–}1)$$

where p is choice probability, and x_1, x_2 and x_3 in this equation are normalised values of the variables for resource conservation, pollution prevention and price increase, respectively, and also shown in Table 24.2.

Note that in Figure 24.3, the preference for a product (or virtual product) is measured on a 7-point rating scale. Therefore, the higher the value of the part worth on the vertical axis, the higher the power is to push up the product preference, and a value of 1 higher on the vertical axis can be regarded as having the effect of pushing up one level on the 7-point rating.

The comparison between the observed and estimated choice probabilities is also shown in Table 24.2. The contribution R^2 of this regression was not so high, that is 0.34, but this result is approximately consistent with the OLS conjoint analytical result, as shown in Okamoto and Suzuki (1998a).

24.3.3 Causal Model Analysis on Purchase of Green Products

To identify, construct and quantitatively estimate the relation between concerned variables, SEM may be one of the most useful tools available (Lilien et al., 1992). Consumers' interest in the green product may accelerate by eco-labels and other sales promotions such as advertising; however, the substantial environmental-consciousness may be more important for a higher-involvement product such as automobiles.

SEM is often used to analyse the relationship between consumer understanding and purchase intent of a product, and it is also used in a case study presented here. The data were collected by presenting respondents with additional questions added to the questionnaire for the conjoint analysis described in the previous section (see Table 24.2). For the SEM analysis, it is first necessary to describe the relationship of each variable including the latent factor with a model of structural formula. The path diagram is a visualisation of the model. It is necessary to correspond to the measurement method for obtaining the value of the observed variable.

In this SEM analysis, the interest is assumed to be measured by asking the attitude toward related data such as advertising materials and exhibitions. The intention to purchase each product is measured by asking the attitude toward the purchase directly. In this causal model, the intention is assumed to be affected by the interest, and the price increase also directly affects the intention.

Given environmentally friendly automobiles, the key attributes that should be adopted in the trade-off analysis are two environmental aspects and price increases, as explained in the previous section. Automobile attributes that affect consumer attitudes are assumed to be the ratio of reusable parts, emission reductions and prices, while other attributes should be at the same level. The relationship between the observed and unobserved (latent) variables concerning an environmentally conscious automobile can be expressed by the path diagram, as shown in Figure 24.4. Although this figure also shows the estimated values of path coefficients, these values of the path coefficients in the structural equation are unknown values at the model construction stage. The path coefficients were estimated from the survey data and added to Figure 24.4 after the analysis. The structural equations are shown in Equations (24–2) to (24–6), where, y_1 is an attitude for collecting data, y_2 is an attitude for attending the exhibition and y_3 is a stated intention of purchase. The variables x_1, x_2 and x_3 are the same as those in Table 24.2 and can be determined without error terms. Although the level of pollution prevention and resource conservation may be highly correlated with price increases, each level of the attribute was allocated in the orthogonal form to maximise the estimation efficiency for the path coefficient. Therefore, the correlation coefficients between these attributes are zero, and, of course, the assigned path coefficients between them are also zero. In this causal model, the latent variables are η_1, the interest in the product and η_2, the intention of purchase.

To prepare the input data for the SEM analysis, the respondents were asked to determine the level of attitude for each product from the designated viewpoint. The product set presented to the respondents was the same as that used for the conjoint analysis shown in Table 24.2. The respondent's attitude was measured by a 7-point rating scale. The estimation of path

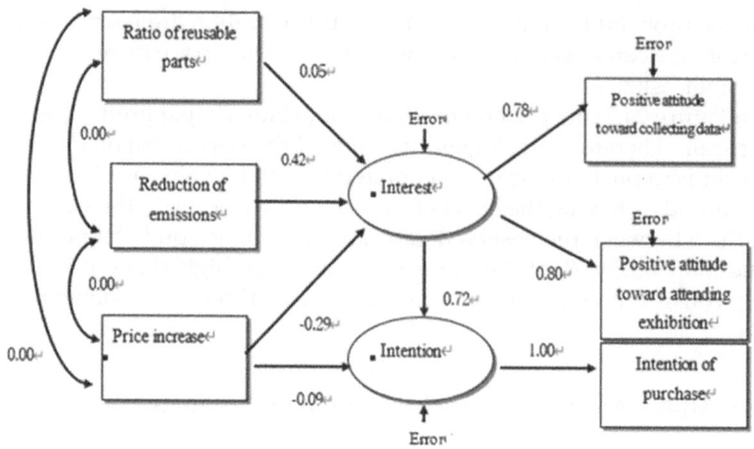

FIGURE 24.4 Path diagram of the model.

Source: Adapted and modified from Okamoto and Suzuki (1998a).

coefficients was carried out by using the AMOS software package (Small Water Corporation, 1997). The derived structural equation model is:

$$\eta_1 = 0.05x_1 + 0.42x_2 - 0.29x_3 + \zeta_1 \quad (24\text{–}2)$$

$$\eta_2 = 0.72\eta_1 - 0.09x_3 + \zeta_2 \quad (24\text{–}3)$$

$$y_1 = 0.78\eta_1 + \varepsilon_1 \quad (24\text{–}4)$$

$$y_2 = 0.80\eta_1 + \varepsilon_2 \quad (24\text{–}5)$$

$$y_3 = \eta_2 \quad (24\text{–}6)$$

where ζ_1, ζ_2, ε_1 and ε_2 are error terms.

The conformity of the model can be evaluated by the goodness-of-fit index (GFI) and adjusted goodness-of-fit index (AGFI) (Toyoda et al., 1992). The obtained GFI and AGFI values were 0.993 and 0.970, respectively. This result suggested that the derived structural equation model seemed to express the relationship between the variables. From empirical consideration compared with the results obtained by the conjoint analysis, this inference seemed to be appropriate.

It was found that the intention to purchase the green product was much affected by interest. The direct effect of higher price increases on the intention of purchase was not so strong compared with the path via the interest of the product. Since the structural equation concerning the interest was similar to the part worth function derived by conjoint analysis, this interest derived by this causal analysis seemed to be consistent with the preference derived by conjoint analysis.

The strength of the psychological desire for a commodity cannot be measured without words. One also does not know whether the respondents are telling the truth. Meanwhile, if the behaviour of the person purchasing the commodity is observed, the result can be measured. Here, the "strength of wanting" is a latent variable, and the "result of the action of purchasing" is an observed variable. This makes it possible to formulate an equation between the "strength of wanting" and the "result of the action of purchasing". This approach has evolved in such fields as psychology and behavioural science, where many variables are difficult to measure. Even in the field of natural science, it seems to be one of the methods that may be expected to be applied when dealing with variables that are difficult to measure.

24.4 Estimating the Economic Value of the Environment

There have been numerous attempts to assess the economic value of the environment. Especially in the field of economics, various discussions have been held from early on, and the field of environmental economics has been established. The economic valuation of the environment discussed here, especially the method of estimating the willingness to pay for the environment, is a part of the results of these efforts. Recently, a similar trend has been observed in various fields related to the environment. For example, attempts to use monetary measures as a criterion for evaluating environmental performance in product life cycle assessments have been introduced (Weidema et al., 2013).

The International Organisation for Standardisation has also published an international standard for the evaluation of the environment in monetary units. The abstract of this ISO 14008 standard (ISO, 2019) states as follows: "This document specifies a methodological framework for the monetary valuation of environmental impacts and related environmental aspects. The monetary valuation methods in this document can also be used to understand organisations' dependencies on the environment better".

Now consider the issues of the preferences of goods, willingness to pay for them and attitudes towards related products and their attributes including the environment.

24.4.1 Preference for Environmental Goods and Attitude toward Environment-Related Products

If one recognises that the environment is important, it may be desirable to measure how well the person feels about it. In this case, it is necessary to answer the question of how to measure the cost of maintaining a good environment, and how to evaluate the products needed to achieve a good environment. In economics, attempts have been made for quite some time to evaluate the preference of goods, and also the environment as its utility function. In marketing, researchers are actively seeking to understand consumers' attitudes toward environmentally friendly products and the environment-related attributes of a product that is a subject of discussion in this section.

Preferences are discussed in a variety of academic disciplines. For example, in economics, psychology, marketing and behavioural sciences. In particular, it is a central theme in our purchasing behaviour and product selection. It is also relevant to the evaluation of the economic value of the environment that is discussed here. Regarding the term "preference", it should be noted that different academic disciplines seem to have slightly different definitions.

- In psychology, "preference" refers to an individual's attitude toward an object, usually reflected in an explicit decision-making process. It is also used in the sense of a judgment about whether one likes or dislikes the subject.
- In economics, "preference" refers to a set of assumptions associated with ordering certain alternatives based on the degree of utility they provide. Human behaviour can be explained by the theory that individuals make decisions based on rational preferences consistent with their self-interests to achieve optimal outcomes. For preferences to be rational, they must satisfy the appropriate axioms.

This preference is conscious in our minds and cannot be directly observed. For this reason, researchers can learn about the preferences by asking respondents certain questions or asking them to answer questionnaires. This is called **stated preference**. On the other hand, the fact that a respondent likes product B more than product A, i.e., that the respondent has a greater preference for product B than A, can be confirmed if one can observe that the respondent chose product B and purchased it. This is called **revealed preference**.

Here, it is possible to estimate not only the preference of the product but also the individual preference of various attributes forming the value of the product. Some of the attributes cannot be purchased by themselves from the market, for example, the environmental performance of the product. By utilising these characteristics, it is also possible to measure the degree of preferences, such as maintaining a good landscape.

The revealed preference theory was introduced by the American economist Paul Samuelson in 1938, which states that consumers' preferences can be revealed by what they purchase (Roper and Zin, 2023). In the field of economics, consumer behaviour (e.g., purchasing behaviour) concerning a product or related product is sometimes observed to estimate "preferences" that cannot be directly observed. Preferences estimated in this way are called revealed preferences, and estimation methods that include observation for this purpose are called revealed preference methods. Here is a quote from the Author of the Britannica (2022) as follows: "the revealed-preferences method involves determining the value that consumers hold for an environmental good by observing their purchase of goods in the market that directly (or indirectly) relate to environmental quality. . . . This revealed-preferences method is also called the hedonic approach".

In the field of economics, there is ongoing debate about how to estimate the preferences, which are not directly observed. When it comes to marketing research, there has always been a dilemma that people do not always behave exactly as they respond to the survey. Some of these issues will be discussed again in the next section on measuring willingness to pay.

Phillips et al. (2002) consider how individuals make decisions and whether "attitude" and "preference" are essentially similar or different, based on a case study of health examinations. In the conclusion, they stated that although the overall findings for attitudes and preferences were similar, that the two approaches resulted in some different conclusions, and that attitude and preference surveys do not result in identical findings. However, they also found evidence that respondents' decision-making processes appear to violate assumptions of economic theory, that is, some respondents were inconsistent in their preferences and appeared to have used cognitive heuristics to simplify decision-making. They explain the attitude survey and the preference survey as follows.

- ***Attitude surveys:*** The concept of an "attitude" comes from social psychology, with "attitude" defined as a psychological tendency that is expressed by evaluating a particular entity with some degree of favour or disfavour. The theoretical literature on attitudes is broad, and approaches to measuring attitudes range from simple approaches that use straightforward ranking questions, to more complex approaches that distinguish attitudes and perceptions.
- ***Preference surveys:*** The concept of preference comes from economic theory, where preference is defined as an individual's utility. In economics, the argument is based on the assumption that individuals maximise a preference (or utility) function. The most useful measure of the strength of preferences is a monetary measure (i.e., the amount willing to pay) that leaves people indifferent between having and not having a given change in utility. Advancing this discussion of preferences requires a variety of assumptions (axioms) about preferences. One powerful tool for this is conjoint analysis, which elicits respondents' preferences by asking them to evaluate options consisting of different combinations of attributes.

The reader can refer to additional literature. One of the key issues in such discussions is how to obtain and manage data in questionnaires and other surveys needed to measure preferences and attitudes. This is illustrated by examples in the next section.

24.4.2 Measuring Willingness-to-Pay

Willingness-to-pay is the maximum amount consumers are willing to pay for a good (e.g., product or service). This is sometimes abbreviated as WTP and is usually expressed as a number in monetary units. Potential customers may be willing to pay less than this value, but it is also important to understand that in most cases they will not want to pay a higher price. It is very important for marketers and/or managers involved in product development and its pricing to know the amount of the purchaser's (or potential consumer's) willingness to pay for the target product. Willingness to pay can vary greatly from customer to customer. This difference is caused by a variety of factors. According to Stobierski (2020), it can usually be classified as either extrinsic or intrinsic.

- Extrinsic differences are observable. These are the factors that make a general judgment about a person without asking a direct question – for example, customer age, gender, income, education history etc. In many questionnaire surveys, it is the content that is written on the face sheet.
- Intrinsic differences are a person's characteristics that become known only by asking them directly. They are difficult to observe and are often referred to as "unobserved differences" – for example, personal risk tolerance and degree of passion for a particular subject.

Stobierski (2020) also lists the following four methods as means of estimating WTP.

1. Surveys and focus groups
2. Conjoint analysis
3. Auctions
4. Experiments and revealed preference

A comprehensive review report on how WTP is measured can be found in Breidert et al. (2006), who explained the classification framework based on data collection methods for measuring WTP, as shown in Table 24.3. The first level of classification is distinguished by whether they are based on actual or simulated price response data or use data derived from surveys. The former involves using market observations or running experiments to generate data; auctions are also included here. Results from price responses are often referred to as revealed preference data. Looking at survey-based methods, there are both direct and indirect surveys to gather relevant data. In contrast to revealed preferences, preference data obtained from surveys are often referred to as stated preferences.

TABLE 24.3

Classification Framework for Methods to Measure Willingness-to-Pay by Breidert et al. (2006)

WTP measurement	Revealed preference	Market data	
		Experiments	Field or laboratory experiment
			Auction
	Stated preference	Direct surveys	Customer survey etc.
		Indirect surveys	Conjoint analysis etc.

Schmidt and Bijmolt (2020) distinguish between hypothetical and realistic contexts for WTP estimation methods depending on whether the measurement includes an obligation to pay. It is HWTP (hypothetical context), or RWTP (real context). Another classification is based on the type of measurement, direct or indirect. Direct methods such as open questioning and indirect methods such as ordinary conjoint analysis are almost hypothetical, but an incentive-aligned conjoint analysis entails a real context. Most measures of RWTP (real context) rely on incentive-compatible methods, which require the participant to buy it if he/she wins the auction. Thus this method cannot be used in the product development process before the final product is completed. Furthermore, measuring RWTP can be difficult and expensive for both practitioners and researchers. They also mentioned that a prevalent assumption in marketing science is that indirect methods measure WTP more accurately than direct methods; however, the deviation primarily depends on the use of a direct or indirect method for measuring HWTP (hypothetical context).

In the field of commerce and marketing, the measurement of WTP is an important issue in product development, pricing strategy and so on. When discussing its measurement, what preferences it pertains to may not be a very important topic. In the field of economics, on the other hand, any discussion of preferences for goods must be based on a rigorous mathematical description, which requires that the preferences satisfy the "axiom of preference".

For example, suppose there are three commodities A, B and C. Suppose researchers ask a question about a person's preference (degree of willingness to buy) for these commodities. The questions were asked three times using the paired comparison method. The respondent was asked to rate the degree of his/her willingness to buy on a 5-point scale. The result was A<B, B<C and C<A. Looking at this result, it is not clear which one this person likes the most. In other words, this choice does not satisfy the "axiom of transitivity". In practical market research, such an academic argument would not be made but would be excluded as inappropriate respondent data in the preliminary screening of the raw data.

In conjoint analysis, researchers can ask respondents whether they like or whether they want to buy options (products or virtual products). In the choice-based conjoint (CBC) analysis, the researchers can ask them which of the options shown on the display they want to buy. If one can set up the appropriate situation at that time, it is possible to reproduce the situation as close as possible to the experiment in the simulated store. By recording the buying or selecting behaviour there and analysing it, the results obtained by this method may be considered a revealed preference. Regarding such issues, the discussion in the following literature may be helpful. Ben-Akiva et al. (2019) stated that preference elicitation methods collect data on consumers by "just asking" about perceptions, intended choices and so on; the CBC analysis asks subjects to make choices from hypothetical menus in experiments that are designed to mimic market experiences. Stated preference experiments promise that they can provide deeper and broader data on the structure of consumer preferences than is obtainable from revealed market observations, with experimental control of the choice environment. It is important to learn about the performance of stated preference elicitations and the conditions under which they can contribute to understanding consumer behaviour and forecasting market demand.

Adamowicz et al. (1994) analysed and evaluated the economic value of recreational sites, particularly near water bodies such as streams and lakes, where activities like fishing and swimming are enjoyed. They used both a stated preference model and a revealed preference model, which are based on random utility theory. The data for both models were obtained from the same individuals. Attributes in the stated preference model are based on the actual levels of attributes in the revealed preference choice set. They contacted randomly selected residents in the study area and inquired whether they had taken any recreational trips during the study period and whether they were willing to complete a more detailed questionnaire. Information about actual trips taken represents revealed preference, whereas responses to options in questionnaire represent stated preference. The results indicated that, although independently estimated models appear to reflect different underlying preferences, joint estimation of the parameters provides evidence that the underlying preferences are actually similar. Additionally, they stated that combining the revealed and stated preference information offers benefits in estimation.

24.4.3 Hedonic Pricing

Hedonic pricing, also called the hedonic approach or hedonic method, is based on the concept that the price of a good is explained by the attributes that make up the good. Specifically, the price of a good (product or commodity) is considered to be an aggregate of the values (part worth) of various attributes of the good, and each attribute worth (in monetary value) can be estimated using regression analysis (hedonic regression).

Rosen (1974), who organised this theory in earnest, defined hedonic prices as the implicit prices of attributes that are revealed to economic agents from observed prices of differentiated products and the specific amounts of characteristics associated with them. Econometrically, implicit prices are estimated by the first step regression analysis (product price regressed on characteristics) in the construction of hedonic price indexes.

The hedonic pricing allows the value (worth) of a characteristic (an attribute) that is not directly traded in the market to be evaluated for that characteristic. According to this model, the total price of a commodity is determined by the bundle of the attributes of the commodity that cannot be sold separately, and it is possible to include the environment-related attributes. Therefore, the value of environmental services that affect the price of a commodity can be estimated. The price model is also used to estimate how characteristics (attributes) influence a consumer's decision to purchase a particular commodity and how much the consumer is willing to pay for those attributes.

The hedonic method can also be used to estimate the price of goods that cannot be purchased directly in the market. It is also used for statistical processing, such as estimating price trends for models or brands that were not marketed in some regions or some years. Such cases are explained in the OECD (2005). According to the OECD definition, the hedonic method is a regression technique used to estimate the prices of qualities or models that are not available on the market in particular periods, but whose prices in those periods are needed to be able to construct price relatives. It is based on the hypothesis that the prices of different models on sale on the market at the same time are functions of certain measurable characteristics such as size, weight, power, speed etc., and so regression methods can be used to estimate by how much the price varies about each of the characteristics.

The method is called the hedonic method perhaps because the price is called "hedonic" for the "pleasure" (in economic terms, utility) that the buyer obtains for the quality of the attributes of a specific good.

24.4.4 Economic Evaluation of Environmental Projects, Policies and Related Initiatives

In the previous sections, the major theme is the economic value of environmentally related products and services. In Sections 24.4.1 and 24.4.2, the discussion mainly focused on products that we can easily imagine. Next, let us consider policies, concepts and other entities such as projects and businesses rather than tangible creations. For instance, let us examine the appropriateness of an energy policy aimed at reducing the use of fossil fuels and transitioning to wind power generation. In such a case, we must confront the question of how much of an increase in the economic burden we are willing to accept in order to adopt alternative energy. This entails evaluating the costs and benefits of implementing this energy policy.

Evaluating the costs and benefits of implementing various policies is called ***cost-benefit analysis (CBA)***. It is an essential step in the planning stages of urban development, land use planning and large-scale public works. If a project has any undesirable effects on the environment, these must be made clear at the environmental impact assessment stage. Furthermore, if the benefits of the project are significant, it may also be possible to calculate an appropriate amount to compensate those who are inconvenienced by the project or its consequence.

Atkinson et al. (2018) provide a summary of the concepts and methodologies involved in the CBA of environmental issues, as well as relevant examples. In this OECD document, they stated:

> The CBA compares the costs and benefits of a particular change to a common good, usually in monetary terms, reflecting how the change is valued by those affected by the project and those who evaluate it. In this process, the net sum of all Willingness to Pay (WTP) and Willingness to Accept (WTA) for the project or policy outcome defines the economic value of the change to the status quo due to the project or policy. The use value relates to the actual (e.g., visits to a national park), planned (future planned visits) or possible use of the goods in question. They state that CBA is a practical tool that can be used to support policy making in practice in the OECD document, and defines it as follows:
>
> - The preferences of individuals are to be taken as the source of value. To say that an individuals' well-being, welfare or utility is higher in state A than in state B is to say that he/she prefers A to B.
> - Preferences are measured by a willingness-to-pay (WTP) for a benefit and a willingness-to-accept compensation (WTA compensation) for a cost.
> - It is assumed that individuals' preferences can be aggregated, with social benefit being the sum of all individuals' benefits and social cost being the sum of all individuals' costs. This implies some degree of cardinalisation of utility.
> - If beneficiaries from a change can hypothetically compensate the losers from a change and have some net gains left over, then the basic test that benefits exceed costs is met.
>
> (Retrieved from Box 2.1. " The theory of CBA" p. 37 in Atkinson et al (2018).)

In the economic evaluation of various projects, including environmental measures, it is necessary to compare their effects and costs. Since the environment does not have an inherent price, it is challenging to evaluate it. Therefore, the economic value is quantitatively assessed using the opinions of many people and the subjective judgment of the respondents. One of the methods for estimating WTP and WTA is the well-known contingent valuation method (CVM). Efforts in this area have grown rapidly since the 1980s, but numerous issues have emerged, generating considerable debate. Many articles and books discuss these issues, including Bateman and Willis (eds.) (2001).

The contingent valuation method in this field is a method of assessing the value of an environment by presenting the current state of the environment and a post-change state of the environment and then asking respondents how much they are willing to pay for the change in the environment. Therefore, if the state of the environment cannot be properly communicated to respondents, they will not be able to properly answer how much they are willing to pay or willing to accept. It has also been pointed out that if the designer of the questionnaire tries to induce respondents to under- (or over-) estimate, it is easy to create such a questionnaire. It is impossible to completely eliminate such bias. However, it is possible to reduce the bias by designing the questionnaire in an appropriate way.

In these discussions, the topic of expenditure on environmental conservation in wildlife preservation areas and scenic spots frequently arose. In this context, if many people wish to visit, the amount of travel expenses incurred can be considered an evaluation of the source that generates the attraction of the place. This method of estimating the conservation costs of a location is known as the travel cost method. Such discussions are particularly prevalent in the field of environmental economics.

Travel cost method (TCM): TCM of economic valuation is a revealed preference approach used in cost-benefit analysis determine the value of non-market goods (e.g., national parks, tourist sites, ecosystems). Initially proposed to measure the benefits that national parks provide to the public, the method has since been used extensively in the field of environmental economics and further refined (Douglas and Taylor, 1998).

Contingent valuation methods (CVM): CVM is a technique for eliciting values by directly asking individuals their willingness to pay (WTP) or willingness to accept (WTA) for a change in a specific target (e.g., environmental level) through surveys or interviews. In theory, it can measure the value of any environmental service. From the perspective of classifying values, it can also measure non-use values such as bequest value and existence value. There are numerous methods to collect data, ranging from directly asking subjects, "How much would you pay for ____?" to having them choose from a list of multiple-choice options, binary choice questions, and other techniques. Originating in the United States, the method is now applied worldwide, with a significant proportion of current research being conducted in developing countries. Simultaneously, the method has transitioned from pure academic speculation to institutional decision analysis. However, the 1990s saw growing criticism of the CVM, with many questioning the fundamental validity of the estimates derived from it, in heated debate. One triggering example was a new law in the United States that required the establishment of procedures to evaluate resource damage and loss due to oil spills. As a result, the National Oceanic and Atmospheric Administration (NOAA) formed a panel to discuss the applicability of the CVM to this problem and published guidelines (Arrow et al., 1993).

24.4.5 Case Study

Recently, the environmental attributes of products and services have attracted a great deal of attention in various fields. As mentioned previously, conjoint analysis can be used for this type of analysis. To do this, take the price (or price increase/decrease) as one of the product attributes and calculate the trade-offs with other attributes. The following is an example of applying conjoint analysis to such a problem. Again, the automobile was chosen as the product of interest for the survey. This was because it is one of the most popular industrial products, has a relatively large environmental impact, and seemed familiar and easy to respond to for many respondents. For this reason, a case study of automobiles was conducted that was detailed in Horie et al. (2005, 2006).

An additional price increase that consumers are willing to tolerate due to saving the environment is called the WTP for the environment, as explained in the previous section. The WTP can be obtained by direct and indirect methods. The former method utilised here is a direct question survey known as the contingent evaluation method. Conversely, as one of the latter methods, conjoint analysis is adopted, which determines the WTP by estimating the part-worth value of each attribute (Horie et al., 2005). In other words, if the decline in the attractiveness of products due to price increases can be offset by the improvement in the attractiveness of environmental attributes, then it can be said that the consideration equivalent to price increases is the WTP for the environmental attributes.

24.4.5.1 Contingent Valuation Method

In this instance, the contingent valuation method (CVM) initially employed directly inquires respondents about their WTP. This method involves asking subjects directly about the amount they are willing to pay. Questions pertaining to the following two issues were prepared:

1. 20% reduction of toxic substances in exhaust gas
2, 5% increase of reusable parts

In the direct questioning survey, the following questions are asked directly. "Suppose a car is sold with 20% less toxic substances in the exhaust gas. Up to how much of a price difference from an ordinary car do you intend to buy it?"

24.4.5.2 Conjoint Analysis, Profile Card Generation and Data Collection

Next, the consumer's preference for environmental attributes of an automobile was analysed with conjoint analysis. The question in conjoint analysis is to ask about the attitude (degree of intention to buy) to the target product. The target product is a combination of attributes of a predetermined level. This allows us to estimate the part worth of the attributes. A trade-off relation between the price and degrees of environmentally conscious attributes can be evaluated based on the part worth of the attributes obtained by the conjoint analysis. In these analyses, the other attributes were assumed to be constant. The selected attributes and their levels are shown in Table 24.4. Profile cards were generated by the procedure of an orthogonal experimental design. This procedure was conducted by the SPSS Conjoint software, and 16 pieces of profile cards were generated. Furthermore, four pieces of holdout cards were added to evaluate the consistency of the answers of the respondents. Therefore, the number of the profiles (or conjoint cards) became 20.

The most notable feature of SPSS Conjoint analysis software is the efficient conjoint card generation method using orthogonal array algorithms. Orthogonal array experimental design has been outlined in Chapter 5 of Part I. More details are available in advanced textbooks. The method adopted by SPSS can provide a smaller number of experiments that can efficiently estimate only the main effect when most interactions are negligible. A specific application example is given in Table 24.4. In this example, there were three factors of four levels, so $4 \times 4 \times 4 = 64$ experiments were required in the factorial design as shown in Section 5.4.2, Chapter 5. However, SPSS Conjoint analysis software specified 16 cards. Respondents' preference for the products displayed on each profile card was measured by a 7-point rating score (7: Definitely will buy, 6: Will buy, 5: Probably will buy, 4: Cannot decide either, 3: Probably will not buy, 2: Will not buy, 1: Definitely will not buy). In this survey, respondents were asked not only to give a rating of 7 levels but also to put an ordinal response for 20 profiles (candidate products). This questionnaire sheet contains 20 profiles as shown in Table 24.5. A detailed explanation for this profile card appeared in Horie et al. (2005).

The part worth of each attribute corresponding to each level and the importance of every attribute were calculated by SPSS Conjoint. Attribute importance was defined as the range of part worth corresponding to the level of the attribute (range, R = max-min). The attribute importance (%) is calculated by dividing the specified attribute range by the sum of all attribute ranges.

TABLE 24.4
Attributes and Their Associated Levels

Attribute	Level
Percentage of reusable parts	Same as the present condition
	Increase 5% above present condition
	Increase 10% above present condition
	Increase 20% above present condition
Toxic substance reduction in exhaust gas	Same as the present condition
	Reduce 20% from the present condition
	Reduce 50% from present condition
	Reduce 100% from present condition
Price increases due to environmental measures	Standard price
	Increase USD1000 above standard price
	Increase USD2000 above the standard price
	Increase USD5000 above the standard price

TABLE 24.5

Level of Each Attribute in the Questionnaire Sheet Used for Conjoint Analysis

No.	Item card	Increased percentage of reusable parts (%)	Toxic substance reduction in exhaust gas (%)	Price increase (USD)	Score (rating scale)	Score (ordinal scale)
1	a	10	ZE	1000		
2	b	5	ZE	2000		
3	c	5	20%	0		
4	d	20	20%	2000		
5	e	Present	50%	2000		
6	f	20	Present	5000		
7	g	5	Present	1000		
8	h	5	50%	5000		
9	i	Present	20%	1000		
10	j	20	50%	1000		
11	k	10	Present	2000		
12	l	Present	Present	0		
13	m	10	20%	5000		
14	n	20	ZE	0		
15	o	10	50%	0		
16	p	Present	ZE	5000		
17	q	5	Present	2000		
18	r	10	Present	1000		
19	s	20	ZE	1000		
20	t	20	Present	1000		

ZE: zero emission (100% reduction).
Note: The survey was conducted in Thailand and Japan. The monetary value was set as follows: 1 USD was equivalent to 100 yen, and 1 baht was equivalent to 3 yen.

24.4.5.3 Results of the Conjoint Analysis

The questionnaire survey was conducted at the Asian Institute of Technology (AIT), Thailand and Tokyo University of Information Sciences, Japan. The respondents were Thai and Japanese students, and the number of respondents was 22 from Thailand and 123 from Japan, (Horie et al., 2006). A few respondents whose answers did not have sufficient consistency in the validation with a holdout card were screened out. The part worth for each level of every attribute was calculated by SPSS Conjoint software. The importance of the attribute is shown in Figure 24.5, which indicates that the most important factor was a price increase due to environmental measures. As for Thai students, the importance of reusable parts was 17.8%, and that of toxic substance reduction in exhaust gas was 33.3%. As for Japanese students, the importance of reusable parts was 16.3% and that of toxic substance reduction was 34%.

If a part worth is the highest at a certain level, that level is favoured by many consumers over other levels of the attribute. Figure 24.6 demonstrates the part worth for each level across all attributes, showing that an increased use of recycled parts and cleaner exhaust emissions are more desirable. It also reveals that the difference between Thai and Japanese students is minimal.

This result was compared with the previous study (Okamoto et al., 1996) in which the preference for two environmental attributes and the price of an automobile were analysed among Thai and Japanese students. The difference between Thai and Japanese students was large in the previous study conducted in 1995 but not in the results obtained in the survey conducted in 2005 as seen here.

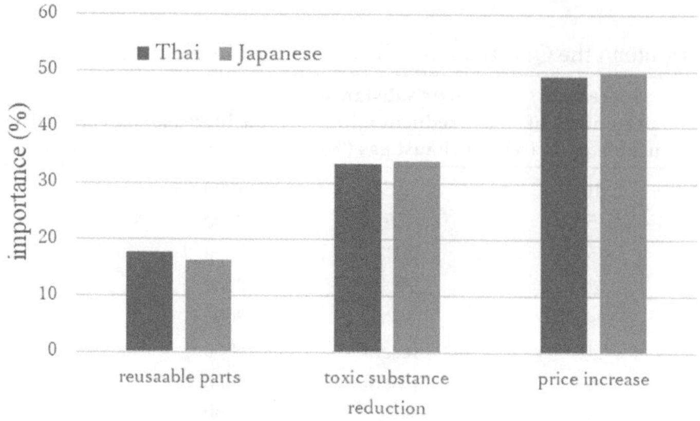

FIGURE 24.5 Importance of each attribute.

Source: Adapted and modified from Horie et al. (2006).

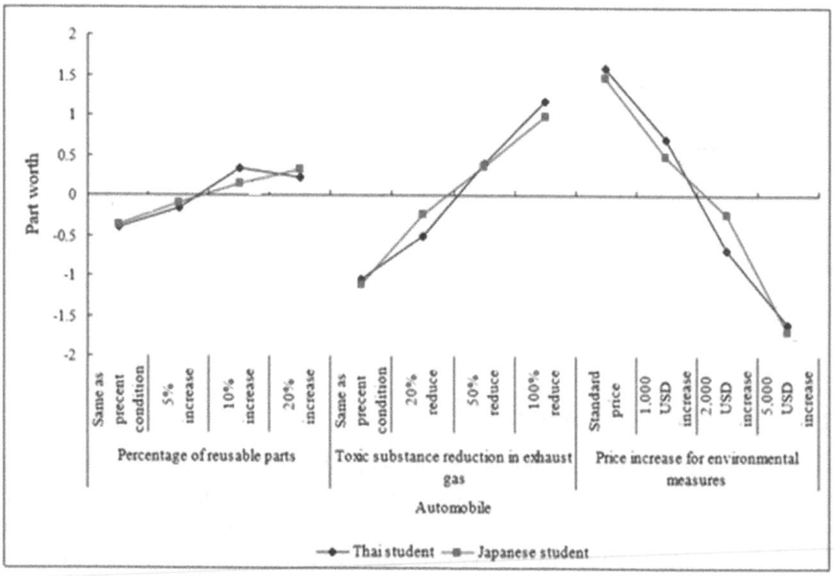

FIGURE 24.6 Part worth for each level.

Source: Adapted and modified from Horie et al. (2006).

24.4.5.4 Estimating the WTP

The results of this conjoint analysis were used to calculate the WTP. The WTP can be calculated from the slope of the formula for the part value graph shown in Figure 24.6. The part worth value tends to increase upward for attributes that are favoured by many consumers. On the other hand, the part worth of the price declines downward, so the coefficient of slope of the linear equation becomes negative. If this empirical equation can be approximated by a straight line, the price increase that negates (offsets) the part worth value of a 20% reduction of toxic substances in exhaust gas can be calculated. It is possible to consider this price value to be the WTP calculated by conjoint analysis, i.e., by an indirect method. The WTP obtained by the direct and indirect methods is shown in Table 24.6.

TABLE 24.6

Amount of Payment Intention (USD) by the Direct and Indirect Methods

Amount of payment intention (WTP)		Japanese student	Thai student
CVM (Direct questioning)	Toxic substance 20% reduction in exhaust gas	690	610
	Reusable parts 5% increase	340	220
Indirect method (Conjoint analysis)	Toxic substance 20% reduction in exhaust gas	880	590
	Reusable parts 5% increase	280	260

Note: Unit: USD (1 USD = 100 yen, 3 baht = 1 yen).

The WTP by CVM was calculated as the average of the declared amounts of all respondents. The results showed that for Thai students, WTP for a 5% increase in reusable parts was about 7330 baht (equivalent to about USD220). The WTP for a 5% increase in reusable parts calculated by the indirect method (i.e., conjoint analysis) was about 8,680 baht (equivalent to about USD260).

The WTP to the 20% reduction of toxic substances in automobile exhaust was about 19,600 baht (about USD590), and it was about 2.5 times higher than that for the increase of reusable parts in monetary units. In the meantime, the WTP for the increase of reusable parts for Japanese students was about 28,000 yen (about USD280). The WTP for the reduction of toxic substances in automobile exhaust was about 88,000 yen (about USD880), i.e., about three times higher than the former. Table 24.6 also showed that the WTP by direct method was in roughly the same range as that by indirect method.

24.4.5.5 Analysis of Individual Data

In ordinary conjoint analysis, where respondents' attitudes (or preferences) are measured on an ordinal or rating scale and parameters are estimated by the least squares method, the number of profiles (conjoint cards) presented to respondents is usually greater than the number of attributes of the target product, so parameters (part worth values) can be estimated from data obtained from each person. This means that parameters (part worth values) can be estimated from a single respondent's data. Of course, data from all respondents (or groups of the respondents by segments) can also be analysed together. However, the method of asking respondents which of the profiles (conjoint cards) presented to them would be their choice would require a considerable number of respondent data to be aligned for such analysis. The SPSS analysis software used here is the former; it can also analyse the data of a single respondent. The personalised analysis presented by Horie et al. (2017) also takes advantage of this feature.

In this case study, the holdout card was verified individually. Holdout cards in SPSS Conjoint analysis software are considered supplementary cards for verification. In the case shown here, in addition to the 16 cards generated by the orthogonal array, four cards with random levels for each factor were added. Respondents provided ratings for 20 cards, but data from only 16 cards were used for estimating part-worth values (equivalent to regression coefficients in regression analysis), which were similar to the description in Section 22.4, Chapter 22. This is similar to the relationship between training data and testing data. According to the obtained part worth value, the rating score value of the respondents for the four holdout cards could be estimated, and it was confirmed how closely it agreed with the actual measured rating value in this case study. However, in such analysis, if the results show a large difference, for example, sometimes the respondent's response and estimate score have a negative correlation coefficient (Kendall's rank correlation is also used), then it may perhaps be because a particular respondent might not fully understand the meaning of the question or might not answer in good faith.

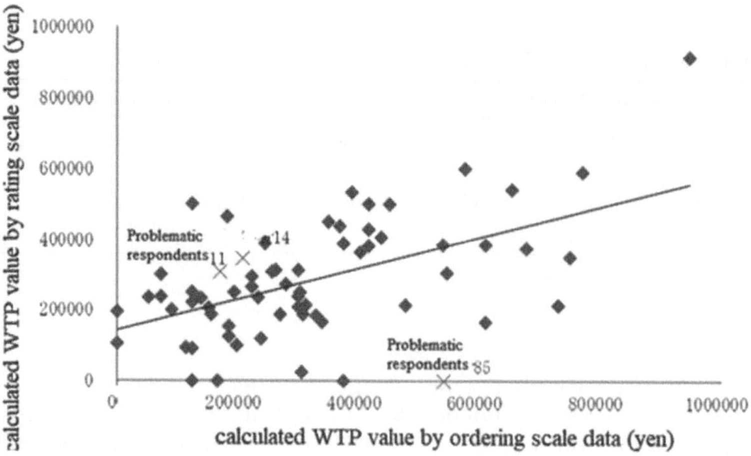

FIGURE 24.7 Comparison of WTP for the zero-emission vehicle obtained from the data of the ranking (ordering) method and the rating scale method. The number in the scatter plot is the respondent number whose answer was questionable, e.g., 11, 14 or 85. The regression line for estimating WTP by rating scale data is $y = 0.44x + 140{,}000$, and the correlation coefficient is 0.54.

Source: Adapted and modified from Horie et al. (2017).

Note that Kendall's rank correlation coefficient is a non-parametric statistic that measures the strength of dependence between two variables. In the case of Kendall's rank correlation coefficient, the correlation is calculated from the amount of data by quantifying the relationship between the pairs to be compared. For more information, see the references, such as Abdi (2007).

The value of WTP shown in Table 24.6 is the result of analysing all of the respondent's data together. In this case, number of Japanese students was 123. However, when looking at the evaluations on the holdout cards for each of the respondents, some respondents gave suspicious answers. Here, the respondents' evaluation values for the four holdout cards and the estimated values from the part worth values were compared. Based on the comparison, it was decided to exclude those respondents whose correlation coefficient was less than 0.5. As a result, only 90 respondents (both of ranking method and rating scale method) were retained for further analyses.

Further, the results were compared for these reliable respondents when the candidate products were evaluated by the rating scale method and the ranking method. Here, let us show the evaluation result of the amount of willingness to pay for 100% reduction of pollutants in the exhaust gas, namely the zero-emission vehicles.

When the part worth value and the attribute level are fitted to a regression equation, the estimated WTP value is consistent with the ratio of the slope of this regression equation. Note that the estimates of WTP values for respondents with very large slopes or too small slopes are unstable. Therefore, some respondents were excluded and the final available data that can be used was 65. The WTP of each of these 65 persons was calculated individually. A comparison of WTP obtained from the data of the ranking method and the rating scale method was conducted. For illustration purposes, one set of the results is shown in Figure 24.7.

Here, the problematic respondents were identified as those who have a median on the ordinal scale located at either end of the rating scale, even though they were not excluded from the first respondent screening. There is also a problem when the observed data by the ranking method is treated simply as metric quantity data. That is, since it means that the measured data by the ranking (or ordering) scale are evaluated at equal intervals, there is a risk of amplifying the slight difference in preference or attitude. Note that this issue has already been pointed out in Chapter 20. Therefore, the results of this analysis suggest that WTPs calculated from ranking data tend to be larger than WTPs calculated from rating scale data.

In summary, the result of the conjoint analysis introduced in this case study showed that the attribute of the highest importance was the price, the second was the reduction of toxic substances in automobile exhaust, and the least important was the increase of reusable parts for both Thai and Japanese students. The willingness-to-pay (WTP) of the consumer was also calculated based on the estimated preference. As far as the estimated preference for environmental aspects of an automobile, a large difference could not be seen between the Thai and Japanese students, and they seemed to emphasize the environmental attribute related to health. While this analysis produced some interesting results, there were few discussions of the target population when selecting respondents, so some challenges remain in this regard.

24.5 Summary

This chapter has introduced some examples of statistical methods used for the analysis of questionnaire data obtained from various evaluation scales. Using such data properly, one may be able to learn as much as or even more than quantitatively measured data. The key message to convey here is not what is stated in this concluding remark mentioned in the case studies but the usefulness of these analytical methods. Even linguistic or categorical data, if obtained in the right way and used appropriately, may be able to provide more valuable insights that cannot be directly obtained from quantitative data measured by conventional methods alone.

These case studies suggest that analytical methods such as conjoint analysis and SEM can extract valuable information. However, in real-life decision-making, it is important to ensure that the selected respondents are representative of the target population. In this case, the importance of sampling theory should not be underestimated, such as selecting respondents based on the purpose of the survey. Knowledge of experimental design is also essential to design questionnaires in conjoint analysis. In addition, it is important not to create questionnaires that intentionally lead respondents. It is also important to understand that in the evaluation of environmental values using CVM and other methods, the analysis results are easily influenced by the questionnaire creator's intentional manipulation.

REFERENCES

Abdi, H., 2007, *The Kendall rank correlation coefficient, "encyclopaedia of measurement and statistics"* (ed. by Salkind, N.), https://personal.utdallas.edu/~herve/Abdi-KendallCorrelation2007-pretty.pdf

Adamowicz, W., Louviere, J. and Williams, M., 1994, Combining revealed and stated preference methods for valuing environmental amenities, *J Environmental Economics and Management*, 26(3), 271–292.

Arrow, K., Solow, R., Portney, P.R., Learner, E.E., Radner, R. and Schuman, H., 1993 January 11, *Report of NOAA panel on contingent valuation*, US. National Oceanic and Atmospheric Administration, https://repository.library.noaa.gov/view/noaa/60900

Atkinson, G., Braathen, N.A., Groom, B. and Mourato, S., 2018, *Cost-benefit analysis and the environment: Further developments and policy use*, OECD Publishing, Paris, http://dx.doi.org/10.1787/9789264085169-en

Author of the Britannica, 2022, Revealed-preferences method, in environmental economics, *Britannica*, viewed on 6 June 2022, https://www.britannica.com/topic/environmental-economics/Revealed-preferences-method

Author of Oxford Reference, 2025, *Causal modelling*, viewed on 20 January 2025, https://www.oxfordreference.com/display/10.1093/oi/authority.20111005143410337

Bateman, I.J. and Willis, K.G. (ed.), 2001, *Valuing environmental preferences: Theory and practice of the contingent valuation method in the US, EU, and developing countries*, Oxford University Press, ISBN 9780199248919, https://doi.org/10.1093/0199248915.001.0001, https://academic.oup.com/book/25949

Beggs, S., Cardell, S. and Hausman, J., 1981, Assessing the potential demand for electric cars, *Journal of Econometrics*, 16, 1–19.

Ben-Akiva, M., McFadden, D. and Train, K., 2019, Foundations of stated preference elicitation: Consumer behavior and choice-based conjoint analysis, *Foundations and Trends(R) in Econometrics*, 10(1–2), 1–144, http://dx.doi.org/10.1561/0800000036

Breidert, C., Hahsler, M. and Reutterer, T., 2006, A review of methods for measuring willingness-to-pay, *Innovative Marketing*, 2(4), 8–32.

Douglas, A.J. and Taylor, J.G., 1998, A new model for the travel cost method: The total expenses approach, *Environmental Modelling & Software*, 14(1), 81–92, https://doi.org/10.1016/S1364-8152(98)00060-7

Greenland, S. and Brumback, B., 2002, An overview of relations among causal modelling methods, *International Journal of Epidemiology*, 31(5), 1030–1037, https://doi.org/10.1093/ije/31.5.1030

Horie, N., Iwamoto, T. and Okamoto, S., 2005 November 5, A study on consumer's WTP toward environmental-consciousness of a product, *Proceedings of 5th meeting of the association of product development and management*, 90–95, Shizuoka, Japan (In Japanese).

Horie, N., Ooki, T., Garivait, H. and Okamoto, S., 2006 December 11–12, Measurement of willingness-to-pay of Japanese and Thai consumer for an environmental-conscious automobile, *Proceedings of ecodesign 2006 Asia Pacific symposium*, Tokyo, Japan.

Horie, N., Shimamura, Y. and Okamoto, S., 2017, Consideration on the willingness-to-pay of the environmental attributes of a product by conjoint analysis, *Journal of Japan Association for Management Systems*, 34(2), 225–233 (In Japanese).

Hox, J.J. and Bechger, T.M., 1999, An introduction to structural equation modelling, *Family Science Review*, 11, 354–373, https://www.joophox.net/publist/semfamre.pdf

International Organization for Standardisation, 2019, *ISO 14008:2019 monetary valuation of environmental impacts and related environmental aspects*, https://www.iso.org/standard/43243.html

Jöreskog, K.G. and Sorbom, D., 1988, *LISREL 7 a guide to the program and applications*, 2nd edition, SPSS Inc., Chicago, USA.

Lancaster, K.J., 1966, A new approach to consumer theory, *J Political Economy*, 74(2), 132–157.

Lilien, G.L., Kotler, P. and Moorthy, K.S., 1992, *Marketing models* (pp. 694–697), Prentice-Hall, Inc., Englewood Cliffs, NJ, USA.

Louviere, J.J., 1994, *Conjoint analysis, chapter 7 of advanced methods of marketing research* (ed. by Bagozzi, R.P.), Basil Blackwell Ltd.

Louviere, J.J. and Woodworth, G., 1986, Design and analysis of simulated consumer choice or allocation experiments: An approach based on aggregate data, *J Marketing Research*, 20, 350–367.

Luce, R.D. and Tukey, J.W., 1964, Simultaneous conjoint measurement: A new type of fundamental measurement, *J Mathematical Psychology*, 1(1), 1–27.

Matsueda, R.L., 2011, *Key advances in the history of structural equation modeling, working paper no. 114*, Center for Statistics and the Social Sciences, University of Washington, https://csss.uw.edu/files/working-papers/2012/wp114.pdf

OECD, 2005, *Hedonic method, glossary of statistical terms*, https://stats.oecd.org/glossary/detail.asp?ID=1225

Okamoto, S., 1999, *Conjoint analysis, marketing research by SPSS*, Nakanishiya Publishing Co., Kyoto, Japan (In Japanese), ISBN 978-4888484862

Okamoto, S., Hishiki, C. and Israngkura, A., 1996 October 15–18, Consumer's attitude toward environment-conscious goods – a case study for a vehicle, *Proceedings, international conference on quality 1996-Yokohama, "quality – key for the 21st century"*, 833–836, Yokohama, Japan.

Okamoto, S. and Suzuki, M., 1998a November 18, A study on the consumer's perceptual evaluation for environment-conscious products, *Proceedings international symposium on manufacturing strategy'98 "strategy-driven manufacturing: A key for the new millennium"*, 199–204, Tokyo, Japan.

Okamoto, S. and Suzuki, M., 1998b May 30, A study on the consumer's evaluation toward the environment-conscious goods, *Proceedings 58th meeting the Japanese society for quality control*, 31–34, Tokyo, Japan (In Japanese).

Phillips, K.A., Johnson, F.R. and Maddala, T., 2002, Measuring what people value: A comparison of "attitude" and "preference" surveys, *Health Services Research*, 37(6), 1659–1679, PMCID: PMC1464045, https://doi.org/10.1111/1475-6773.01116

Roper, J.E. and Zin, D.M., 2023, Revealed preference theory, *Britannica Money*, viewed on 2 October 2023, https://www.britannica.com/money/revealed-preference-theory

Rosen, S., 1974, Hedonic prices and implicit markets: Product differentiation in pure competition, *Journal of Political Economy*, 82, 34–55, https://doi.org/10.1086/260169

Schmidt, J. and Bijmolt, T.H.A., 2020, Accurately measuring willingness to pay for consumer goods: A meta-analysis of the hypothetical bias, *Journal of the Academy of Marketing Science*, 48, 499–518, https://doi.org/10.1007/s11747-019-00666-6

Small Water Corporation, 1997, *Amos user's guide, version 3.6*, Small Water Co., Chicago, USA.

SPSS Incorporated, 1991, *SPSS statistical algorithms*, 2nd edition, SPSS Inc., Chicago, USA.

SPSS Incorporated, 1997, *SPSS Conjoint™ 8.0*, SPSS Inc., Chicago, USA.

Stobierski, T., 2020, Willingness to pay: What it is & how to calculate, *Business Insights, Harvard Business School Online*, https://online.hbs.edu/blog/post/willingness-to-pay

Tarka, P., 2018, An overview of structural equation modeling: Its beginnings, historical development, usefulness and controversies in the social sciences, *Quality & Quantity, International J of Methodology*, 52, 313–354, https://doi.org/10.1007/s11135-017-0469-8

Toyoda, H., Maeda, T. and Yanai, H., 1992, *Statistics for searching the cause, an introduction to covariance structural analysis, blue backs series* (pp. 174–177), Kodansha publishing Co., Tokyo, Japan (In Japanese).

Weidema, B.P., Pizzol, M. and Brandão, M., 2013 October, *The use of monetary valuation of environmental impacts in Life cycle assessment: State of the art, strengths and weaknesses*, Publication INFO SCORE-LCA Report Nb 2012-03, https://lca-net.com/publications/show/the-use-of-monetary-valuation-of-environmental-impacts-in-life-cycle-assessment-state-of-the-art-strengths-and-weaknesses/

Index

Note: Page numbers in *italics* indicate a figure and page numbers in **bold** indicate a table on the corresponding page.

A

accuracy, 118–119, *see also* prediction accuracy
acid rain, IPCAJ's study, quality assurance, 248, 255–256
  data screening and evaluation, 254–255
  inter-laboratory comparison, 251–252, 252 253
  precipitation sampler and instrumental error, 253–254
  storage condition, rain samples, 249, **250**, *251*
active sampling technique, 146
adaptive cluster sampling design, 140–141, *141*
adjusted goodness-of-fit index (AGFI), 436
adsorbable organic halogens (AOX), 169, 170, 175, 176
AERMOD model, 282–283, 293, 296, 297
air-dried pulp (ADP), 156, 158, 169–170
  chemical balance, *159*
  material balance, 158, *159*
air pollution, 2, 3, 57, 182–183, *see also* air quality model; air quality predictions; Asian Regional Air Pollution Research Network (AIRPET); Kashima SPM study; receptor models; spatial resolution, air pollution; temporal resolution/averaging time, air pollution
  asthma case study, 318–320
  automobiles, 25
  chlorine-bleached Kraft pulp and paper mills, 156–157
  composting, 148
  PCA, roadside analysis, 341–342
  monitoring parameters, 111–112
  Northern Thailand, 336–341
  photochemical, 350–352, 377–381
  pollutant units, 113
  real time measurement, 117
  satellite data, 303–304
  SO_2 concentrations, Tokyo, 1960s, 381–383
  spatial distribution of photochemical, 350–352
  spatiotemporal scope, 218, *219*
  Tokyo 1960s, 335–336, 381–383
air quality model, 195, 219–221, 281–283, *see also* air tracer dispersion experiments, roadways
  concentration values, 283–284
  Eulerian models, 242, 282
  evaluation criteria, inter-comparison of, 305–307
  highway line source modelling, 266–267
  initial plume spreads, 274
  in Japan, 294–295, **295**
  Kashima area, 285
    annual averages, 287, 290
    case study, 285–286
    hourly maximum concentration, 290
    long-term and short-term models, 286–287
    SO_2 concentrations, **288–289**, *291*, *292*
    target, average *vs.* maximum values, 290–293
  Lagrangian models, 282
  NAAQS, **284**
  performance evaluation, 305
  pollutants behaviour, *281*
  principal component analysis, *307*
  representativeness, 240
  summer and winter datasets, **306**
  in United States, 293–294, *294*
air quality model performance evaluation, 300
  3D photochemical smog models, 302–304
  challenges in, 300–301
  fractional bias, 302
  framework, *302*
  fundamentals, 301
  robust highest concentrations, 302
  ten-day weather forecast, **300**
air quality predictions, 366, 386–387; *see also* canonical correlation analysis; cluster analysis
  challenges, 366–368
  haze prediction, stepwise regression modelling, 384–386
  maximum hourly ozone (*see* artificial neural networks (ANNs) *vs.* linear regression model)
  photochemical air pollution, discriminant analysis, 378–381, **380**
  SO_2 concentrations, Tokyo, 381
    explanatory variables, 381–382, **382**
    modelling results and evaluation, 382–383, *383*
air tracer dispersion experiments, roadways, 258–259, 277
  auxiliary tracer released from both sides of road, 261
  auxiliary tracer released from one side of road, 261
  concentration attenuation tendency, 273–274
  database development, 266–267
    experimental method and design, 267–270
  plume spreads
    comparison, 274
    initial plume spreads, 274–275, *276*
    traffic conditions comparison, 276–277
  quality assurance, flux calculations, 270, 271, 272, 273
  releasing/disseminating plan, 259–260, *261*

sampling grid, 264–266
sampling plan, roadway experiment, 261–264, *262*, *263*
turbulence measurement, 264–266
vehicle speed *vs.* wind turbulence, *266*
alternative hypothesis, 44–49, 324, 325
American Society for Quality (ASQ), 11, 13–14
ammonia (NH_3), 111, 112
analysis of variance (ANOVA), 60
acid rain, **252**
air quality monitoring, 232
factorial design, 68–70, **70**
mixed models, 72–73
repeated measures data, 73
vs. two-way ANOVA, 71–72, **72**
Kashima SPM study, **207**
local control, 74
one-way ANOVA, 62–68
orthogonal design, 74–76, **75**
vs. population means, 60–61
randomisation, 74
randomised block design, 62–68
regression analysis, **86**, **91–92**
replication, 74
split design, 74
vs. two-way ANOVA, 71–72, **72**
analytical quality control (AQC), 12–13
anomaly detection framework (ADF), 123
artificial intelligence (AI), 8, 107, 122, 333
artificial neural networks (ANNs) *vs.* linear regression model, 373–374
architecture, *373*
input data, 374–375
MLP vs. LR, *375*, 375–377, **376**
Asian Regional Air Pollution Research Network (AIRPET), 182
definition, 182–184
framework, *183*
monitoring programme, 190–193
$PM_{2.5}$, 192
QA/QC procedure, 184
chemical analyses, 187–189
data quality, 189–190
filter pretreatment, 187
filter selection, 187
frequency, 186–187
monitoring sites, 184–186
PM, 186
sampled filter retrieval, 187
sampling time, 186–187
visual check, 187
attitude surveys, 438
averaging time, *see* temporal resolution/averaging time, air pollution
automatic sampling, 147, 154

B

backward elimination method, 100
Bayesian statistics, 8, 28
Beer-Lambert absorption law, 126
benzene, 111, 341–342
benzene, toluene, ethylbenzene and xylenes (BTEX), 116, 184, 341–342, *342*
Big Data Technology, 7, 8, 106, 122, 333
biochemical oxygen demand (BOD), 112, 157, 158, 170–171
bioconcentration, 179
biological monitoring parameters, 113
biostatistics, 8, 9
black carbon (BC), 112, 113
monitoring equipment, 188–189
Bradley–Terry–Luce (BTL) model, 434
butterfly distribution, Thailand, *see also* principal component analysis (PCA)
principal component analysis (PCA), 353
correlation matrix, **356**
dataset, 354–356, **355**
results, 356–357
varimax rotation, variables, **357**
zoogeographic zones, *353*, 353–354

C

canonical correlation analysis, 367, 368, *see also* air quality predictions
algorithms, 368–369
application, 369
weighting coefficient vector, **370**
carbon dioxide (CO_2), 8, 25, 112, 218
carbon monoxide (CO), 111–112, 117
3D photochemical smog models, 304
atmospheric concentration, 222–223, *223*
representativeness, 119
categorical data, 315–317, 326, 334, 449
air quality model, 286–287
categorical variables, 316
contingency table, **324**, **325**
health survey results, **323**
proportion, 56
RBD, 70
causal model, 361, 428, 430
purchase of green products, 435–437
central limit theorem, 27–28, 36, 142
chemical mass balance (CMB), 192
calculations, 399, 403–405
abnormal values, 402–403, **403**
aluminium and sodium, *407*
data, consistency of, 405–407, **406**
element, consistency for, 407–408
procedure, 399–402
result evaluation, 402–403, **403**
samarium and arsenic, *408*
source profiles, **405**
receptor models, 390–393, *393*
effective variance model, 397
least square model, 395
linear programming model, 398
with multivariate models, 394
ridge regression model, 397
chemical oxygen demand (COD), 112, 170–171
chemistry transport model (CTM), 302–303

chi-square distribution (χ^2 distribution), 38–39
relationship with other distributions, 39–40
chlorine-alkali extraction-hypochlorite-hypochlorite (C-E-H1-H2), 156, 170, 173–174
chlorine-bleached Kraft pulp and paper mill
chlorinated compounds, 173–174
bleaching effluents and bleached pulp, **178**
heavy metals samples, **179**
levels of, 176–179
organochlorinated compounds and metal analyses, **175**
sample analyses, 175
sampling, 174
environmental discharges, 169–170
mill effluents and water, test results, **173**
physicochemical parameters of effluents, 170–171, **171**
toxicity test results, 172–173
toxicity tests, 171
wastewater sampling, 172
water in receiving bodies, 172
monitoring program, 156–157
air pollution, 160–166
material balance analysis, 158–160
performance auditing, 157–158
cluster analysis, 370–371, *see also* air quality predictions
hierarchical method, 370–371, *371*
non-hierarchical method, *372*, 372–373
spatial homogeneity, 349–350, *350*
cluster sampling, 137–138
communality, 331–332
comparability/harmonisation, 120
composite sample, 148–149
compositing protocol, 148–149
confirmatory factor analysis (CFA), 104–105
conjoint analysis, 426–428, 432–435, **433**, *434*, *see also* willingness to pay (WTP)
individual data, 447–449
market share simulation, 434
profile card generation and data collection, 444
purchase of green products, 435–437
results, 445
continuous automatic monitors, 124
continuous data, 32, 144, 315, 317
continuous probability distribution, 31–32
control charts, 12
industrial pollution, 164–166, *165*
consumer perception, green products
case study, 443–444 (*see also* conjoint analysis; willingness to pay (WTP))
conjoint analysis (*see* conjoint analysis)
economic value, 437–439
economic evaluation, projects, policies, initiatives, 441–443
hedonic pricing, 441
PCA, 431–433
purchase of green products, 435–437
SEM, 435–437
willingness-to-pay (*see* willingness to pay)
convenience sampling, 134
correlation analysis, 78
correlation coefficient, 78
vs. regression analysis, 84–85
typical and characteristic scatter plots, *79*
variable relationship, 78–79
correspondence analysis, 324, 326, 426
cosine similarity, 345–346
cost-benefit analysis (CBA), 441–442

D

data, 315, 329
analytical methods, 317–320
categorical, 315–317
discrete, 315
environment data, **319**
GLM, 322
interrelation among categorical variables, 323–326
logistic regression, 320–321
as prospective study, 321
as retrospective study, 321
measuring scale (*see* measuring scale)
morbidity and air pollutant concentration, *318*
perceptron, 322–323
pollutant concentration, **318**
quantitative, 316
respiratory morbidity, **318**
sigmoid-type experimental model, *319*
type of, 315–316, **316**
variables and, 316–317
data management, 1, 7, 10
quality issues, 11–13
data quality, *see* quality assurance and quality control (QA/QC)
data system, 120, *121*
data analysis, 121–122
data validation, 120–121
dissemination, 122
raw data, 120
reporting, 122
degrees of freedom (df), 34–35, *35*, 52
decomposition of, 84
factorial design, 69–70
one-way ANOVA, 63
prediction accuracy, 83
t distribution with, *45*, *56*
descriptive statistics, 20, 30, 42
centre of data, 21–22
distribution, 20–21
moment, 23–24, *24*
spatial and temporal resolutions of data, 24–26, *25*
variability, 22–23
detectability, 119
dioxins, 7, 112, 176–177, 180
in bleaching effluents and bleached pulp, **178**
discrete data, 144, 315, 317
discrete probability distribution, 31
discrete samples, 147–148
discriminant analysis, 102, 104, 105, 300, 317
geometric description, *105*
Mahalanobis distance, 105

multiple correlation ratio, 105
photochemical air pollution, 378–381, **380**
as retrospective study, 321
dissolved oxygen (DO), 112, 146

E

economic value of environment, 437
conjoint analysis, profile card generation and data collection, 444, **445**
CVM, 443
goods and attitude, 437–439
hedonic pricing, 441
projects, policies and related initiatives, 441–443
willingness-to-pay, 439–441
ecosystem diversity, 9
effective variance model, 397
end-of-pipe control, 152
environment, definition of, 1–2
environmental monitoring, 1, 111, 123, 127
continuous automatic monitors, 124
data management methods, 122–123
data system, 120, *121*
data analysis, 121–122
data validation, 120–121
dissemination, 122
raw data, 120
reporting, 122
development trend in, 6–7
framework, 4–6, *5*
low-cost sensors, 124–125
manual monitors, 123–124
parameters (*see* monitoring parameters)
QA/QC (*see* quality assurance and quality control QA/QC)
quality management, 1–4
remote sensing, 125–126
environmental pollution, definition of, 3
environmental population, *see* population
environmental
quality, 2–3
standards, 4
statistics, 7–10
error monitoring, *see* measurement errors
estimation, 45, 49, 59
interval, 50–51
point, 50
Eulerian models, 282–283
experimental design, 8, 20, 60, 76
factorial design, 68
and multivariate analysis, 106
randomisation, 74
replication, 74
exploratory factor analysis (EFA), 104–105
factor extraction process, 332
factors (components) to retain, 332
particulate matter (PM), 357–359
conceptual diagram, *359*
FA, 360–362
non-detected data points, 359–360
spatial distributions of provinces, *358*
varimax rotated factor loadings, *358*
vs. PCA, 333–334
rotation, 332–333
terminology, 331
extractable organochlorines (EOCl), 169, 170, 175, 176

F

factor analysis (FA), 8, 104, 121, 332
CFA, 104–105
EFA, 104–105
model, 390, 395, 398, 409
PM source factors, 360–362, **361**
factor analysis–multiple regression (FA-MR), 408
data used, 410, **411**
evaluation and discussion, 413–415, *414*, *415*
formulation, 408–410
source apportionment, 411–413, **412**, **413**
factorial design
analysis of variance (ANOVA), 68–70, **70**
mixed models, 72–73
repeated measures data, 73
vs. two-way ANOVA, 71–72, **72**
water quality survey, **69**
with/without interaction, *68*
F distribution, 39–40, 60, 86
relationship with other distributions, 39–40
F test (variance test), 58–59
finite population, 30, 131
sample size, 143–144
fractional bias (FB), 302, 305

G

gaseous pollutants, 113–114, 213, 413
continuous automatic monitors, 124
grab air sampling, 147
low-cost sensors, 125
remote sensing, 125
Gaussian-type diffusion equation, 282
goodness-of-fit test, 224
analysis method, 224–225
application, 225–226
concept, 224–225
frequency distribution, **225**
histogram of, *225–226*
result, **226**
grab sampling, 147–148
greenhouse gases (GHGs), 8–9, 156
grid sampling, 136

H

haze prediction, stepwise regression modelling, 384
early warning, 386
statistical prediction model, 384–386
synoptic patterns, 384

heavy metals, 111–113, 117, 140, 168
and bioconcentration, 179
hierarchical method, 349, 370–371, *371*
hypothesis testing, 45–46, 59
power of, 47–48, *48*
regression coefficient, 87–88
terminology, 43–44

I

image analysis, 333
independence, 26–27
inductively coupled plasma emission spectroscopy (ICP), 209–210
industrial discharges, 124, 152, 168, 180, *see also* chlorine-bleached Kraft pulp and paper mill
harms of, 168–169
regulations, 155
industrial pollution, 151, 166, *see also* chlorine-bleached Kraft pulp and paper mill
auditing, 153
permanent monitoring, 154
preliminary survey, 153–154
monitoring categories, 152–153
normal discharge level *vs.* actual discharges, *155*
operating conditions, 154–155
reduction of, 151–152
regulatory monitoring, 155–156
Industrial Pollution Control Association of Japan (IPCAJ), *see* acid rain, IPCAJ's study, quality assurance; Kashima SPM study
inferential statistics, 30, 59, *see also* estimation; hypothesis testing; population means; statistical test
infinite population, 30, 131
sample size, 142–143
instrumental neutron activation analysis (INAA), 197, 210–212, *211*
specifications, 217
International Organization for Standardization (ISO), 153
ISO9000 standard, 13–14
ISO9001 standard, 13–14
ISO14000, 153
ISO14001, 153
interpretability, 332
interval estimation, 50–51
interval scale, 327

J

judgmental sampling, 130, 134, **138**

K

Kaiser criterion, 332, 360
Kashima SPM study, 195, 212–215
analytical plan, 197–198
chromium concentrations, spatial distribution of, *213*
composite sampling, 197–198
cutting design of filter, *198*
flow rate, 202–204
flow volume calculation, 202–204
gaseous and particulate pollutants, **214**
humidity, 205–206
instrumental differences effects, 198–202
MMD, 213, *214*
QA/QC, 198
Anderson cascade impactor, **200**, *201*, *202*
ANOVA, **207**
chemical analysis, 209–212
concentration of instrument, **201**
cyclone, efficiency of, *200*
element, measurement for, **212**
filter media, 208
filter paper, 209
instrumental differences, 207–208
monitoring instrument for, 204–205
sampling location, device and elements, **199**
sampling design
collection conditions and sampling durations, **197**
equipment, 196–197
frequency, 196–197
monitoring sites, 196
time, 196–197
sampling site, sampling device and element, **199**
sizing efficiency, 198–202

L

Lagrangian models, 282
Large Power Plant Effluent Study (LAPPES), 293–294, *294*
law of large numbers, 26, 27, 30, 50
time-series analysis of concentration data and 227–228
lead (Pb), 111, 402
least square model, 395–397
light detection and ranging (LiDAR), 126
linear programming model, 398
linear regression (LR), *375*, 375–377, **376**
location parameters, 21–22
logistic regression, 318–321
GLM, 322
logit model, 434
long-term and short-term models, 286–287
low-cost sensors (LCS), 122–125, 303

M

machine learning, 28, 122–123, 322, 333
Mahalanobis distance, 105, 317, 371
manual monitors, 123–124
manual sampling, 147
market-based instruments (MBI), 151–152
mass median diameter (MMD), 213, *214*, 379

mass to mass (mg/kg), 113–114
mass to volume (mg/m^3), 113–114
Matlab, 107
maximum utility model, 434
mean fractional error (MFE), 306, 307
mean squares (MS), 63
measurement errors, 117
  random, 117–118
  schematic presentation, *118*
  systematic, 117–118
measurement protocol, 146
measuring scale, 326
  how to, 327–329
  level of measurement, 326–327
  rating and ranking scores, *327*
  rating score value, *328*
microplastics (MP), 6, 123
Microtox test, 171–173, **173**
Model Verification Kit, 266
molar to volume (mol/m^3), 113–114
moment, 23–24, *24*, 147
monitoring parameters, 111
  air pollution, 111–112
  biological, 113
  definition, 3–4
  measurement units, 113–114
  sediment pollution, 113
  soil pollution, 113
  water pollution, 112
multicollinearity, 100, 104, 383, 397
  ridge regression model, 397
  VIF value, 401
multilayer perceptron (MLP), 373–375
  performance, *375*, 375–377
multiple regression analysis, 94, 102, 104, 378–379
  atmospheric concentrations, **94**
  equation, 97–100, **98**, **99**
  partial regression coefficient, 95–97, **96**
  photochemical air pollution, 378–381, **380**
  regression coefficients, **94**
  scatter diagrams, *95*
  variable selection method, 100
multistage mixed sampling designs, 130, 139–140
  adaptive cluster sampling, 140–141, *141*
  ranked set sampling, 141–142
multistage sampling design, 131–133, 138–139
  adaptive cluster sampling, 140–141, *141*
  multistage mixed sampling, 139–140
  multistage probability sampling, 139
  ranked set sampling, 141–142
multivariate analysis, 102
  classification, **103**
  discriminant analysis, 105
  experimental design, 106
  factor analysis, 104–105
  PCA, 103–104
  statistical analysis software, 106–107

N

nanoplastics (NP), 6, 123
National Ambient Air Quality Standards (NAAQS), 4, 111, 114, 283
  in Japan, **284**
  in United States, **284**
nitric oxide (NO), 231, 234
  3D photochemical smog models, 304
  concentration variability, **235**, **236**, *237*
  FA-MR model, 413–414
nitrogen dioxide (NO_2), 94, 111, 213, 221
  air quality
    analysis of differences, **233**
    concentration variability, 234–238, **235**, **236**
  representativeness, **243**
nitrogen oxides (NO_x), 2, 121, 230
  air tracer experiments, *260*
  plume spreads, 274
nominal scale, 316, 326–327
non-hierarchical method, 372–373
non-ortho PCBs, 175, 177, 179, 180
non-probability sampling, 134, 137, 139
  convenience, 134
  judgmental, 134
  quota, 134–135
  snowball, 135
non-random sampling, 132–133
normal distribution, 37, *37–38*
  relationship with other distributions, 39–40
Northern Thailand
  air quality in, 336–341
    Chiangmai's monitoring stations, *340*
    first and second principal components, *337*
    regional surface synoptic charts, *339*
    varimax rotation, **338**
null hypothesis, 43–49, 161
number to volume (number/m^3), 113–114

O

oblique rotation, 332–333
one-tailed test, 48–49, *49*, 57, 325
one-way ANOVA, 62, 317, *see also* analysis of variance (ANOVA)
  data structure, **62–65**, *65*
  factorial design, 70–71, **71**
  and randomised block design, 62–68
  regression analysis, **91**
ordinal scale, 327, 328, 426–428, 448
ordinary least squares (OLS), 433, 435
organic carbon (OC), 112, 187
organic ions, 112, 148
organic matter (OM), 113, 190
orthogonal design, 74–76, **75**
orthogonal rotation, 332–333
ozone (O_3), 2, 111, 218, **303**
  maximum hourly (*see* artificial neural networks (ANNs) *vs.* linear regression model)
  photochemical air pollution, 351

P

paired comparison method, 327, 440
paper mill, *see* chlorine-bleached Kraft pulp and paper mill
partial regression coefficient, 95–97, **96**
particulate matter (PM), 111–112, 146, 148
 PCA, 357–359
  conceptual diagram, *359*
  FA, 360–362
  non-detected data points, 359–360
  spatial distributions of provinces, *358*
  varimax rotated factor loadings, *358*
 QA/QC procedure, in AIRPET, 184
  chemical analyses, 187–189
  data quality, 189–190
  filter pretreatment, 187
  filter samples for composition analyses, *188*
  filter selection, 187
  frequency, 186–187
  monitoring sites, 184–186
  receptor modelling study, 186
  sampled filter retrieval, 187
  sampling time, 186–187
  visual check, 187
 QA/QC procedure, in Kashima SPM study
  Anderson cascade impactor, **200**, *201*, *202*
  chemical analysis, 209–212
  concentration of instrument, **201**
  cyclone, efficiency of, *200*
  filter media, 208
  filter paper, 209
  instrumental differences, 207–208
  mass and reconstructed mass groups, *192*
  monitoring instrument for, 204–205
  sampling location, device and elements, **199**
 receptor model, 390
Pasquill-Gifford algorithm, 294, 295
passive sampling technique, 146–147
path analysis, 428–431, *430*
perception map, 431–432
perceptron, 320, 322–323
pH, 112, 113, 124, 146
 fluctuations in, 253
 rain samples, 249
 soil quality, 113
 water quality, 124
phosphorus, 2, 112, 169
photochemical air pollution, 350–352, **352**
 Tokyo, 377
  discriminant analysis model, 378–381, **380**
  input data, 377–378
  multiple regression model, 378–381, **380**
Plume Model Validation and Development (PMV&D) project, 295–296
$PM_{2.5}$, 3, 4, 6, 111–112, *see also* AIRPET
 LCS network, 122
 representativeness, *241*
PM_{10}, 111, 122, 126, *see also* AIRPET
point estimation, 50
polycyclic aromatic hydrocarbons (PAHs), 111–112, 115, 410
population, 19, 130–131, 145, *see also* sampling design
 finite, 30, 143–144
 infinite, 30, 142–143
 mean, 30–31
 vs. sample, **30**, 30–31, 36, 132
population means, 51
 ANOVA, 60–61, *60–61*
 paired data, 54–55, **55**
 pooled value, 53
 variances equal), 52–54
 variances unequal, 54
positive matrix factorisation (PMF), 192, 362, 395, 398, 400
 algorithm, 415–416
 Bangkok dataset, 419–421, **420**
 concept, 415–416
 Kashima dataset, 421–423, **422**, *423*
 multi-site dataset, 416–418
 source profile estimation results, 418–419, **419**
precision, 6, 12, 117–119, 187
 chemical analysis, 209–212, *211*
 element measurement, **212**
 finite population, 143–144
 infinite population, 142–143
prediction accuracy, 80, 83, 285, 297, 378
preference surveys, 438
principal component analysis (PCA), 103–105, 331, 333
 consumer perception, green products, 431–432
  conjoint analysis, 432–435
  purchase of, 435–437
 roadside air quality, 341
  BTEX measurements and traffic flows in urban area, 341–343, *342*, **343**
  particulate matter (PM) in Tokyo, 343–344, **344**, *345*
 spatial homogeneity, 345
  application, 346–349
  cluster analysis, 349–350, *350*
  correlation matrix, **348**
  cosine similarity, 345–346
  first and second principal components, plotting, *349*
  Kashima coastal area, 346
  result, **349**
  wind vector correlation matrix, *347*
 waste water discharge, 162–164, *163*
principal component analysis (PCA) *vs.* exploratory factor analysis (EFA), 331, 333–334, 362
 air quality
  data, Tokyo, **335**, 335–336, **336**
  meteorological patterns, Northern Thailand, 336–341
 butterfly distribution, binary dataset, Thailand, 353
  correlation matrix, **356**
  dataset, 354–356, **355**
  results, 356–357

varimax rotation, variables, **357**
zoogeographic zones, *353*, 353–354
factor extraction process, 332
number of factor to retain, 332
rotation, 332–333
terminology, 331
photochemical air pollution, 350–352, **352**
particulate matter (PM), 357–359
conceptual diagram, *359*
FA, 360–362
non-detected data points, 359–360
spatial distributions of provinces, *358*
varimax rotated factor loadings, *358*
principal components (PCs), 331, 332, 335
plotting, *349*
scatter plots, *337*
variables, **343**
probability density distributions, 32–33, *32–33*
probability sampling, 132, 135, 149
cluster, 137–138
judgmental, **138**
simple random sampling, 135–136
stratified, 137
systematic, 136
proportion test, 56–58
proton-induced X-ray emission spectroscopy (PIXE), 188
vs. XRF, *189*
puff modelling approach, 282–283

Q

quality assurance and quality control (QA/QC), 10–14, 114–115; *see also* acid rain, IPCAJ's study, quality assurance
accuracy, trueness and precision, 118–119
AIRPET, 184
chemical analyses, 187–189
data quality, 189–190
filter pretreatment, 187
filter selection, 187
frequency, 186–187
monitoring sites, 184–186
PM, 186
sampled filter retrieval, 187
sampling time, 186–187
visual check, 187
air tracer, flux calculations, 270, 271, 272, 273
comparability/harmonisation, 120
data system, 120, *121*
data analysis, 121–122
data validation, 120–121
dissemination, 122
raw data, 120
reporting, 122
detectability, 119
errors, monitoring, 117–118, *118*
international quality standards, 13–14
Kashima SPM study, 198
Anderson cascade impactor, **200**, *201*, *202*
chemical analysis, 209–212
concentration of instrument, **201**
cyclone, efficiency of, *200*
element, measurement for, **212**
filter media, 208
filter paper, 209
instrumental differences, 207–208
monitoring instrument for, 204–205
sampling location, device and elements, **199**
representativeness, 119
samples, 146
spatial coverage, 119
temporal completeness, 119
uncertainty of measurement, 115–117
quantiles-quantiles (Q-Q) diagrams, 220, 308, 310
quantitative data, 20, 315–317, 449
questionnaire survey data, 426, *see also* consumer perception, green products
conjoint analysis (*see* conjoint analysis)
structural equation modelling (SEM), 428–431
quota sampling, 132, 134–137, 140

R

random errors, 117–118
random factors, 72–73, 117
randomisation, 74
randomised block design (RBD), 62–68, 70
random sampling, 132–133
random variable, 31, 47, 234, 316, 317
continuous probability distribution, 31–32
degrees of freedom, 34–35, *35*
discrete probability distribution, 31
expected value, 33–34
probability density distributions, 32–33, *32–33*
sum of squares, 34–35, *35*
variance, 33–34
ranked set sampling, 141–142, 149
ranking method, 327, 329, 448
rating scale method, 327–329, 448
ratio scale, 327
receptor models, 100, 112, 183, 283, 389
CMB, 390–393, *393*
concept, 389–390
effective variance model, 397
factor analysis model, 398
least square model, 395–397
linear programming model, 398
particulate matter, *390*
PM, 186
PMF, 398
ridge regression model, 397
SEM, 398
regression analysis, 76
ANOVA, **91–92**
calculation auxiliary table, **89**
cause–effect relationship, 79–80
vs. correlation analysis, 84–85
dataset, 92–93, *93*
degrees of freedom, 84
one-way ANOVA, **91**
prediction accuracy, 80, 83, *84*

regression coefficients, *80*, 80–82, **82**
test, 85–88, **86**
for repetitive data, 88–92
scatter diagram, *80*
scatter plots, *90*
variance, 84
regression coefficients
air pollutant concentrations, **82**
coefficients and residuals, **82**
estimation of, 80–82
hypothesis test, 87–88
regression equation, *82*
test for, 85–87
regulatory monitoring, 155–156
rejection area, 44, 53–55, *56*
remote sensing techniques, 125–126, 141, 303
repeated measures data, 73
replication, 74
representativeness, 20, 111, 119, 134, 141
concentration spatial distribution in target area, 241–242
concept, 240
correlation coefficient, distribution of, *243*
criterion setting condition, **244**
in Himeji region, *244*
network monitoring, 245–246
NO_2 monitoring station, **243**
$PM_{2.5}$ annual average, *241*
station monitoring, 242–245, *245*
residual sum of squares, 81, 92, 97, 381, 405
ridge regression model, 397
Ringelmann Smoke chart, 315
roadside air quality, PCA analysis
principal component analysis (PCA), 341
BTEX measurements and traffic flows in urban area, 341–343, *342*, **343**
particulate matter (PM) in Tokyo, 343–344, **344**, *345*
root mean square errors (RMSE), 122–123, 208, 238, 386
rotation, 103, 105
oblique, 332–333
orthogonal, 332–333

S

sampling design, 116, 130, 144, 149
environmental measurements
collection techniques, 146–149
context, 144–146
types of samples, 146–149
methods (*see* sampling methods)
multistage, 138–139
probability, 139
mixed, 139–142
population, 130–132
vs. population, 30–31, 36, 132
sample size, 32, 47–48, 132
finite population, 143–144
infinite population, 142–143
sampling distributions, 19, 36
chi-square distribution, 38–39
F distribution, 39–40
normal distribution, *37*, *37–38*
population, 36
population and, **30**, 30–31
and statistical inference, 42
t distribution, 38, *39*
sampling methods, 132–133, *133*
non-probability sampling, 134
convenience, 134
judgmental, 134
quota, 134–135
snowball, 135
non-random sampling, 132–133
probability sampling, 135
cluster, 137–138
judgmental, **138**
simple random sampling, 135–136
stratified, 137
systematic, 136
random sampling, 132–133
satellite
monitoring, 125–126, 126
products, 6, 122, 125, 241
scoring method, 327
significance level, 44, 45, 53, 55, 57
significance testing, 45–47
snowball sampling, 134, 135
software, statistical analysis, 106–107
soil pollution, 7, 113, 140
solid waste, 111, 140, 151, 156, 170
spatial resolution, air pollution, 230, 246, *see also* temporal resolution/averaging time, air pollution
additive nature of variance, 233–234
analysis of difference,
measuring instruments, 231–232
measuring sites, 232–233
concentration measures, 230–231
concentration variability
same building, 234–238
residential area, 238–240
monitoring stations, optimal placement, 240–241
concentration spatial distribution, 241–242
network evaluation, 245–246
representativeness of, 242–245
species diversity, 9
split design, 74, 90
standard deviation (SD), 12–13, 30–31
statistical inference, 20, 132, 139
sampling and, 42
statistical test, 42
concept of, 42–43
hypothesis testing, **43**, 43–46
power of, 47–48, *48*
one-tailed test, 48–49, *49*
procedure for, 44–45
sample size, 47–48
significance testing, 46–47
two-tailed test, 48–49, *49*
statistics, 19

descriptive problems, 20
inferential problems, 20
stepwise method, 100
stratified sampling, 137–139
structural equation modelling (SEM), 105, 398, 428, 435
causal modelling, 428
concept, 428–429
path analysis, 429–431, *430*
sulphur dioxide (SO_2), 2, *21*, 94, 111, 112
hourly concentrations, reproducibility of, 308
Kashima SPM study, 195
Kashima study, air quality model, **288–289**, *291*, *292*
multiple regression analysis, 94
NAAQS
in Japan, **284**
in United States, **284**
Tokyo, 381
explanatory variables, 381–382, **382**
modelling results and evaluation, 382–383, *383*
sulphur hexafluoride (SF_6), 223, 236, 258
concentration data, *224*, *225*, *226*
concentration variability, *237*, *238*
highway line source modelling, 267
plume spreads, 274
vertical distribution, *265*
sum of squared errors (SSE), 81, 376
sum of squares (SS), 34–35, *35*, 170
one-way ANOVA, 63
suspended particulate matter (SPM), *see* Kashima SPM study
Sustainable Development Goals (SDGs), 7
systematic errors, 117–118
systematic sampling method, 136, 137, 139

T

target population, 6, 130, 131, 133, 449
cluster sampling, 137–138
convenience sampling, 134
multistage probability sampling, 139
quota sampling, 134–135
sampled population, 145
simple random sampling, 135
stratified sampling, 137
t distribution, 22, 38, *39*, 61, 79
relationship with other distributions, 39–40
variance test, 58–59
temporal resolution/averaging time, air pollution, 217, 228, *see also* spatial resolution
averaging time, 219–222
concentration, averaging time shorter than one hour, 222–223
concentration data, statistical analysis, 223–224
goodness-of fit, 224–226
time-series analysis and law of large numbers, 227–228
spatiotemporal scope of air pollution, 218, *219*, **219**
thermo-optical transmittance (TOT), 188–189
time-series analysis, 121
concentration data, *227*, 227–228
Tokyo
air quality in, **335**, 335–336, **336**
photochemical air pollution, 377
analysis model, 378–381, **380**
input data, 377–378
multiple regression model, 378–381, **380**
SO_2 concentration, 381
explanatory variables, 381–382, **382**
modelling results and evaluation, 382–383, *383*
trueness, 6, 117–119, 210
turbidity, 112, 124, 155, 161, 166
turbulence, 116, 228, 259–260, *263*
air tracer sampling grid, 264–266, *266*
two-tailed test, 48–49, *49*
two-way ANOVA, 69, **70**, 71–72, **72**
type I error, 12, 44, 45, 48
type II error, 13, 44, 45, 48

U

ultrafine particles (UFPs), 112
uniqueness, 331
United Nations Statistics Division (UNSD), 7
United States
air quality model, 293–294
EPA, 301
vs. Japan data, 295–296, **296**
SIP, 284
SO_2 NAAQS in, **284**
United States Geological Survey (USGS), 11–12

V

variables, 8, 22–23, 308
categorical, 323–326
and data, 316–317
selection method, 100
variance, 33–34
decomposition of, 84
variance inflation factor (VIF), 100, 401, 402
volatile organic compounds (VOCs), 111–112, 124, 146
AIRPET monitoring, 183–184
volume to volume (m^3/m^3), 113–114

W

wastewater (WW), 151, 153–154, 168, **171**
biological effects, 172
chlorine-bleached Kraft pulp, 157, 166, 169, 170, 174
composite sampling, 149
convenience sampling, 134
discrete samples, 147–148
ecotoxicological tests, 171
judgmental sampling, 134
PCA, 162–164, *163*

regulatory monitoring, 155–156
sampling, 172
sampling design, 145–146
water pollution, 24, 113, 125
monitoring parameters, 112
water-soluble ions, 112, 157, 187, 343
willingness to pay (WTP), 426, 439–441, 448–449
amount of payment intention, **447**
CVM, 443
estimating, 446–447
for zero-emission vehicle, *448*

X

X-ray fluorescence spectrometers (XRF), 188, 197, 210
chemical composition for PM, *211*
PIXE *vs.*, *189*